Dae Mann Kim

Introductory Quantum Mechanics for Applied Nanotechnology

Related Titles

Hommelhoff, P., Kling, M. (Eds.)

Attosecond Nanophysics
From Basic Science to Applications

2015
Print ISBN: 978-3-527-41171-9

Caironi, M., Noh, Y.-Y. (Eds.)

Large Area and Flexible Electronics

2015
Print ISBN: 978-3-527-33639-5

Fayngold, M., Fayngold, V.

Quantum Mechanics and Quantum Information
A Guide through the Quantum World

2013
Print ISBN: 978-3-527-40647-0

Wolf, E.L.

Nanophysics of Solar and Renewable Energy

2012
Print ISBN: 978-3-527-41052-1

Sullivan, D.M.

Quantum Mechanics for Electrical Engineers

2012
Print ISBN: 978-0-470-87409-7

Ahn, D., Park, S.

Engineering Quantum Mechanics

2011
Print ISBN: 978-0-470-10763-8

Sanghera, P.

Quantum Physics for Scientists and Technologists
Fundamental Principles and Applications for Biologists, Chemists, Computer Scientists, and Nanotech

2011
Print ISBN: 978-0-470-29452-9

Rubahn, H.

Basics of Nanotechnology

3rd Edition
2008
Print ISBN: 978-3-527-40800-9

Dae Mann Kim

Introductory Quantum Mechanics for Applied Nanotechnology

WILEY-VCH
Verlag GmbH & Co. KGaA

Author

Prof. Dae Mann Kim
Korea Inst. f. Advanced Study
Cheongnyangni 2-dong
130-722 Seoul
South Korea

Cover
© Istockphoto/*kynny*

Library of Congress Card No.: applied for

British Library Cataloguing-in-Publication Data
A catalogue record for this book is available from the British Library.

Bibliographic information published by the Deutsche Nationalbibliothek
The Deutsche Nationalbibliothek lists this publication in the Deutsche Nationalbibliografie; detailed bibliographic data are available on the Internet at <http://dnb.d-nb.de>.

Print ISBN: 978-3-527-41245-7
ePdf ISBN: 978-3-527-67718-4
ePub ISBN: 978-3-527-67717-7
Mobi ISBN: 978-3-527-67715-3
oBook ISBN: 978-3-527-67719-1

Typesetting Laserwords Private Limited, Chennai, India
Printing and Binding Markono Print Media Pte Ltd, Singapore

Printed on acid-free paper

To my grandma and family

Contents

Preface

The multidisciplinary science education has been prompted by the rapid advancement and utilization of IT/BT/NT, and the quantum mechanics is the basic science supporting the technologies. It further provides the platform on which to bridge different disciplines in science and engineering. This introductory textbook is intended for the undergraduate seniors and beginning graduate students and is focused on the application and multidisciplinary aspects of the quantum mechanics.

The applications have been chosen primarily from the semiconductor and optoelectronic devices to make the discussion practical. The p-n junction diode is first singled out for the discussion as the simplest solid state switch and also as photodiode, light-emitting and laser diodes and solar cells. Moreover, the field effect transistors are treated in some detail. The well-known theory of MOSFET is first compactly presented to serve as the general background for considering other kinds of novel FETs such as nanowire and spin field-effect transistors. The working principles of these devices are treated from a unified standpoint of the equilibrium and nonequilibrium statistics and device physics in conjunction with the quantum mechanical concepts. Additionally, these FETs as the nonvolatile memory cells, biosensor, and solar cells are highlighted. As an extension of the discussion of the spin FET the quantum computing is briefly touched upon.

The organization of the book is as follows. The classical and statistical mechanics and the electromagnetic fields are compactly summarized as a general background. After a short visit to the milestones leading to quantum mechanics, the Schrödinger equation is applied immediately to problems of practical interests, involving the quantum wells and subbands, 1D, 2D, and 3D densities of states. In particular, the tunneling and its applications are highlighted. Two key bound systems are treated in some detail. Specifically, the harmonic oscillator is analyzed based on the quantum mechanical and operator treatments. In addition, the hydrogen atom is considered as the simplest atomic system and as an essential ingredient for analyzing the atomic spectroscopy, multielectron atoms, paramagnetic electron resonance and molecules.

The chemical bond for the molecular formation is included in the discussion list. In particular, the molecular spectroscopy is treated as an extension of the atomic spectroscopy by utilizing the time-independent perturbation theory and focused on the rotational and vibrational motions of diatomic molecules. The nuclear spin, hyperfine structure, and nuclear magnetic resonance for molecular imaging are briefly introduced. Moreover, the interaction of light with matter is highlighted, based on the time-dependent perturbation theory, and the operation principle of the laser is elucidated. Finally, the semiconductor statistics and the transport of the charge carriers are discussed as an essential background for modeling the semiconductor devices. An effort has been expended to make the presentation and discussion brief and clear by simplifying the mathematics and by making use of the analogies existing between different dynamic systems.

The contents of this book have evolved from the courses offered in the Department of Electrical and Computer engineering, Rice University, Houston, TX., USA; POSTECH, Pohang, Korea; and the College of Engineering, Seoul National University, Seoul, Korea. The active and enthusiastic participation of the attending students made it a joyful experience to teach the courses. My thanks are due to those students. I would also like to express my sincere thanks to Miss You-Na Hwang for her tireless cooperation in preparing the figures for the book. Finally, it is my pleasure and honor to express my heartfelt gratitude to Professor Willis E. Lamb, whose courses on quantum mechanics and laser physics were most inspiring.

Seoul, Korea *Dae Mann Kim*

1
Review of Classical Theories

A compact review of classical theories is presented, including the classical and statistical mechanics and electromagnetism. These theories are inherently intertwined with quantum mechanics and provide the general background from which to understand the quantum mechanics in a proper perspective.

1.1
Harmonic Oscillator

The harmonic oscillator (HO) is one of the simplest, yet ubiquitous dynamical systems appearing in a variety of physical and chemical systems such as electromagnetic waves and molecules. The HO is a particle attached to a spring, executing oscillatory motion. When the spring is compressed or stretched, the spring provides a restoring force for putting the particle back to the equilibrium position (Figure 1.1). In the process, an oscillatory motion ensues, and the motion represents a variety of important natural phenomena such as molecular vibrations and electromagnetic waves.

Newton's equation of motion of the HO reads as

$$m\ddot{x} = -kx \tag{1.1}$$

where m is the mass of the oscillator, x the displacement from the equilibrium position, and k the spring constant. The double dots denote the second-order differentiation with respect to time, and $-kx$ is Hook's restoring force. The equation can be put into a form

$$\ddot{x} + \omega^2 x = 0, \quad \omega^2 \equiv \frac{k}{m} \tag{1.2}$$

where ω is the characteristic frequency. Trigonometric functions, for example, $\sin \omega t$, $\cos \omega t$ are well-known solutions of Eq. (1.2). When the oscillator is pulled by x_0 and gently released, for instance, the displacement $x(t)$ and the velocity $v(t)$ are given by

$$x(t) = x_0 \cos \omega t, \quad v(t) \equiv \dot{x}(t) = -\omega x_0 \sin \omega t \tag{1.3}$$

and $x(t)$, $v(t)$ oscillate in time in quadrature (Figure 1.2) with the period $T = 2\pi/\omega$.

Introductory Quantum Mechanics for Applied Nanotechnology, First Edition. Dae Mann Kim.

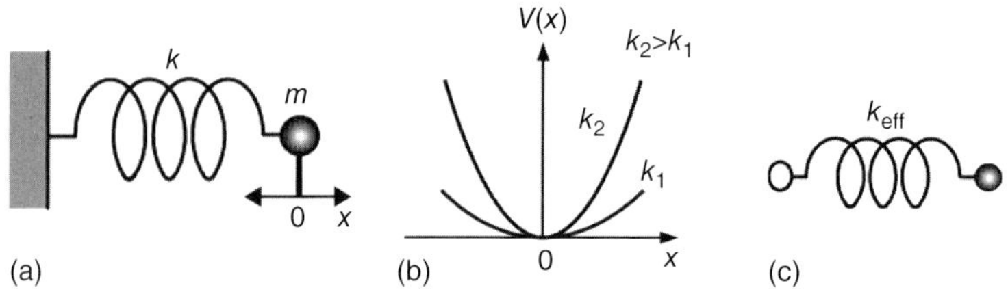

Figure 1.1 The harmonic oscillator, a particle of mass m attached to a spring with the spring constant k (a); the potential energy of HO (b); a diatomic molecule as represented by two atoms coupled via an effective spring constant (c).

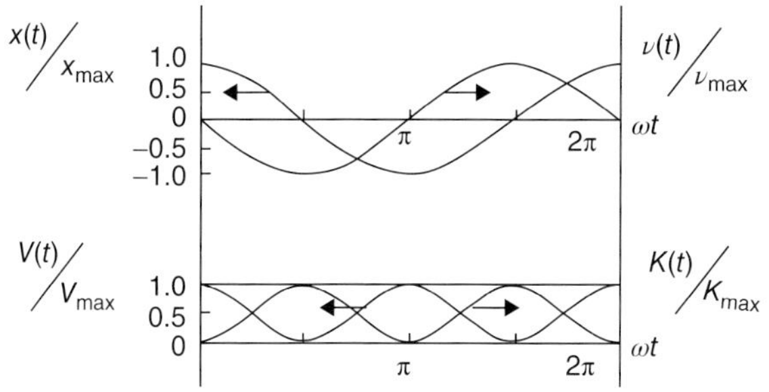

Figure 1.2 The displacement x, velocity v, and kinetic K and potential V energies versus ωt, all scaled with respective maximum values. The total energy $K + V$ is constant in time, and HO is a conservative system.

The potential energy of the HO is obtained by integrating the work done for displacing the HO from the equilibrium position to x against the restoring force:

$$V(x) \equiv -\int_0^x dx\,(-kx) = \frac{1}{2}kx^2 \tag{1.4}$$

The total energy is often denoted by Hamiltonian H and is expressed in terms of the linear momentum p_x and the displacement x as

$$H \equiv K + V = \frac{p_x^2}{2m} + \frac{1}{2}kx^2 \tag{1.5}$$

Given H, Hamilton's equations of motion read as

$$\dot{x} \equiv \frac{\partial H}{\partial p_x} = \frac{p_x}{m} \quad \dot{p}_x \equiv -\frac{\partial H}{\partial x} = -kx \tag{1.6}$$

The pair of equations in (Eq. (1.6)), when combined, reduces to Newton's equation of motion, and the variables x, p_x are known as *canonically conjugate variables*. The essence of classical mechanics is to solve the equation of motion and to precisely specify the position and momentum of a particle or a system of particles.

1.2
Boltzmann Distribution Function

The properties of macroscopic quantities are derived from the dynamics of an ensemble of microscopic objects such as electrons, holes, atoms, and molecules. Statistical mechanics describes such an ensemble of particles by means of the distribution function, $f(\boldsymbol{r}, \boldsymbol{v}, t)$. The function represents the probability of finding the particles in the phase space volume element $d\boldsymbol{r}d\boldsymbol{v}$ at $\boldsymbol{r}$, $\boldsymbol{v}$, and t. Thus, when multiplied by density n of the particle $f(\boldsymbol{r}, \boldsymbol{v}, t)\, d\boldsymbol{r}d\boldsymbol{v}$ represents the number of particles in the volume element at t.

The change in time of $f(\boldsymbol{r}, \boldsymbol{v}, t)$ is given from the chain rule by

$$\frac{df(\underline{r}, \underline{v}, t)}{dt} = \frac{\partial f}{\partial t} + \frac{\partial f}{\partial x}\frac{\partial x}{\partial t} + \cdots + \frac{\partial f}{\partial v_x}\frac{\partial v_x}{\partial t} + \cdots = \frac{\partial f}{\partial t} + \underline{v} \cdot \nabla f + \underline{a} \cdot \nabla_v f \quad (1.7a)$$

where the operators

$$\nabla \equiv \hat{x}\frac{\partial}{\partial x} + \hat{y}\frac{\partial}{\partial y} + \hat{z}\frac{\partial}{\partial z}, \quad \nabla_v \equiv \hat{x}\frac{\partial}{\partial v_x} + \hat{y}\frac{\partial}{\partial v_y} + \hat{z}\frac{\partial}{\partial v_z} \quad (1.7b)$$

are the gradient operators with respect to $\boldsymbol{r}$, $\boldsymbol{v}$, and $\boldsymbol{a}$ is the acceleration. The distribution function also changes in time due to collisions by which the particles are pushed out of or pulled into the volume element. Hence, the transport equation is given by

$$\frac{\partial f}{\partial t} + \underline{v} \cdot \nabla f + \frac{\underline{F}}{m} \cdot \nabla_v f = \left.\frac{\delta f}{\delta t}\right|_{\text{coll}}, \quad \underline{a} = \frac{\underline{F}}{m} \quad (1.8)$$

with $\boldsymbol{F}$ denoting the force.

Equilibrium

In the thermodynamic equilibrium, the distribution function f_0 is independent of time, that is, $(\partial/\partial t)f_0 = 0$, and the collision term should also be put to zero. This is because every process is balanced by its inverse process in equilibrium (detailed balancing). Consequently, the number of particles pushed out of and pulled into the phase space volume element due to collision is the same. Thus, the one-dimensional transport equation in equilibrium is given from Eq. (1.8) by

$$v_x \cdot \frac{\partial f_0}{\partial x} - \frac{1}{m}\frac{\partial \varphi}{\partial x}\frac{\partial f_0}{\partial v_x} = 0, \quad F_x \equiv -\frac{\partial \varphi}{\partial x} \quad (1.9)$$

where the force has been expressed in terms of the potential φ.

We may look for the solution in the form

$$f_0(x, v_x) = N e^{-E(x)/k_B T}, \quad E(x) = \frac{m v_x^2}{2 k_B T} + \varphi(x) \quad (1.10)$$

where N is the constant of integration and k_B the Boltzmann constant having the value $1.381 \times 10^{-23}\,\mathrm{J\,K^{-1}}$ or $8.617 \times 10^{-5}\,\mathrm{eV\,K^{-1}}$, and $E(x)$ is the total energy at x, consisting of kinetic and potential energies. By inserting Eq. (1.10) into Eq. (1.9)

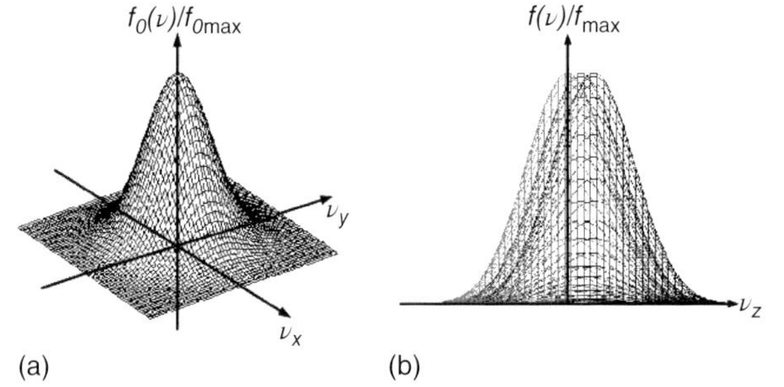

Figure 1.3 The distribution function of an ensemble of free particles in equilibrium (a) and under an electric field in the z-direction (b), all scaled with the maximum values; $f_0(\boldsymbol{v})$ is symmetric in $\boldsymbol{v}$, while $f(v)$ is not in the z-direction.

and carrying out the differentiation with respect to x and v_x, we find that Eq. (1.10) is indeed the solution. Also N can be used for normalizing $f_0(x, v_x)$. For a system of free particles in which $\varphi = 0$, the normalized equilibrium distribution function is given by

$$f_0\left(v_x\right) = \left(\frac{m}{2\pi k_B T}\right)^{1/2} e^{-mv_x^2/2k_B T} \tag{1.11}$$

where N has been found from the normalization condition,

$$N\int_{-\infty}^{\infty} dv_x e^{-mv_x^2/2k_B T} = N\left(\frac{2\pi k_B T}{m}\right)^{1/2} = 1$$

Naturally, $f_0(v_x)$ can be generalized to three dimensions as

$$f_0\left(\underline{v}\right) = \left(\frac{m}{2\pi k_B T}\right)^{3/2} e^{-mv^2/2k_B T}, \quad v^2 = v_x^2 + v_y^2 + v_z^2 \tag{1.12}$$

The function f_0 is the celebrated Boltzmann distribution function for a system of free particles, and the exponential factor appearing therein is called the *Boltzmann probability factor*. Clearly, $f_0(\boldsymbol{v})$ is symmetric in $\boldsymbol{v}$ and represents the fact that there is no preferred direction, a well-known property of the equilibrium (Figure 1.3).

Equipartition Theorem

In equilibrium, the probability of a particle moving from left to right is the same as that of moving from right to left (Eq. (1.11)). Therefore, the average velocity is zero, but the average value of v_x^2 is not zero and can be found as

$$\begin{aligned}\left\langle v_x^2\right\rangle &\equiv \left(\frac{m}{2\pi k_B T}\right)^{3/2} \int_{-\infty}^{\infty} dv_x v_x^2 e^{-mv_x^2/2k_B T} \int_{-\infty}^{\infty} dv_y e^{-mv_y^2/2k_B T} \int_{-\infty}^{\infty} dv_z e^{-mv_z^2/2k_B T} \\ &= \frac{k_B T}{m}\end{aligned} \tag{1.13}$$

By inspection, we can write

$$\langle v_x^2 \rangle = \langle v_y^2 \rangle = \langle v_z^2 \rangle = \frac{k_B T}{m} \tag{1.14}$$

Hence, the total average kinetic equation is given by

$$\frac{1}{2} m \langle v^2 \rangle = \frac{1}{2} m \left(\langle v_x^2 \rangle + \langle v_y^2 \rangle + \langle v_z^2 \rangle \right) = \frac{3}{2} k_B T \tag{1.15}$$

which represents the equipartition theorem, namely, that the average kinetic energy of a free particle is equally divided into x-, y-, and z-directions, respectively, in equilibrium.

Nonequilibrium Distribution Function

Let us next consider an ensemble of electrons uniformly distributed in space and subjected to an electric field in the z-direction, $\hat{z}\mathrm{E}_0$. In this case, f is independent of $\boldsymbol{r}$ and at the steady state $\partial f/\partial t = 0$; hence, Eq. (1.8) reads in relaxation approach as

$$\frac{(-q\mathrm{E}_0)}{m_n} \frac{\partial f}{\partial v_z} = -\frac{f - f_0}{\tau}; \quad \left. \frac{\delta f}{\delta t} \right|_{\mathrm{coll}} = -\frac{f - f_0}{\tau} \tag{1.16}$$

where $-q\mathrm{E}_0$ is the force acting on an electron with charge $-q$ and mass m_n. The collision term used describes the system relaxing back to the equilibrium in a time scale determined by τ called the *longitudinal relaxation time*, and f_0 and f are the equilibrium and nonequilibrium distribution functions, respectively. Let us assume for simplicity that f does not depart very much from f_0, that is, $f - f_0 \ll f, f_0$. In this case, we can find f iteratively by putting $f = f_0$ on the left-hand side, obtaining

$$f \approx f_0 + \frac{q\mathrm{E}_0 \tau}{m_n} \frac{\partial f_0}{\partial v_z} = f_0 \left(1 - \frac{q\mathrm{E}_0 \tau v_z}{k_B T} \right) \tag{1.17}$$

where Eq. (1.12) has been used for f_0. Clearly, f is asymmetric in v_z due to the electric field applied, while symmetric in v_x, v_y as shown in Figure 1.3.

Mobility and Conductivity

Once f is found, the physical quantities of interest can be specified explicitly. For example, consider the average velocity of electrons. As f is still symmetric with respect to v_x, v_y, $\langle v_x \rangle = \langle v_y \rangle = 0$ but $\langle v_z \rangle$ is not zero and is given by

$$\langle v_z \rangle \equiv \frac{\int_{-\infty}^{\infty} dv_x \int_{-\infty}^{\infty} dv_y \int_{-\infty}^{\infty} dv_z v_z \, f_0 \left(1 - \frac{q\mathrm{E}_0 \tau(v) v_z}{k_B T} \right)}{\int_{-\infty}^{\infty} dv_x \int_{-\infty}^{\infty} dv_y \int_{-\infty}^{\infty} dv_z \, f_0 \left(1 - \frac{q\mathrm{E}_0 \tau(v) v_z}{k_B T} \right)} = -\frac{q\mathrm{E}_0}{m_n} \langle \tau_n \rangle \tag{1.18a}$$

where

$$\langle \tau_n \rangle \equiv \frac{m_n}{k_B T} \int_{-\infty}^{\infty} dv_x \int_{-\infty}^{\infty} dv_y \int_{-\infty}^{\infty} dv_z v_z^2 \tau(v) f \tag{1.18b}$$

denotes the effective relaxation time. Note in Eq. (1.18) that the first integral in the numerator and the second integral in the denominator vanish because the integrands therein are odd functions of v_z. This renders the denominator equal to unity because f_0 is a normalized distribution function (Eq. (1.12)). The relaxation time depends in general on the velocity $\boldsymbol{v}$ and has been averaged over.

The average velocity $\langle v_z \rangle$ derived in Eq. (1.18) represents the drift velocity with which all electrons move uniformly on top of their random thermal motion. The drift velocity is driven by E_0 exerting force on the electrons and can be viewed as the output of E_0:

$$v_{dn} \equiv \langle v_z \rangle = -\frac{q\mathrm{E}_0 \langle \tau_n \rangle}{m_n} \equiv -\mu_n \mathrm{E}_0, \quad \mu_n \equiv \frac{q\langle \tau_n \rangle}{m_n} \tag{1.19}$$

The response function μ_n connecting the input field and the output drift velocity is called the *mobility*. The current density of electrons due to drift is therefore given from Eq. (1.19) by

$$J_\mathrm{D} \equiv -q\sum_{j=1}^{n}(\underline{v}_{jth} + v_{dn}) = -\sigma_n \mathrm{E}_0, \quad \sigma_n \equiv q\mu_n n \tag{1.20}$$

where n is the electron density, and the random thermal velocities v_{jth} sum up to zero. The quantity σ_n connecting E_0 to J_D is known as the *conductivity*. The mobility μ_n and conductivity σ_n are the key transport coefficients.

1.3 Maxwell's Equations and EM Waves

Maxwell's equations are the foundations of the electromagnetism and are summarized as follows. When the charge and current density ρ and $\boldsymbol{J}$ are spatially distributed and vary in time, the electric $\mathbf{E}(\boldsymbol{r},t)$ and magnetic $\boldsymbol{B}(\boldsymbol{r},t)$ fields are generated and coupled to each other according to Maxwell's equations:

$$\nabla \times \underline{\mathrm{E}} = -\frac{\partial \underline{B}}{\partial t} \tag{1.21}$$

$$\nabla \times \underline{H} = \underline{J} + \frac{\partial \underline{D}}{\partial t} \tag{1.22}$$

$$\nabla \cdot \underline{\mathrm{E}} = \frac{\rho}{\varepsilon} \tag{1.23}$$

$$\nabla \cdot \underline{B} = 0 \tag{1.24}$$

The displacement vectors $\boldsymbol{D}$ and $\boldsymbol{B}$ are correlated to $\mathbf{E}$ and the magnetic field intensity $\boldsymbol{H}$ via the permittivity ε and the permeability μ of the medium as

$$\underline{D} = \varepsilon \underline{\mathrm{E}}, \quad \underline{B} = \mu \underline{H} \tag{1.25}$$

The addition of the continuity or charge conservation equation renders Maxwell's equations self-contained:

$$\frac{\partial \rho}{\partial t} + \nabla \cdot \underline{J} = 0 \tag{1.26}$$

As well known, Eq. (1.21) is Faraday's law of induction, specifying $\boldsymbol{B}(\boldsymbol{r},t)$ as the source of generating **E**, while Eq. (1.22) is Ampere's circuital law describing $\boldsymbol{J}$ as the source for generating $\boldsymbol{B}$. Also Eq. (1.23) represents Coulomb's law and Eq. (1.24) is the theoretical statement of the fact that no magnetic monopole has been observed. Ampere's circuital law, Eq. (1.22), was complemented by Maxwell, who introduced $\partial \underline{D}/\partial t$, called the *displacement current*. The modification was necessitated by the fact that the curl of any vector, $\nabla \times \underline{A}$, should be solenoidal, that is, $\nabla \cdot \nabla \times \underline{A} \equiv 0$, as can be readily verified. With $\boldsymbol{D}$ thus introduced, the requirement that $\boldsymbol{H}$ in Eq. (1.22) is solenoidal is satisfied, because the divergence operation on the right-hand side of Eq. (1.22) reduces the equation to the continuity equation to become zero. Maxwell's equations are rooted in the observed laws of nature and have successfully undergone the test of time and have been the source of unceasing applications.

Wave Equation

The electric and magnetic fields **E** and $\boldsymbol{H}$ coupled inherently via the two laws Eqs. (1.21) and (1.22) can be decoupled and examined separately. Thus, consider a medium free of charge ρ and $\boldsymbol{J}$. Then, the curl operations on both sides of Eq. (1.21) lead to

$$\nabla \times \nabla \times \underline{\mathrm{E}} \equiv \left[\nabla\nabla \cdot - \nabla^2\right] \underline{\mathrm{E}} = -\nabla^2 \underline{\mathrm{E}}; \quad \nabla \cdot \underline{\mathrm{E}} \propto \rho = 0 \tag{1.27a}$$

$$\nabla \times \left(-\frac{\partial \underline{B}}{\partial t}\right) = -\mu \frac{\partial}{\partial t}\left(\underline{J} + \varepsilon \frac{\partial}{\partial t}\underline{\mathrm{E}}\right) = -\mu\varepsilon \frac{\partial^2}{\partial t^2}\underline{\mathrm{E}}, \quad \underline{J} = 0 \tag{1.27b}$$

where a vector identity and Ampere's law have been used in Eqs. (1.27a) and (1.27b), respectively. Hence, by equating Eqs. (1.27a) and (1.27b), there results the wave equation:

$$\nabla^2 \underline{\mathrm{E}} - \frac{1}{v^2}\frac{\partial^2}{\partial t^2}\underline{\mathrm{E}} = 0, \quad \frac{1}{v^2} \equiv \mu\varepsilon = \mu_0\varepsilon_0\mu_r\varepsilon_r = \frac{1}{(c/n)^2} \tag{1.28}$$

Here, v is the velocity of light in the medium in which $\mu_r = 1$ and is specified in terms of the velocity of light in the vacuum $1/\mu_0\varepsilon_0$ and the index of refraction n via $\varepsilon_r = n^2$,with ε_r denoting the dielectric constant. Clearly, $\boldsymbol{D}$ is indispensable in bringing out the wave nature of the electromagnetic field. We can likewise derive the identical wave equation for $\boldsymbol{H}$.

Plane Waves and Wave Packets

A typical solution of the wave equation (1.28) is the plane wave

$$\underline{\mathrm{E}}(z,t) = \hat{x}\mathrm{E}_0 e^{-i(\omega t - kz)}, \quad \omega = \frac{k}{\sqrt{\mu\varepsilon}} \tag{1.29}$$

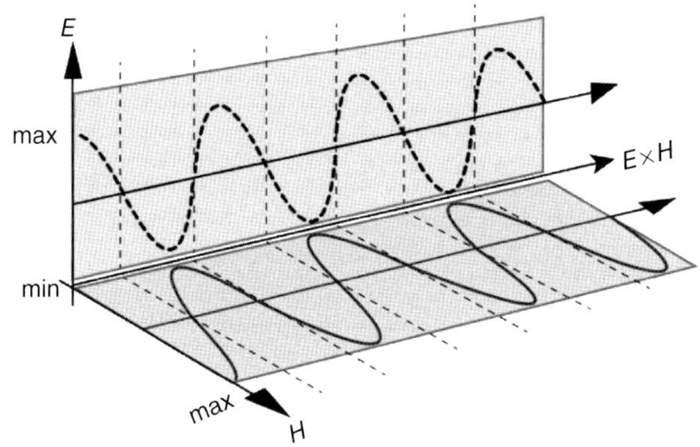

Figure 1.4 Spatial profiles of electric and magnetic fields traveling in the z-direction. Also shown is the Poynting vector, accompanying the propagation with the power.

propagating in the z-direction, for example, with the wave vector $k = 2\pi/\lambda$ obeying the dispersion relation as given in Eq. (1.29). The amplitude $\mathbf{E}_0$ has to be taken perpendicular to $\boldsymbol{k}$, say in the x-direction so that $\nabla \cdot \underline{\mathrm{E}} \propto \hat{z} \cdot \hat{x} = 0$ in accordance with Coulomb's law. In this case, the $\boldsymbol{H}$-field is obtained from Eqs. (1.21) and (1.29) as

$$\underline{H} = \hat{y}\sqrt{\varepsilon/\mu}\mathrm{E}_0 e^{-i(\omega t - kz)}, \quad \hat{y} = \hat{z} \times \hat{x} \tag{1.30}$$

Therefore, $\mathbf{E}$, $\boldsymbol{H}$, and $\boldsymbol{k}$ are mutually perpendicular, and the complex Pointing vector $\mathrm{E} \times H^*$ represents the power flow in the z-direction, as shown in Figure 1.4.

Wave Packets

The wave equation (1.28) is linear, so that the linear superposition of plane waves is also the solution:

$$\underline{\mathrm{E}}(z, t) = \mathrm{Re}\sum_n \underline{\mathrm{E}}_n e^{-i(\omega_n t - k_n z)} = \mathrm{Re}\int_{-\infty}^{\infty} dk \underline{\mathrm{E}}(k)\, e^{-i(\omega t - kz)} \tag{1.31}$$

The wave packet can be put into a compact form by Taylor expanding ω at k_0:

$$\omega(k) = \omega(k_0) + v_g(k - k_0) + \alpha(k - k_0)^2 + \cdots; \quad v_g \equiv \frac{\partial\omega(k_0)}{\partial k} \tag{1.32}$$

In a linear medium $\alpha = 0$, $v_g = c/n$, and by using Eq. (1.32), we can express Eq. (1.31) as

$$\underline{\mathrm{E}}(z, t) = \mathrm{Re}e^{-i(\omega_0 t - k_0 z)} \int_{-\infty}^{\infty} dk \underline{\mathrm{E}}(k)\, e^{i(z - v_g t)(k - k_0)} \tag{1.33}$$

and represent the wave packet in terms of two components: (i) the mode function oscillating with the carrier frequency ω_0 and propagating with the phase velocity ω_0/k_0 and (ii) the envelope contributed by superposed plane waves

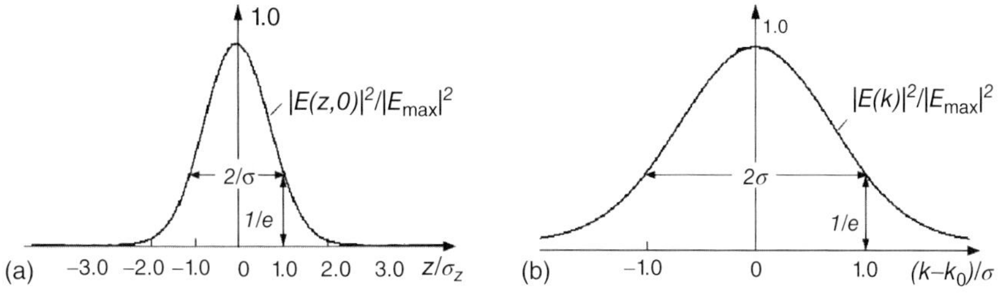

Figure 1.5 Spatial profile of the field intensity in the z-direction (a) and power spectrum versus the wave vector (b).

and propagating with the group velocity v_g. For the Gaussian spectral density centered at k_0,

$$\underline{E}(k) = \frac{\underline{E}_0 e^{-(k-k_0)^2/2\sigma^2}}{\sqrt{2\pi}\sigma} \tag{1.34}$$

the integration of Eq. (1.33) yields

$$\underline{E}(z,t) = \mathrm{Re}\underline{E}_0 e^{-i(\omega_0 t - k_0 z)} e^{[-\sigma^2 (z-v_g t)^2/2]} \tag{1.35}$$

The wave packet in this case consists of a Gaussian envelope propagating with the group velocity v_g, while the mode function rapidly oscillates within the envelope and propagates with the phase velocity ω_0/k_0.

Shown in Figure 1.5 are the spatial profile of the wave packet Eq. (1.35) at $t = 0$ and the power spectrum. The bandwidth of the power spectrum Δk is often defined by the width between two $1/e$ points from its peak, that is, $\Delta k = 2\sigma$. The spatial extent of the intensity envelope is likewise specified by $\Delta z = 2/\sigma = 4/\Delta k$. Given Δk, the frequency band width is given from the dispersion relation by $\Delta\omega = v_g \Delta k = 2v_g\sigma$. Finally, the time duration of the wave packet is given by $\Delta t = \Delta z/v_g = 4/\Delta\omega$. Therefore, the wave packet is characterized by the basic relation

$$\Delta z \propto \frac{1}{\Delta k}, \quad \Delta t \propto \frac{1}{\Delta\omega} \tag{1.36}$$

where the proportionality constants are of the order of unity and depends on the dispersion relation occurring in the power spectrum. The relationship (Eq. (1.36)) is of fundamental importance in quantum mechanics and is followed up in due course.

The Interference

The interference effect is a signature of the wave and was demonstrated by Young with his classic double-slit experiment as shown in Figure 1.6. In this experiment, two plane waves emanating from a distant source are passed through two slits. The two beams are detected on a screen L distance away from the slits. At a point

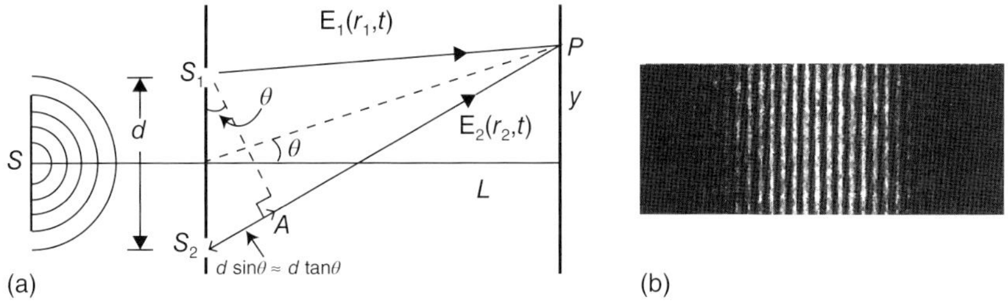

Figure 1.6 (a) Young's double-slit experimental scheme and (b) the observed fringe pattern.

P on the screen, the total field registered consists of the two plane waves:

$$\underline{E}\left(\underline{r}, t\right) = \sum_{j=1}^{2} \mathrm{Re}\underline{E}_0 e^{-i\left(\omega t - \underline{k}_j \cdot \underline{r}_j\right)} \tag{1.37}$$

The detected time-averaged intensity is thus given by

$$I = \left\langle \left(\underline{E}_1 + \underline{E}_2\right) \cdot \left(\underline{E}_1^* + \underline{E}_2^*\right)\right\rangle_t = \left|E_1\right|^2 + \left|\underline{E}_2\right|^2 + \left(\underline{E}_1 \cdot \underline{E}_2^* + \underline{E}_2 \cdot \underline{E}_1^*\right) \tag{1.38}$$

and consists of two background and interference terms, respectively. Naturally, the latter two terms depend on the difference in optical paths the two beams have traversed before reaching P. The resulting phase difference is given in the far-field approximation by $kd \sin\theta$ (Figure 1.6), and therefore I reads as

$$I = 2\left|\underline{E}_0\right|^2 (1 + \cos\varphi), \quad \varphi = kd \sin\theta \simeq \left(\frac{2\pi}{\lambda}\right) d \left(\frac{y}{L}\right) \tag{1.39}$$

where d and y are the space between two slits and the height of P on the screen, respectively. For $L \gg y$, $\sin\theta \simeq \tan\theta \simeq y/L$. Obviously, the interference term adds to or subtracts from the background, depending on the relative phase between the two beams. The maximum and minimum intensities are attained for $\varphi = 2n\pi$ and $\varphi = 2\pi\,(n + 1/2)$, respectively, with n denoting an integer. Therefore, bright and dark strips appear at $y_n = (\lambda L/d)\,n$ and $y_n = (\lambda L/d)\,(n + 1/2)$, respectively.

Problems

1.1 The H_2 molecule consists of two protons coupled via an effective spring with the spring constant k. The 1D Hamiltonian is given by (Figure 1.7)

$$H = \frac{1}{2} m_1 \dot{x}_1^2 + \frac{1}{2} m_2 \dot{x}_2^2 + \frac{1}{2} k \left(x_1 - x_2\right)^2$$

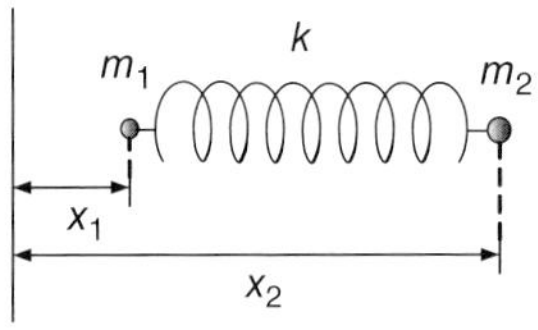

Figure 1.7 Two particles coupled via a spring with spring constant k.

(a) Introduce the center of mass and relative coordinates as

$$X = x_1 + x_2, \quad x = x_1 - x_2$$

and express the Hamiltonian in terms of X and x and interpret the result.

(b) Write down the equations of motion for the center of mass X and relative displacement x and interpret the equations of motion.

1.2 Find the thermal velocity of (a) electron, (b) proton, (c) H_2 molecule, and (d) particle of mass 1 g at $T = 10$, 300, and 1000 K.

1.3 (a) Show that the electric field given in Eq. (1.29) is the solution of the wave equation, provided ω, k satisfy the dispersion relation, $\omega^2 = v^2k^2$ with k denoting the wave vector.

(b) Show that the magnetic field intensity H given in Eq. (1.30) and $\mathbf{E}$ in Eq. (1.29) satisfy Faraday's law of induction and Ampere's circuital law in a medium free of charge and current.

(c) Derive the wave equation of $\boldsymbol{H}$.

1.4 Given the wave packet Eq. (1.35), find variance of $|\underline{\mathrm{E}}(z,t)|^2$ at $t = 0$

$$(\Delta z)^2 = \left\langle (z - \langle z \rangle)^2 \right\rangle; \quad \langle a \rangle \equiv \int_{-\infty}^{\infty} dz a |\mathrm{E}(z,0)|^2$$

1.5 By using the relations

$$\underline{\hat{x}} \cdot \underline{\hat{x}} = \underline{\hat{y}} \cdot \underline{\hat{y}} = \underline{\hat{z}} \cdot \underline{\hat{z}} = 1, \quad \underline{\hat{x}} \cdot \underline{\hat{y}} = \underline{\hat{y}} \cdot \underline{\hat{z}} = \underline{\hat{z}} \cdot \underline{\hat{x}} = 0,$$

$$\underline{\hat{x}} \times \underline{\hat{y}} = \underline{\hat{z}}, \underline{\hat{y}} \times \underline{\hat{z}} = \underline{\hat{x}}, \quad \underline{\hat{z}} \times \underline{\hat{x}} = \underline{\hat{y}}$$

show that all vectors are solenoidal, that is, $\nabla \cdot \nabla \times \underline{A} \equiv 0$.

1.6 By combining Eqs. (1.23), (1.25), and (1.26), show that $\boldsymbol{H}$ in Eq. (1.22) is solenoidal.

Suggested Readings

1. D. M. Kim, Introductory Quantum Mechanics for Semiconductor Nanotechnology, Wiley-VCH, 2010.
2. R. A. Serway, C. J. Moses, and C. A. Moyer, Modern Physics, Third Edition, Brooks Cole, 2004.
3. D. Halliday, R. Resnick, and J. Walker, Fundamentals of Physics Extended, Eighth edn, John Wiley & Sons, 2007.
4. L. C. Shen and J. A. Kong, Applied Electromagnetism, Second edn, PWS Publishing Company, 1987.

2
Milestones Leading to Quantum Mechanics

The milestone discoveries leading to the advent of quantum mechanics are discussed in conjunction with the concepts of the quantized energy level, photon, matter wave, and spectroscopy.

2.1
Blackbody Radiation and Quantum of Energy

The puzzling data confronting the classical theory was the spectral energy density ρ observed from a blackbody. A blackbody is a material that absorbs all radiations incident on its surface. A cavity with a small hole is a good implementation. Once the light passes through the hole into the cavity, it undergoes multiple reflections until it is absorbed by atoms on the surface of the wall. The equilibrium is established, and the atoms constantly absorb and emit the same amount of radiation. Figure 2.1 shows the observed $\rho(\nu)$, which rises and falls with increasing frequency ν at a given temperature T.

Rayleigh and Jeans partially explained the data by multiplying the number of standing-wave modes in the cavity in the frequency interval from ν to $\nu + d\nu$ and the average field energy $k_B T$ therein:

$$\rho(\nu) = (8\pi\nu^2/c^3)k_B T \tag{2.1}$$

The theory agrees with the data for small ν, but at high ν, the data exponentially fall down, while the theoretical curve increases without any upper bound. The disagreement between the theory and the experiment is known as the *ultraviolet catastrophe*.

To resolve the problem, Planck introduced the novel concept of the quantum of energy. He postulated that a system oscillating with frequency ν is inherently associated with the quantum of energy $\varepsilon = h\nu$ that cannot be divided. The constant h is called the *universal Planck constant* and has the value 6.626×10^{-34} J s or 4.136×10^{-15} eV s. By using the postulate, we can now find the average energy as

Introductory Quantum Mechanics for Applied Nanotechnology, First Edition. Dae Mann Kim.

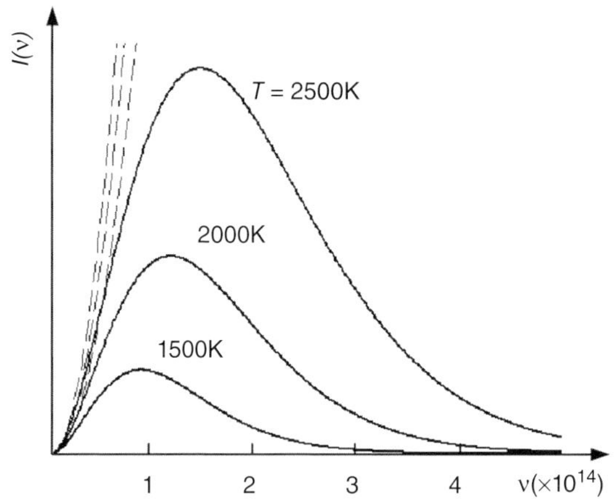

Figure 2.1 The spectral intensity of the blackbody radiation versus frequency at different temperatures. Also shown are the Rayleigh–Jeans's theoretical curves (broken lines).

$$\langle \varepsilon \rangle = \frac{\sum_{n=0}^{\infty} \varepsilon_n e^{-\beta\varepsilon_n}}{\sum_{n=0}^{\infty} e^{-\beta\varepsilon_n}} = -\frac{\partial}{\partial\beta} \ln \sum_{n=0}^{\infty} e^{-\beta\varepsilon_n}, \quad \beta \equiv \frac{1}{k_B T} \tag{2.2a}$$

where $\exp-(\beta\varepsilon_n)$ is the Boltzmann probability factor, as discussed. As the energy ε varies digitally in units of $h\nu$, that is, $\varepsilon_n = nh\nu$, we can sum up the probability factors as

$$\sum_{n=0}^{\infty} e^{-\beta\varepsilon_n} = \frac{1}{1 - e^{-\beta h\nu}} \tag{2.2b}$$

Thus, by inserting Eq. (2.2b) in Eq. (2.2a) and multiplying the resulting average energy by the mode density, which will be further detailed later on, we obtain

$$\rho(\nu) = \frac{8\pi\nu^2}{c^3}\langle\varepsilon\rangle = \frac{8\pi\nu^2}{c^3}\frac{h\nu}{\left(e^{h\nu/k_B T} - 1\right)} \tag{2.3}$$

Equation (2.3) is the celebrated Planck theory and quantitatively accounts for the data. For small ν, $h\nu \ll k_B T$ and Eq. (2.3) reduces to Eq. (2.1) and for large ν, $\rho(\nu)$ decreases exponentially with increasing ν in agreement with the data. The cornerstone of the theory is the quantum of energy.

2.2
Photoelectric Effect and Photon

The cathode-ray tube has been instrumental in bringing out key discoveries and concepts in the history of physics, and the photoelectric effect is one of such

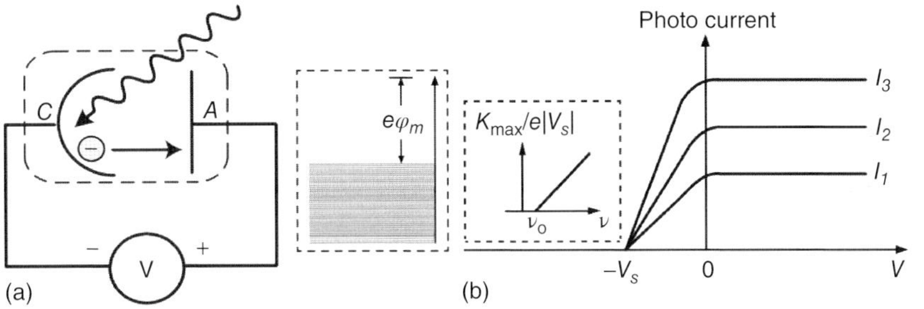

Figure 2.2 A vacuum tube with the cathode and the anode. The cathode modeled by a sea of electrons confined by potential barrier at the surface (a). The photocurrent versus the anode voltage (b). Also shown is the stopping power eV_S versus frequency of the incident light.

examples. The tube is made of glass, filled with a rarefied gas, and the anode and cathode inserted therein (Figure 2.2). The photoelectric effect consists of the input light on the cathode producing the output photocurrent, I_P. A striking feature of I_P is that it flows only when the light frequency ν is greater than a critical value for given cathode. Also, I_P terminates at a negative bias at the anode $-V_S$, and the stopping power eV_S increases linearly with ν. Naturally, I_P is contributed by electrons, emitted from the cathode and pulled toward the anode by the positive bias. In the classical theory, the energy gained by the electrons from the light is proportional to the light intensity. This suggests that eV_S should increase with the intensity of light in contrast with the observed data.

Einstein resolved the puzzling features of I_P by invoking Planck's concept of the quantum of energy. Specifically, the light of frequency ν was taken to consist of photons, with each photon carrying the undividable quantum of energy $h\nu$ and traveling with the velocity of light. In this corpuscular picture of the light, the intensity I is given by the flux of photons $I/h\nu$ crossing unit area per unit time. As the energy $h\nu$ cannot be divided, a photon interacts digitally with an electron and is either absorbed completely, imparting all of its energy to the electron, or not absorbed at all. Hence, the energy of the photon absorbed is used in part for the electron to overcome the surface barrier potential, while the remainder is converted to the kinetic energy of the emitted electron,

$$h\nu = e\varphi + \frac{mv^2}{2} \tag{2.4}$$

where the surface barrier of the cathode $e\varphi$ is called the *work function* (Figure 2.2).

Equation (2.4) explains the observed behavior of the I_P data. The higher light intensity or the greater photon flux should emit more electrons from the cathode, increasing I_P. Also, with increasing bias V at the anode, the electrons are guided more efficiently, and I_P should increase with V. But the current saturates when the guiding efficiency reaches unity. Also electrons are emitted only when $h\nu \geq e\varphi$, which accounts for the critical frequency required for I_P. Finally, eV_S is determined by the kinetic energy given by Eq. (2.4) and therefore depends only on ν, in agreement with the data.

2.3 Compton Scattering

The photon picture of light was also demonstrated by Compton, who performed the X-ray scattering experiments as sketched in Figure 2.3. An X-ray beam irradiating an electron is scattered off at an angle θ, while the target electron recoils back. The wavelength of the scattered X-ray is shifted by an amount depending on θ. This is in contradiction with the classical theory, which predicts that the shift is caused by the intensity and illumination time of the radiation.

Compton interpreted the data by modeling the X-ray to consist of photons, streaming with the velocity of light c with the quantum of energy $h\nu$ and zero rest mass. Hence, the energy and momentum of the photon are specified from the special theory of relativity as

$$h\nu = [m_{\mathrm{ph}}^2 c^4 + c^2 p^2]^{1/2} = cp, \quad m_{\mathrm{ph}} = 0 \tag{2.5}$$

During the collision, both the energy and momentum are conserved:

$$p_i c + mc^2 = p_f c + (m^2 c^4 + p_e^2 c^2)^{1/2} \tag{2.6}$$

$$\mathbf{p}_i = \mathbf{p}_f + \mathbf{p}_e \tag{2.7}$$

where $\mathbf{p}_i$, $\mathbf{p}_f$ are the photon momenta before and after the scattering, $\mathbf{p}_e$ the momentum of the electron after the scattering due to recoil, and m, mc^2 the rest mass and rest energy of electron. Hence, by finding p_e^2 from Eq. (2.7) as

$$p_e^2 = \mathbf{p}_e \cdot \mathbf{p}_e = (\mathbf{p}_i - \mathbf{p}_f) \cdot (\mathbf{p}_i - \mathbf{p}_f) = p_i^2 + p_f^2 - 2p_i p_f \cos\theta$$

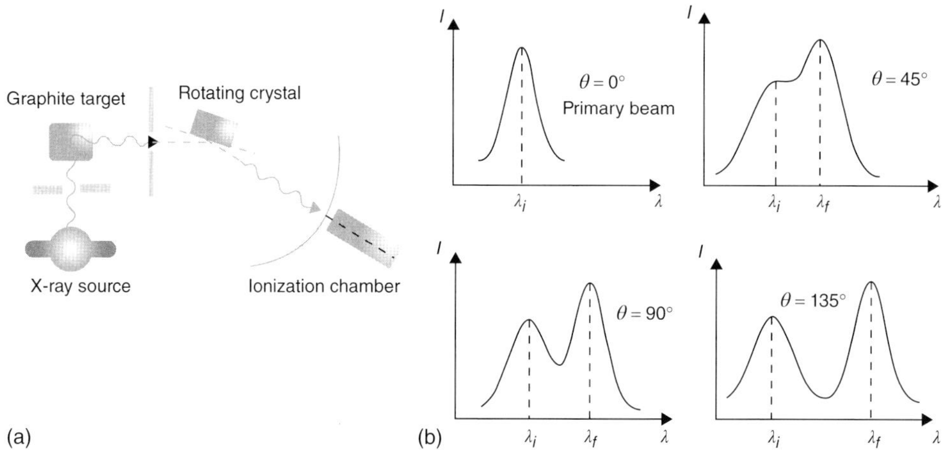

Figure 2.3 The experimental setup of the X-ray scattering (a) and the measured shifts of the wavelengths versus the scattering angle (b).

and equating it to p_e^2 as it appears in Eq. (2.6), we find after a straightforward algebra

$$\frac{1}{p_f} - \frac{1}{p_i} = \frac{2}{mc}\sin^2\left(\frac{\theta}{2}\right) \tag{2.8}$$

Hence, the shift in wavelength due to scattering can be found by expressing p_f, p_i in terms of the corresponding wavelength via the relation $p = h\nu/c = h/\lambda$ (see Eq. (2.5)):

$$\Delta\lambda \equiv \lambda_f - \lambda_i = 4\pi\lambda_e \sin^2\left(\frac{\theta}{2}\right); \quad \lambda_e = \frac{2\hbar}{mc}, \quad \hbar \equiv \frac{h}{2\pi} \tag{2.9}$$

The constant λ_e is the Compton wavelength having the value 4×10^{-4} nm, and Eq. (2.9) is in agreement with the data, confirming thereby the photon picture of light. As λ_e is independent of the wavelength, the relative shift, $\Delta\lambda/\lambda_i$, is more readily observed in the X-ray wavelength region. The binding energy of the electron bound to an atom is small compared with the energies involved in the scattering and has been neglected in Eq. (2.6).

2.4 de Broglie Wavelength and Duality of Matter

The wave nature of light has been firmly built into the classical optics as evidenced, for example, by Young's interference experiment, but the corpuscular nature of light has also been demonstrated experimentally. The two different tracks point to the duality of light, namely, that the light exhibits both the wave-like and particle-like natures.

On the material side, the particle nature of electrons, atoms, molecules, and so on, has been taken for granted. But de Broglie introduced a daring concept of the matter wave and postulated that a particle also behaves as a wave with wavelength λ given by

$$\lambda = \frac{h}{p} = \frac{h}{[2mE]^{1/2}}; \quad E = \frac{p^2}{2m} \tag{2.10}$$

where h, p, and E denote Planck constant, the linear momentum, and kinetic energy of the particle, respectively. The wavelength λ thus introduced is called the *de Broglie wavelength*. The matter wave was experimentally confirmed by Davisson and Germer, who obtained the diffraction pattern of electrons just like that of the X-ray (Figure 2.4).

Thus, the duality of matter was also established, and a particle has to be taken to exhibit both the particle-like and wave-like natures. Although abstract in concept, the matter wave has become an integral part of everyday life. The electron microscope, for example, utilizes the wave nature of electrons just as the optical microscope uses visible light for imaging the object. In the electron microscope, the wavelength λ can be tuned by varying p via the kinetic energy. Specifically, the

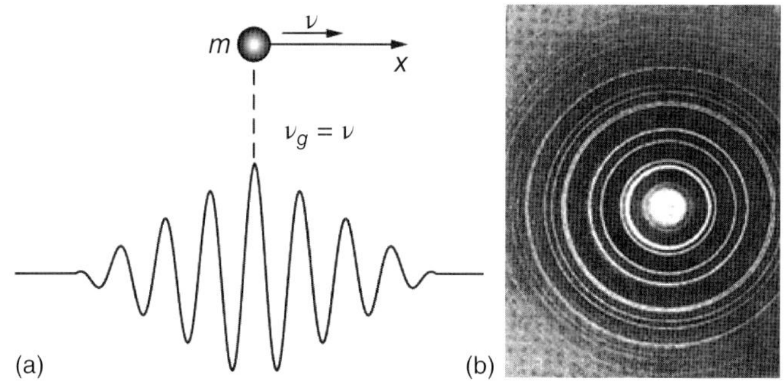

Figure 2.4 The graphical representation of de Broglie matter wave (a) and the diffraction pattern of 50 keV electrons from Cu_3Au film (b). (Courtesy of the late Dr. L. H. Germer.)

electron accelerated by the voltage V possesses the de Broglie wavelength given in nanometers by

$$\lambda = \frac{1.226}{\sqrt{E(\mathrm{eV})}}\,\mathrm{nm} \tag{2.11}$$

2.5
Hydrogen Atom and Spectroscopy

Bohr's theory of the hydrogen atom is a landmark achievement, and it culminates the old quantum theory. Bohr modeled the H-atom to consist of an electron revolving in the circular orbit around the proton. The model is derived from the α particle scattering experiment by Rutherford, which pointed to the existence of a nucleus at the center of an atom. The atomic model brought out the issue of the stability of matter. An electron in a circular orbit is subjected to acceleration and should therefore emit the radiation, according to the electrodynamics. Therefore, the electron should lose energy constantly while circling around the nucleus and spiral into the nucleus.

Additionally, the radiation emitted from the H-atom was observed to consist of several sets of infinite number of discrete lines instead of the continuous spectra, as predicted by the electrodynamics. The observed spectral lines were shown empirically fitted by the Ritz combination rule:

$$\frac{1}{\lambda} = R\left(\frac{1}{n^2} - \frac{1}{m^2}\right), \quad n < m \tag{2.12}$$

where λ is the wavelength, n, m positive integers, and R the Rydberg constant with the value $R = 0.010973732$ per nm. Each infinite series of discrete lines can be fitted by fixing n while varying m: Lyman series, $n = 1$ and $m \geq 2$; Balmer series, $n = 2$ and $m \geq 3$; Paschen series, $n = 3$ and $m \geq 4$; Brackett series, $n = 4$ and $m \geq 5$; Pfund series, $n = 5$ and $m \geq 6$.

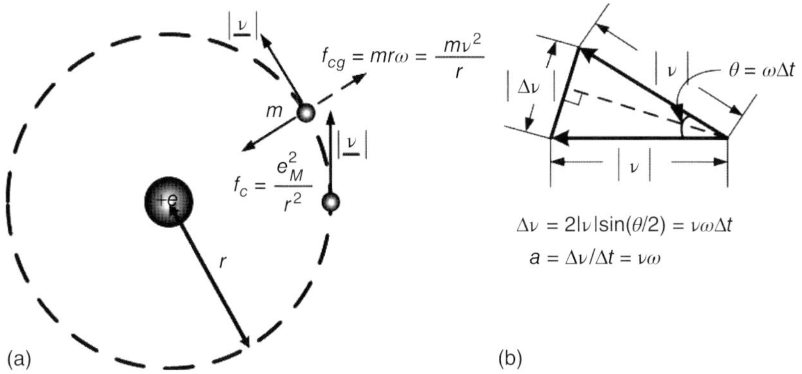

Figure 2.5 (a,b) The circular orbit of the electron around the proton formed by the balance between the centripetal and centrifugal forces. Also shown is the graphical description of the centrifugal force.

Bohr introduced a few basic postulates in his theory of the H-atom.

1) *Quantized orbits*: The electron resides in stable, non-radiating circular orbits whose angular momentum L with respect to proton has a discrete set of values:

$$L_n \equiv mv_n r_n = n\hbar, \quad \hbar \equiv \frac{h}{2\pi}, \quad n = 1, 2, 3, \dots \tag{2.13}$$

2) *Quantum transition*: The electron can make transitions between two orbits, emitting or absorbing the radiation to conserve the energy.

The circular orbit is maintained in general by the balance of two forces, that is,

$$\frac{e_M^2}{r_n^2} = \frac{mv_n^2}{r_n}, \quad e_M^2 \equiv \frac{e^2}{4\pi\varepsilon_0} \tag{2.14}$$

where ε_0 is the vacuum permittivity. The left-hand side is the centripetal force resulting from the attractive Coulomb force between the proton and the electron. The right-hand side is the centrifugal force associated with a circular motion (Figure 2.5). We can combine Eqs. (2.13) and (2.14) and eliminate v_n, obtaining

$$r_n = r_B n^2; \quad r_B \equiv \frac{\hbar^2}{me_M^2} \tag{2.15}$$

where r_B is known as the *Bohr radius* and has the value $r_B = 0.053\,\text{nm}$.

We can also combine Eqs. (2.14) and (2.15) and find the kinetic energy K_n and the total energy E_n of the electron in the nth orbit as

$$K_n \equiv \frac{1}{2}mv_n^2 = \frac{1}{2}\frac{e_M^2}{r_n} = -\frac{1}{2}V_n, \quad V_n \equiv -\frac{e_M^2}{r_n} \tag{2.16}$$

$$E_n = K_n + V_n = -E_0\frac{1}{n^2}, \quad E_0 = \frac{e^4 m}{2(4\pi\varepsilon_0)^2\hbar^2} \tag{2.17}$$

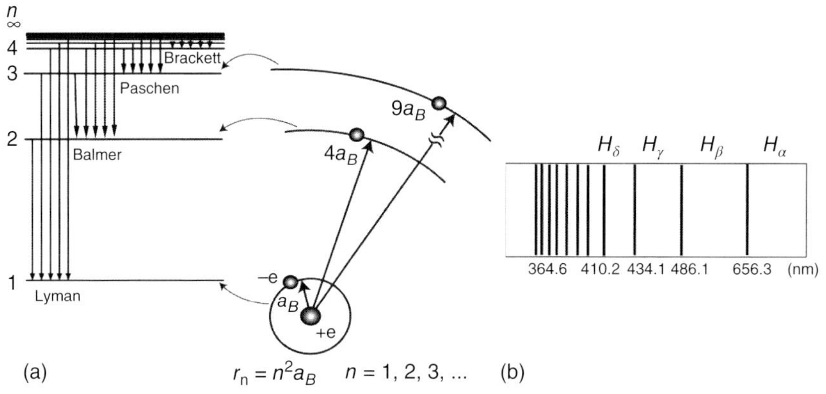

Figure 2.6 The quantized energy level and the corresponding electron orbits of the H-atom (a). The Balmer series of the emission spectral lines (b).

and $E_0 = 13.6\,\text{eV}$ is called the *ionization energy of the H-atom*, and the integer n is known as the *principal quantum number*.

As mentioned, the electron can make the transition from higher (n_i) to lower (n_f) orbits, emitting a photon of frequency ν or wavelength λ to conserve energy:

$$h\nu = \frac{hc}{\lambda} = E_0\left(\frac{1}{n_f^2} - \frac{1}{n_i^2}\right) \tag{2.18}$$

By comparing Eq. (2.18) with Ritz's combination rule Eq. (2.12), the Rydberg constant is theoretically specified as

$$R = \frac{E_0}{hc} = \frac{e^4 m}{4\pi(4\pi\varepsilon_0)^2\hbar^3 c} \tag{2.19}$$

The agreement of the empirical constant with its theoretical expression is an outstanding highlight of the theoretical physics. The lowest energy level for $n = 1$ is called the *ground state*, and E_0 is the energy required to ionize the atom by knocking out the electron from the ground state to the vacuum level, hence the name ionization energy. The discrete orbits, energy levels, and the quantum transitions are shown in Figure 2.6.

Atomic Orbits and Standing Waves

The key point of Bohr's theory is to quantize the angular momentum and introduce the electron orbits. The electron in these orbits has the de Broglie wavelength given by

$$\lambda_n \equiv \frac{h}{p_n} = \frac{h}{mv_n}, \quad n = 1, 2, \ldots \tag{2.20}$$

When v_n is replaced by r_n by using the quantization condition (Eq. (2.13)), Eq. (2.20) reduces to

$$2\pi r_n = n\lambda_n \tag{2.21}$$

and states that the circumference of the quantized orbit is an integer multiple of de Broglie wavelength of the electron therein. This means that the optical path of the orbit satisfies the standing-wave condition. If the condition is not met, the wave interferes with itself destructively, and the electron cannot reside in the orbit.

Problems

2.1 (a) Calculate the energy of a photon in electron volt units with the wavelengths 10 m (radiowave), 1 m (microwave), 10 000 nm (infrared), 600 nm (visible), 200 nm (ultraviolet), 50 nm (EUV), and 1 nm (X-ray).
(b) Find the corresponding wave numbers $1/\lambda$ per centimeter.

2.2 Find the de Broglie wavelengths of
(a) the electron, proton, and H-atom moving at room temperature.
(b) The electron with kinetic energy 200 eV, 100 keV, and 1 MeV.
(c) The electron in the ground state of the H-atom.

2.3 (a) Show that the Planck theory (Eq. (2.3)) reduces to R–J theory (Eq. (2.1)) for $h\nu \ll k_B T$.
(b) Fill in the algebra and derive the theoretical description of the X-ray scattering (Eq. (2.9)) from Eqs. (2.6) and (2.7).
(c) Combine Eqs. (2.13) and (2.14) and derive the Bohr radius (Eq. (2.15)).

2.4 (a) Lithium, beryllium, and mercury have the work functions of 2.3, 3.9, and 4.5 eV, respectively. Which metal will exhibit the photoelectric effect and find the stopping power therein when the light of wavelength 300 nm is incident.
(b) The stopping powers of photoelectrons from aluminum are −2.3 and −0.9 V when emitted by light of wave length of 194 and 248 nm, respectively. Find from these data Planck's constant and the work function of the aluminum.

2.5 The ionized helium atom He^+ is a hydrogen-like atom consisting of two protons in the nucleus and one electron revolving around the nucleus. Find the ionization energy in electron volt unit, the atomic radius of the ground state with $n = 1$, and the shortest and longest wavelengths of Balmer series.

2.6 The X-ray with energy 200 keV is scattered off an electron at rest. If the scattered beam is detected at 90° with respect to the incident direction, find
(a) the shift in wavelength and energy of the scattered X-ray and
(b) the kinetic energy of the recoiling electron.

2.7 (a) Find the radius, kinetic, potential, and total energies of an electron in the ground ($n = 1$) and first excited ($n = 2$) states of the H-atom.
Find the transition wavelength between the first excited state and the ground state.

Suggested Readings

1. D. M. Kim, Introductory Quantum Mechanics for Semiconductor Nanotechnology, Wiley-VCH, 2010.
2. R. A. Serway, C. J. Moses, and C. A. Moyer, Modern Physics, Third Edition, Brooks Cole, 2004.
3. D. Halliday, R. Resnick, and J. Walker, Fundamentals of Physics Extended, Eighth Edition, John Wiley & Sons, 2007.
4. J. Singh, Quantum Mechanics, Fundamentals and Applications to Technology, John Wiley & Sons, 1996.
5. R. L. Liboff, Introductory Quantum Mechanics, Fourth Edition, Addison Wesley Publishing Company, Reading, MA, 2002.
6. S. Gasiorowics, Quantum Physics, Third Edition, John Wiley & Sons, 2003.
7. A. I. M. Rae, Quantum Mechanics, Fourth Edition, Taylor & Francis, 2002.

3
Schrödinger Wave Equation

The Schrödinger wave equation is to the quantum mechanics what Newton's equation of motion is to the classical mechanics. Both equations represent the basic postulates, the validity of which can be proven solely by the agreement between the theoretical results derived from it and experimental data. The Schrödinger equation of a particle reads as

$$i\hbar \frac{\partial \psi(\underline{r}, t)}{\partial t} = \hat{H}\psi(\underline{r}, t), \quad \hbar \equiv \frac{h}{2\pi} \tag{3.1a}$$

where the Hamiltonian $\hat{H}$ is the total energy consisting of the kinetic K and potential V energies and Ψ the wavefunction of the particle.

3.1
Operator Algebra and Basic Postulates

In quantum mechanics, a physical quantity is inherently associated with an operator. For instance, the momentum and energy are represented by the operators

$$\underline{p} \to -i\hbar\nabla, \quad E \to i\hbar\frac{\partial}{\partial t} \tag{3.1b}$$

so that the Hamiltonian operator reads as

$$\hat{H} = \frac{p^2}{2m} + V = -\frac{\hbar^2}{2m}\nabla^2 + V(\underline{r}); \quad \nabla^2 = (\frac{\partial^2}{\partial x^2} + \frac{\partial^2}{\partial y^2} + \frac{\partial^2}{\partial z^2}) \tag{3.1c}$$

The wave equation is a linear, second-order partial differential equation. The essence of quantum mechanics is to find Ψ and extract the dynamical information from it based on a few postulates as summarized below.

Postulates

(i) A dynamical system is associated with a wavefunction $\psi(\underline{r}, t)$ that contains all possible information of the system, (ii) ψ evolves in time according to the Schrödinger equation, and (iii) the quantity $\psi^*\psi\, d\boldsymbol{r}$ represents the probability of finding the system in the volume element $d\boldsymbol{r}$ at $\boldsymbol{r}$ and at time t. Hence, the

Introductory Quantum Mechanics for Applied Nanotechnology, First Edition. Dae Mann Kim.

expectation or average value of a physical quantity as represented by an operator $\widehat{A}$ is theoretically described by

$$\langle \widehat{A} \rangle = \frac{\int_{-\infty}^{\infty} d\underline{r}\psi^*(\underline{r},t)\widehat{A}\psi(\underline{r},t)}{\int_{-\infty}^{\infty} d\underline{r}\psi^*(\underline{r},t)\psi(\underline{r},t)} \tag{3.2}$$

Thus, $\psi^*\psi$ plays the role of a distribution function except that the operator is inserted in between the wavefunctions.

Bra and Ket Vectors

The spatial integration involving a product of two functions f^* and g is called the *inner product* and is compactly expressed as

$$\int d\underline{r}f^*(\underline{r})g(\underline{r}) \equiv \langle f|g\rangle; \quad f^* \to \langle f|, g \to |g\rangle \tag{3.3}$$

Here f^* ($\langle f|$) and g ($|g\rangle$) are called the *bra and ket vectors*, and the expectation value of $\widehat{A}$ is also compactly expressed as

$$\langle \widehat{A} \rangle = \frac{\langle \Psi|\widehat{A}|\Psi\rangle}{\langle \Psi|\Psi\rangle} \tag{3.4}$$

3.2
Eigenequation, Eigenfuntion and Eigenvalue

Energy Eigenfunction

The time-dependent Schrödinger equation (3.1) is analyzed in general by looking for the solution in the form

$$\psi(\underline{r},t) = T(t)u(\underline{r}) = e^{-i(Et/\hbar)}u(\underline{r}) \tag{3.5}$$

where E is the total energy of the system. Upon inserting Eq. (3.5) into Eq. (3.1) and canceling the exponential factor from both sides, there results

$$\widehat{H}u(\underline{r}) = Eu(\underline{r}) \tag{3.6}$$

The time-independent equation (3.6) is called an *eigenequation,* in which an operator, for example, $\widehat{H}$, acting on u reproduces the same function, multiplied by a constant E. In this case, u and E are called the *energy eigenfunction* and *eigenvalue,* respectively. The eigenequation generates a set of eigenfunctions $\{u_n\}$ and eigenvalues $\{E_n\}$, and the wavefunction can be generally expanded in terms of the eigenfunctions as

$$\psi(\underline{r},t) = \sum_n a_n e^{-i\omega_n t}u_n(\underline{r}), \quad \omega_n \equiv \frac{E_n}{\hbar} \tag{3.7}$$

where a_n is the expansion coefficient.

Momentum Eigenfunction

Let us consider a free particle moving in the x-direction. The momentum eigenequation reads then from Eq. (3.1b) as

$$-i\hbar\frac{\partial}{\partial x}u(x) = p_x u(x), \quad \hat{p}_x \equiv -i\hbar\frac{\partial}{\partial x} \tag{3.8}$$

By dividing both sides with the eigenfunction $u(x)$, Eq. (3.8) is rearranged as

$$\frac{\partial u(x)}{u(x)} = ik_x \partial x, \quad k_x \equiv \frac{p_x}{\hbar}$$

Thus, by integrating both sides with respect to x $u(x)$ is readily found as

$$u(x) = Ne^{ik_x x}, \quad p_x = \hbar k_x \tag{3.9}$$

where N is the constant of integration, $\hbar k_x$ the momentum eigenvalue, and k_x the wave vector, which plays essentially the same role as the optical wave vector.

The wave vector k_x is determined by the boundary conditions imposed. For example, when a periodic boundary condition is imposed in the interval from 0 to L, that is, $u(0) = u(L)$, then k_x is found from Eq. (3.9) by $k_{xn}L = 2\pi n$ with n denoting an integer.

Also, the constant N can be used for normalizing $u(x)$:

$$1 = \int_0^L dxu^* u = N^2 L$$

Therefore, the normalized eigenfunction of the momentum is given by

$$u_n(x) = \left(\frac{1}{L}\right)^{1/2} e^{in(2\pi/L)x}, \quad n = \pm 1, \pm 2, \ldots \tag{3.10}$$

The 1D momentum eigenequation (Eq. (3.8)) is straightforwardly extended to 3D as

$$-i\hbar\nabla u(\underline{r}) = \underline{p}u(\underline{r}), \quad \nabla = \left[\hat{x}\frac{\partial}{\partial x} + \hat{y}\frac{\partial}{\partial y} + \hat{z}\frac{\partial}{\partial z}\right] \tag{3.11}$$

and the eigenfunction can likewise be found in analogy with Eq. (3.10) as

$$u(\underline{r}) = \frac{1}{(L_x L_y L_z)^{1/2}} e^{i\underline{k}\cdot\underline{r}}, \quad \underline{k} = \hat{x}k_x + \hat{y}k_y + \hat{z}k_z \tag{3.12}$$

3.3 Properties of Eigenfunctions

A few basic properties of the eigenfunction are presented as follows:

Hermitian Operator

An operator $\hat{A}$ of a physical quantity should satisfy the Hermitian condition given by

$$\int d\underline{r} f^*(\underline{r})\hat{A}g(\underline{r}) = \int d\underline{r}[\hat{A}f(\underline{r})]^* g(\underline{r}); \quad \langle f|\hat{A}g\rangle = \langle \hat{A}f|g\rangle \tag{3.13}$$

where f and g are arbitrary well-behaving functions, differentiable and vanishing at infinity. The condition has also been expressed in terms of the bra – ket notation. The Hemiticity of the1D momentum operator, for example, can be shown with the use of integration by parts as follows:

$$\int_{-\infty}^{\infty} dx f^*(-i\hbar\frac{\partial}{\partial x})g = -i\hbar\left[f^*g|_{-\infty}^{\infty} - \int_{-\infty}^{\infty} dx g\frac{\partial f^*}{\partial x}\right] = \int_{-\infty}^{\infty} dx(-i\hbar\frac{\partial}{\partial x}f)^* g$$

Orthogonality of Eigenfunctions

The eigenfunctions of a Hermitian operator are orthogonal to each other, and the eigenvalues are real. To prove the theorem, let us consider the eigenequation

$$\hat{A}u_n(\underline{r}) = a_n u_n(\underline{r}); \quad \hat{A}|n\rangle = a_n|n\rangle, \quad |u_n(\underline{r})\rangle \equiv |n\rangle \tag{3.14}$$

where n is an integer called the *quantum number*. By performing the inner product on both sides, we obtain

$$\int_{-\infty}^{\infty} d\underline{r}u_m^*\hat{A}u_n = a_n\int_{-\infty}^{\infty} d\underline{r}u_m^*u_n; \quad \langle u_m|\hat{A}u_n\rangle = a_n\langle u_m|u_n\rangle \tag{3.15}$$

Upon interchanging u_n and u_m, Eq. (3.15) reads as

$$\int_{-\infty}^{\infty} d\underline{r}u_n^*\hat{A}u_m = a_m\int_{-\infty}^{\infty} d\underline{r}u_n^*u_m; \quad \langle u_n|\hat{A}u_m\rangle = a_m\langle u_n|u_m\rangle \tag{3.16}$$

As $\hat{A}$ is Hermitian, Eq. (3.16) can also be expressed as

$$\int_{-\infty}^{\infty} d\underline{r}(\hat{A}u_n)^* u_m = a_m\int_{-\infty}^{\infty} d\underline{r}u_n^*u_m; \quad \langle \hat{A}u_n|u_m\rangle = a_m\langle u_n|u_m\rangle \tag{3.17}$$

and its complex conjugate is given by

$$\int_{-\infty}^{\infty} d\underline{r}u_m^*(\hat{A}u_n) = a_m^*\int_{-\infty}^{\infty} d\underline{r}u_m^*u_n; \quad \langle u_m|\hat{A}u_n\rangle = a_m^*\langle u_m|u_n\rangle \tag{3.18}$$

By subtracting Eq. (3.18) from Eq. (3.15), there results

$$(a_n - a_m^*)\int_{-\infty}^{\infty} dru_m^*u_n = 0; \quad (a_n - a_m^*)\langle u_m|u_n\rangle = 0 \tag{3.19}$$

For $n = m$, $u_n^*u_n$ is positive definite, and the inner product does not vanish, hence $a_n = a_n^*$, that is, the eigenvalue is real. For $n \neq m$, the eigenvalues are not the same, that is, $a_n \neq a_m$, in the nondegenerate system. Hence, the eigenfunctions should be orthogonal, that is,

$$\int_{-\infty}^{\infty} dru_m^*u_n = 0; \quad \langle u_m|u_n\rangle = 0, \quad n \neq m \tag{3.20}$$

For the degenerate case in which the eigenvalues can be the same even if $n \neq m$, the present proof does not apply. However, the degenerate eigenfunctions can be made orthogonal by devising appropriate linear combinations of the eigenfunctions.

The Completeness of Eigenfunctions

The wavefunction can be expanded in terms of a set of eigenfunctions that are orthogonal and normalized or orthonormal for short:

$$|\varphi(\underline{r})\rangle = \sum_{n=0}^{\infty} c_n|u_n(\underline{r})\rangle = \sum_{n=0}^{\infty} c_n|n\rangle \tag{3.21}$$

where the expansion coefficient c_n is specified by means of the inner product as

$$\langle k|\varphi(\underline{r})\rangle = \sum_{n=0}^{\infty} c_n\langle k|n\rangle = \sum_{n=0}^{\infty} c_n\delta_{kn} = c_k \tag{3.22a}$$

where δ_{kn} is called the *Kronecker delta function* and is defined as

$$\delta_{kn} = \begin{cases} 1 & k = n \\ 0 & k \neq n \end{cases} \tag{3.22b}$$

Hence, by inserting Eq. (3.22) into Eq. (3.21), the eigenfunction is represented by

$$|\varphi(\underline{r})\rangle = \sum_{n=0}^{\infty}\langle n|\varphi(\underline{r})\rangle|n\rangle = \sum_{n=0}^{\infty}|n\rangle\langle n|\varphi(\underline{r})\rangle \tag{3.23}$$

In Eq. (3.23), the constant c_n has been slipped past the ket vector. Hence, it is clear that

$$\sum_{n=0}^{\infty}|n\rangle\langle n| = I \tag{3.24}$$

The identity Eq. (3.24) is known as the *closure property* and represents the completeness of the eigenfunctions of the Hermitian operators.

It is interesting to note the similarity existing between the expansion scheme of Eq. (3.21) and the representation of a 3D vector:

$$\underline{A} = \hat{x}A_x + \hat{y}A_y + \hat{z}A_z$$

In this representation, the unit vectors are orthonormal, that is, $\hat{x}\cdot\hat{x} = 1$, $\hat{x}\cdot\hat{y} = 0$, and so on, and the three components are extracted by performing the scalar product $\hat{x}\cdot\underline{A} = A_x$, $\hat{y}\cdot\underline{A} = A_y$, and $\hat{z}\cdot\underline{A} = A_z$. In a similar context, $|\varphi(\underline{r})\rangle$ is to be viewed as a vector in infinite orthogonal Hilbert space and can be expanded in terms of $\{u_n\}$ with the set of expansion coefficient $\{c_n\}$ specified by means of the inner product.

3.4 Commutation Relation and Conjugate Variables

The motion of a particle can be described precisely in classical mechanics, as discussed. The theory presupposes that the act of measurement does not disturb the dynamical system under investigation. In contrast, the quantum mechanical description is based on the premise that measurement process itself disturbs and modifies the system. The modifications are pronounced in microsystems such as atoms, molecules, and electrons.

Commutation Relation

A thought experiment for measuring the size of the H-atom makes the point clear. To resolve its diameter, the probing light should have wavelength $\lambda < 0.1$ nm (see Eq. (2.15)) or the frequency $\nu\ (= c/\lambda)$ greater than 3×10^{18} Hz. Thus, the probing photons should have the energy $h\nu$ greater than $\sim 1.23 \times 10^4$ eV, a value larger than the binding energy of the H-atom ~ 13.6 eV by orders of magnitude. Hence, the measurement would end up by ionizing the H-atom.

The thought experiment implies that the consecutive measurements of two physical quantities, or operators, $\hat{A}$, $\hat{B}$ do not necessarily yield the same results, when performed in reverse order. That is to say, the respective theoretical values are not necessarily the same:

$$\langle\psi|\hat{A}\hat{B}|\psi\rangle \neq \langle\psi|\hat{B}\hat{A}|\psi\rangle \tag{3.25}$$

Equivalently, the commutator of two operators is not necessarily zero:

$$[\hat{A}, \hat{B}] \equiv \hat{A}\hat{B} - \hat{B}\hat{A} \neq 0 \tag{3.26}$$

Conjugate Variables

The typical examples of the noncommuting operators are the canonically conjugate variables appearing in pairs in Hamilton's equation of motion (Eq. (1.6)). Specifically, the conjugate pairs obey the relation

$$[x, p_x] = [y, p_y] = [z, p_z] = i\hbar \tag{3.27}$$

The relation (Eq. (3.27)) can be proven for x and p_x, for example, as

$$[x, p_x]f(x) \equiv x\left(-i\hbar\frac{\partial}{\partial x}\right)f(x) - \left(-i\hbar\frac{\partial}{\partial x}\right)[xf(x)] = i\hbar f(x) \quad \text{q.e.d.}$$

where $f(x)$ is an arbitrary function. The combinations of the position and momentum operators other than those in Eq. (3.27) commute, however.

Commuting Operators and Common Eigenfunction

It is important to point out that the commutation relation carries important consequences. For example, if the two operators commute, they can share a common eigenfunction. To prove it, let us consider the eigenfunction of $\hat{B}$:

$$\hat{B}|u_n\rangle = b_n|u_n\rangle \tag{3.28}$$

As $\hat{A}\hat{B} = \hat{B}\hat{A}$ in this case, it follows from Eq. (3.28) that

$$\hat{A}\hat{B}|u_n\rangle = b_n\hat{A}|u_n\rangle \equiv \hat{B}\hat{A}|u_n\rangle \tag{3.29}$$

Therefore, the new function $|v_n\rangle \equiv \hat{A}|u_n\rangle$ is also an eigenfunction of $\hat{B}$. As an eigenfunction is determined to within a constant, one can put

$$|v_n\rangle \equiv \hat{A}|u_n\rangle \propto |u_n\rangle = a_n|u_n\rangle \tag{3.30}$$

proving thereby that $|u_n\rangle$ is also the eigenfunction of $\hat{A}$. Also, if $\hat{A}$, $\hat{B}$ share a common eigenfunction, we can write by definition

$$\langle u_n|\hat{A}\hat{B}|u_n\rangle = a_n b_n \langle u_n|u_n\rangle = a_n b_n, \quad \langle u_n|\hat{B}\hat{A}|u_n\rangle = a_n b_n \tag{3.31}$$

Hence, $\hat{A}$ and $\hat{B}$ are shown to commute. An additional implication of Eq. (3.31) is that it is possible to measure two commuting observables simultaneously.

3.5
Uncertainty Relation

Uncertainty in Position and Momentum

The fact that x and p_x do not commute carries an important consequence, namely, that it is not possible to precisely measure x and p_x simultaneously. Rather the uncertainty in $\boldsymbol{r}$ and $\boldsymbol{p}$ is specified by

$$\Delta x \Delta p_x \approx \hbar, \quad \Delta y \Delta p_y \approx \hbar, \quad \Delta z \Delta p_z \approx \hbar \tag{3.32}$$

The relations (Eq. (3.32)) constitute the crux of Heisenberg's uncertainty principle, and the principle is rooted in the wave nature of particles. The uncertainty in x, p_x, for instance, can be shown explicitly by considering the wavefunction of a free particle as represented by a Gaussian wave packet:

$$\psi(x,t) \propto e^{-(i\omega_0 t - k_0 x)} e^{-(x - v_g t)^2/2\sigma^2}; \quad \frac{E}{\hbar} = \omega_0, \quad \frac{p}{\hbar} = k_0 \tag{3.33}$$

The spatial profile of the probability density then reads as

$$|\psi(x,t=0)|^2 \propto e^{-(x^2/\sigma^2)} \tag{3.34}$$

and the uncertainty or variance Δx is obtained by evaluating the average values

$$(\Delta x)^2 \equiv \langle (x - \langle x\rangle)^2\rangle = \langle x^2 - 2x\langle x\rangle + \langle x\rangle^2\rangle = \langle x^2\rangle - \langle x\rangle^2 \tag{3.35a}$$

As the probability density is an even function of x, $\langle x\rangle = 0$ and $\langle x^2\rangle$ is evaluated as

$$\langle x^2\rangle \equiv \frac{\displaystyle\int_{-\infty}^{\infty} dx x^2 e^{-(x^2/\sigma^2)}}{\displaystyle\int_{-\infty}^{\infty} dx e^{-(x^2/\sigma^2)}} = \frac{\sigma^2}{2} \tag{3.35b}$$

Hence, the uncertainty in position is given by

$$\Delta x = \frac{\sigma}{\sqrt{2}} \tag{3.36}$$

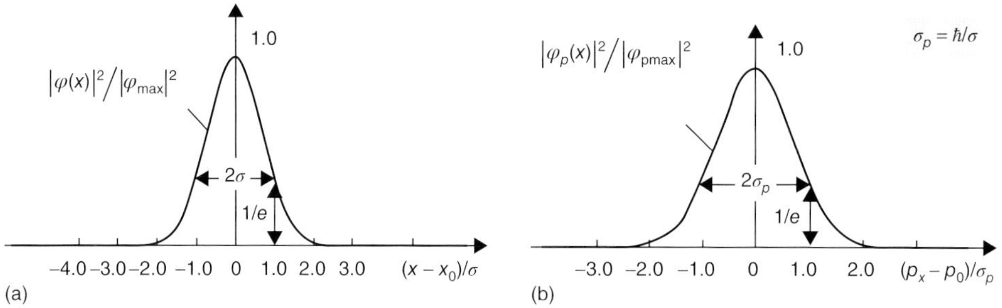

Figure 3.1 The spatial profile of the intensity of a Gaussian wave packet (a) and the distribution of the momentum expansion coefficients associated (b).

Also, at $x = 0$, for example, the temporal profile of Ψ is given from Eq. (3.33) by

$$|\psi(t, x = 0)|^2 \propto e^{-(t^2/\sigma_t^2)}, \quad \sigma_t^2 \equiv \frac{\sigma^2}{v_g^2} \tag{3.37}$$

The variance in time Δt can likewise be calculated as

$$\Delta t = \frac{\sigma}{\sqrt{2}v_g} = \frac{\Delta x}{v_g} \tag{3.38}$$

The result indicates that the uncertainty in time of detecting the wave packet is determined by its transit time $\Delta x/v_g$.

Now, the wave packet Eq. (3.34) can be expanded in terms of a complete set of the momentum eigenfunctions of a free particle derived in Eq. (3.12):

$$\psi(x, t = 0) = e^{[ik_0x-(x^2/2\sigma^2)]} = \int_{-\infty}^{\infty} dp\varphi_p e^{ikx}, \quad k \equiv \frac{p_x}{\hbar} \tag{3.39}$$

with the expansion coefficient given by

$$\varphi_p \equiv \int_{-\infty}^{\infty} dx e^{-ikx} e^{k_0x} e^{-x^2/2\sigma^2} = \langle e^{ikx}|\psi(x, t = 0)\rangle \propto e^{-\sigma^2(k-k_0)^2/2} \tag{3.40}$$

(see Eq. (3.22)). Thus, φ_p is centered at p_0 and Gaussian distributed as shown in Figure 3.1. Therefore, the variance in momentum can likewise be calculated as

$$\Delta p_x \equiv \hbar\Delta k_x = \frac{1}{\sqrt{2}}\frac{\hbar}{\sigma} \tag{3.41}$$

and by combining Eqs. (3.36) and (3.41), we find

$$\Delta x \Delta p_x \approx \hbar \tag{3.42}$$

It is thus clear that the smaller Δx, the larger Δp_x or vice versa.

Uncertainty in Energy and Time

The uncertainty in energy and time can also be shown as follows. Given a free particle with kinetic energy $E = p_x^2/2m$, we can differentiate both sides, obtaining

$$\Delta E = \frac{p_x}{m}\Delta p_x = v_g \Delta p_x \tag{3.43}$$

Therefore, by combining Eqs. (3.38), (3.41), and (3.43), we also find the relation

$$\Delta E \Delta t \approx \Delta p_x \Delta x \approx \hbar \tag{3.44}$$

Clearly, Eq. (3.44) states that it is not possible to precisely measure both E and t simultaneously. Rather the accuracy of measuring E depends on the measurement time Δt. Moreover, in view of $\Delta E \approx \hbar\Delta\omega$, Eq. (3.44) is consistent with the basic relationship between the time duration and the frequency bandwidth in electromagnetic pulses (Eq. (1.36)).

Problems

3.1 Given the 3D momentum eigenequation (3.11), derive the normalized eigenfunction as given in Eq. (3.12).
Hint: Look for the eigenfunction in the form

$$u(\underline{r}) = f_x(x)f_y(y)f_z(z) \tag{A}$$

and insert (A) into Eq. (3.11). By dividing both sides with (A), reduce the equation to three 1D momentum eigenequations with respect to p_x, p_y, and p_z.

3.2 The laser pulses have been continually compressed from nano (10^{-9}s), pico (10^{-12}s), and femto (10^{-15}s) seconds. Find the corresponding frequency band widths.

3.3 The diameter of the nucleus is $\sim 1 \times 10^{-5}$ nm. Use the uncertainty relation to estimate the minimum kinetic energy for the electron and the proton to have within the nucleus. The binding energy per nucleon is $\sim 5 \times 10^6$ eV. Can the proton reside in the nucleus? Can the electron reside in the nucleus?

3.4 Estimate the minimum energy in eV unit of an electron and proton, which are spatially confined in a cube with the edge lengths of 1, 0.5, 0.05 nm, and compare the results with the thermal energy at room temperature.

3.5 When the electron in the H-atom is promoted from the ground state with $n = 1$ to the first excited state with $n = 2$, the electron stays in the excited state typically 10 ns before returning to the ground state. Find the center wavelength and spread of wavelengths resulting from the finite lifetime of the electron when the electron makes the transition from $n = 2$ state to $n = 1$ state.

3.6 Show by using the integration by parts that the Hamiltonian

$$\hat{H} = -\frac{\hbar^2}{2m}\nabla^2 + V(r)$$

is a Hermitian operator.

Suggested Readings

1. J. Singh, Quantum Mechanics, Fundamentals and Applications to Technology, John Wiley & Sons, 1996.
2. A. Yariv, An Introduction to Theory and Applications of Quantum Mechanics, John Wiley & Sons, 1982.

3. R. L. Liboff, Introductory Quantum Mechanics, Fourth Edition, Addison Wesley Publishing Company, Reading, MA, 2002.
4. S. Gasiorowics, Quantum Physics, Third Edition, John Wiley & Sons, 2003.
5. R. W. Robinett, Quantum Mechanics, Classical Results, Modern Systems and Visualized Examples, Oxford University Press, 2006.
6. H. Kroemer, Quantum Mechanics for Engineering, Materials Science, and Applied Physics, International Edition, Prentice Hall, 1994.

4
Bound States in Quantum Well and Wire

A particle in a simple potential well is an interesting dynamic system and provides valuable insights for the bound states. In particular, the energy quantization of a particle is naturally brought out from the self-evident fact of the wavefunction physically well behaving. Moreover, the results obtained are pertinent to the problems of practical interest and provide useful backgrounds for designing and analyzing the semiconductor devices. The quantum well and wire are discussed together with the density of states in one, two, and three dimensions.

4.1
Electrons in Solids

An electron in solids is often modeled as a free particle in a box, which in turn is taken as the 3D infinite square well potential. To analyze the motion of the electron therein, let us first consider a particle in 1D infinite square well potential of width L (Figure 4.1). The potential is then given by

$$V(x) = \begin{cases} 0 & 0 \leq x \leq L \\ \infty & \text{otherwise} \end{cases} \tag{4.1}$$

The electron therein is a free particle, and the energy eigenequation is given by

$$-\frac{\hbar^2}{2m}\frac{\partial^2}{\partial x^2}u(x) = Eu(x); \quad \frac{p^2}{2m} = -\frac{\hbar^2}{2m}\frac{\partial^2}{\partial x^2} \tag{4.2a}$$

or equivalently by

$$u'' + k^2 u = 0, \quad k^2 \equiv \frac{2mE}{\hbar^2} = \frac{p^2}{\hbar^2} \tag{4.2b}$$

Equation (4.2b) is identical to that of the harmonic oscillator, when t is replaced by x, and we can thus take sinusoidal functions $\sin kx$ or $\cos kx$ as the solution. As the probability of finding the particle outside the infinite potential well has to be zero, $u(x)$ should vanish at the two edges of the well. Moreover, the probability density should sum up to unity. Hence, the normalized eigenfunctions are given by

$$u_n(x) = \left(\frac{2}{L}\right)^{1/2} \sin k_n x; \quad k_n = \frac{n\pi}{L}, \quad n = 1, 2, \ldots \tag{4.3}$$

Introductory Quantum Mechanics for Applied Nanotechnology, First Edition. Dae Mann Kim.

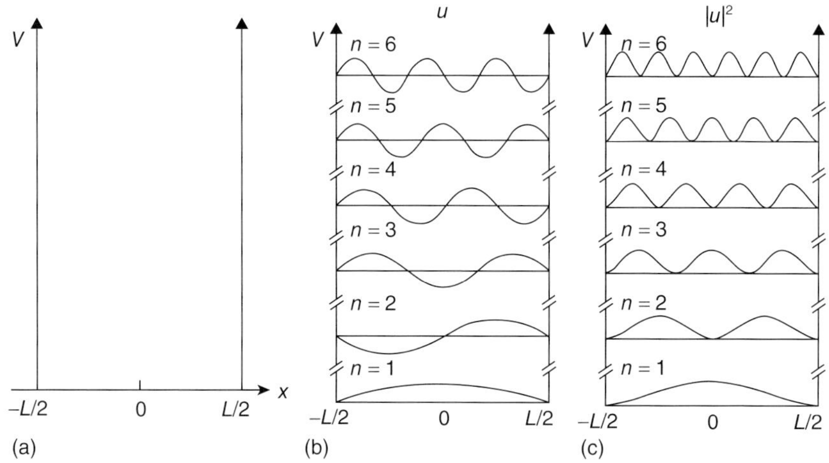

Figure 4.1 The infinite square well potential with width L (a) and typical eigenfunctions (b) and the probability densities (c) and the subbands.

and satisfy the required boundary conditions $u_n(x = 0) = u_n(x = L) = 0$. The condition is identical to the standing-wave condition. The associated eigenenergies are therefore naturally quantized and given by

$$E_n = \frac{p^2}{2m} = \frac{\hbar^2 k_n^2}{2m} = \frac{\hbar^2 \pi^2}{2mL^2} n^2, \quad n = 1, 2, 3, \dots \tag{4.4}$$

The integer n is known as the *quantum number*, and the quantized energy levels are called the *sublevels* or *subbands*. Typical probability densities and subbands are shown in Figure 4.1. It is interesting to note that the ground state energy E_1 is not zero but is finite. Classically, a particle can be completely at rest in the potential well at a position precisely known so that $\Delta p_x = 0$ and $\Delta x = 0$, in contradiction with the uncertainty principle. Herein lies a fundamental difference between the classical and quantum theories.

Particle in 3D Box

Let us consider a particle in 3D box and model the potential as

$$V(\underline{r}) = \begin{cases} 0 & 0 \leq x, y, z \leq L \\ \infty & \text{otherwise} \end{cases} \tag{4.5}$$

The energy eigenequation of the particle inside the box then reads as

$$-\frac{\hbar^2}{2m}\left(\frac{\partial^2}{\partial x^2} + \frac{\partial^2}{\partial y^2} + \frac{\partial^2}{\partial z^2}\right) u(x, y, z) = Eu(x, y, z) \tag{4.6}$$

We may use the separation of variable technique and look for the solution in the form

$$u(x, y, z) = X(x)Y(y)Z(z) \tag{4.7}$$

and insert Eq. (4.7) into Eq. (4.6) and divide both sides with Eq. (4.7), obtaining

$$-\frac{\hbar^2}{2m}\left(\frac{X''}{X}+\frac{Y''}{Y}+\frac{Z''}{Z}\right)=E \tag{4.8}$$

The double primes denote the second-order differentiations with respect to x, y, and z, respectively.

Each term on the left-hand side depends solely on x, y, and z, so that we can put each term to a constant, and let the three constants add up to the total energy E. In this manner, Eq. (4.8) is reduced to three independent 1D equations with each one identical to Eq. (4.2). Therefore, we can express the eigenfunction and eigenenergy by extending Eqs. (4.3) and (4.4) as

$$u_n(x,y,z)=\left(\frac{2}{L}\right)^{3/2}\sin\left(\frac{n_x\pi}{L}x\right)\sin\left(\frac{n_y\pi}{L}y\right)\sin\left(\frac{n_z\pi}{L}z\right) \tag{4.9a}$$

$$E_n=\frac{\hbar^2\pi^2}{2mL^2}(n_x^2+n_y^2+n_z^2) \tag{4.9b}$$

Evidently, the ground state corresponds to $n_x=n_y=n_z=1$, while the first excited state is associated with $n_x=2,\quad n_y=n_z=1,\ n_y=2,\quad n_x=n_z=1,\ n_z=2,\quad n_x=n_y=1$. The three quantum states share a common eigenvalue; hence, there is the threefold degeneracy in the first exited state. The degree of degeneracy increases in higher-lying energy levels.

4.2
1D, 2D, and 3D Densities of States

The electron in solids is to be modeled as a free particle in 3D box with its wavefunction satisfying the stationary boundary condition, namely, that the wavefunction vanishes at the edges of the box. The boundary condition ensures that the electron is well confined in the solid. In addition, the periodic boundary condition is also utilized to describe the electron freely propagating in the bulk solid. Such propagation is represented by the traveling wavefunction given in Eqs. (3.10) and (3.12) by

$$\Psi(\underline{r},t)=e^{-i\omega t}u(\underline{r})=\frac{1}{L^{3/2}}e^{-i(\omega t-\underline{k}\cdot\underline{r})},\quad E=\hbar\omega=\frac{\hbar^2k^2}{2m} \tag{4.10}$$

where $u(\boldsymbol{r})$ satisfies the 3D energy eigenequation of a free particle Eq. (4.6). When $u(r)$ is combined with the exponential time factor, it provides a mode function of a free particle traveling in the $\boldsymbol{k}$-direction as a matter wave.

The periodic boundary condition states that a particle exiting at $x+L$, for example, reenters at x and is thus specified by

$$u(x,y,z)=u(x+L,y,z)=u(x,y+L,z)=u(x,y,z+L) \tag{4.11}$$

(see Figure 4.2). The condition forces the wave vector $\boldsymbol{k}$ in Eq. (4.10) to satisfy

$$k_xL=2\pi n_x,\quad k_yL=2\pi n_y,\quad k_zL=2\pi n_z \tag{4.12}$$

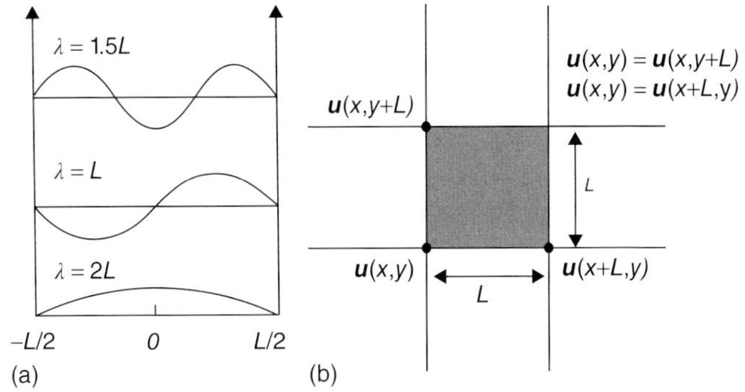

Figure 4.2 Stationary (a) and periodic (b) boundary conditions.

where the quantum numbers, n_x, n_y, n_z are positive or negative integers, describing the particle traveling in $\boldsymbol{k}$- or $-\boldsymbol{k}$-directions. The eigenenergy is then given by

$$E_{\underline{n}} = \frac{\hbar^2}{2m}(k_x^2 + k_y^2 + k_z^2) = \frac{\hbar^2}{2m}\left(\frac{2\pi}{L}\right)^2 (n_x^2 + n_y^2 + n_z^2) \tag{4.13}$$

It is therefore clear that there is the one-to-one correspondence between $\boldsymbol{k}$ (k_x, k_y, k_z) and $\boldsymbol{n}$ (n_x, n_y, n_z), and each $\boldsymbol{k}$ represents a single quantum state.

A key quantity of interest is the number of quantum states in the interval from $\boldsymbol{k}$ to $\boldsymbol{k} + \mathrm{d}\boldsymbol{k}$ or equivalently from E to $E + \mathrm{d}E$ in 1D, 2D, and 3D environments. Such number of states is readily found by considering 1D, 2D, and 3D $\boldsymbol{k}$-spaces, which are scaled with the unit length $2\pi/L$ (Figure 4.3). The respective unit cell containing a single dot, that is, a single quantum state is given by

$$\left(\frac{2\pi}{L}\right)^j, \quad j = 3, 2, 1$$

and the differential volume elements between k and $k + \mathrm{d}k$ are given, respectively, by

$$4\pi k^2 dk, \quad 2\pi k dk, \quad 2dk$$

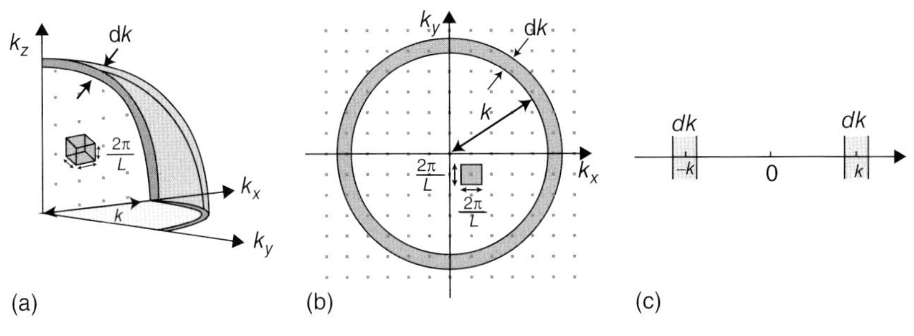

Figure 4.3 3D (a), 2D (b), and 1D (c) volume elements in the $\boldsymbol{k}$-space with each dot representing a quantum state.

3D Density of States

The number of quantum states in 3D space can be found by dividing the differential volume element with the unit cell. However, for each quantum state for given $\boldsymbol{k}$, there are two independent quantum states, corresponding to the spin-up and spin-down states of the electron. Therefore, the number of the states per unit volume is given by

$$g_{3D}(k)dk = 2 \times \frac{4\pi k^2 dk}{(2\pi/L)^3} \times \frac{1}{L^3} = \frac{k^2 dk}{\pi^2} \tag{4.14a}$$

yielding thereby the 3D density of states in $\boldsymbol{k}$-space

$$g_{3D}(k) = \frac{k^2}{\pi^2} \tag{4.14b}$$

We can transcribe Eq. (4.14b) in the E-space via the dispersion relation $E = \hbar^2 k^2/2m$ as

$$g_{3D}(k)dk \equiv g_{3D}(E)dE = \frac{\sqrt{2}m^{3/2}E^{1/2}}{\pi^2\hbar^3}dE \tag{4.15a}$$

Therefore, the 3D density of states is given in terms of E by

$$g_{3D}(E) = \frac{\sqrt{2}m^{3/2}E^{1/2}}{\pi^2\hbar^3} \propto E^{1/2} \tag{4.15b}$$

2D and 1D Densities of States

We can likewise divide the 2D volume element by the 2D unit cell, obtaining

$$g_{2D}(k)dk = 2\frac{2\pi k dk}{(2\pi/L)^2}\frac{1}{L^2} = \frac{kdk}{\pi} \tag{4.16a}$$

The resulting $\boldsymbol{k}$-space density of states $g_{2D}(k) = k/\pi$ is likewise transcribed into E as

$$g_{2D}(E) = \frac{m}{\pi\hbar^2} \propto E^0 \tag{4.16b}$$

The 1D counterpart of Eqs. (4.14a) and (4.16a) is given by

$$g_{1D}(k) = 2\frac{2dk}{(2\pi/L)}\frac{1}{L} = \frac{2}{\pi}dk \tag{4.17a}$$

and is transcribed into E as

$$g_{1D}(E) = \frac{\sqrt{2}m^{1/2}}{\pi\hbar}\frac{1}{E^{1/2}} \propto E^{-1/2} \tag{4.17b}$$

The 3D density of states $g_{3D}(E)$ is a key factor for analyzing the bulk semiconductor devices such as the metal oxide semiconductor field-effect transistor (MOSFET), while $g_{2D}(E)$ and $g_{1D}(E)$ are essential for modeling nanoelectronic devices, such as FinFET and nanowire field-effect transistors (FETs). Figure 4.4 shows g_{1D}, g_{2D}, and g_{3D} versus energy.

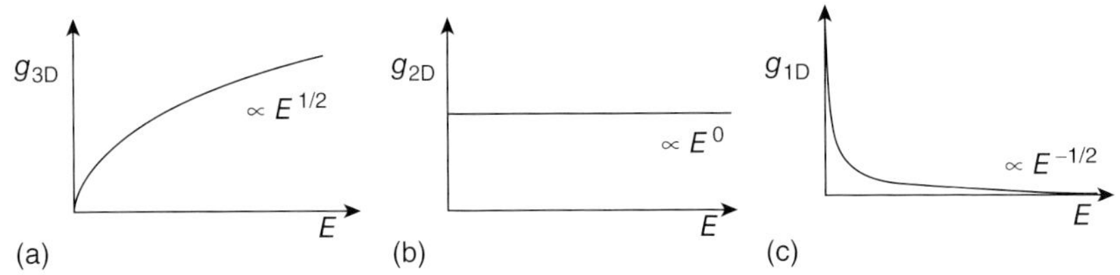

Figure 4.4 The E-space density of states: 3D (a), 2D (b), and 1D (c).

4.3 Particle in Quantum Well

The potential well with a finite barrier height V is called the *quantum well* and has become an essential part of semiconductor and optoelectronic device structures, for example, laser diodes, bipolar junction transistors, FinFETs, and nanowire FETs. Thus, consider a particle in the quantum well of height V and width W, as shown in Figure 4.5:

$$V(x) = \begin{cases} 0 & |x| \le W/2 \\ V & |x| \ge W/2 \end{cases} \tag{4.18}$$

Inside the well, $V = 0$ and the eigenequation of a free particle is given as usual by

$$u'' + k^2 u = 0; \quad k^2 \equiv \frac{2mE}{\hbar^2}, \quad |x| \le \frac{W}{2} \tag{4.19a}$$

The analysis is confined to the bound state, that is, $E \le V$; hence, the eigenequation outside the well reads as

$$u'' - \kappa^2 u = 0; \quad \kappa^2 \equiv \frac{2m(V-E)}{\hbar^2}, \quad |x| > \frac{W}{2} \tag{4.19b}$$

Obviously, $u(x)$ should assume the sinusoidal ($\sin kx$, $\cos kx$) and exponential ($\exp \pm\kappa x$) functions inside and outside the well, respectively. We can therefore construct the even and odd eigenfunctions to expedite the analysis as

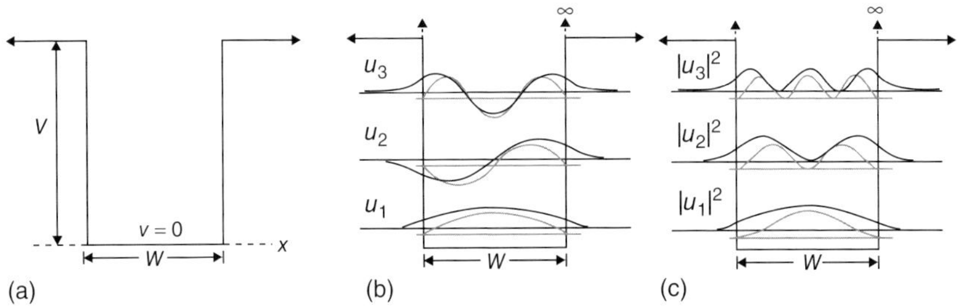

Figure 4.5 The quantum well with a finite potential depth V and width W (a). Typical eigenfunctions (b) and corresponding probability densities (c) and subbands. Also shown for comparison are the eigenfunctions, probability densities, and subbands in the infinite square well potential (thin lines).

$$u_e(x) = N\begin{cases} Ae^{\kappa x} & x < -W/2 \\ \cos kx & |x| \le W/2 \\ Ae^{-\kappa x} & x > W/2 \end{cases}; \quad u_o(x) = N\begin{cases} -Ae^{\kappa x}; & x < -W/2 \\ \sin kx & |x| \le W/2 \\ Ae^{-\kappa x} & x > W/2 \end{cases} \quad (4.20)$$

where the exponential functions chosen vanish for $x \to \pm\infty$ as it should, and the constants of integration A and N are used for satisfying the boundary and normalization conditions.

Boundary Conditions

The conditions for the eigenfunctions to satisfy are that $u(x)$ and its derivative $\partial u(x)/\partial x$ must be continuous everywhere. These two conditions are required to render the probability density $u^*(x)u(x)$ and the momentum $\propto \partial u(x)/\partial x$ continuous everywhere. Clearly, these conditions are automatically satisfied inside and outside the well as $u(x)$ is described by analytical functions in the two regions. Therefore, the conditions need to be applied only at the two edges of the well where two different solutions meet. However, as $u(x)$ is even or odd in x, when the condition is satisfied at one edge, it is also satisfied at the other edge.

For $u_e(x)$, the two boundary conditions at $W/2$ are specified by

$$\cos\xi = Ae^{-\eta}, \quad \xi \equiv \frac{kW}{2}, \quad \eta \equiv \frac{\kappa W}{2} \qquad (4.21a)$$

$$-k\sin\xi = -\kappa Ae^{-\eta} \qquad (4.21b)$$

The two equations can be combined into one by multiplying both sides of Eq. (4.21b) by $W/2$ and dividing it with Eq. (4.21a):

$$\xi\tan\xi = \eta \qquad (4.22)$$

The boundary conditions for $u_o(x)$ are likewise compacted as

$$-\xi\cot\xi = \eta \qquad (4.23)$$

Also the parameters ξ and η introduced in Eq. (4.21a) are constrained by Eqs. (4.19a) and (4.19b) as

$$\xi^2 + \eta^2 \equiv \left(\frac{kW}{2}\right)^2 + \left(\frac{\kappa W}{2}\right)^2 = \frac{mVW^2}{2\hbar^2} \qquad (4.24)$$

Therefore, the problem is reduced to finding k and κ, such that the pair of boundary conditions (Eq. (4.22)) for $u_e(x)$ and (Eq. (4.23)) for $u_o(x)$ are satisfied. The unknown values k and κ can be found by numerical or graphical means, and let us resort to the latter means. For this purpose, η in Eqs. (4.22) and (4.23) is plotted versus ξ in Figure 4.6. Also plotted in the figure is a family of circles (Eq. (4.24)) corresponding to different potential depths, V, and widths, W. Thus, finding the values of ξ and η or k and κ consists of reading off the coordinates of the cross points of the two curves Eqs. (4.22) and (4.24) for $u_e(x)$ and Eqs. (4.23) and (4.24) for $u_o(x)$. Once k and κ are thus determined, we can find the eigenfunctions and energy eigenvalues from Eq. (4.19).

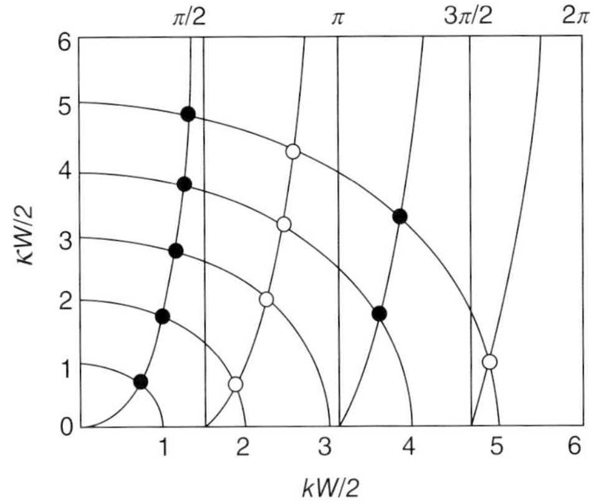

Figure 4.6 The graphical scheme for finding $\boldsymbol{k}$ and $\boldsymbol{\kappa}$. Plotted are the two sets of curves Eq. (4.22) (thick lines) and Eq. (4.23) (thin lines) and a family of circles Eq. (4.24). Also shown are the intersection points for finding u_e (filled circles) and for finding u_o (open circles).

The typical eigenfunctions thus found and its probability densities are plotted in Figure 4.5 together with those corresponding to the infinite square well potential, for comparison. The sublevels in the latter are consistently lower than those in the former, indicating the tighter binding of the particle. Also the eigenfunctions in the quantum well are shown to penetrate into the classically forbidden region outside the well, the significance of which will become clear soon. We can also observe a few additional features of the bound states from Figure 4.6. When the radius of the circle becomes large with deeper well depth V for given W, more bound states exist in the well. Also, the lowest ground state is always associated with $u_e(x)$, and higher-lying states alternate between $u_e(x)$ and $u_o(x)$, and at least one bound state exists, regardless of the well depth. Finally, in the limit of infinite V, there are two sets of infinite number of cross points for $u_e(x)$ and $u_o(x)$, respectively, determined by

$$\xi_n \equiv \frac{k_n W}{2} = \frac{\pi}{2}(2n+1), \quad \xi_n \equiv \frac{k_n W}{2} = n\pi, \quad n = 0, 1, 2, \ldots$$

When combined, these two conditions lead to the energy eigenvalues derived in Eq. (4.4), as it should.

4.4 Quantum Well and Wire

Quantum Well

It has become possible to grow atomic layers of varying thicknesses by using the molecular beam epitaxy or metal organic chemical vapor deposition techniques.

As a result, superlattice structures containing multiple quantum wells are routinely fabricated. Figure 4.7 shows a typical example consisting of AlGaAs and GaAs layers. An electron in the semiconductor moves freely in certain energy ranges, called the *conduction* and *valence bands*. These two bands are separated by the energy gap called *bandgap*, and the electrons are forbidden to propagate in such a bandgap. Hence, the quantum wells are formed by two semiconducting materials with different bandgaps in equilibrium contact with the larger bandgap providing the potential barrier.

Let us consider a quantum well in which electrons are confined, say in the z-direction, while propagating freely in the x-, y-directions, forming thereby the 2D electron gas. The energy eigenequation then reads as

$$\left[-\frac{\hbar^2}{2m_x}\frac{\partial^2}{\partial x^2}-\frac{\hbar^2}{2m_y}\frac{\partial^2}{\partial y^2}-\frac{\hbar^2}{2m_z}\frac{\partial^2}{\partial z^2}+V(z)\right]u(x,y,z)=Eu(x,y,z) \quad (4.25a)$$

where the potential is given by

$$V(z)=\begin{cases}0 & |z|\le W/2\\ V & |z|\ge W/2\end{cases} \quad (4.25b)$$

and m_x, m_y, and m_z denote the effective masses of the electron with which it moves in x-, y-, and z-directions, respectively. The effective mass of the electron in solids is different from its rest mass and depends on the crystallographic directions.

We can as usual use the separation of variable technique and decompose (Eq. (4.25a)) into three separate equations involving x, y, z variables and obtain the sublevels as

$$E_n=\frac{\hbar^2k_x^2}{2m_x}+\frac{\hbar^2k_y^2}{2m_y}+\frac{\hbar^2\pi^2}{2m_zW^2}n^2, \quad n=1,2,\ \ldots \quad (4.26)$$

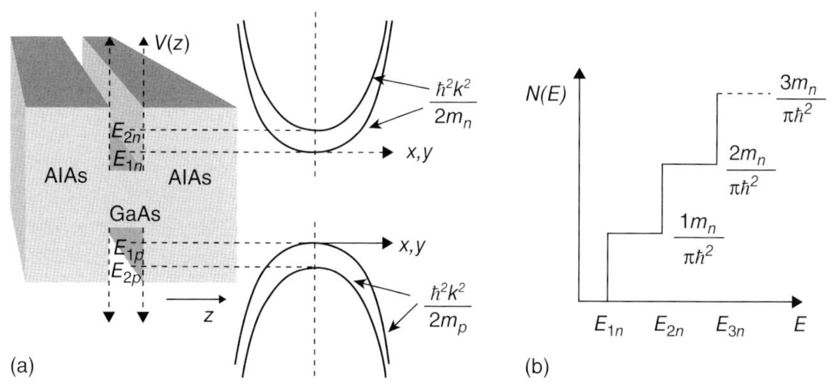

Figure 4.7 The quantum well of electrons and holes, the respective subbands, and dispersion curves (a). The 2D density of states versus energy (b).

The total eigenenergy thus consists of the kinetic energies in the x-, y-directions, and the sublevels resulting from the confinement in the quantum well of width W. For simplicity, the well depth has been taken as infinity in Eq. (4.26). Plotted in Figure 4.7 are the subbands and the density of states. As the 2D density of states is constant, independent of energy (Eq. (4.16b)), the number of quantum states increases stepwise whenever E crosses the discrete subband with the energy E_{zn}. Also, each E_{zn} is associated with the kinetic energy associated with the free propagations in x-, y-directions.

Quantum Wire

The quantum wires with nanoscale cross-sections are fast becoming essential elements of the nanoelectronic devices. Thus, consider the electrons in such nanowires. A particle therein is confined in, for example, y-, z-directions while free to move in the x-direction (Figure 4.8). The energy spectrum therefore consists of two sets of sublevels resulting from the confinement in y-, z-directions and the kinetic energy with which the particle freely moves in the x-direction:

$$E_{n,m} = \frac{\hbar^2 k_x^2}{2m_x} + \frac{\hbar^2 \pi^2}{2m_y W_y^2} n^2 + \frac{\hbar^2 \pi^2}{2m_z W_z^2} m^2, \quad n, m = 1, 2, \ldots \tag{4.27}$$

Again, for simplicity, the well depth has been taken infinite in Eq. (4.27). Shown in Figure 4.8 are the subbands and the density of states. As the 1D density of states follows the power law, $E^{-1/2}$ (Eq. (4.17b)), the density of states exhibits a sawtooth-like characteristics versus E.

The quantum wells and wires have become essential elements of various semiconductor devices. For example, in high-efficiency laser diodes, electrons and holes are injected into the respective quantum wells and are allowed to have longer radiative recombination time while confined in the well. Additionally, the operation of MOSFET is based on injecting 2D electrons or holes into the gate voltage-induced quantum well. Moreover, nanowire FETs enjoy the prospect of becoming one of the mainstream drivers of nanoelectronics.

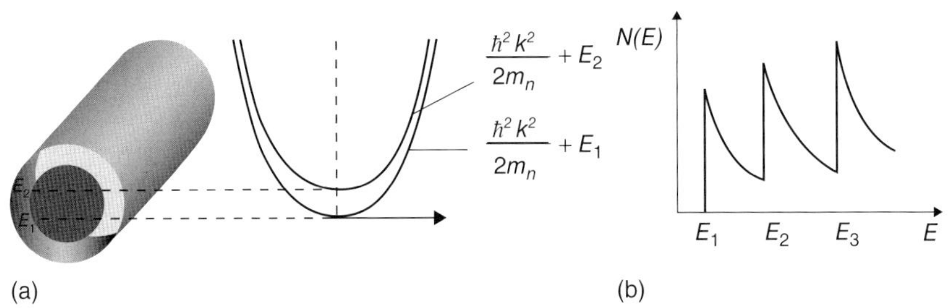

Figure 4.8 The quantum wire, subbands, and dispersion curves of the electron (a) and 1D density of states versus energy (b).

Problems

4.1 (a) Show that the traveling wave eigenfunction given in Eq. (4.10) satisfies the energy eigenequation of a free particle with eigenvalue (Eq. (4.13)) under the periodic boundary condition.
(b) Fill in the algebra and obtain the 3D, 2D, and 1D densities of states in the E-space from those in the k-space (Eqs. (4.14b), (4.16a), and (4.17a)) by using the dispersion relation E versus k.

4.2 (a) Derive the 3D density of states in the cubic box of width W by using the stationary boundary conditions, that is, the energy eigenfunctions vanish at the edges of the box and show that the result is same as Eq. (4.14b).
(b) Express the density of states in terms of the frequency ν and show that it reduces precisely to the number of standing-wave modes in the cavity that was used in Rayleigh–Jeans theory (Eq. (2.1)).

4.3 An electron is contained in two cubic quantum dots of dimension 0.1 and 1.0 nm, respectively.
(a) Find the lowest three energy levels in eV units corresponding to $n = 1, 2, 3$ and the degeneracy of each level.
(b) Calculate the wavelengths of photons emitted when the electron cascades down the energy levels from 3 to 2 and 2 to 1.
(c) Compare the ground state energies of the two quantum dots with the thermal energy at room temperature $k_B T$.

4.4 An electron in silicon bounded by two SiO_2 layers is to be taken confined in the quantum well given by

$$V(z) = \begin{cases} 3.1\,\text{eV} & z \leq -W/2, z \geq W/2 \\ 0\,\text{eV} & -W/2 \leq z \leq W/2 \end{cases}$$

(a) Calculate numerically the bound state energy in electron volt unit by taking $W = 2$ nm and the effective mass of electron $m_n = 0.1\, m_0$ with m_0 denoting the rest mass.
(b) Write a short program enabling the analysis of bound states for varying well width W and plot the energy eigenfunction and eigenvalue versus W.

4.5 (a) Starting from the energy eigenequation of a particle in a quantum well, fill in the algebra and find the energy eigenfunction and eigenvalue in Eq. (4.26).
(b) Repeat the analysis and find the eigenfunction and eigenvalue Eq. (4.27) in a quantum wire. Take the barrier height to be infinite for simplicity.

Suggested Readings

1. D. M. Kim, Introductory Quantum Mechanics for Semiconductor Nanotechnology, Wiley-VCH, 2010.
2. R. W. Robinett, Quantum Mechanics, Classical results, Modern Systems and Visualized

Examples, Oxford University Press, 2006.

3. R. L. Liboff, Introductory Quantum Mechanics, Fourth Edition, Addison Wesley Publishing Company, Reading, MA, 2002.
4. J. Singh, Quantum Mechanics, Fundamentals and Applications to Technology, John Wiley & Sons, 1996.
5. H. Kroemer, Quantum Mechanics for Engineering, Materials Science, and Applied Physics, International Edition, Prentice Hall, 1994.
6. A. Yariv, An Introduction to Theory and Applications of Quantum Mechanics, John Wiley & Sons, 1982.

5
Scattering and Tunneling of 1D Particle

The scattering of the 1D particle from the potential barrier or well is discussed in terms of reflection, transmission, and resonant transmission. Additionally, the tunneling of a particle through the potential barrier, a feature unique in quantum mechanics, is highlighted, and its applications in memory, display, nanometrology, and single-electron transistor (SET) are discussed.

5.1
Scattering at the Step Potential

Consider a particle incident on a step potential with height V (Figure 5.1). Classically, if the incident particle has a kinetic energy E greater than V, it flies over the barrier with diminished velocity. If E is less than V, it bounces back from the barrier. Quantum mechanically, however, both transmission and reflection occur with probabilities depending on E and V.

Let us first consider the case in which $E > V$. Since the step potential is given by

$$V(x) = \begin{cases} 0 & x \leq 0 \\ V & x > 0 \end{cases} \tag{5.1}$$

the energy eigenequation reads as

$$u(x)'' + \alpha^2 u(x) = 0 \tag{5.2a}$$

with the wave vector given by

$$\alpha^2 = \begin{cases} k_0^2 & k_0^2 = \frac{2mE}{\hbar^2}, \quad x \leq 0 \\ k^2 & k^2 = \frac{2m(E-V)}{\hbar^2}, \quad x > 0 \end{cases} \tag{5.2b}$$

The eigenequation (Eq. (5.2)) has been dealt with, and let us use the solution given by

$$\psi(x,t) \propto e^{-i\omega t} u(x) \propto e^{-i(\omega t \mp \alpha x)}, \quad \omega = \frac{E}{\hbar} \tag{5.3}$$

Introductory Quantum Mechanics for Applied Nanotechnology, First Edition. Dae Mann Kim.

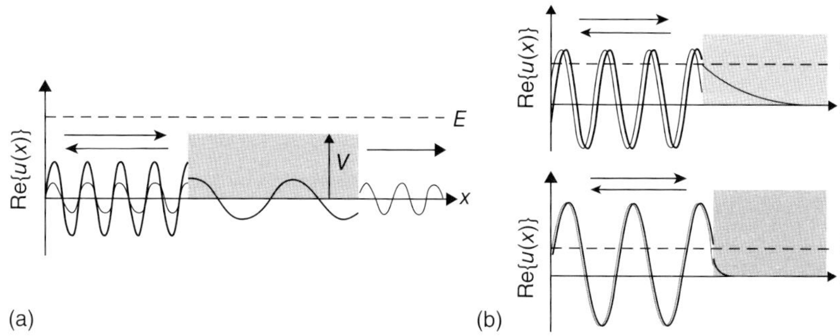

Figure 5.1 A particle incident on a step potential with height V, undergoing both reflection or transmission for $E > V$ (a) and total reflection for $E < V$ (b). Also shown is the penetration of the particle into the potential barrier before total reflection.

(see Eqs. (4.2) and (4.10)). Evidently, Eq. (5.3) describes a particle propagating in $\pm x$-directions, and $u(x)$ is given by

$$u(x) = \begin{cases} i_0 e^{ik_0x} + re^{-ik_0x}, & x \leq 0 \\ te^{ikx}, & x > 0 \end{cases} \tag{5.4}$$

In Eq. (5.4), the terms associated with i_0, r, and t account for the incident and reflected beams in the region $x \leq 0$ and the transmitted beam in the region $x \geq 0$, respectively. Once the particle is transmitted, there is no barrier to reflect it back; hence, only the forward component needs to be retained for $x > 0$.The constants of integration, i_0, r, and t can be used for satisfying the boundary conditions. The boundary conditions, namely, $u(x)$ and its derivate should be continuous need to be applied at $x = 0$, where the two solutions meet. The two conditions read as

$$i_0 + r = t, \quad k_0(i_0 - r) = kt \tag{5.5}$$

We can find r and t from two conditions in Eq. (5.5) with i_0 taken as the input parameter:

$$\frac{r}{i_0} = \frac{k_0 - k}{k_0 + k}, \quad \frac{t}{i_0} = \frac{2k_0}{k_0 + k} \tag{5.6}$$

The Probability Current Density

To proceed further, it is necessary to introduce the probability current density. Thus, consider the change in time of the probability density

$$\frac{\partial}{\partial t}(\psi^*\psi) = \left(\frac{\partial}{\partial t}\psi^*\right)\psi + \psi^*\left(\frac{\partial}{\partial t}\psi\right) \tag{5.7}$$

Upon using the time-dependent Schrödinger equation

$$i\hbar\frac{\partial\psi(\underline{r}, t)}{\partial t} = -\frac{\hbar^2}{2m}\nabla^2\psi(\underline{r}, t) + V(\underline{r})\psi(\underline{r}, t)$$

and its complex conjugate and the well-known vector identity,

$$\psi^*\nabla^2\psi - \psi\nabla^2\psi^* \equiv \nabla\cdot(\psi^*\nabla\psi - \text{c.c.})$$

in Eq. (5.7), we can recast Eq. (5.7) in a straightforward manner as

$$\frac{\partial}{\partial t}\psi^*\psi = -\nabla\cdot\underline{S} \tag{5.8a}$$

$$\underline{S} \equiv \frac{\hbar}{2mi}(\psi^*\nabla\psi - \text{c.c.}) = \frac{\hbar}{2mi}\left(u^*(x)\frac{\partial}{\partial x}u(x) - \text{c.c.}\right) \tag{5.8b}$$

The quantity $\boldsymbol{S}$ is called the *probability current density*, and c.c. denotes the complex conjugate. Equation (5.8) represents the conservation of matter and is analogous to the charge conservation equation.

Reflection and Transmission

With $u(x)$ in Eq. (5.4) used in Eq. (5.8b), there results

$$S_\text{I}(x) = \frac{\hbar k_0}{m}|i_0|^2 - \frac{\hbar k_0}{m}|r|^2, \quad x \le 0 \tag{5.9a}$$

$$S_\text{II}(x) = \frac{\hbar k}{m}|t|^2, \quad x > 0 \tag{5.9b}$$

The first term in S_I represents the incident flux specified in terms of the probability density $|i_0|^2$ and the velocity of propagation $\hbar k_0/m$. Likewise, the two terms $\propto |r|^2$ and $\propto |t|^2$ describe the reflected and transmitted fluxes propagating with the velocities $\hbar k_0/m$ and $\hbar k/m$, respectively. Thus, the reflection R and transmission T coefficients are given by

$$R \equiv \frac{(\hbar k_0/m)|r|^2}{(\hbar k_0/m)|i_0|^2} = \frac{(k_0-k)^2}{(k_0+k)^2}, \qquad T \equiv \frac{(\hbar k/m)|t|^2}{(\hbar k_0/m)|i_0|^2} = \frac{4kk_0}{(k_0+k)^2} \tag{5.10}$$

Therefore, the incoming particle with $E > V$ is either reflected or transmitted with the probabilities given by Eq. (5.10). This is in apparent contrast with the classical description. It also follows from Eq. (5.10) that R and T add up to unity, as it should,

$$R + T = 1 \tag{5.11}$$

Evidently, the quantum treatment is analogous with the reflection and transmission of a light beam, incident on a dielectric interface.

The Total Reflection

For $E < V$, the analysis can be done in a similar manner. In this case, $k^2 < 0$ for $x \ge 0$ (see Eq. (5.2b)), and k is turned into an imaginary wave vector

$$k \to i\kappa, \quad \kappa^2 \equiv \frac{2m(V-E)}{\hbar^2} \tag{5.12}$$

Therefore, the reflection coefficient is obtained from Eq. (5.10) by replacing k by $i\kappa$ as

$$R \equiv \frac{(\hbar k_0/m)|r|^2}{(\hbar k_0/m)|t|^2} = \frac{k_0 - i\kappa}{k_0 + i\kappa} \cdot \frac{k_0 + i\kappa}{k_0 - i\kappa} = 1 \tag{5.13}$$

Equation (5.13) states that the particle is bound to be reflected back. Also, because $u(x)$ is real for $x > 0$, in this case (see Eq. (5.4)), $S_{II}(x)$ is zero, and there is no transmission. The result is in agreement with the classical description, which predicts 100% reflection for $E < V$. However, there is an important difference, namely, that the particle penetrates into the classically forbidden barrier region by an amount, $\delta \approx 1/(2\kappa)$ before undergoing the total reflection (Figure 5.1).

5.2 Scattering from a Quantum Well

When a particle is incident on a quantum well (Figure 5.2), the particle undergoes both reflection and transmission with the nonzero probabilities, again in contradiction with the classical description. The energy eigenequation is split in this case into two regimes, inside and outside the well, and is identical to Eq. (5.2) but with the wave vectors given by

$$\alpha^2 = \begin{cases} k_0^2, & k_0^2 = \frac{2mE}{\hbar^2}, \quad |x| \geq \frac{W}{2} \\ k^2, & k^2 = \frac{2m(E+V)}{\hbar^2}, \quad |x| \leq \frac{W}{2} \end{cases} \tag{5.14}$$

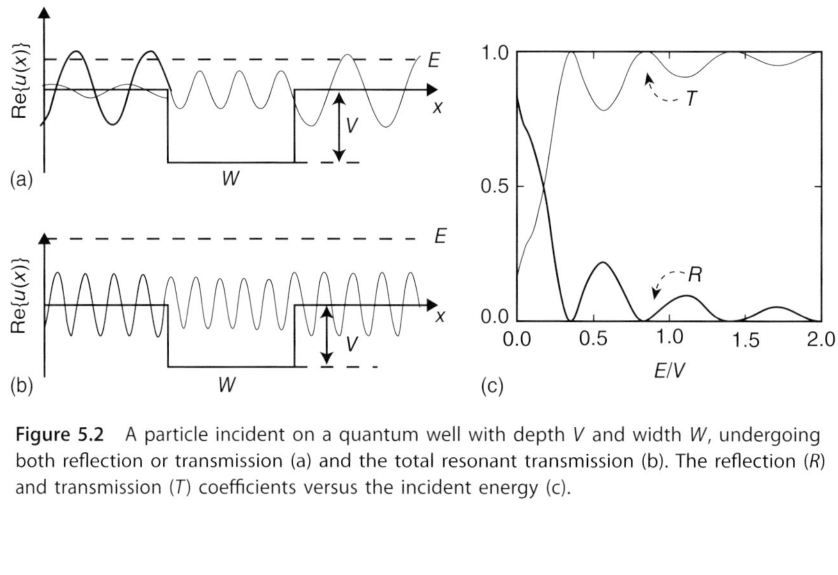

Figure 5.2 A particle incident on a quantum well with depth V and width W, undergoing both reflection or transmission (a) and the total resonant transmission (b). The reflection (R) and transmission (T) coefficients versus the incident energy (c).

We can thus express $u(x)$ in analogy with Eq. (5.4) as

$$u(x) = \begin{cases} i_0 e^{ik_0 x} + r e^{-ik_0 x} & x < -W/2 \\ A e^{ikx} + B e^{-ikx} & |x| \leq W/2 \\ t e^{ik_0 x} & x > W/2 \end{cases} \tag{5.15}$$

and account for the incident and reflected beams to the left of the well and the transmitted beam to the right of the well. Inside the well, two counter-running beams should be used as the solution.

The constants of integration are again determined from the boundary conditions, namely, $u(x)$, and its derivatives be continuous at the two edges of the well, $x = \mp W/2$:

$$i_0 e^{-ik_0 W/2} + r e^{ik_0 W/2} = A e^{-ikW/2} + B e^{ikW/2} \tag{5.16a}$$

$$A e^{ikW/2} + B e^{-ikW/2} = t e^{ik_0 W/2} \tag{5.16b}$$

$$i_0 k_0 e^{-ik_0 W/2} - r k_0 e^{ik_0 W/2} = A k e^{-ikW/2} - B k e^{ikW/2} \tag{5.16c}$$

$$A k e^{ikW/2} - B k e^{-ikW/2} = t k_0 e^{ik_0 W/2} \tag{5.16d}$$

There are five constants with which to satisfy four conditions, and we can again take i_0 as an input parameter and determine the rest in a straightforward manner, obtaining

$$\frac{t}{i_0} = \frac{e^{-ik_0 W} 2k_0 k}{2k_0 k \cos(kW) - i(k_0^2 + k^2) \sin kW} \tag{5.17a}$$

$$\frac{r}{i_0} = \frac{i e^{-ik_0 W}(k^2 - k_0^2) \sin(kW)}{2k_0 k \cos(kW) - i(k_0^2 + k^2) \sin kW} \tag{5.17b}$$

Hence R and T are specified with the use of Eq. (5.17) and in analogy with Eq. (5.10) as

$$T = \frac{1}{1 + \Lambda(E, V, W)}, \qquad R = \frac{\Lambda(E, V, W)}{1 + \Lambda(E, V, W)} \tag{5.18a}$$

$$\Lambda(E, V, W) \equiv \frac{V^2}{4E(E + V)} \sin^2 \left[W \sqrt{\frac{2m}{\hbar^2}(E + V)} \right] \tag{5.18b}$$

and R and T thus found add up to unity, as they should.

Resonant Transmission

As clear from Eq. (5.18), R and T are again nonzero, in contrast with the classical theory. However, for $E \gg V$, $\Lambda \to 0$, and $T \approx 1$, in agreement with the classical theory. Moreover, even for E comparable with V, Eq. (5.18) indicates that 100% transmission ensues, that is, $R = 0$ and $T = 1$, when the incident energy satisfies

$$Wk_n \equiv W\sqrt{\frac{2m}{\hbar^2}(E_n + V)} = n\pi, \quad n = 1, 2, \ldots \tag{5.19}$$

Equation (5.19) can be interpreted in light of de Broglie wavelength λ. As $k_n = 2\pi/\lambda_n$, Eq. (5.19) is equivalent to $2W = n\lambda_n$, which indicates that the round-trip distance of the quantum well is an integer multiple of de Broglie wavelength of the particle. This is precisely the condition for 100% transmission of light in Fabry–Perot etalon or Bragg diffraction. The total transmission of a particle is known as the *resonant transmission*. The R and T are plotted in Figure 5.2 versus the energy of the incident particle. The resonant condition lends to an alternative interpretation, when expressed as

$$E_n + V = \frac{\pi^2\hbar^2}{2mW^2}n^2, \quad n = 1, 2, \ldots \tag{5.20}$$

Equation (5.20) indicates that if the incident energy of the particle as viewed from the bottom of the quantum well corresponds to one of the possible energy eigenvalues of the infinite square well potential, there ensues the total transmission (see Eq. (4.4)). This carries an important bearing in the band theory of solids, as will be discussed.

5.3 Tunneling

A particle incident on a potential barrier with height V greater than its kinetic energy E has a finite probability of transmitting through the barrier. Such transmission, a feature unique in quantum mechanics, is called the *tunneling*. Thus, consider a particle incident on a potential barrier with height V and thickness d (Figure 5.3). The tunneling can be analyzed in parallel with the transmission of a particle through a quantum well (see Eqs. (5.14)–(5.18)). The only modification required is to change k in Eq. (5.14) as

$$k = \sqrt{\frac{2m(E - V)}{\hbar^2}}, \quad E \geq V; \quad k = i\kappa, \quad \kappa \equiv \sqrt{\frac{2m(V - E)}{\hbar^2}}, \quad E \leq V \tag{5.21}$$

Thus, for $E > V$, the expressions of R and T in Eq. (5.18) can be used directly, provided W is replaced by d and the new k is used as defined in Eq. (5.21). For $E < V$, we can again use Eqs. (5.17) and (5.18), with k replaced by $i\kappa$ as defined in Eq. (5.21). The algebra is lengthy but simple and straightforward, and we can find

$$T = \frac{1}{1 + \Lambda(E, V, d)}, \quad R = \frac{\Lambda(E, V, d)}{1 + \Lambda(E, V, d)} \tag{5.22a}$$

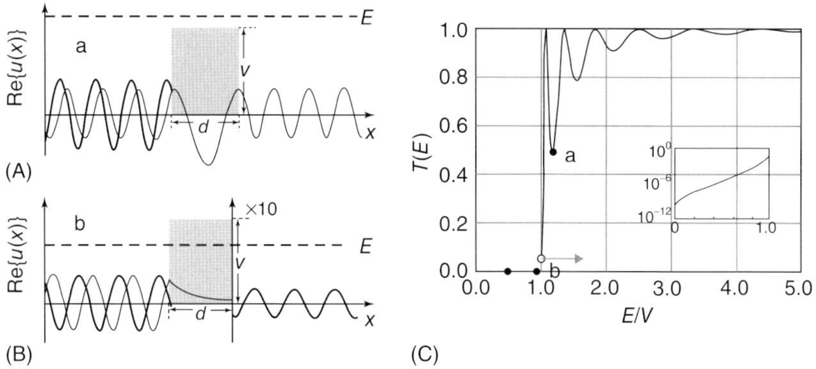

Figure 5.3 A particle incident on a potential barrier with height V and thickness d, undergoing both reflection and transmission for $E > V$ (A), reflection and tunneling for $E < V$ (B). The transmission coefficient versus the incident energy E/V for $E \geq V$ and tunneling probability versus E/V for $E \leq V$ (C).

$$\Lambda(E, V, d) \equiv \frac{V^2}{4E(V-E)} \sinh^2\left[d\sqrt{\frac{2m}{\hbar^2}(V-E)}\right] \tag{5.22b}$$

In deriving Eq. (5.22), the trigonometric identities have been used, that is,

$$\sin ix = i \sinh x, \quad \cos ix = \cosh x, \quad \cosh^2 x = 1 + \sinh^2 x$$

Figure 5.3 shows the typical eigenfunctions of the incident, reflected, and transmitted beams for $E \geq V$ and $E \leq V$. For the latter case, T decreases exponentially with decreasing incident energy for given V. For the former case, there is a clear trace of resonant transmissions occurring for potential barrier as well. The tunneling can be understood in light of the finite penetration the particle makes before undergoing the total reflection off the step potential (Figure 5.1). The total reflection occurs at the step potential because of the infinite width of the barrier. When the barrier width is cut to a finite value d, there is a finite probability density for $x \geq d$ as clear from the figure. This means that the particle has a finite probability of penetrating beyond d, that is, tunneling through the barrier.

The penetration depth is analogous to the skin depth of the light at the metallic surface, and the tunneling has the optical analog as well. The light propagates in the waveguide or optical fiber by means of the total internal reflection. But if another waveguide or optical fiber is placed nearby as in a directional coupler (Figure 5.4), the light leaks into the other, thereby modulating and switching the light. The coupling of power between the waveguides is due to the guided electromagnetic waves tailing out of the waveguide. Likewise, tunneling is due to the finite penetration of the wavefunction into the classically forbidden region.

The tunneling analysis can be extended to an arbitrary-shaped potential barrier $V(x)$. Given $V(x)$, it can be decomposed into a juxtaposition of square barrier elements with infinitesimal thickness Δx and height $V(n\Delta x)$ (Figure 5.5). We can then take the tunneling through each barrier element as statistically independent

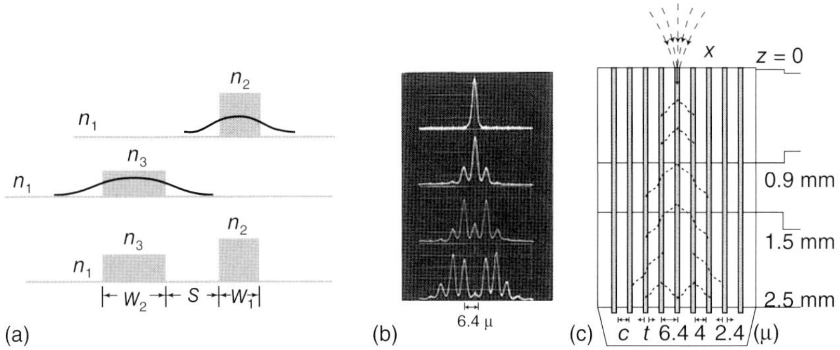

Figure 5.4 The index of refraction profiles of a directional coupler (a) and the observed light switching (b). Also shown is the tailing of the electric field amplitude out of the waveguide, inducing the coupling between the two waveguides (c). (Taken from Optical Electronics, A. Yariv, Holt, Rinehart, and Winston, 1985.)

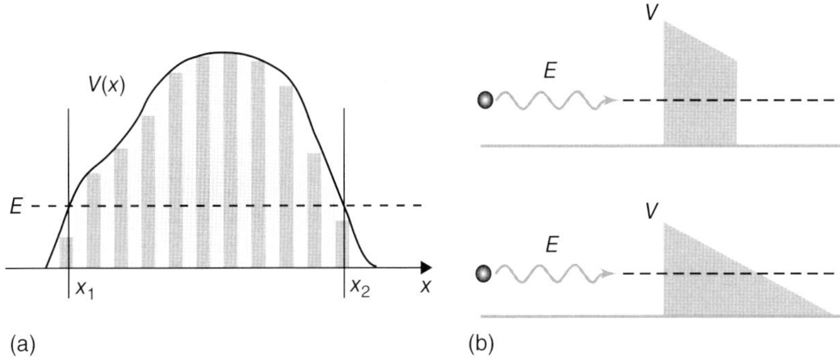

Figure 5.5 The tunneling of a particle through an arbitrary-shaped potential barrier (a) and the direct tunneling through the trapezoidal and the F–N tunneling through the triangular potential barriers (b).

and multiply the differential tunneling probabilities T_j to obtain the net tunneling probability:

$$T = \prod_j T_j \approx \exp - \frac{2\sqrt{2m}}{\hbar} \int_{x_1}^{x_2} dx[(V(x) - E)]^{1/2} \tag{5.23}$$

where for T_j, the dominant exponential factor derived in Eq. (5.22) has been used and the two limits x_1 and x_2 are determined by the condition $V(x) = E$ (Figure 5.5).

5.3.1
Direct and Fowler–Nordheim Tunneling

We next apply the tunneling probability Eq. (5.23) to a trapezoidal potential barrier, as shown in Figure 5.5. This kind of potential barrier is encountered by an electron or a hole incident on a dielectric layer in the presence of an external electric field **E**. In this case, $V(x)$ is given by

$$V(x) = V - q\text{E}x \tag{5.24}$$

where V is the barrier height at the dielectric interface, q the magnitude of the electron charge, and **E** the applied electric field. Upon using Eq. (5.24) in Eq. (5.23), we find

$$T = \exp - \left\{ \frac{4(2m)^{1/2}}{3q\text{E}\hbar} \left[(V - E)^{3/2} - (V - E - q\text{E}d)^{3/2} \right] \right\} \tag{5.25}$$

where d is the width of the barrier. The tunneling through the trapezoidal barrier is known as *direct tunneling* and accounts for one of the limiting processes hindering the downscaling of the metal oxide semiconductor field-effect transistor (MOSFETs). When the potential barrier is of a triangular shape, the second term in Eq. (5.25) drops out, and the tunneling probability reduces to

$$T = \exp - \frac{4(2m)^{1/2}}{3q\text{E}\hbar}(V - E)^{3/2} \tag{5.26}$$

and is known as the *Fowler–Nordheim* (*F–N*) tunneling. The F–N tunneling is utilized extensively for various semiconductor device operations. Figure 5.6 shows the direct and F–N tunneling probabilities versus the incident electron energy E for a different electric field E. The two parameters critically affect the tunneling probabilities as clear from the figure.

5.3.2 Resonant Tunneling

The superlattice structure is composed of a series of quantum wells with each well formed by two potential barriers and is an important element in optoelectronic devices. The electrons in such structures undergo resonant tunneling. To examine it, let us consider an electron incident on two potential barriers with height V, thickness d, and distance W apart as shown in Figure 5.7. An electron incident on

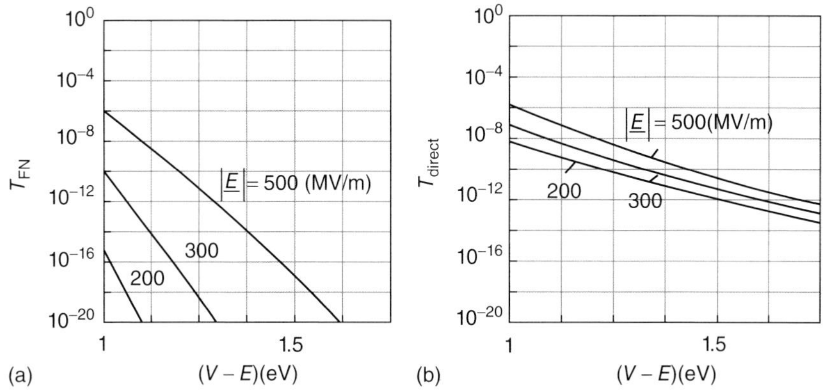

Figure 5.6 The F–N (a) and direct (b) tunneling probabilities of an electron versus the effective barrier height $V - E$ for a different electric field E.

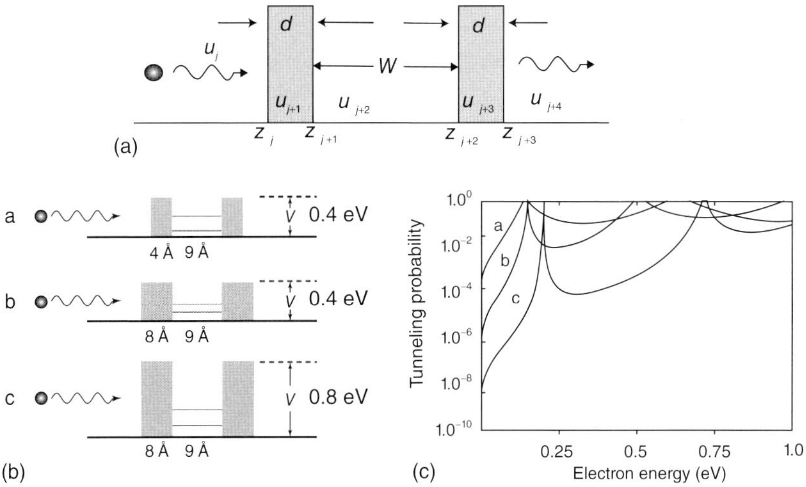

Figure 5.7 A particle incident on two potential barriers of height V and thickness d and distance W apart (a). A particle incident on the two potential barriers with different height and thickness (b) and corresponding tunneling probability versus incident energy (c).

the input plane at z_j with $E < V$ may exit at the output plane at z_{j+3} after undergoing two successive tunneling through the two barriers. The eigenfunctions in the regions j and $j+1$ are given in analogy with Eq. (5.4) by

$$u_j(z) = A_j e^{ikz} + B_j e^{-ikz}; \quad k \equiv \left(\frac{2mE}{\hbar^2}\right)^{1/2}, \quad z \le z_j \tag{5.27a}$$

$$u_{j+1}(z) = A_{j+1} e^{-\kappa z} + B_{j+1} e^{\kappa z}; \quad \kappa \equiv \left[\frac{2m(V-E)}{\hbar^2}\right]^{1/2}, \quad z_j \le z \le z_{j+1} \tag{5.27b}$$

Naturally, $u_j(z)$ consists of the incident and reflected beams, while $u_{j+1}(z)$ is composed of two exponential terms, $\exp \pm\kappa x$, because the width of the barrier is finite. The boundary conditions at z_j

$$u_j(z_j) = u_{j+1}(z_j), \quad u_j'(z_j) = u_{j+1}'(z_j) \tag{5.28}$$

yields coupled equations relating A_j, B_j to A_{j+1}, B_{j+1} as in Eq. (5.16), however, with k_0 and k replaced by k and $i\kappa$, respectively. Thus, by expressing A_j, B_j in terms of A_{j+1}, B_{j+1}, we can write

$$\begin{pmatrix} A_j \\ B_j \end{pmatrix} = M(z_j, i\kappa, k) \begin{pmatrix} A_{j+1} \\ B_{j+1} \end{pmatrix} \tag{5.29a}$$

where the 2×2 transfer matrix elements are given by

$$M(z_j; i\kappa, k) = \frac{1}{2}\begin{pmatrix} \left(1 + \frac{i\kappa}{k}\right) e^{i(i\kappa - k)z_j} & \left(1 - \frac{i\kappa}{k}\right) e^{-i(i\kappa + k)z_j} \\ \left(1 - \frac{i\kappa}{k}\right) e^{i(i\kappa + k)z_j} & \left(1 + \frac{i\kappa}{k}\right) e^{-i(i\kappa - k)z_j} \end{pmatrix} \tag{5.29b}$$

We can likewise express A_{j+1}, B_{j+1} in terms of A_{j+2}, B_{j+2} by imposing the usual boundary conditions at $z_j + d$ as

$$\begin{pmatrix} A_{j+1} \\ B_{j+1} \end{pmatrix} = M(z_j + d, k, i\kappa) \begin{pmatrix} A_{j+2} \\ B_{j+2} \end{pmatrix} \tag{5.30}$$

In fact, the transfer matrix in Eq. (5.30) is obtained from Eq. (5.29b) by simply interchanging k and κ and replacing z_j by $z_j + d$. In this manner, A_j, B_j are coupled to A_{j+2}, B_{j+2} as

$$\begin{pmatrix} A_j \\ B_j \end{pmatrix} = M(z_j, z_j + d) \begin{pmatrix} A_{j+2} \\ B_{j+2} \end{pmatrix} \tag{5.31a}$$

where the net transfer matrix is obtained by multiplying the matrices in Eqs. (5.29) and (5.30):

$$\begin{aligned} M(z_j; d) &\equiv M(z_j, i\kappa, k) M(z_j + d, k, i\kappa) \\ &= \begin{pmatrix} m_{11}(d) & m_{12}(z_j, d) \\ m_{12}^*(z_j, d) & m_{11}^*(d) \end{pmatrix} \end{aligned} \tag{5.31b}$$

with the matrix elements given by

$$m_{11}(d) = e^{ikd} \left(\cosh \kappa d - i \frac{k^2 - \kappa^2}{2k\kappa} \sinh \kappa d \right) \tag{5.31c}$$

$$m_{12}(z_j, d) = i e^{-ik(2z_j + d)} \frac{k^2 + \kappa^2}{2k\kappa} \sinh \kappa d \tag{5.31d}$$

The matrix Eq. (5.31) is the unit transfer matrix by which to describe the multiple tunneling.

Let us revisit the tunneling through a single barrier by using Eq. (5.31). In this case, constants, A_j, B_j, and A_{j+2} in Eq. (5.30) represent the incident, reflected, and transmitted beams, respectively. Once the particle is transmitted, there is no reflection, hence $B_{j+2} = 0$, and the tunneling probability can be found in analogy with Eqs. (5.17a) and (5.18a) as

$$T_{1B} \propto \left| \frac{A_{j+2}}{A_j} \right|^2 = \left| \frac{1}{m_{11}} \right|^2 \tag{5.32}$$

When k, κ are expressed in terms of E by using Eq. (5.27), Eq. (5.32) leads to the same results as obtained in Eq. (5.22).

We next treat the tunneling through two successive barriers. The extension of a single-barrier tunneling Eq. (5.32) to tunneling through two barriers in succession is straightforward and can be done by coupling A_{j+2}, B_{j+2} to A_{j+4}, B_{j+4} via the transfer matrix Eq. (5.31) with appropriate changes of z_j's. The transfer matrix connecting A_j, B_j to A_{j+4}, B_{j+4} is then obtained by multiplying the two unit transfer matrices:

$$\begin{pmatrix} A_j \\ B_j \end{pmatrix} = M(z_j, d) M(z_{j+2}, d) \begin{pmatrix} A_{j+4} \\ 0 \end{pmatrix}, \quad z_{j+2} = z_j + W + d, \quad B_{j+4} = 0 \tag{5.33}$$

Thus, after performing the matrix multiplication, we find

$$\frac{A_{j+4}}{A_j} = \frac{1}{m_{11}(d)m_{11}(d) + m_{12}(z_j, d)m_{12}^*(z_{j+2}, d)}$$
$$= \frac{-e^{-2ikd}4k^2\kappa^2}{[(k^2-\kappa^2)\sinh\kappa d + 2ik\kappa\cosh\kappa d]^2 - e^{2ikW}(k_1^2+\kappa^2)^2\sinh^2\kappa d} \tag{5.34}$$

The ratio (Eq. (5.34)) can be put into a simple mathematical form by noting that the first bracket in the denominator gives the tunneling probability for a single barrier T_{1B} when combined with the numerator (see Eqs. (5.31) and (5.32)). Thus, we can rewrite Eq. (5.34) as

$$\frac{A_{j+4}}{A_j} = \frac{-e^{-2ikd-2i\theta}T_{1B}}{1-e^{2i(kW-\theta)}R_{1B}}, \quad R_{1B} = 1 - T_{1B} \tag{5.35a}$$

$$\tan\theta = \frac{2k\kappa\cosh\kappa d}{(k^2-\kappa^2)\sinh\kappa d} \tag{5.35b}$$

Therefore, the probability of tunneling the two successive barriers is obtained as

$$T_{2B} = \frac{(\hbar k_0/m)|A_{j+4}|^2}{(\hbar k_0/m)|A_j|^2} = \frac{1}{1+4(R_{1B}/T_{1B}^2)\sin^2(kW-\theta)} \tag{5.36}$$

In deriving Eq. (5.36), use has been made of the identities

$$|1 - f\exp i\chi|^2 = 1 + f^2 - 2f\cos\chi, \cos x = 1 - 2\sin^2\left(\frac{x}{2}\right)$$

It is thus clear from Eq. (5.36) that the resonant tunneling can occur when $kW \approx n\pi$. The condition can be expressed in terms of E via the relation $E = (\hbar k)^2/2m$ as

$$E_n = \frac{\hbar^2\pi^2 n^2}{2mW^2}, \quad n = 1, 2, \ldots \tag{5.37}$$

and points clearly to the fact that the resonant tunneling occurs when the energy of the incident electron coincides with the bound state energies of the quantum well formed in between the two barriers (see Eq. (4.4)). This fact carries an important bearing on the band theory of solids, as will be discussed. Plotted in Figure 5.7 is the tunneling probability through two successive barriers T_{2B} versus the incident energy for various barrier heights and widths. Indeed, T_{2B} is drastically reduced with increasing V and d, but the general features of the resonant tunneling are preserved.

5.4 The Applications of Tunneling

The tunneling is utilized extensively in semiconductor and optoelectronic devices, and the list of applications is fast increasing. Some of the applications are briefly discussed.

5.4.1 Metrology and Display

Figure 5.8 shows the schematics of the scanning tunneling microscope (STM). In this scheme, the high sensitivity of the F–N tunneling probability on the thickness of the barrier potential is utilized for probing the surface morphology with atomic-scale resolution. The probing is done by fixing the tunnel current I_T flowing between the probe tip and the surface atoms. Keeping I_T fixed necessitates the adjustment of the height of the probe tip so that the distance between the tip and surface atoms is kept constant. The required adjustment of the height of the tip versus the $x-y$ scan reveals the surface morphology with about 0.1 nm in accuracy. Alternatively, the height of the tip is fixed at a constant level while scanning. In this case, I_T should vary depending on the varying distance between the tip and atoms, which can be translated into the surface morphology.

Field Emission Display

The schematics for the display are shown in Figure 5.9. The image information is transmitted by the driver circuitry via the strings of voltages applied to the array of metallic tips, forming the pixels. The signal voltages then induce the field

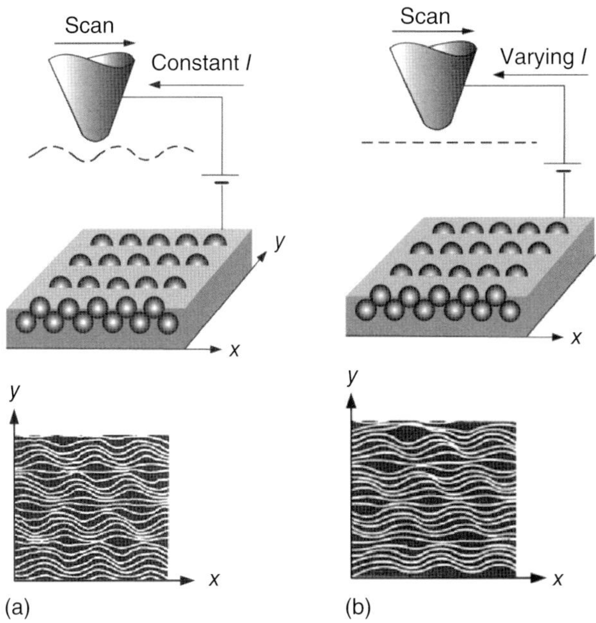

Figure 5.8 The schematics of the scanning tunneling microscopy: adjusting the probe tip distance from the sample surface at a fixed current level while scanning the surface (a) and fixing the probe tip height and monitoring the tunnel current while scanning the surface (b).

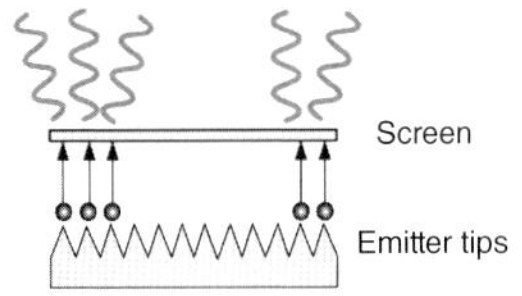

Figure 5.9 The schematics of the flat panel display; the electrons emitted from pixel tips via the field crowding assisted tunneling and transferring the image information to the display screen.

crowding at the metallic tip, enhancing the electric field amplitude and enabling F–N tunneling to occur at the tip. The electrons thus tunneled out from each pixel transmit the image information to the screen for display.

5.4.2
Single-Electron Transistor

The SET is based on the tunneling of a single electron and utilizes a quantum dot as the channel island. The dot is made of a metal or semiconductor and insulated from the two electrodes, called the *source* and *drain* (Figure 5.10). With shrinking size of the dot, the electron potential energy therein varies significantly depending on the presence or absence of a single electron. This effect is used for the controlled tunneling of a single electron for the transistor action.

The size effect can be discussed by taking the junction between the electrodes and quantum dot as the parallel-plate capacitor for simplicity. The capacitance is then given by the area A and thickness d of the junction as

$$C = \frac{\varepsilon A}{d} \tag{5.38}$$

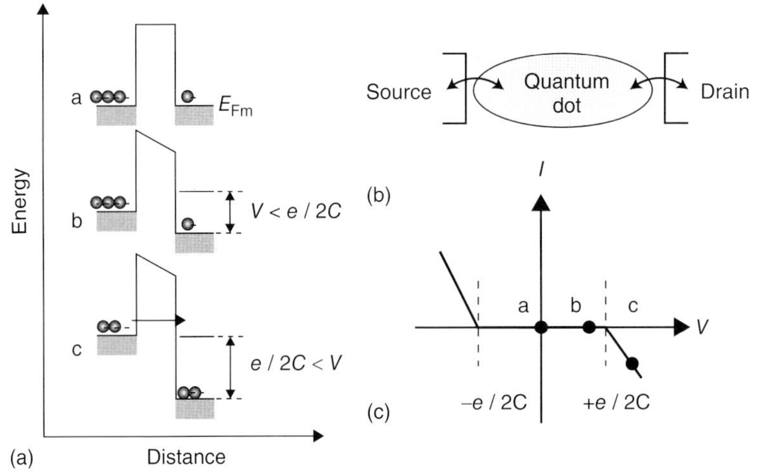

Figure 5.10 The single-electron transistor consisting of a quantum dot as the channel insulated from the source and drain electrodes (b). The static I–V behavior and the Coulomb blockade (c). The energy-level diagrams operative for a single electron tunneling (a); the blocked tunneling for the drain bias V or eV less than the charging energy E_C (b) and the tunneling with sufficient V to compensate for E_C.

where ε is the permittivity of the junction. Now, the charging energy of a single electron in the quantum dot is given from the well-known electromagnetic theory by

$$E_c = \frac{e^2}{2C} \tag{5.39}$$

and E_c can dictate the tunneling, provided it far exceeds the thermal energy, that is,

$$\frac{e^2}{2C} \gg k_B T \tag{5.40}$$

Otherwise, E_c is simply buried in the thermal fluctuations. For SiO_2 and for $A \approx 10\,nm \times 10\,nm$ and $d \approx 2\,nm$, E_c is $\sim 15\,meV$ and is comparable with the thermal energy at room temperature $25\,meV$. It is therefore possible to satisfy the condition of Eq. (5.40) by further downsizing the quantum dot.

Additionally, E_c should exceed the uncertainty ΔE in the energy of the single electron, which is inherently associated with its finite lifetime in the dot. The lifetime can be estimated by $\tau = R_T C$, where R_T is the tunneling resistance inversely proportional to its probability. Thus, τ is analogous to the RC time constant of a capacitor connected to a series resistance, in this case, the tunneling resistance. The condition that E_c is much greater than ΔE can be expressed by using the uncertainty relation as

$$\Delta E \approx \frac{h}{\tau} = \frac{h}{R_T C} \ll \frac{e^2}{2C} \tag{5.41a}$$

Or equivalently,

$$R_K \ll R_T, \quad R_K \equiv \frac{h}{e^2} \simeq 25.8\,\mathrm{k\Omega} \tag{5.41b}$$

The resistance R_K is called the *quantum resistance,* and the Eq. (5.41) ensures that the electron is localized in the dot in a quantum state therein.

Once Eqs. (5.40) and (5.41) are satisfied, E_c plays the critical role for the SET operation. A single electron when tunneling into the quantum dot from the source electrode raises the electron energy level therein by E_c, which should hinder the tunneling event. This is because the tunneling is an elastic process, and the energy level of the dot after the tunneling should not exceed the initial energy level of the tunneling electron. However, when the drain voltage V_D in excess of $e/2C$ is applied, the electron potential energy in the dot is lowered by E_c or more via the capacitive coupling between the drain electrode and the quantum dot. Hence, the electron can now tunnel from the source into the quantum dot, contributing to the drain current I_D. Note that a positive V applied to the drain lowers the electron potential energy by $-qV$. By the same token, if a negative V_D is applied below $-e/2C$ to the drain, the electron energy level in the dot is lowered by more than E_c compared with that of the drain electrode. Therefore, an electron can tunnel into the channel from the drain and contribute to I_D flowing in the opposite direction.

It is therefore clear that tunneling of a single electron is prohibited in the range of V_D

$$-\frac{e}{2C} < V_{\mathrm{D}} < \frac{e}{2C} \tag{5.42}$$

This is known as the *Coulomb blockade.* Outside of this V_D range, however, the drain current flows contributed by the tunneling events of a single electron. Figure 5.10 shows the resulting current–voltage characteristics. In summary, the operation of SET is based on the Coulomb blockade caused by the charging energy in the quantum dot, but the blockade is overcome by means of the capacitive coupling of V_D to the channel island.

Problems

5.1 (a) A particle of mass m is incident on a two-step potential barrier with E greater than V_2 and in the direction normal to the barrier. Find (i) the energy eigenfunctions in the regions $x \le 0$, $0 \le x \le d_1$, and $x > d_1$, (ii) R and T by imposing the boundary conditions at $x = 0, d_1$, and (iii) V_1 and d_1 at which 100% transmission occurs.
(b) A particle is incident on a potential barrier V_2 from the region V_1 at an angle θ_i with respect to the z-direction. Write down the incident, reflected, and transmitted wavefunctions and find the angle of reflection and transmission by using the boundary conditions at the potential boundary. Interpret the result in light of the reflection and refraction of light at a dielectric interface (Figure 5.11).

5.2 (a) Starting from Eq. (5.7), fill in the algebra and derive the expression of the probability current density $\boldsymbol{S}$ (Eq. (5.8)).
(b) Use the eigenfunction Eq. (5.4) in Eq. (5.8) and derive Eq. (5.9).

5.3 (a) Starting from four coupled equations (Eq. (5.16)), find the ratios t/i_0, r/i_0 (Eq. (5.17)), and T, R given (Eq. (5.18)).
(b) Carry out a parallel analysis and derive the tunneling probability T (Eq. (5.22)) with the use of Eq. (5.21).

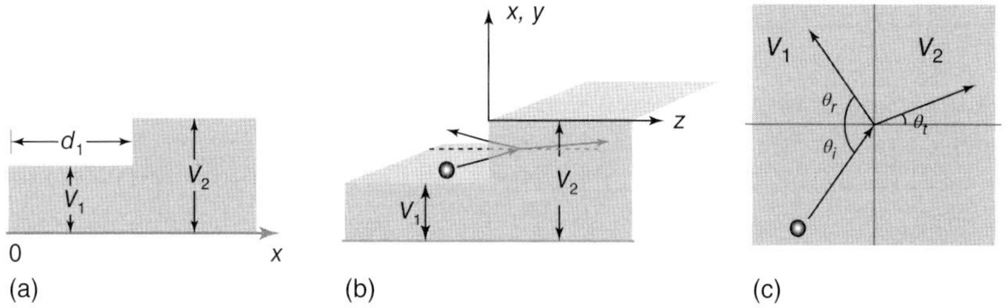

Figure 5.11 A particle incident on a two-step potential barrier with $E > V_2$ (a) and incident on V_2 (b) with $V_1 < E < V_2$ from the region V_1. The incident, reflected, and transmitted angles (c).

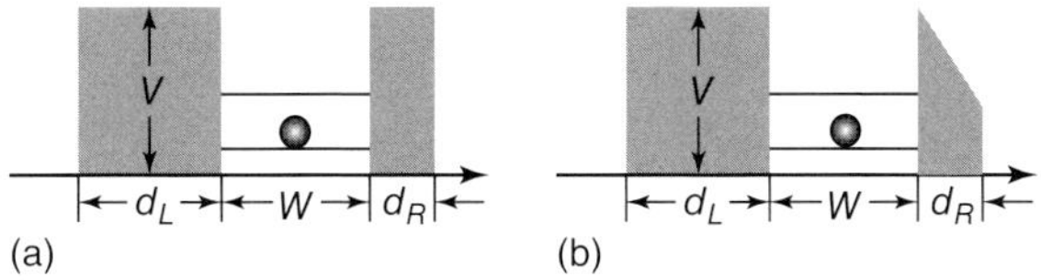

Figure 5.12 An electron in the ground state of a quantum well formed by two square barrier potential (a) and with one barrier subjected to an electric field (b).

5.4 Consider a quantum well formed in between two potential barriers of height V, thicknesses d_L and d_R, and distance W apart with $d_L > d_R$.
 (a) Estimate the ground state energy of an electron in electron volt unit by assuming infinite barrier height, for widths of 1, 10 nm.
 (b) Find the lifetime of the electron in the ground state.
 Hint: The lifetime is defined by $T\widetilde{n} = 1$, where T is the tunneling probability while $\widetilde{n} = 2W/v_T$ is the number for the electron to encounter the barrier, with v_T denoting the thermal velocity of the electron.

5.5 Consider the quantum well shown in Figure 5.12. When subjected to an electric field E, the barrier potential is transformed to the trapezoidal shape as shown in Figure 5.12.
 (a) Express the trapezoidal shape in terms of E and find the electron lifetime versus E.
 (b) Find E necessary to shorten the lifetime to 1 μs. (Take the infinite barrier height for estimating the ground state energy for simplicity.)

5.6 The metal tip of the STM has the work function of 4.5 eV.
 (a) Find the electric field at which the electron tunneling probability is 10^{-4} if the distance between the tip and the sample is 5 *nm*.
 (b) If 5 V is applied between the tip and the sample, estimate the distance between them to attain the same tunneling probability of 10^{-4}.
 Hint: Use a triangular barrier with the height given by the work function.

5.7 (a) Starting from the wavefunction given in Eq. (5.27), fill in the algebra and derive the results (Eqs. (5.29), (5.31), and (5.32)).
 (b) Extend the analysis and derive Eq. (5.36).
 Hint: Use the matrix algebra.

$$\begin{pmatrix} a_{11} & a_{12} \\ a_{21} & a_{22} \end{pmatrix} \begin{pmatrix} A \\ B \end{pmatrix} = \begin{pmatrix} a_{11}A + a_{12}B \\ a_{21}A + a_{22}B \end{pmatrix}$$

$$\begin{pmatrix} a_{11} & a_{12} \\ a_{21} & a_{22} \end{pmatrix} \begin{pmatrix} b_{11} & b_{12} \\ b_{21} & b_{22} \end{pmatrix} = \begin{pmatrix} a_{11}b_{11} + a_{12}b_{21} & a_{11}b_{12} + a_{12}b_{22} \\ a_{21}b_{11} + a_{22}b_{21} & a_{21}b_{12} + a_{22}b_{22} \end{pmatrix}$$

Suggested Readings

1. J. Singh, Quantum Mechanics, Fundamentals and Applications to Technology, John Wiley & Sons, 1996.
2. D. M. Kim, Introductory Quantum Mechanics for Semiconductor Nanotechnology, Wiley-VCH, 2010.

3. A. Yariv, An Introduction to Theory and Applications of Quantum Mechanics, John Wiley & Sons, 1982.
4. R. L. Liboff, Introductory Quantum Mechanics, Fourth Edition, Addison Wesley Publishing Company, Reading, MA, 2002.
5. S. Gasiorowics, Quantum Physics, Third Edition, John Wiley & Sons, 2003.
6. R. W. Robinett, Quantum Mechanics, Classical Results, Modern Systems and Visualized Examples, Oxford University Press, 2006.
7. H. Kroemer, Quantum Mechanics for Engineering, Materials Science, and Applied Physics, International Edition, Prentice Hall, 1994.

6
Energy Bands in Solids

The energy band is a natural consequence of the wave nature of the particle and provides the basic foundation for analyzing the condensed matters and the semiconductor devices. The energy band is discussed based on the Kronig–Penney (K–P) potential, Bloch wavefunction, and the resonant tunneling. Additionally, the motion of electrons in solids is discussed with the use of the dispersion relation operative in conduction and valence bands.

6.1
Bloch Wavefunction in Kronig–Penney Potential

The 1D crystal is often modeled as a linear array of positive ions, located periodically (Figure 6.1). An electron in the crystal interacts with ions via the attractive Coulomb force, and the periodic Coulomb potential can be approximated by a string of square barrier potentials, separated by identical quantum wells. This simplified version of the 1D potential is known as *Kronig–Penney potential*, and it brings out the concept of the energy band in a simple manner.

The unit cell of the K–P potential consists of the quantum well of width a and the barrier potential of thickness b, so that the length d of the unit cell is the sum of a and b. A focal point of the discussion is the electron wavefunction in the periodic potential given by

$$V(x) = V(x + d), \quad d = a + b \tag{6.1}$$

For such potential, the Hamiltonian is also periodic

$$\hat{H}(x) = -\frac{\hbar^2}{2m}\frac{\partial^2}{\partial x^2} + V(x) = \hat{H}(x + d) \tag{6.2}$$

In this case, the Bloch theorem states that the energy eigenfunction is specified by the modulated plain wave

$$\varphi_k(x) = e^{ikx}u(x) \tag{6.3a}$$

with the modulating envelop satisfying the periodic boundary condition

$$u(x) = u(x + d) \tag{6.3b}$$

Introductory Quantum Mechanics for Applied Nanotechnology, First Edition. Dae Mann Kim.

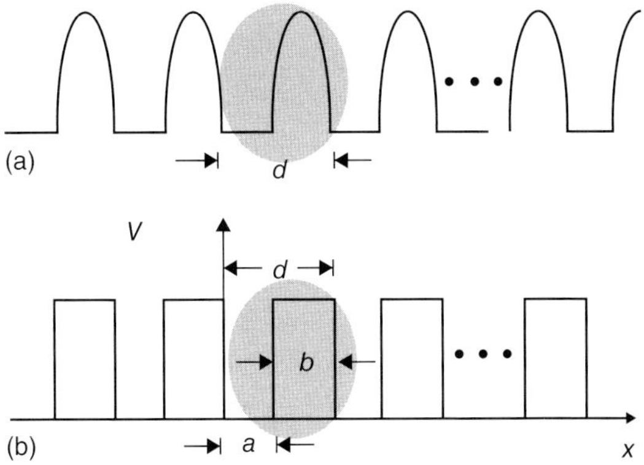

Figure 6.1 The Coulomb potential of an electron in 1D crystal (a) and the Kronig–Penney potential mimicking it via a string of unit cells composed of a quantum well and barrier potential (b).

The wavefunction Eq. (6.3) is known as the *Bloch wavefunction*.

To prove the Bloch theorem, let us introduce the displacement operator

$$\hat{D}f(x) \equiv f(x+d) \tag{6.4}$$

where $f(x)$ is an arbitrary function. When $\hat{D}$ operates on Eq. (6.3), there results

$$\hat{D}\varphi_k(x) = \hat{D}e^{ikx}u(x) \equiv e^{ik(x+d)}u(x+d) = e^{ikd}\varphi_k(x) \tag{6.5}$$

Therefore, $\varphi_k(x)$ is the eigenfunction of $\hat{D}$ with the eigenvalue $\exp(ikd)$. Moreover, because $\hat{H}$ is periodic, we can write

$$\hat{D}\hat{H}(x)f(x) = \hat{H}(x+d)f(x+d) = \hat{H}(x)\hat{D}f(x)$$

where $f(x)$ is an arbitrary function and therefore $\hat{D}$ and $\hat{H}$ commute. Since the commuting operators can share a common eigenfunction (see Eq. (3.30)) and since $\varphi_k(x)$ is an eigenfunction of $\hat{D}$, $\varphi_k(x)$ is also an eigenfunction of $\hat{H}$, thus proving the Bloch theorem. The probability density of $\varphi_k(x)$ is given by

$$|\varphi_k(x+nd)|^2 = |e^{ik(x+nd)}u(x+nd)|^2 = |e^{iknd}\varphi_k(x)|^2 = |\varphi(x)|^2, \quad n = 1, 2, \ldots \tag{6.6}$$

and is consistent with the premise of a periodic system, namely, that the electron is found in all unit cells with equal probability.

We next specify the Bloch wavefunction in the K–P potential by using the ring boundary condition (Figure 6.2). The ring consists in this case of a large number N of unit cells, and the periodic boundary condition is equivalent to stating that the electron leaving the last cell in the ring reenters into the first one. The condition is conveniently used for describing the motion of electrons in the bulk crystal, free of edge effects. The wave vector k in Eq. (6.3a) should then satisfy the condition

$$e^{ikdN} = 1 \equiv e^{i2\pi n}, \quad n = 1, 2, \ldots \tag{6.7}$$

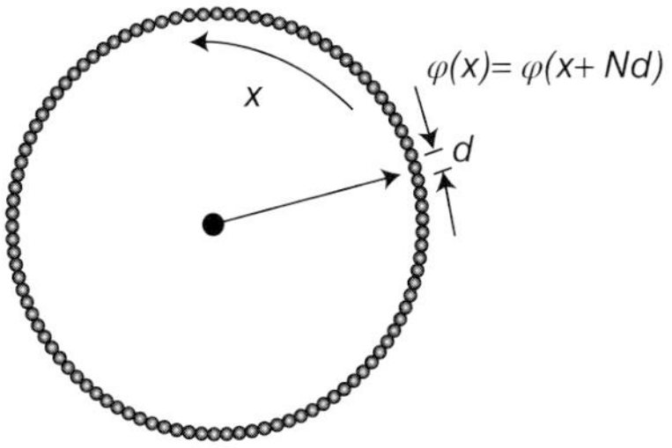

Figure 6.2 The ring boundary condition in 1D crystal.

and therefore should be specified by

$$k_n = \frac{2\pi n}{\mathrm{d}N} = \frac{2\pi n}{L}, \quad n = 1, 2, \ldots \tag{6.8}$$

where L is the length of the crystal. Thus, k becomes quasi-continuous in the limit of large N, and the momentum associated $\hbar k_n$ is called *crystal momentum*. In the limit of infinite well width, the envelope function $u(x)$ in Eq. (6.3) should be put to unity, in which case $\varphi_k(x)$ reduces to the wavefunction of a free particle, as it should. For a finite well width, however, $\varphi_k(x)$ is modulated by $u(x)$, which should be an identical function in each unit cell. With this general fact in mind, we can represent the energy eigenfunction in a unit cell in the usual manner as

$$\varphi_k(x) = \begin{cases} Ae^{ik_1x} + Be^{-ik_1x}, & k_1 = \left[\frac{2mE}{\hbar^2}\right]^{1/2}, \quad 0 \le x \le a \\ Ce^{-\kappa x} + De^{\kappa x}, & \kappa = \left[\frac{2m(V-E)}{\hbar^2}\right]^{1/2}, \quad a \le x \le d \end{cases} \tag{6.9}$$

Here, the analysis is confined to the bound state, in which $E < V$.

Boundary Conditions

As noted, $u(x)$ should be periodic and therefore should satisfy the conditions $u(0^+) = u(d^-)$ and $u'(0^+) = u'(d^-)$. Or in terms of $\varphi_k(x)$, the conditions read as

$$\varphi(0^+) = e^{-ikd}\varphi(d^-) \tag{6.10a}$$

$$\varphi'(0^+) = e^{-ikd}\varphi'(d^-) \tag{6.10b}$$

In Eq. (6.10b), use has been made of $u'(x) = \varphi'(x)[\exp -(ikx)] - iku(x)$, and the condition to be satisfied by the second term $iku(x)$ has already taken into account in Eq. (6.10a). Upon inserting Eq. (6.9) into Eq. (6.10), there result

$$A + B = e^{-ikd}(Ce^{-\kappa d} + De^{\kappa d}) \tag{6.11a}$$

$$ik_1(A - B) = -\kappa e^{-ikd}(C^{-\kappa d} - De^{\kappa d}) \tag{6.11b}$$

Additionally, the usual boundary condition, namely, that $\varphi(x)$, $\varphi'(x)$ be continuous everywhere should be applied at $x = a$ where $V(x)$ is discontinuous. These conditions read as

$$Ae^{ik_1a} + Be^{-ik_1a} = Ce^{-\kappa a} + De^{\kappa a} \tag{6.12a}$$

$$ik_1(Ae^{ik_1a} - Be^{-ik_1a}) = -\kappa(C^{-\kappa a} - De^{\kappa a}) \tag{6.12b}$$

Thus, finding the Bloch wavefunction has been reduced to determining the constants of integration A, B, C, and D from the four boundary conditions Eqs. (6.11) and (6.12). For this purpose, let us first find A, B in terms of C, D from Eq. (6.11) as

$$A = \frac{1}{2}e^{-ikd}(\alpha e^{-\kappa d}C + \alpha^* e^{\kappa d}D), \quad \alpha \equiv 1 + \frac{i\kappa}{k_1} \tag{6.13a}$$

$$B = \frac{1}{2}e^{-ikd}(\alpha^* e^{-\kappa d}C + \alpha e^{\kappa d}D) \tag{6.13b}$$

We can likewise express A, B in terms of C, D from Eq. (6.12) as

$$A = \frac{1}{2}e^{-ik_1a}(\alpha e^{-\kappa a}C + \alpha^* e^{\kappa a}D) \tag{6.14a}$$

$$B = \frac{1}{2}e^{ik_1a}(\alpha^* e^{-\kappa a}C + \alpha e^{\kappa a}D) \tag{6.14b}$$

Hence, we can eliminate A, B by equating the right-hand sides of Eqs. (6.13a), (6.14a) and (6.13b), (6.14b), respectively, and write the coupled equation for C, D as

$$\begin{cases} a_{11}C + a_{12}D = 0 \\ a_{21}C + a_{22}D = 0 \end{cases} \quad \text{or} \quad \begin{pmatrix} a_{11} & a_{12} \\ a_{21} & a_{22} \end{pmatrix}\begin{pmatrix} C \\ D \end{pmatrix} = 0 \tag{6.15a}$$

where the matrix elements are given by

$$\begin{aligned} a_{11} &= \alpha(e^{-ikd-\kappa d} - e^{-ik_1a-\kappa a}), \quad & a_{12} &= \alpha^*(e^{-ikd+\kappa d} - e^{-ik_1a+\kappa a}) \\ a_{21} &= \alpha^*(e^{-ikd-\kappa d} - e^{ik_1a-\kappa a}), \quad & a_{22} &= \alpha(e^{-ikd+\kappa d} - e^{ik_1a+\kappa a}) \end{aligned} \tag{6.15b}$$

Secular Equation and Dispersion Relation

Since the coupled equation (6.15) is homogeneous, that is, the right-hand side is zero, C, D will be zero, unless the secular equation is satisfied, that is, the determinant of the coupling matrix is zero

$$\begin{vmatrix} a_{11} & a_{12} \\ a_{21} & a_{22} \end{vmatrix} = 0, \quad \text{or} \quad a_{11}a_{22} - a_{12}a_{21} = 0 \tag{6.16}$$

If Eq. (6.16) is not met, it can be readily shown that $C = D = 0$, in which case $A = B = 0$, as clearly follows from Eq. (6.13) and Eq. (6.14). Therefore, the Bloch wavefunction becomes trivial. Thus, the secular equation (6.16) is a critical condition to be satisfied for obtaining the nontrivial wavefunction. We can spell out the determinant Eq. (6.16) explicitly by using Eq. (6.15b) and obtain after a lengthy but straightforward algebra

$$\cos k_1a \cosh \kappa b - \frac{k_1^2 - \kappa^2}{2k_1\kappa} \sin k_1a \sinh \kappa b = \cos kd \tag{6.17}$$

As k_1 and κ are given functions of E (see Eq. (6.9)), Eq. (6.17) implicitly relates E with the wave vector k. Once E is found as an explicit function of k and the dispersion relation is obtained, the Bloch wavefunction $\varphi_k(x)$ given in Eq. (6.3) is completely specified.

6.2
E–k Dispersion and Energy Bands

To find the dispersion relation specifying E explicitly in terms of k from the transcendental equation (6.17), we can resort to a graphical means. For this purpose, the left-hand side of Eq. (6.17) is plotted versus E/V in Figure 6.3. The resulting curve is clearly shown to oscillate with diminishing amplitudes with increasing E/V. Also shown in the figure are the values of $\cos kd$ appearing on the right-hand side of Eq. (6.17). From these plots, it is possible to find E as a function of k and to bring out the key features of the energy band.

Allowed Bands and Forbidden Gaps

As $|\cos kd| \leq 1$, it is clear from Figure 6.3 that only for those ranges of E for which the left-hand side of Eq. (6.17) falls within the bounds of $\cos kd$, we can find the relationship between real k and real E. In these energy regimes, the electron can propagate in the crystal with a real propagation vector k, and these ranges are called *energy bands*. On the other hand, for E values in which the magnitude of the left-hand side of Eq. (6.17) is greater than unity, k therein should be a complex quantity. In this case, the electrons cannot propagate, and such energy ranges are called *forbidden gaps*. In summary, the spectrum of the electron energy in 1D crystal consists of a series of allowed bands, separated by forbidden gaps. Also the allowed band broadens with increasing E, while it decreases with increasing V and tighter binding of electrons.

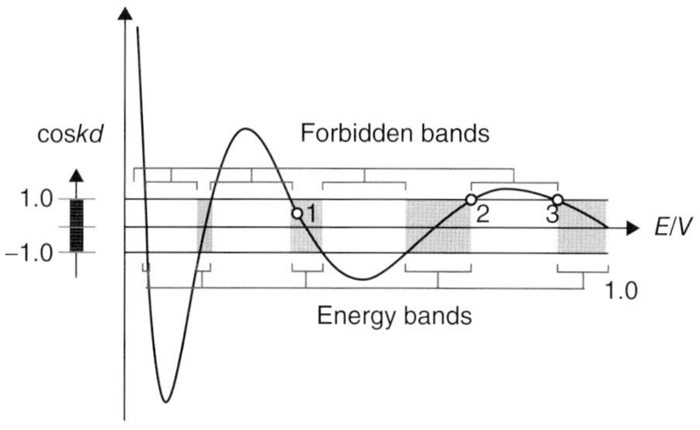

Figure 6.3 The graphical analysis of the dispersion relation: the left-hand side of Eq. (6.17) is plotted versus E/V, and values of $\cos kd$ on the right-hand side are indicated. Also shown are the allowed energy bands and forbidden gaps.

Bloch Wavefunction in Allowed Bands

We next specify the Bloch wavefunction in the allowed energy bands. For this purpose, let us choose from the plot a pair of E, k values in the energy band and insert the pair in the coupled equation (6.15). In this case, the secular equation (6.17) is automatically satisfied by the pair, and therefore the two equations become redundant. That is, C and D are coupled with each other via either $C = -(a_{12}/a_{11})D$ or $C = -(a_{22}/a_{21})D$. Once C is expressed in terms of D, for instance, A and B can also be found in terms of D (see Eq. (6.13) or (6.14)). In this way, $\varphi_k(x)$ is specified in terms of E, k and other crystal parameters with D serving as the normalization constant. In Figure 6.4 are plotted typical wavefunctions thus found, together with probability densities. The wavefunctions are similar in shape to the bound state wavefunctions in the quantum well and are periodic over the unit cells.

Characteristics of $E-k$ Relationship

It is clear from Eq. (6.17) and Figure 6.3 that a given E can be matched by a string of k values $k + 2\pi n/d$ with n denoting an integer. Also, a given k is matched by multiple E values. However, we can set the one-to-one correspondence between E and k by allowing k to increase continually in steps of $2\pi/d$. The resulting E$-k$ curves are shown in Figure 6.5 in which E is shown as an even function of kd. This is expected because $\cos kd$ is even in kd, so that a given E can be matched by both

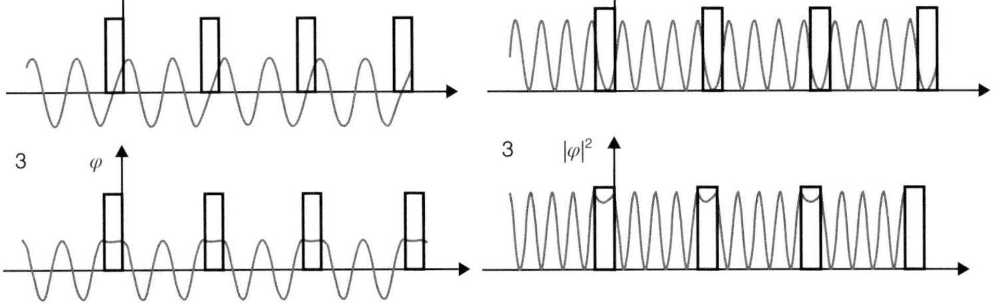

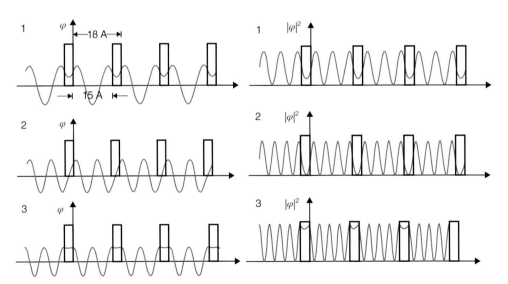

Figure 6.4 Typical Bloch wavefunctions and probability densities for different E, k pairs: E chosen from within the band (1), at the top of given band (2), and at the bottom of next higher lying band (3).

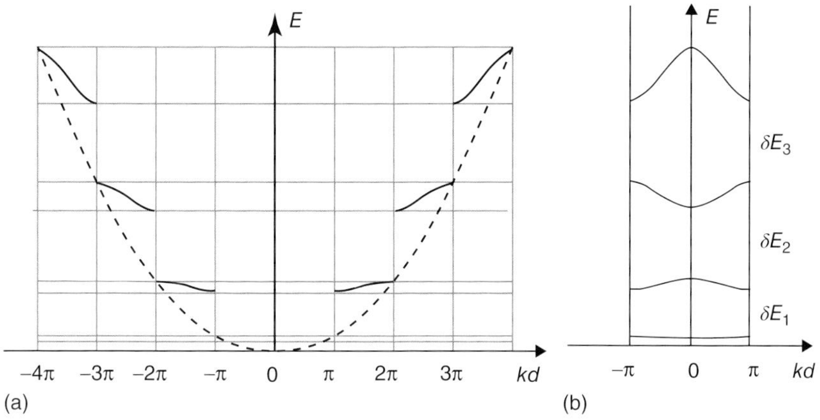

Figure 6.5 The dispersion curves: E versus $k + 2\pi n/d$, $n = 0, 1, 2, \ldots$ (a) and in reduced Brillouin zone (b). Also shown for comparison is the dispersion curve for a free particle $E = (\hbar^2/2m)k^2$ (broken line).

kd and $-kd$. For comparison, the equivalent free particle dispersion relation

$$E = \frac{[\hbar(k + 2\pi n/d)]^2}{2m}$$

is plotted by varying continuously the momentum equivalent to $\hbar(k + 2\pi n/d)$. Clearly, the two curves look alike, but differ considerably near the band edges.

Moreover, because k is determined to within an integer multiple of $2\pi/d$, it suffices to collect all of the E–k curves in the entire energy bands in a single interval, $-\pi \leq kd \leq \pi$, called the *first Brillouin zone*. This can be done by sliding the E–k curves in other Brillouin zones by $\pm 2\pi n/d$, and the resulting dispersion curves in the reduced zone are shown in Figure 6.4. From the figure, we can note a few revealing features of the E–k characteristics. Near the band edges, E is flat with respect to k, that is, $\partial E(k)/\partial k = 0$. This can be seen on a general ground by differentiating both sides of Eq. (6.17) with k and find that $\partial E(k)/\partial k \propto \sin kd$. At the band edges, $kd = n\pi$, hence $\sin kd = 0$. This clearly suggests that near the band edges, $E \propto k^2$ and the kinetic energy E of the electron is well represented by that of a free particle. Therefore, the electron is shown to behave as a free particle near the band edges.

Also, at the edges of the forbidden gap, $\cos kd = \pm 1$, so that $kd \equiv (2\pi/\lambda)d = n\pi$ or equivalently $2d = n\lambda$. The relationship indicates that the round-trip distance of the unit cell is an integer multiple of the de Broglie wavelength of the electron at band edges. This condition is precisely the 1D Bragg reflection condition, representing the constructive interference of reflected waves. Hence, the wave cannot penetrate into the next cell and propagate but becomes evanescent. In this case, the wavefunction degenerates into a standing wave, consisting of both forward and backward components with equal amplitudes. Naturally, there are two ways of forming standing waves, even and odd parity waves or wavefunctions with two different energy eigenvalues. The resulting splitting of energy levels accounts for the top and bottom of the energy gap at the band edge. This can be clearly seen from Figure 6.4, in which $\varphi_k(x)$ at the top of an energy band and that at the bottom of the next higher-lying energy band are plotted. The main difference between the two wavefunctions consists of high- and low-probability amplitudes near the potential barrier, as clearly shown in Figure 6.4. The resulting difference in average energy of the electron in the unit cell accounts for the energy gap.

Number of Quantum States per Band

We next consider the total number of quantum states in a band by using the ring boundary condition Eq. (6.8). Clearly, the number of k values or equivalently the number of wavefunctions in the range from k to $k + dk$ is given from Eq. (6.8) by

$$dn = \frac{L}{2\pi}dk \tag{6.18}$$

But we have to multiply Eq. (6.18) by 2 to account for the two electron spin states, spin-up and -down, for each k and integrate Eq. (6.18) over the Brillouin zone, obtaining

$$n = 2\frac{L}{2\pi}\int_{-\pi/d}^{\pi/d} dk = \frac{2L}{d} = 2N \tag{6.19}$$

where N is the total number of unit cells in the crystal. Hence, the total number of the quantum states per band is given by the number of unit cells constituting the 1D crystal multiplied by factor 2.

6.3 The Motion of Electrons in Energy Bands

The dispersion relation shown in Figure 6.5 provides the basis by which to describe the motion of electrons in 1D crystal. Thus, let us revisit the $E - k$ dispersion of a free particle

$$E = \frac{\hbar^2 k^2}{2m}, \quad \hbar k \equiv p \tag{6.20}$$

The velocity of the particle is given in this case by

$$v \equiv \frac{1}{\hbar}\frac{dE}{dk} = \frac{\hbar k}{m} = \frac{p}{m} \tag{6.21}$$

and is identical to the group velocity of the wave packet (Eq. (1.32)):

$$v_g \equiv \frac{d\omega}{dk}, \quad \omega = \frac{E}{\hbar} \tag{6.22}$$

Since $E \propto k^2$ near the band edge, the propagation velocity of the electron can also be represented by the slope of the E–k curve.

When an external electric field E is applied, the energy gained by an electron in δt from the field is given by

$$\delta E = -eEv_g\delta t \tag{6.23}$$

where $-e\text{E}$ is the force acting on the electron and $v_g\delta t$ the displacement the electron makes in time δt. The energy gain can also be expressed from Eqs. (6.20) and (6.21) as

$$\delta E \equiv \frac{dE}{dk}\delta k = \hbar v_g \delta k \tag{6.24}$$

Hence, by equating the right-hand sides of Eqs. (6.23) and (6.24), we obtain the equation of motion of the electron as

$$\frac{\hbar\delta k}{\delta t} \equiv \frac{dp}{dt} = -e\text{E} \tag{6.25}$$

Also the acceleration of the electron can be expressed from Eq. (6.22) as

$$a \equiv \frac{dv_g}{dt} = \frac{1}{\hbar}\frac{d}{dt}\left(\frac{\partial E(k)}{\partial k}\right) = \frac{1}{\hbar}\frac{\partial^2 E(k)}{\partial k^2}\frac{\partial k}{\partial t} \tag{6.26}$$

Hence, by equating $\partial k/\partial t$ in Eqs. (6.25) and (6.26), there results

$$\left[\frac{1}{\hbar^2}\frac{\partial^2 E(k)}{\partial k^2}\right]^{-1} a = -e\text{E} \tag{6.27}$$

Clearly, Eq. (6.27) is the well-known equation of motion, relating the force to the acceleration and the mass. The effective mass of the electron in the crystal can therefore be represented by

$$m_n = \left(\frac{1}{\hbar^2}\frac{\partial^2 E(k)}{\partial k^2}\right)^{-1} \tag{6.28}$$

In this manner, the dynamic parameters of the electron in the crystal are specified with the use of the dispersion relation in the energy bands.

6.4
Energy Bands and Resonant Tunneling

The energy band in solids has been discussed thus far in conjunction with the Bloch wavefunction and the dispersion relation. It can also be understood from an alternative standpoint of the resonant tunneling of electrons in the periodic potential. Thus, consider the limiting case in which the barrier height V is infinite. In this case, the parameters κ, $\sinh \kappa b$, and $\cosh \kappa b$ in Eq. (6.17) all diverge (see Eq. (6.9)), and the second term on the left-hand side of Eq. (6.17) becomes dominant, and other terms can be put to zero. Hence, the condition for Eq. (6.17) to hold true is given by

$$\sin k_1 a = 0, \quad \text{or} \quad k_1 a \equiv \left[\frac{2mE}{\hbar^2}\right]^{1/2} a = n\pi \tag{6.29}$$

Consequently, the energy levels associated are given by

$$E_n = \frac{\hbar^2\pi^2}{2ma^2}n^2, \quad n = 1, 2, \ldots$$

and are identical to those in an infinite square well potential of width a (see Eq. (4.4)). This is expected because the electron in this case is strictly confined to one unit cell independent of other cells.

Another limiting case to consider is the infinite potential barrier width b, and in this limit, $\sinh \kappa b = \cosh \kappa b \to \infty$. Thus, when both sides of Eq. (6.17) are divided by $\cosh \kappa b$ and $\cos k_1 a$, it reduces to

$$\tan 2\xi = \frac{2\xi\eta}{\xi^2 - \eta^2}, \quad k_1 a \equiv 2\xi, \quad \kappa a \equiv 2\eta \tag{6.30}$$

Or equivalently, with the use of a well-known trigonometric identity,

$$\tan 2\xi = \frac{2\tan\xi}{(1 - \tan^2\xi)}$$

Equation (6.30) is further reduced to a quadratic equation for $\tan\xi$

$$\tan^2\xi + \frac{\xi^2 - \eta^2}{\xi\eta}\tan\xi - 1 = 0 \tag{6.31}$$

Hence, we can solve for $\tan\xi$, obtaining

$$\xi\tan\xi = \eta, \quad \xi\cot\xi = -\eta \tag{6.32}$$

as the positive and negative branches of the solution. Clearly, Eq. (6.32) is the reduced version of the dispersion relation (6.17) and is identical to the quantization condition of the bound state energy for even and odd parity eigenfunctions in the quantum well (see Eqs. (4.22) and (4.23)).

In light of these two limiting cases, it is clear that the energy bands originate from the same energy quantization conditions as those discrete energy levels in isolated quantum wells. The only difference between the two cases consists of the discrete energy levels of the quantum well being broadened into bands due to the coupling between unit cells via the overlap of the wavefunctions in adjacent cells (see Figure 6.4). Therefore, an electron behaving as a free particle in allowed bands can be understood in light of the resonant tunneling. That is, an electron in allowed bands automatically satisfies the condition of the resonant tunneling by residing in energy eigenstates of the quantum well. Therefore, the electron can tunnel through

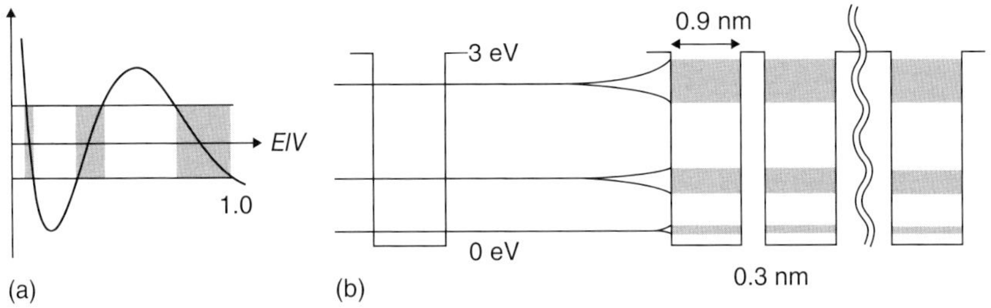

Figure 6.6 The discrete bound state energy levels in individual quantum well being broadened into energy bands (b) and the equivalent energy bands resulting from the dispersion relation (a).

the string of potential barriers with 100% probability. In this context, the potential barriers become transparent, and the electron moves as a free particle.

Problems

6.1 (a) Given the coupled equation

$$a_{11}x + a_{12}y = c_1$$
$$a_{21}x + a_{22}y = c_2$$

show that the solution can be expressed by Kramer's rule:

$$x = \frac{\begin{vmatrix} c_1 & a_{12} \\ c_2 & a_{22} \end{vmatrix}}{\begin{vmatrix} a_{11} & a_{12} \\ a_{21} & a_{22} \end{vmatrix}} = \frac{c_1a_{22} - c_2a_{12}}{a_{11}a_{22} - a_{12}a_{21}}, \quad y = \frac{\begin{vmatrix} a_{11} & c_1 \\ a_{21} & c_2 \end{vmatrix}}{\begin{vmatrix} a_{11} & a_{12} \\ a_{21} & a_{22} \end{vmatrix}} = \frac{c_2a_{11} - c_1a_{21}}{a_{11}a_{22} - a_{12}a_{21}}$$

(b) Given the characteristic matrix equation

$$\begin{pmatrix} 1 & 2 \\ 2 & 1 \end{pmatrix} \begin{pmatrix} x_1 \\ x_2 \end{pmatrix} = \lambda \begin{pmatrix} x_1 \\ x_2 \end{pmatrix} \quad \text{or} \quad \begin{cases} (1-\lambda)x_1 + 2x_2 = 0 \\ 2x_1 + (1-\lambda)x_2 = 0 \end{cases} \tag{A}$$

Show that the solution of the coupled equation is trivial unless the secular equation is satisfied,

$$\begin{vmatrix} 1-\lambda & 2 \\ 2 & 1-\lambda \end{vmatrix} = 0 \quad \text{or} \quad (1-\lambda)^2 - 4 = 0$$

(c) Show that the two characteristic roots $\lambda_{\pm}$ when inserted into (A) yields the infinite number of solutions as long as x_1 and x_2 are related by $x_2 = \pm x_1$.
Show that the condition $x_2 = \pm x_1$ can be found from any one of two equations in (A).
Show that if the normalization condition is imposed $x_1^2 + x_2^2 = 1$, the solution is given by

$$X_1 = \begin{pmatrix} x_1 \\ x_2 \end{pmatrix} = \frac{1}{\sqrt{2}} \begin{pmatrix} 1 \\ 1 \end{pmatrix}, \quad X_2 = \frac{1}{\sqrt{2}} \begin{pmatrix} 1 \\ -1 \end{pmatrix}$$

6.2 (a) Starting from boundary conditions Eqs. (6.11) and (6.12), eliminate A, B and derive Eq. (6.15).
(b) Starting from the secular equation (6.16), derive the dispersion relation (6.17).
Hint: Use the identity $e^{-2ikd} + 1 = 2e^{-ikd} \cos kd$.

6.3 The superlattice structure consists of a series of quantum wells and barrier potentials for both electrons and holes, as shown.
(a) Take the barrier height to be infinite and design the well width a such that the first two subbands of the electrons are separated by 40 meV. Use the effective electron mass of $m_n \sim 0.07\,m_0$ with m_0 denoting the rest mass.

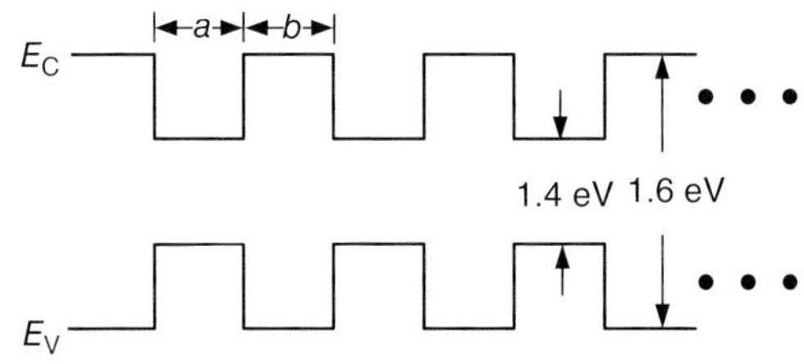

Figure 6.7 A superlattice structure consisting of a string of the quantum wells.

(b) Repeat the analysis numerically or by graphical means using the potential barrier of 0.1 V and estimate the widths of the two subbands (Figure 6.7).

Suggested Readings

1. J. Singh, Quantum Mechanics, Fundamentals and Applications to Technology, John Wiley & Sons, 1996.
2. A. Yariv, An Introduction to Theory and Applications of Quantum Mechanics, John Wiley & Sons, 1982.
3. H. Kroemer, Quantum Mechanics for Engineering, Materials Science, and Applied Physics, International Edition, Prentice Hall, 1994.
4. J. S. Blakemore, Solid State Physics, Second Edition, Cambridge University Press, 1985.
5. D. M. Kim, Introductory Quantum Mechanics for Semiconductor Nanotechnology, Wiley-VCH, 2010.
6. R. L. Liboff, Introductory Quantum Mechanics, Fourth Edition, Addison Wesley Publishing Company, Reading, MA, 2002.

7
The Quantum Treatment of Harmonic Oscillator

The harmonic oscillator (HO) is treated quantum mechanically. The HO is a key component in various kinds of dynamical systems, and it is thus essential to comprehend the physics of the HO. For instance, the dynamics of HO provides the general background for the quantum treatment of electromagnetic (EM) field, the molecular vibrations, chemical bonds, atomic oscillations in condensed matter, and so on. Moreover, the quantization of EM field is carried out in analogy with the operator treatment of HO.

7.1
Energy Eigenfunction and Energy Quantization

Consider a particle of mass m attached to a spring with spring constant k as shown in Figure 7.1. The energy eigenequation of the HO is given by

$$\left[-\frac{\hbar^2}{2m}\frac{\partial^2}{\partial x^2}+\frac{1}{2}kx^2\right]u(x)=Eu(x),\quad \frac{k}{m}\equiv\omega^2 \tag{7.1}$$

where the Hamiltonian consists of the kinetic and potential energies and ω is the characteristic frequency. In treating the differential equation, it is convenient to introduce the dimensionless displacement ξ and the energy parameter λ as

$$\xi=\alpha x,\quad \alpha\equiv\left(\frac{mk}{\hbar^2}\right)^{1/4}=\left(\frac{m\omega}{\hbar}\right)^{1/2};\quad \lambda\equiv\frac{2E}{\hbar\omega} \tag{7.2}$$

and recast Eq. (7.1) by multiplying both sides with $2(m/k)^{1/2}/\hbar=2/\hbar\omega$ as

$$\frac{\mathrm{d}^2}{\mathrm{d}\xi^2}u(\xi)+(\lambda-\xi^2)u(\xi)=0 \tag{7.3}$$

In the asymptotic limit $\xi\to\infty$, λ can be neglected, and the solution is given by $u(\xi)\approx\exp-\xi^2/2$, as can be easily verified. Thus, we may try the solution in the form

$$u(\xi)=e^{-\xi^2/2}H(\xi) \tag{7.4}$$

and insert it into Eq. (7.3), obtaining

$$H''-2\xi H'+(\lambda-1)H=0 \tag{7.5}$$

Introductory Quantum Mechanics for Applied Nanotechnology, First Edition. Dae Mann Kim.

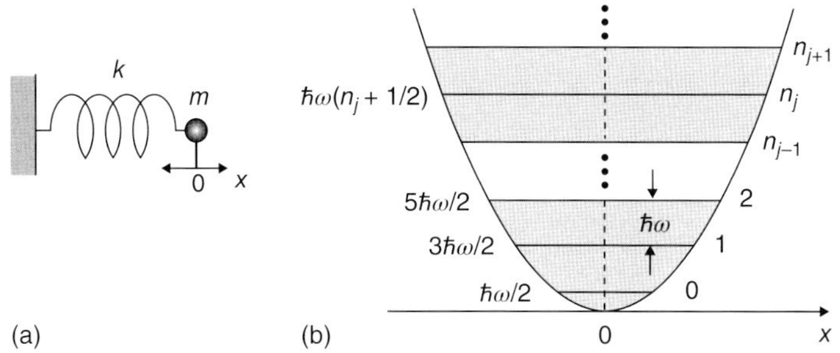

Figure 7.1 (a,b) A harmonic oscillator and its quantized energy spectrum consisting of the discrete levels separated by the quantum of energy $\hbar\omega$. Also indicated is the continuous energy spectrum of the classical harmonic oscillator.

where the primes denote the differentiation with respect to ξ.

Series Solution

We can use the method of the series solution and expand $H(\xi)$ as

$$H(\xi) = \xi^s \sum_{n=0}^{\infty} a_n \xi^n, \quad a_0 \neq 0 \tag{7.6}$$

By inserting Eq. (7.6) into Eq. (7.5) and carrying out the differentiations involved, we can write

$$s(s-1)a_0\xi^{s-2} + (s+1)sa_1\xi^{s-1}$$
$$+\sum_{n=0}^{\infty}\{[(s+n+2)(s+n+1)a_{n+2} - [2(s+n)+1-\lambda]a_n\}\xi^{s+n} = 0 \tag{7.7}$$

In Eq. (7.7), the first two terms resulting from $H(\xi)''$ have been taken out of the summation, while all other terms are combined into two groups by appropriate adjustments of dummy index n. Thus, solving Eq. (7.5) is reduced to satisfying Eq. (7.7) for arbitrary powers and values of ξ, which can be done by putting,

$$s(s-1)a_0 = 0, \quad (s+1)sa_1 = 0 \tag{7.8a}$$

$$(s+n+2)(s+n+1)a_{n+2} = [2(s+n)+1-\lambda]a_n \tag{7.8b}$$

The two conditions in Eq. (7.8a) are known as the *indicial equations*, while Eq. (7.8b) is the recurrence relation specifying higher-order coefficients recursively in terms of a_0, a_1.

As a_0 is taken nonzero (see Eq. (7.6)), the indicial equations are satisfied with the choice of $s = 0$ regardless of whether or not a_1 is zero. Hence, $H(\xi)$ is obtained in terms of two infinite order polynomials, with a_0, a_1 specifying higher-order coefficients:

$$H(\xi) = a_0\left(1 + \frac{a_2}{a_0}\xi^2 + \frac{a_4}{a_2}\frac{a_2}{a_0}\xi^4 + \cdots\right) + a_1\xi\left(1 + \frac{a_3}{a_1}\xi^2 + \cdots\right) \tag{7.9}$$

Thus, the eigenfunction $u(x)$ is found with the insertion of Eq. (7.9) into Eq. (7.4).

Energy Quantization

Naturally, $u(\xi)$ should be physically well behaving, in particular for large ξ. To examine the asymptotic behavior, let us consider the Taylor expansion of the exponential function, namely,

$$\exp \xi^2 = \sum_{n=0}^{\infty} b_n \xi^{2n}, \quad b_n = \frac{1}{n!}$$

In this series, the ratio between two successive expansion coefficients is given by $b_{n+1}/b_n \approx 1/n$ for large n and is identical to the corresponding ratio as appears in Eq. (7.8b). This indicates that $H(\xi)$ diverges as $H(\xi) \approx \exp \xi^2$, so that $u(\xi)$ also diverges as $u(x) \approx \exp(\xi^2/2)$ (see Eq. (7.4)). Therefore, the appropriate modifications of $H(\xi)$ are in order to make the eigenfunction physically well behaving.

The modification can be made by terminating the a_0-series in Eq. (7.9) at a finite order while eliminating the other series by putting $a_1 = 0$. The termination at the nth order can be made by constraining λ in Eq. (7.8b) by the condition

$$0 \equiv a_{n+2} = \frac{2n + 1 - \lambda}{(n + 1)(n + 2)} a_n, \quad s = 0 \tag{7.10a}$$

The requirement of Eq. (7.10a), when combined with Eq. (7.2), provides the natural ground for the quantization of energy:

$$\lambda_n \equiv \frac{2E_n}{\hbar\omega} = 2n + 1, \quad n = 0, 2, 4, \ldots \tag{7.10b}$$

In this case, the polynomial $H(\xi)$ consists of the even powers of ξ, which renders $u(\xi)$ an even function of ξ (see Eq. (7.4)).

For $s = 1$, we can again satisfy Eq. (7.8a) by putting $a_1 = 0$ and obtain the finite-order polynomial $H(\xi)$ with the use of the recurrence relation (7.8b):

$$0 \equiv a_{n+2} = \frac{2n + 3 - \lambda}{(n + 2)(n + 3)} a_n, \quad s = 1 \tag{7.11a}$$

Thus, the energy is naturally quantized from Eq. (7.11a) as

$$\lambda_n \equiv \frac{2E_n}{\hbar\omega} = 2n + 3, \quad n = 0, 2, 4, \ldots \tag{7.11b}$$

Moreover, $H(\xi)$ consists in this case of odd powers of ξ so that $u(\xi)$ is an odd function of ξ. The two energy quantization equations (7.10b) and (7.11b) can be combined into one as

$$E_n = \hbar\omega\left(n + \frac{1}{2}\right); \quad \lambda_n = 2n + 1, \quad n = 0, 1, 2, 3, \ldots \tag{7.12}$$

(see Figure 7.1). In this manner, the energy eigenvalues are naturally quantized by the obvious requirement that the eigenfunctions should be physically well behaving. The quantized energy level of HO is in distinct contrast with that of the classical HO, for which E varies continuously by any infinitesimal amount, as discussed in Chapter 1. The discrete energy levels are equally spaced and separated by $\hbar\omega$, and the integer n in Eq. (7.12) is the quantum number specifying the discrete energy levels.

Ground State

The ground state energy for $n = 0$ is given from Eq. (7.12) by

$$E_0 = \frac{\hbar\omega}{2} \tag{7.13}$$

and is not zero but has a finite value, called *zero-point energy*. This is again in contrast with the classical oscillator whose total energy at rest in the equilibrium position has zero value. The zero-point energy originates from to the wave-like behavior of the particle and the uncertainty principle just like the finite ground state energy in the quantum well.

7.2 The Properties of Eigenfunctions

Hermite Polynomials

The nth order polynomial $H_n(\xi)$ thus found is the well-known Hermite polynomial, obeying the differential equation

$$H_n'' - 2\xi H_n' + 2nH_n = 0 \tag{7.14}$$

The Hermite differential equation is identical to Eq. (7.5), when λ is replaced by λ_n in Eq. (7.11b). The properties of $H_n(\xi)$ can be conveniently discussed by using its generating function

$$G(\xi, s) \equiv e^{\xi^2-(s-\xi)^2} = e^{-s^2+2s\xi} \equiv \sum_{n=0}^{\infty} \frac{H_n(\xi)s^n}{n!} \tag{7.15}$$

The generating function yields various useful recurrence relations. For instance, by differentiating both sides of Eq. (7.15) with respect to ξ, there results

$$\frac{\partial}{\partial\xi}e^{-s^2+2s\xi} = 2se^{-s^2+2s\xi} = 2\sum_{n=0}^{\infty} \frac{H_n(\xi)s^{n+1}}{n!} = \sum_{n=0}^{\infty} \frac{H_n'(\xi)s^n}{n!} \tag{7.16}$$

with the primes denoting the differentiation with respect to ξ. We can thus single out the coefficients of the equal powers of s from both sides of Eq. (7.16), obtaining

$$H_n' = 2nH_{n-1} \tag{7.17}$$

Also by differentiating Eq. (7.15) with respect to s, we can similarly find

$$\xi H_n = \frac{1}{2}H_{n+1} + nH_{n-1} \tag{7.18}$$

In addition, we can obtain $H_n(\xi)$ by differentiating the generating function $G(\xi,s)$ with respect to s n times and putting $s = 0$. In this case, the terms with powers in s less than n vanish while getting differentiated n times, while the terms with powers in s greater than n also vanish in the limit $s \to 0$. In this manner, $H_n(x)$ is singled out from the summation to be given by

$$H_n(\xi) \equiv \frac{\partial^n}{\partial s^n} e^{\xi^2-(s-\xi)^2}\Big|_{s=0}$$
$$= e^{\xi^2}(-)^n \frac{\partial^n}{\partial \xi^n} e^{-(s-\xi)^2}\Big|_{s=0} = (-)^n e^{\xi^2} \frac{\partial^n}{\partial \xi^n} e^{-\xi^2} \tag{7.19}$$

The operational representation of $H_n(\xi)$ is known as *Rodrigues's formula* and is useful for generating $H_n(\xi)$. For example, we can easily obtain by mere differentiations

$$H_0 = 1, \quad H_1 = 2\xi, \quad H_2 = 4\xi^2 - 2, \ \ldots$$

The Orthogonality of Eigenfunctions

The energy eigenfunction of HO is given from Eq. (7.4) by

$$u_n(\xi) = N_n e^{-\xi^2/2} H_n(\xi) \tag{7.20}$$

where N_n is the normalization constant. To find N_n and also to examine the orthogonality of $u_n(x)$, let us integrate the product of two generating functions:

$$\int_{-\infty}^{\infty} d\xi e^{-s^2+2s\xi} e^{-t^2+2t\xi} e^{-\xi^2} \equiv \sum_{n=0}^{\infty} \frac{s^n}{n!} \sum_{m=0}^{\infty} \frac{t^m}{m!} \int_{-\infty}^{\infty} d\xi H_n(\xi) H_m(\xi) e^{-\xi^2} \tag{7.21}$$

We can perform the integration on the left-hand side by using the table, obtaining

$$e^{-s^2-t^2} \int_{-\infty}^{\infty} d\xi e^{-\xi^2} e^{2(s+t)\xi} = \sqrt{\pi} e^{2ts} = \sqrt{\pi} \sum_n \frac{2^n t^n s^n}{n!} \tag{7.22}$$

In Eq. (7.22), the exponential function $\exp(2ts)$ has been Taylor expanded.

Hence, the double sum on the right-hand side of Eq. (7.21) has to be reduced to the single sum. Therefore, the coefficients of the terms proportional to $s^n t^m$ on the right-hand side of Eq. (7.21) should satisfy the relation

$$\int_{-\infty}^{\infty} d\xi H_n H_m e^{-\xi^2} = \sqrt{\pi} 2^n n! \delta_{nm} \tag{7.23a}$$

where δ_{nm} is the Kronecker delta function defined as

$$\delta_{nm} = \begin{cases} 1 & \text{for } n = m \\ 0 & \text{for } n \neq m \end{cases} \tag{7.23b}$$

In this manner, the orthogonality of $\{u_n(\xi)\}$ is shown explicitly. At the same time, the normalization constant N_n can also be found from Eqs. (7.2), (7.20), and (7.23) as

$$1 \equiv \int_{-\infty}^{\infty} dx [u_n(x)]^2 = \frac{N_n^2}{\alpha} \int_{-\infty}^{\infty} d\xi e^{-\xi^2} H_n^2 = \frac{N_n^2}{\alpha} \sqrt{\pi} 2^n n! \tag{7.24}$$

The normalized eigenfunctions of HO are thus given by

$$u_n(x) = \left(\frac{\alpha}{\sqrt{\pi} 2^n n!} \right)^{1/2} e^{-\xi^2/2} H_n(\xi), \quad \xi \equiv \alpha x, \quad \alpha \equiv \left[\frac{m\omega}{\hbar} \right]^{1/2} \tag{7.25}$$

Typical eigenfunctions and probability densities are plotted in Figure 7.2. Note that there is a small but finite probability of finding the HO in the classically forbidden region. Also the penetration depth increases with increasing n, that is, with increasing energy. Moreover, the peak of the probability densities shifts from the center at $x \approx 0$ for small n to the edges at $x \approx x_0$ for large n. Classically, the probability P of finding the HO in the interval from x to $x + dx$ is inversely proportional to its dwell time therein. Therefore, P attains the minimum value near the origin $x \approx 0$, where the velocity of the HO is at its maximum. On the other hand, P attains the maximum value at $x \approx x_0$, where the HO is momentarily at rest before reversing its direction. The classical probability P is also plotted in Figure 7.2, for comparison. Clearly, the profile of P is in general agreement with the trace of the sub-peaks of the probability density $|u_n(x)|^2$ for large n. The agreement of the probability profile for large n is referred to as the *correspondence principle*.

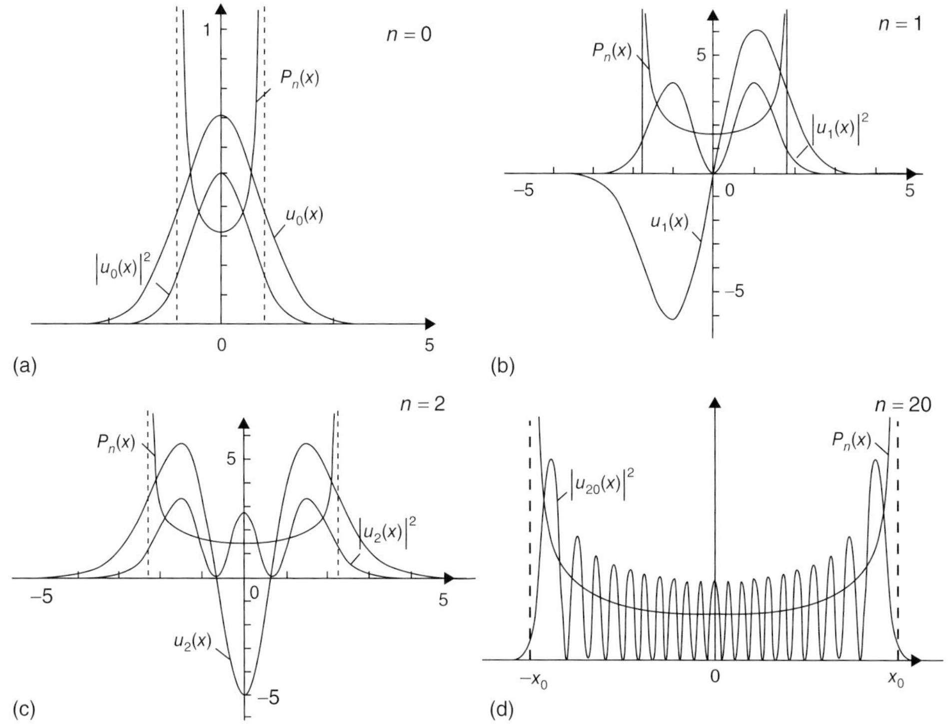

Figure 7.2 (a–d) Typical eigenfunctions and probability densities of the HO. Also shown are the classical turning points (broken lines) and the classical analog of the probability density (thin lines).

The Uncertainty Relation

We next examine the uncertainties in x and p_x in the nth eigenstate. The evaluation of Δx, Δp_x can be done conveniently by using the recurrence relations given in Eqs. (7.17) and (7.18). Obviously, the average value of x is zero because $u_n^*(x)u_n(x)$ is even in x regardless of whether $u_n(x)$ is even or odd, so that the parity of $u_n^*(x)xu_n(x)$ is odd. But the average value of x^2 is not zero and given by

$$\langle x^2\rangle \equiv \langle u_n|x^2|u_n\rangle$$
$$\equiv \int_{-\infty}^{\infty} dx u_n^* x^2 u_n = \frac{N_n^2}{\alpha^3}\int_{-\infty}^{\infty} d\xi e^{-\xi^2/2}H_n(\xi)\xi^2 e^{-\xi^2/2}H_n(\xi), \quad \xi = \alpha x \tag{7.26a}$$

At this point, we can make use of the recurrence relation Eq. (7.18) and the orthonormality of the eigenfunctions in performing the integration, obtaining

$$\langle x^2\rangle = \frac{\hbar}{m\omega}\left(n+\frac{1}{2}\right) \tag{7.26b}$$

Once $\langle x\rangle$, $\langle x^2\rangle$ are known, the variance Δx is obtained as

$$\Delta x^2 \equiv \langle (x-\langle x\rangle)^2\rangle = \langle x^2 - 2x\langle x\rangle + \langle x\rangle^2\rangle = \langle x^2\rangle - \langle x\rangle^2 = \frac{\hbar}{m\omega}\left(n+\frac{1}{2}\right) \tag{7.27}$$

We can likewise obtain the variance of p_x by using the recurrence relation Eq. (7.17) as

$$\Delta p^2 \equiv \langle (p_x - \langle p_x\rangle)^2\rangle = \langle p_x^2\rangle - \langle p_x\rangle^2 = m\omega\hbar\left(n+\frac{1}{2}\right), \quad p_x = -i\hbar\frac{\partial}{\partial x} \tag{7.28}$$

Therefore, the uncertainty relation between x and p_x in the nth eigenstate is given by

$$\Delta x \Delta p_x = \hbar\left(n+\frac{1}{2}\right) \tag{7.29}$$

and is shown to increase with increasing energy level. It is also clear from Eq. (7.29) that the ground state for $n = 0$ has the minimum uncertainty limit $\hbar/2$.

7.3 HO in Linearly Superposed State

Useful Matrix Elements

As mentioned, the HO is a key element in a number of important dynamical systems, and a few matrix elements involving the eigenfunctions are extensively utilized. These matrix elements can be simply evaluated with the use of recurrence relations Eqs. (7.17) and (7.18). For instance, the matrix element

$$\langle u_n|x|u_m\rangle = \frac{N_n N_m}{\alpha^2}\int_{-\infty}^{\infty} d\xi e^{-\xi^2/2}H_n(\xi)\xi e^{-\xi^2/2}H_m(\xi) \tag{7.30}$$

plays an essential role for describing the interaction of light and matter. We can again use the recurrence relation Eq. (7.18) and the orthonormality of eigenfunctions and obtain

$$\langle u_l|x|u_{l'}\rangle = \begin{cases} (l+1)^{1/2}/(2m\omega/\hbar)^{1/2}, & l' = l+1 \\ l^{1/2}/(2m\omega/\hbar)^{1/2}, & l' = l-1 \\ 0, & \text{otherwise} \end{cases} \tag{7.31}$$

We can also evaluate the matrix elements involving the momentum by using the recurrence relation Eq. (7.17) and obtain

$$\langle p_x\rangle \propto \langle u_l|\frac{\partial}{\partial x}|u_{l'}\rangle = \begin{cases} (m\omega/2\hbar)^{1/2}(l+1)^{1/2}, & l' = l+1 \\ -(m\omega/2\hbar)^{1/2}l^{1/2}, & l' = l-1 \\ 0, & \text{otherwise} \end{cases} \tag{7.32}$$

When the HO is in a superposed state, consisting of the ground and first excited states with equal probability, for example, the wavefunction is given by

$$\psi(x,t) = \frac{1}{\sqrt{2}}\left(e^{-i\omega t/2}u_0 + e^{-i3\omega t/2}u_1\right), \quad \omega = \frac{E}{\hbar} \tag{7.33}$$

In Eq. (7.33), the oscillatory time components $\exp -iEt/\hbar$ have been added to each eigenstate, and the factor $1/\sqrt{2}$ is introduced for normalizing the wavefunction. Then, the probability density

$$\psi^*\psi = \frac{1}{2}(u_0^2 + u_1^2 + 2u_0u_1\cos\omega t) \tag{7.34}$$

consists of the time-independent background terms u_0^2, u_1^2, and an oscillatory term, as shown in Figure 7.3. The oscillatory behavior of the HO can be seen

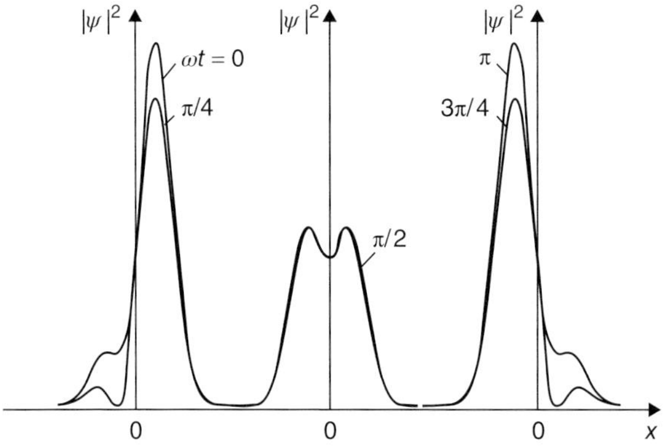

Figure 7.3 The probability density at different times of the superposed state consisting of the ground and first excited states with equal probability. Also shown is the oscillation in time of the probability density profile.

more clearly by considering the expectation values of x and p_x:

$$\langle x\rangle = \frac{1}{2}\left\langle e^{-i\omega t/2}u_0 + e^{-i3\omega t/2}u_1\right| x \left| e^{-i\omega t/2}u_0 + e^{-i3\omega t/2}u_1\right\rangle$$
$$= x_{12}\cos\omega t, \quad x_{12} \equiv \langle u_0|x|u_1\rangle = \frac{1}{(2m\omega/\hbar)^{1/2}} \tag{7.35}$$

$$\langle p_x\rangle = \frac{1}{2}\left\langle e^{-i\omega t/2}u_0 + e^{-i3\omega t/2}u_1\right| -i\hbar\frac{\partial}{\partial x} \left| e^{-i\omega t/2}u_0 + e^{-i3\omega t/2}u_1\right\rangle$$
$$= -p_{12}\sin\omega t, \quad p_{12} \equiv \hbar\langle u_0|\frac{\partial}{\partial x}|u_1\rangle = \left(\frac{m\omega\hbar}{2}\right)^{1/2} \tag{7.36}$$

In deriving Eqs. (7.35) and (7.36), the amplitudes of oscillation x_{12}, p_{12} have been obtained by using the matrix elements derived in Eqs. (7.31) and (7.32). Indeed, these average quantities oscillate in time in a manner similar to the classical HO.

7.4 The Operator Treatment of HO

The operator treatment of the HO is important by itself, but it also paves the way for the quantum treatment of EM fields and molecular vibrations. Moreover, the concept of phonons is naturally brought out. Thus, consider the operators defined as

$$\begin{pmatrix} a \\ a^+ \end{pmatrix} = \frac{1}{\sqrt{2}}\left[\alpha x \pm i\frac{1}{\hbar\alpha}p_x\right] = \frac{1}{\sqrt{2}}\left(\xi \pm \frac{\partial}{\partial\xi}\right) \tag{7.37}$$

where ξ is the dimensionless variable specified in Eq. (7.2), and a and a^+ are called the *lowering and raising operators*, respectively. We can find the commutation relation of a and a^+ by using the commutation relations $[x,p] = i\hbar$, $[x,x] = [p,p] = 0$, obtaining

$$[a,a^+] = \frac{1}{2}\left(\frac{-i}{\hbar}\right)[x,p_x] + \frac{1}{2}\left(\frac{i}{\hbar}\right)[p_x,x] = 1 \tag{7.38}$$

Also, we can invert Eq. (7.37) and find x and p in terms of a, a^+ as

$$x = \frac{1}{\sqrt{2}\alpha}(a + a^+) \tag{7.39a}$$

$$p_x = \frac{i\hbar\alpha}{\sqrt{2}}(a^+ - a) \tag{7.39b}$$

and express the Hamiltonian of HO in terms of a and a^+ as

$$\hat{H} = \frac{p_x^2}{2m} + \frac{1}{2}kx^2 = \hbar\omega\left(a^+a + \frac{1}{2}\right) \tag{7.40}$$

In obtaining Eq. (7.40), Eq. (7.38) has been used, that is, $aa^+ = a^+a + 1$ together with the identities $k/\alpha^2 = (\hbar\alpha)^2/m = \hbar\omega$ (see Eq. (7.2)). The operator a^+a is known as the *number operator*, and it commutes with $\hat{H}$, as clear from Eq. (7.40). Since two commuting operators share a common eigenfunction, the energy eigenfunction $u_n(x)$ can also be used as the eigenfunction of the number operator.

Raising, Lowering, and Number Operators

The operators a, a^+, and a^+a yield interesting results when they operate on $u_n(x)$. Thus, consider a operating on $u_n(x)$:

$$au_n = \frac{1}{\sqrt{2}}\left(\xi + \frac{\partial}{\partial \xi}\right)\left[N_n H_n e^{-\xi^2/2}\right]$$
$$= \frac{1}{\sqrt{2}} N_n H_n' e^{-\xi^2/2} = \frac{1}{\sqrt{2}} N_n 2n H_{n-1} e^{-\xi^2/2} = \sqrt{n} u_{n-1} \tag{7.41}$$

In obtaining Eq. (7.41), Eq. (7.25) was used for $u_n(x)$, and the recurrence relation Eq. (7.17) was used for converting H_n' to H_{n-1}. It is therefore clear that the operator a lowers the eigenstate from n to $n-1$ and is called the *lowering operator*. On the other hand, a^+ raises the eigenstate from n to $n+1$ and is known as the *raising operator*. The raising operation can be shown with the use of two recurrence relations (7.17) and (7.18) as

$$a^+ u_n = \frac{N_n}{\sqrt{2}}\left(\xi - \frac{\partial}{\partial \xi}\right) H_n e^{-\xi^2/2} = \frac{N_n}{\sqrt{2}}(2\xi H_n - H_n')e^{-\xi^2/2} = \sqrt{n+1}u_{n+1} \tag{7.42}$$

Therefore, the two consecutive operations of a and a^+ on $u_n(x)$ yield

$$a^+ a u_n = a^+[\sqrt{n} u_{n-1}] = \sqrt{n}\sqrt{(n-1+1)} u_{n-1+1} = n u_n \tag{7.43}$$

and indeed $u_n(x)$ is also the eigenfunction of a^+a with n as the eigenvalue.

7.4.1
Creation and Annihilation Operators and Phonons

In view of the roles of a, a^+, and a^+a, the eigenenergy E_n of the HO can be interpreted as consisting of the n number of the quantum of energy $h\nu$ called the *phonon*. In this context, the operator a destroys a phonon and lowers u_n to u_{n-1} and is also called the *annihilation operator*. By the same token, a^+ creates a phonon and raises u_n to u_{n+1} and is called the *creation operator*.

It is also possible to generate the set of eigenfunctions $\{u_n(x)\}$ by creating a series of phonons in succession from the ground state in which $n = 0$. The starting point of this operational approach for finding $u_n(x)$ is the fact that when a operates on the ground state, it pushes the state out of the Hilbert space of the eigenfunction, that is,

$$au_0 = \sqrt{0}u_{0-1} = 0 \tag{7.44}$$

The operation (Eq. (7.44)) can be cast into a differential equation by using Eq. (7.37) as

$$\left(\xi + \frac{\partial}{\partial \xi}\right) u_0 = 0 \tag{7.45}$$

The solution of this simple differential equation can be easily obtained as

$$u_0 \propto e^{-\xi^2/2} = N_0 e^{-\xi^2/2}, \quad N_0 = \left(\frac{\alpha}{\sqrt{\pi}}\right)^{1/2} \tag{7.46}$$

where the constant of integration N_0 has been used as the normalization constant. Once u_0 is found, the higher-lying eigenstates can be systematically generated by creating a single phonon in succession, that is, by applying a^+ on u_0 repeatedly. For instance, the first excited state is obtained by creating one phonon in the ground state:

$$\begin{aligned} u_1 \equiv \frac{1}{\sqrt{1}} a^+ u_0 &= \frac{1}{\sqrt{1}} \frac{1}{\sqrt{2}} \left(\xi - \frac{\partial}{\partial \xi}\right) u_0(\xi) \\ &= \left(\frac{\alpha}{\sqrt{\pi 2!}}\right)^{1/2} e^{-\xi^2/2} H_1(\xi) \end{aligned} \tag{7.47}$$

Here again, recurrence relations (7.17) and (7.18) have been used together with Eq. (7.25) for $u_n(x)$ and Eq. (7.42) for the operation $a^+ u_n$. The eigenfunction $u_n(x)$ is obtained in general by performing the operation

$$u_n(\xi) = \frac{1}{\sqrt{n!}} (a^+)^n u_0(\xi), \quad a^+ = \frac{1}{\sqrt{2}} \left(\xi - \frac{\partial}{\partial \xi}\right) \tag{7.48}$$

Problems

7.1 (a) Starting with the energy eigenequation (Eq. (7.1)), fill in the algebraic steps, and derive the eigenequation given in Eq. (7.3) in terms of ξ.
(b) By looking for the solution of the eigenfunction in the form given in Eq. (7.4), reduce the eigenequation to (7.5).

7.2 Consider the 3D HO with the Hamiltonian

$$\hat{H} = -\frac{\hbar^2}{2m} \nabla^2 + \frac{1}{2} k_x x^2 + \frac{1}{2} k_y y^2 + \frac{1}{2} k_z z^2$$

(a) Set up the energy eigenequation and look for the eigenfunction in the form

$$u(x, y, z) = u_x(x) u_y(y) u_z(z)$$

and reduce the 3D eigenequation to three 1D eigenequations.
(b) Find the eigenfunction and the eigenvalue of the 3D oscillator. For $k_x = k_y = k_z = k$, discuss the energy spectrum and find the degeneracy of the first, second, and third excited states.

7.3 (a) Derive the recurrence relations (7.17) and (7.18) by filling in the algebraic steps described.
(b) By using the recurrence relations and the normalized eigenfunction (Eq. (7.25)), evaluate the variances Δx^2 and Δp_x^2 and derive the uncertainty relation between x and p_x given in Eq. (7.29).

(c) By using the same recurrence relations, derive the matrix element (Eq. (7.31)).

7.4 Consider a classical oscillator with mass m and spring constant k and oscillating with an amplitude x_0.

(a) Find the kinetic and potential energies averaged over one period of oscillation and compare the results with the total energy.
(b) Find the average kinetic and potential energies in the n th eigenstate of the quantum oscillator and compare the results with the total energy of the eigenstate.
(c) Discuss the similarities or differences between the classical and quantum descriptions.

7.5 The vibrational spectra of molecules can be observed by the infrared spectroscopy. The carbon monoxide (CO) molecule can be modeled as C and O atoms coupled via a spring with an effective spring constant k. The energy spacing between the two energy eigenstates is observed to be given by the wavenumber $1/\lambda = 2170\ \text{cm}^{-1}$.

(a) By taking the masses of C and O to be 12 and 16 atomic units, determine k, which is a measure of the bond stiffness.
(b) Find the zero-point energy.

7.6 By using the representation of x, p in terms of raising and lowering operators, a^+ and a (see Eq. (7.39)), derive the Hamiltonian given in terms of the number operator (Eq. (7.40)).

Suggested Readings

1. D. M. Kim, Introductory Quantum Mechanics for Semiconductor Nanotechnology, Wiley-VCH, 2010.
2. A. Yariv, An Introduction to Theory and Applications of Quantum Mechanics, John Wiley & Sons, 1982.
3. J. Singh, Quantum Mechanics, Fundamentals and Applications to Technology, John Wiley & Sons, 1996.
4. S. Gasiorowics, Quantum Physics, Third Edition, John Wiley & Sons, 2003.

8
Schrödinger Treatment of Hydrogen Atom

The quantum treatment of the H-atom is presented. The H-atom is the simplest atomic system, but the theory of H-atom contains the central core of quantum mechanics. Bohr's H-atom theory was the culmination of the old quantum theory, and the Schrödinger treatment of the H-atom demonstrated the versatility of his wave equation. Moreover, the theory is used as the general background for treating the multi-electron atoms and molecules. The topics included for discussion are angular momentum, spatial quantization, atomic orbital, quantized energy level and atomic spectroscopy, Doppler broadening, and so on.

8.1
Angular Momentum Operators

The angular momentum is a key ingredient of quantum mechanics. Bohr's H-atom theory, for instance, starts with quantizing the angular momentum. Understandably, it also plays a central role in the quantum treatment of the H-atom. Moreover, the eigenfunction of the angular momentum offers the tool for treating the atomic orbital, multi-electron atomic system, chemical bonding, molecular structures, and so on.

Thus, consider a particle with mass m and moving in a circular orbit with the linear momentum, $m\boldsymbol{v}$ at $\boldsymbol{r}$ distance from a fixed center (Figure 8.1). The angular momentum operator $\underline{\hat{l}}$ is defined as the vector product of $\boldsymbol{r}$ and $\boldsymbol{p}$, and in Cartesian coordinate frame, it reads as

$$\underline{\hat{l}} = \underline{r} \times \underline{\hat{p}} = \underline{r} \times (-i\hbar\nabla), \quad \nabla = \hat{x}\frac{\partial}{\partial x} + \hat{y}\frac{\partial}{\partial y} + \hat{z}\frac{\partial}{\partial z} \tag{8.1}$$

The three components are then given by

$$\hat{l}_x = y\hat{p}_z - z\hat{p}_y, \quad \hat{l}_y = z\hat{p}_x - x\hat{p}_z, \quad \hat{l}_z = x\hat{p}_y - y\hat{p}_x \tag{8.2}$$

where the cyclic property of the vector products of unit vectors $\hat{x} \times \hat{y} = \hat{z}$, $\hat{y} \times \hat{z} = \hat{x}$, $\hat{z} \times \hat{x} = \hat{y}$ has been used.

Introductory Quantum Mechanics for Applied Nanotechnology, First Edition. Dae Mann Kim.

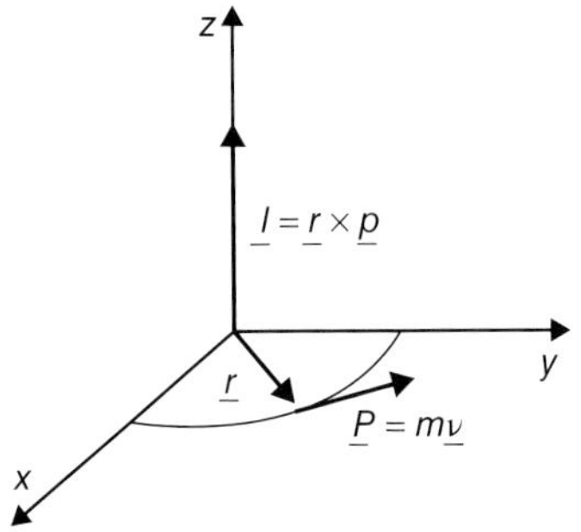

Figure 8.1 The angular momentum as the vector product of ***r*** and ***p***.

It is convenient to treat the angular momentum in the spherical coordinate frame, in which case the angular momentum operator reads as

$$\hat{\underline{l}} = \underline{r} \times (-i\hbar\nabla), \quad \nabla = \left(\hat{r}\frac{\partial}{\partial r} + \hat{\theta}\frac{1}{r}\frac{\partial}{\partial \theta} + \hat{\varphi}\frac{1}{r\sin\theta}\frac{\partial}{\partial\varphi}\right)$$

$$= -i\hbar\hat{\underline{f}}(\theta,\varphi), \quad \hat{\underline{f}}(\theta,\varphi) \equiv \left(\hat{\varphi}\frac{\partial}{\partial\theta} - \hat{\theta}\frac{1}{\sin\theta}\frac{\partial}{\partial\varphi}\right) \tag{8.3}$$

where $\hat{r}$, $\hat{\theta}$, $\hat{\varphi}$ are the unit vectors in the frame (Figure 8.2), and the cyclic property of the vector products of the three unit vectors has also been used. We can single out x, y, and z components of the angular momentum by projecting Eq. (8.3) into the x, y, z axes:

$$\hat{l}_z \equiv \hat{z}\cdot\hat{\underline{l}} = (\hat{a}_r\cos\theta - \hat{a}_\theta\sin\theta)\cdot\hat{\underline{l}} = -i\hbar\frac{\partial}{\partial\varphi} \tag{8.4a}$$

$$\hat{l}_x \equiv \hat{x}\cdot\hat{\underline{l}} = i\hbar\left(\sin\varphi\frac{\partial}{\partial\theta} + \cot\theta\cos\varphi\frac{\partial}{\partial\varphi}\right) \tag{8.4b}$$

$$\hat{l}_y \equiv \hat{y}\cdot\hat{\underline{l}} = i\hbar\left(-\cos\varphi\frac{\partial}{\partial\theta} + \cot\theta\sin\varphi\frac{\partial}{\partial\varphi}\right) \tag{8.4c}$$

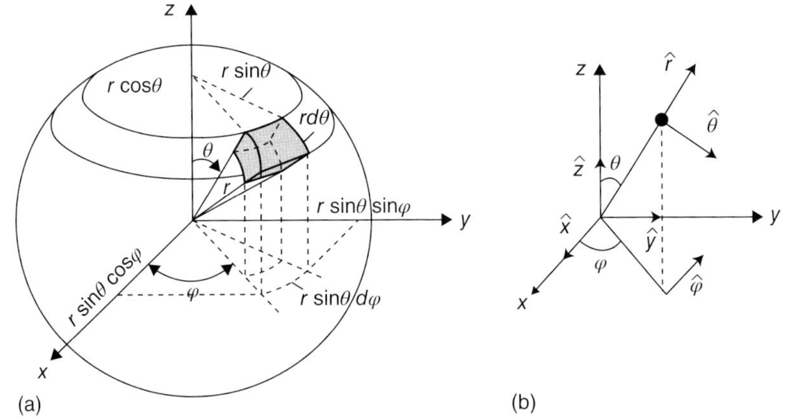

Figure 8.2 The spherical coordinate frame with variables r, θ, and φ (a) and two sets of unit vectors $\hat{r}, \hat{\theta}, \hat{\varphi}$ and $\hat{x}$, $\hat{y}$, $\hat{z}$ (b).

where $\hat{x}, \hat{y}, \hat{z}$ have been expressed in terms of $\hat{a}_r, \hat{a}_\theta, \hat{a}_\varphi$, respectively.

Additionally, the operator expression of $\hat{l}^2$ can be obtained by using the definition (Eq. (8.3)) and the vector identity.

$$\hat{l}^2 \equiv -\hbar^2(\underline{r} \times \nabla) \cdot (\underline{r} \times \nabla) = -\hbar^2 \underline{r} \cdot [\nabla \times (\underline{r} \times \nabla)] \tag{8.5}$$

In Eq. (8.5), we can again use Eq. (8.3) for replacing $\underline{r} \times \nabla$ by $\underline{\hat{f}}(\theta, \varphi)$, obtaining

$$\begin{aligned} \hat{l}^2 &= -\hbar^2 \underline{r} \cdot \nabla \times \underline{\hat{f}}(\theta, \varphi) \\ &= -\hbar^2 \left[\frac{1}{\sin\theta} \frac{\partial}{\partial\theta} \left(\sin\theta \frac{\partial}{\partial\theta} \right) + \frac{1}{\sin^2\theta} \frac{\partial^2}{\partial\varphi^2} \right] \end{aligned} \tag{8.6}$$

where use has been made of

$$[\nabla \times \underline{\hat{f}}(\theta, \varphi)]_r = \frac{1}{r\sin\theta} \frac{\partial}{\partial\theta} (\sin\theta \hat{f}_\varphi) + \frac{1}{r\sin\theta} \frac{\partial \hat{f}_\theta}{\partial\varphi}$$

Also the commutation relations involving $\hat{l}_x$, $\hat{l}_y$, and $\hat{l}_z$ can be derived from those of $\boldsymbol{r}$ and $\boldsymbol{p}$ given in Eq. (3.27). For example,

$$\begin{aligned} [\hat{l}_x, \hat{l}_y] &\equiv [(y\hat{p}_z - z\hat{p}_y), (z\hat{p}_x - x\hat{p}_z)] = [y\hat{p}_z, z\hat{p}_x] + [z\hat{p}_y, x\hat{p}_z] \\ &= y\hat{p}_x[\hat{p}_z, z] + \hat{p}_y x[z, \hat{p}_z] = i\hbar(x\hat{p}_y - y\hat{p}_x) \equiv i\hbar\hat{l}_z \end{aligned} \tag{8.7a}$$

Likewise, we can obtain

$$[\hat{l}_y, \hat{l}_z] = i\hbar\hat{l}_x, \quad [\hat{l}_z, \hat{l}_x] = i\hbar\hat{l}_y \tag{8.7b}$$

and these cyclic relations are compactly summarized as

$$\underline{\hat{l}} \times \underline{\hat{l}} = i\hbar\underline{\hat{l}} \tag{8.8}$$

Finally, let us consider the commutator

$$[\hat{l}^2, \hat{l}_z] = [(\hat{l}_x^2 + \hat{l}_y^2 + \hat{l}_z^2), \hat{l}_z]$$

The first term can be calculated by using Eq. (8.7b) as

$$\begin{aligned} [\hat{l}_x^2, \hat{l}_z] &= \hat{l}_x\hat{l}_x\hat{l}_z - \hat{l}_z\hat{l}_x\hat{l}_x \\ &= \hat{l}_x(\hat{l}_z\hat{l}_x - i\hbar\hat{l}_y) - (\hat{l}_x\hat{l}_z + i\hbar\hat{l}_y)\hat{l}_x \\ &= -i\hbar(\hat{l}_x\hat{l}_y + \hat{l}_y\hat{l}_x) \end{aligned} \tag{8.9a}$$

Likewise, we can obtain

$$[\hat{l}_y^2, \hat{l}_z] = i\hbar(\hat{l}_x\hat{l}_y + \hat{l}_y\hat{l}_x) \tag{8.9b}$$

Hence, it follows from Eq. (8.9) that

$$[\hat{l}^2, \hat{l}_z] = [\hat{l}_x^2, \hat{l}_z] + [\hat{l}_y^2, \hat{l}_z] + [\hat{l}_z^2, \hat{l}_z] = 0 \tag{8.10a}$$

where the last commutator is by definition zero. We can likewise obtain

$$[\hat{l}^2, \hat{l}_x] = [\hat{l}^2, \hat{l}_y] = 0 \tag{8.10b}$$

8.2
Spherical Harmonics and Spatial Quantization

We next consider the eigenequation of $\hat{l}_z$ that is given from Eq. (8.4) by

$$-i\hbar\frac{\partial u(\varphi)}{\partial\varphi} = l_z u(\varphi) \tag{8.11}$$

where $u(\varphi)$ and l_z are the eigenfunction and eigenvalue, respectively. By rearranging the equation as $\partial u(\varphi)/u(\varphi) = i(l_z/\hbar)\partial\varphi$, and integrating both sides, there results

$$u(\varphi) \propto e^{i(l_z/\hbar)\varphi} = \frac{1}{\sqrt{2\pi}}e^{i(l_z/\hbar)\varphi} \tag{8.12}$$

In Eq. (8.12), the constant of integration has been used for normalizing $u(\varphi)$ over the interval from 0 to 2π. Naturally, the eigenfunction $u(\varphi)$ should be single valued, that is,

$$u(\varphi) = u(\varphi + 2\pi) \tag{8.13}$$

Hence l_z should satisfy the condition $(l_z/\hbar)2\pi = 2\pi m$ with m as an integer, that is,

$$l_z = m\hbar, \quad m = 0, \pm 1, \pm 2, \ldots \tag{8.14}$$

Therefore, the normalized eigenfunction of $\hat{l}_z$ reads as

$$u_m(\varphi) = \frac{1}{\sqrt{2\pi}}e^{im\varphi}, \quad m = 0, \pm 1, \pm 2, \ldots \tag{8.15}$$

Next the eigenequation of $\hat{l}^2$ is given from Eq. (8.6) by

$$-\hbar^2\left[\frac{1}{\sin\theta}\frac{\partial}{\partial\theta}\left(\sin\theta\frac{\partial}{\partial\theta}\right) + \frac{1}{\sin^2\theta}\frac{\partial^2}{\partial\varphi^2}\right]Y_\beta(\theta,\varphi) = \beta\hbar^2 Y_\beta(\theta,\varphi) \tag{8.16}$$

where $Y_\beta(\theta,\varphi)$, $\beta\hbar^2$ are the eigenfunction and eigenvalue, respectively. As $\hat{l}^2$ and $\hat{l}_z$ commute, the two operators share a common eigenfunction. We can therefore put

$$Y_{\beta m}(\theta,\varphi) \propto u_m(\varphi) = u_m(\varphi)P_{\beta m}(\theta) \tag{8.17}$$

It thus follows from Eqs. (8.15) and (8.17) that $\partial^2 Y_{\beta m}/\partial\varphi^2 = -m^2 Y_{\beta m}$, so that the eigenequation (8.16) involves only one variable θ and reduces to

$$\frac{d}{dw}(1-w^2)\frac{dP_{\beta m}}{dw} + \left[\beta - \frac{m^2}{1-w^2}\right]P_{\beta m} = 0, \quad w = \cos\theta \tag{8.18}$$

The differential equation (8.18) can again be solved by the series method, but the series solutions diverge and are physically unacceptable unless β and m are constrained by

$$\beta = l(l+1), \quad m = -l, -l+1, \ldots, -1, 0, 1, 2, \ldots, l-1, l \tag{8.19}$$

When the Eq. (8.19) is put into Eq. (8.18), the equation becomes identical to the well-known Legendre differential equation. The solutions are known as the *Legendre* and *associated Legendre polynomials,* and denoted by $P_l^0(\equiv P_l)$ and P_l^m, respectively. The detailed analysis of Eq. (8.18) is found in the reference books listed at the end of this chapter. Suffice it to say here that the Legendre polynomials belong to the list of well-known special functions in mathematical physics and can be generated by the Rodrigues's formula:

$$P_l(w) = \frac{1}{2^l l!}\left(\frac{d}{dw}\right)^l (w^2-1)^l, \quad l = 0, 1, 2, \dots, \quad w = \cos\theta \tag{8.20}$$

Once P_l is found, P_l^m is obtained by the operation

$$P_l^m(w) = (-)^m(1-w^2)^{m/2}\frac{d^m}{dw^m}P_l(w), \quad P_l^{-m}(w) = (-)^m P_l^m(w) \tag{8.21}$$

Also the Legendre polynomials are orthogonal

$$\int_{-1}^{1} dw P_l^m(w) P_{l'}^m = \frac{2}{2l+1}\frac{(l+|m|)!}{(l-|m|)!}\delta_{ll'} \tag{8.22}$$

so that the normalized eigenfunction of both $\hat{l}_z$ and $\hat{l}^2$ is given by

$$Y_l^m(\theta,\varphi) = (-)^m\left[\frac{2l+1}{4\pi}\frac{(l-|m|)!}{(l+|m|)!}\right]^{1/2} P_l^m(\theta)e^{im\varphi} \tag{8.23a}$$

The eigenfunction (8.23a) is the celebrated spherical harmonics and is related to its complex conjugate as

$$Y_l^{-m}(\theta,\varphi) = (-)^m Y_l^m(\theta,\varphi)^* \tag{8.23b}$$

The Spatial Quantization

The spherical harmonics are often denoted by Dirac's ket vector, and the eigenequations are then compactly expressed as (see Eqs.(8.11), (8.16))

$$\hat{l}^2|lm\rangle = \hbar^2 l(l+1)|lm\rangle, \quad |lm\rangle \equiv Y_l^m \equiv Y_{lm}, \quad l = 0, 1, 2, \dots \tag{8.24a}$$

$$\hat{l}_z|lm\rangle = \hbar m|lm\rangle, \quad m = -l, -l+1, \dots, l-1, l \tag{8.24b}$$

Here, the integer l is called the *angular momentum quantum number* and m the magnetic quantum number. The eigenfunctions (8.24) clearly indicate that the angular momentum is specified by a discrete set of the quantum numbers, l, m in units of $\hbar$, as illustrated in Figure 8.3. Given l, the angular momentum precesses around the z-axis in discrete orientations such that its projection onto the z-axis varies in units of $\hbar$. The resulting spatial quantization of $\hat{l}_z$ is a feature again unique in quantum mechanics.

When the angular momentum is in the state $|lm\rangle$, it follows from Eqs. (8.4b), (8.4c), and (8.24b) that

$$\langle lm|\hat{l}_z|lm\rangle = \hbar m \tag{8.25a}$$

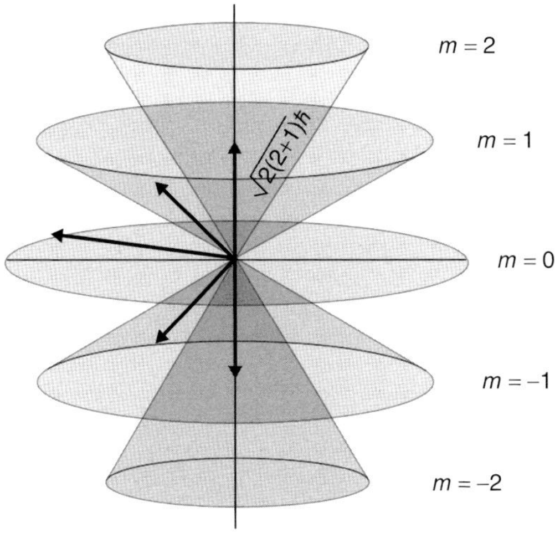

Figure 8.3 The spatial quantization of the z-component of ***l*** for $l = 2$.

$$\langle lm|\hat{l}_x|lm\rangle = \langle lm|i\hbar\left(\sin\varphi\frac{\partial}{\partial\theta} + \cot\theta\cos\varphi\frac{\partial}{\partial\varphi}\right)|lm\rangle = 0 \tag{8.25b}$$

$$\langle lm|\hat{l}_y|lm\rangle = \langle lm|i\hbar\left(-\cos\varphi\frac{\partial}{\partial\theta} + \cot\theta\sin\varphi\frac{\partial}{\partial\varphi}\right)|lm\rangle = 0 \tag{8.25c}$$

It is clear from Eq. (8.25) that the information of l_x, l_y is lost, when l_z is known precisely. This is expected because $\hat{l}_x$, $\hat{l}_y$, and $\hat{l}_z$ do not commute. Typical polar plots of the spherical harmonics are shown in Figure 8.4, and Table 8.1 lists Y_{lm}'s as a function of φ and θ.

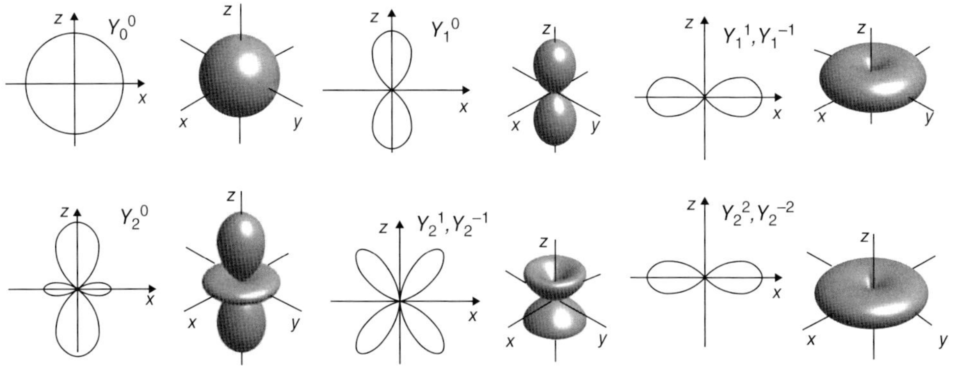

Figure 8.4 The 3D and projected polar plots of Y_0^0, Y_1^0, $Y_1^{\pm 1}$, Y_2^0, $Y_2^{\pm 1}$, $Y_2^{\pm 2}$.

Table 8.1 Typical spherical harmonics.

l	m	$Y_l^m(\theta,\varphi) = Y_{lm}(\theta,\varphi)$
0	0	$Y_{00} = \frac{1}{2\pi^{1/2}}$
1	0	$Y_{10} = \frac{1}{2}(3/\pi)^{1/2}\cos\theta$
	± 1	$Y_{1\pm1} = \mp\frac{1}{2}(3/2\pi)^{1/2}\sin\theta e^{\pm i\varphi}$
2	0	$Y_{20} = \frac{1}{4}(5/\pi)^{1/2}(3\cos^2\theta - 1)$
	± 1	$Y_{2\pm1} = \mp\frac{1}{2}(15/2\pi)^{1/2}\cos\theta\sin\theta e^{\pm i\varphi}$
	± 2	$Y_{2\pm2} = \frac{1}{4}(15/2\pi)^{1/2}\sin^2\theta e^{\pm 2i\varphi}$

8.3
The H-Atom and Electron–Proton Interaction

The H-atom consists of a proton and an electron bound together via the attractive Coulomb potential, and it is a two-body central force system. The equations of motion of the electron and proton read as

$$m_e \ddot{\underline{r}}_e = \underline{f}(r), \quad m_p \ddot{\underline{r}}_p = -\underline{f}(r), \quad \underline{r} \equiv \underline{r}_e - \underline{r}_p \tag{8.26}$$

where $\boldsymbol{r}_e, \boldsymbol{r}_p$ are the coordinates of the electron and proton, and m_e, m_p the respective masses. The central force depends only on the distance between the two and acts on each other in the opposite directions (Figure 8.5). When the two equations are added together, there results

$$m_e \ddot{\underline{r}}_e + m_p \ddot{\underline{r}}_p = 0 \tag{8.27}$$

We can recast Eq. (8.27) into the equation of motion of the center of mass as

$$M\ddot{\underline{R}} = 0; \quad \underline{R} \equiv \frac{m_e \underline{r}_e + m_p \underline{r}_p}{M}, \quad M \equiv m_e + m_p \tag{8.28a}$$

where $\boldsymbol{R}$ is the center of mass coordinate and M the total mass. When the two equations in Eq. (8.26) are divided by m_e, m_p, respectively, and the latter equation

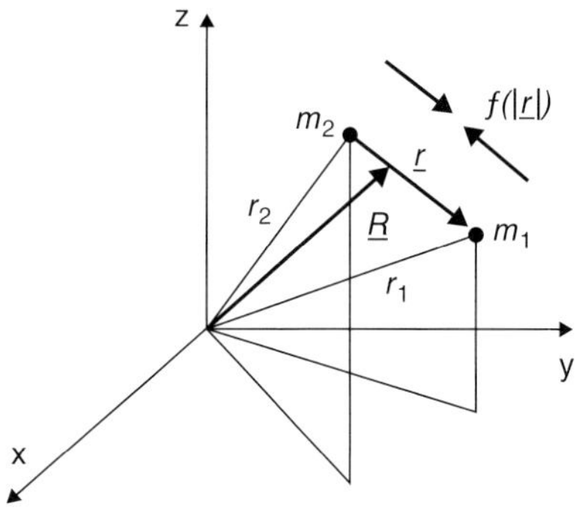

Figure 8.5 Two particles bound by a central force: $\boldsymbol{r}_1$, $\boldsymbol{r}_2$, $\boldsymbol{R}$, and $\boldsymbol{r}$ are the coordinates of particle 1, 2, the center of mass, and the displacement of particle 1 with respect to 2.

is subtracted from the former, there results

$$\ddot{\underline{r}} = \mu \underline{f}(r), \quad \underline{r} = \underline{r}_e - \underline{r}_p, \frac{1}{\mu} = \frac{1}{m_e} + \frac{1}{m_p} \tag{8.28b}$$

where μ is called the *reduced mass*. The motions of the two particles are thus partitioned into (i) the motion of the center of mass, moving as a free particle and (ii) the motion of a fictitious particle with the reduced mass μ with respect to the fixed force center. Since $m_e \ll m_p$, $\mu \approx m_e$, and the relative motion is essentially the motion of the electron with respect to the proton.

The Hydrogenic Atom

Energy Eigenequation and Eigenfunction

The Hamiltonian of the H-atom reads as

$$\hat{H} = -\frac{\hbar^2}{2M}\nabla_{\mathrm{R}}^2 - \frac{\hbar^2}{2\mu}\nabla^2 + V(r), \quad V(r) = -\frac{Ze_M^2}{r}, \quad e_M^2 \equiv \frac{e^2}{4\pi\varepsilon_0} \tag{8.29}$$

where the first two terms represent the kinetic energies of the center of mass and relative motions expressed via the Laplacian operators with respect to $\boldsymbol{R}$ and $\boldsymbol{r}$. The term $V(r)$ is the attractive Coulomb potential in MKS units and ε_0 the vacuum permittivity. The atomic number Z denotes the number of protons in the nucleus and $Z = 1$ in this case. The Schrödinger equation thus reads as

$$i\hbar\frac{\partial}{\partial t}\psi(\underline{R}, \underline{r}, t) = \left[-\frac{\hbar^2}{2M}\nabla_{\mathrm{R}}^2 - \frac{\hbar^2}{2\mu}\nabla^2 + V(r)\right]\psi(\underline{R}, \underline{r}, t) \tag{8.30}$$

We can partition the solution into the center of mass and relative motions as

$$\psi(\underline{R}, \underline{r}, t) = e^{-i(E_{CM}/\hbar)t}u_{CM}(\underline{R}) \cdot e^{-i(E/\hbar)t}u(\underline{r}) \tag{8.31}$$

Upon inserting Eq. (8.31) into Eq. (8.30) and carrying out the differentiations involved and dividing both sides with Eq. (8.31), there results

$$-\frac{\hbar^2}{2M}\nabla_R^2 u_{CM}(\underline{R}) = E_{CM}u(\underline{R}) \tag{8.32a}$$

$$\left[-\frac{\hbar^2}{2\mu}\nabla^2 - \frac{Ze_M^2}{r}\right]u(\underline{r}) = Eu(\underline{r}) \tag{8.32b}$$

where the total energy is the sum of the kinetic energy of the center of mass and the internal energy associated with the relative motion

$$E_T = E_{CM} + E \tag{8.32c}$$

The wavefunction of the center of mass moving as a free particle is given from Eq. (5.3) by

$$\varphi_{CM}(\underline{R}, t) \propto e^{-i(E_{CM}/\hbar)t}e^{i\underline{K}\cdot\underline{R}}, \quad \frac{\hbar^2K^2}{2M} = E_{CM} \tag{8.33}$$

where K is the wave vector. This leaves the bulk of analysis to solving Eq. (8.32b).

The Bound States

We next treat Eq. (8.32b) in the spherical coordinate frame and express the Laplacian as

$$\nabla^2 = \frac{1}{r^2}\frac{\partial}{\partial r}r^2\frac{\partial}{\partial r} + \frac{1}{r^2}\left[\frac{1}{\sin\theta}\frac{\partial}{\partial\theta}\sin\theta\frac{\partial}{\partial\theta} + \frac{1}{\sin^2\theta}\frac{\partial^2}{\partial\varphi^2}\right] \tag{8.34}$$

in which the bracket containing the angular variables is identical to $-\hat{l}^2/\hbar^2$ (see Eq. (8.6)). Hence, Eq. (8.32b) can be expressed as

$$\left[-\frac{\hbar^2}{2\mu}\left(\frac{1}{r^2}\frac{\partial}{\partial r}r^2\frac{\partial}{\partial r} - \frac{1}{r^2}\frac{1}{\hbar^2}\hat{l}^2\right) - \frac{Ze_{\mathrm{M}}^2}{r}\right]u(r,\theta,\varphi) = Eu(r,\theta,\varphi) \tag{8.35}$$

It is thus clear from Eq. (8.35) that $\hat{H}$ and $\hat{l}^2$ commute, and therefore the two operators can share the common eigenfunction, in this case the spherical harmonics:

$$u(r,\theta,\varphi) \propto Y_l^m(\theta,\varphi) = Y_l^m(\theta,\varphi)R(r) \tag{8.36}$$

Upon inserting Eq. (8.36) into Eq. (8.35), using Eq. (8.24a) and dividing both sides by Eq. (8.36), there results

$$\left[-\frac{\hbar^2}{2\mu}\frac{1}{r^2}\frac{\partial}{\partial r}\left(r^2\frac{\partial}{\partial r}\right) + V_{\mathrm{eff}}(r)\right]R(r) = ER(r) \tag{8.37a}$$

where the effective potential

$$V_{\mathrm{eff}}(r) = -\frac{Ze_M^2}{r} + \frac{1}{2\mu r^2}\hbar^2 l(l+1) \tag{8.37b}$$

consists of an attractive Coulomb potential and the repulsive centrifugal potential arising from the rotational motion of the electron. For large r, the attractive term dominates, while for small r, the repulsive term is prevalent. These two potentials combine to form a potential well, as shown in Figure 8.6. It is in this potential well that the bound states of the H-atom are formed. For $l = 0$, however, only the attractive Coulomb force binds the electron to the proton.

Radial Wavefunction

For analyzing Eq. (8.37), we can take E to be at the zero level when the electron is at a large distance from the proton and not bound by it. The bound state energy should then be taken negative. Let us introduce the dimensionless variable

$$\rho = \alpha r, \quad \alpha^2 \equiv \frac{8\mu|E|}{\hbar^2} \tag{8.38}$$

and express Eq. (8.37a) as

$$\left[\frac{1}{\rho^2}\frac{d}{d\rho}\rho^2\frac{d}{d\rho} + \frac{\lambda}{\rho} - \frac{1}{4} - \frac{l(l+1)}{\rho^2}\right]R_{\lambda l}(\rho) = 0; \quad \lambda \equiv \frac{Ze_M^2}{\hbar}\left(\frac{\mu}{2|E|}\right)^{1/2} \tag{8.39}$$

In the asymptotic limit, in which $\rho \to \infty$, Eq. (8.39) reduces to

$$R''_{\lambda l} - \frac{1}{4}R_{\lambda l} = 0 \tag{8.40}$$

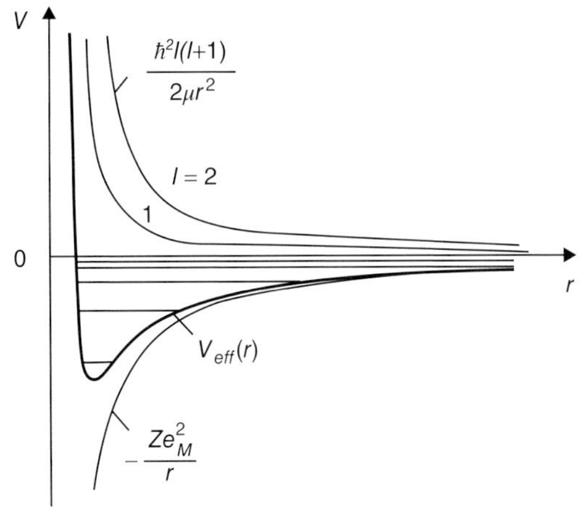

Figure 8.6 The effective potential energy resulting from the attractive Coulomb force and repulsive centrifugal force.

with primes denoting the differentiation with respect to ρ. The solution is then given by $R_{\lambda l} \propto \exp(\pm\rho/2)$, but the positive branch of the exponent has to be discarded to prevent $R_{\lambda l}$ from diverging at large ρ. We thus look for the solution of Eq. (8.39) in the form

$$R_{\lambda l}(\rho) = \rho^l L(\rho) e^{-\rho/2} \tag{8.41}$$

and insert it into Eq. (8.39), obtaining

$$\rho L'' + [2(l+1) - \rho]L' + (\lambda - l - 1)L = 0 \tag{8.42}$$

Equation (8.42) can again be solved by means of the series method, but the solution diverges as usual unless the parameters λ and l are constrained by the condition

$$\lambda = n = m + l + 1, \quad m = 0, 1, 2, \ldots \tag{8.43a}$$

where n is an integer greater than the angular momentum quantum number l:

$$n \geq l + 1 \tag{8.43b}$$

When the conditions (8.43) are inserted into Eq. (8.42), the resulting differential equation

$$\rho L''_{nl} + [2(l+1) - \rho]L'_{nl} + (n - l - 1)L_{nl} = 0 \tag{8.44a}$$

reduces to the well-known Laguerre differential equation

$$\rho L_q^{p\prime\prime} + [p + 1 - \rho]L_q^{p\prime} + (q - p)L_q^p = 0 \tag{8.44b}$$

with the identification $p = 2l + 1$, $q = n + l$. Therefore, the Laguerre polynomial can be used as the solution

$$R_{nl}(\rho) = N_{nl}\rho^l e^{-\rho/2} L_{nl}(\rho); \quad L_{nl}(\rho) \equiv L_{n+l}^{2l+1}(\rho) \tag{8.45}$$

with N_{nl} denoting the normalization constant.

Energy Quantization

Let us revisit the condition (8.43) and examine the bound state energy. The constraint imposed on λ reads with the use of Eq. (8.39) as

$$\lambda_n \equiv \frac{Ze_M^2}{\hbar}\left(\frac{\mu}{2\left|E_n\right|}\right)^{1/2} = n, \quad n = 1, 2, \ldots \tag{8.46a}$$

Therefore, the bound state energy is naturally quantized as

$$E_n \equiv -|E_n| = -E_0\frac{1}{n^2}; \quad E_0 = \frac{\mu Z^2 e_M^4}{2\hbar^2} = \frac{Z^2 e_M^2}{2a_0}, \quad n = 1, 2, \ldots \tag{8.46b}$$

where E_0 (= 13.6 eV) for $Z = 1$ is called the *ionization energy* that is required to ionize the H-atom from its ground state ($n = 1$) and the parameter

$$a_0 \equiv \frac{\hbar^2}{\mu e_M^2} = \frac{\hbar^2}{e_M^2 m_e}\left(1 + \frac{m_e}{m_N}\right) = a_B\left(1 + \frac{m_e}{m_N}\right) \tag{8.46c}$$

represents the Bohr radius a_B scaling the atomic radius (see Eq. (2.15)). As the mass of the proton is much greater than m_e, μ is practically identical to m_e, and the result is in agreement with Bohr's H-atom theory. However, the information contained in the wavefunction is much richer.

8.3.1 Atomic Radius and the Energy Eigenfunction

When the energy eigenvalue in Eq. (8.46b) is inserted into Eq. (8.38), the parameter α is specified as

$$\alpha_n^2 = \frac{8\mu}{\hbar^2}|E_n| = \frac{8\mu}{\hbar^2}\frac{Z^2 e_M^2}{2a_0}\frac{1}{n^2} = \left(\frac{2Z}{a_0 n}\right)^2 \tag{8.47a}$$

Hence, the dimensionless radial variable

$$\rho \equiv \alpha_n r = \left(\frac{2Z}{a_0 n}\right) r \tag{8.47b}$$

naturally scales the atomic radius in terms of the Bohr radius a_B, atomic number Z, and the energy level n. Moreover, the normalization constant N_{nl} in Eq. (8.45) is found from

$$\begin{aligned} 1 &= N_{lm}^2 \int_0^\infty r^2 dr [R_{nl}(r)]^2 = N_{lm}^2 \frac{1}{\alpha_n^3}\int_0^\infty d\rho \rho^2 \left[\rho^l \mathrm{e}^{-\rho/2} L_{n+l}^{2l+1}(\rho)\right]^2 \\ &= N_{lm}^2 \frac{1}{\alpha_n^3}\frac{2n[(n+l)!]^3}{(n-l-1)!} \end{aligned} \tag{8.48}$$

where the well-known integral involving Laguerre polynomials has been used. Therefore, by finding N_{lm} from Eq. (8.48), we obtain the normalized radial wavefunction as

$$R_{nl}(r) = \left[\left(\frac{2Z}{na_0}\right)^3 \frac{(n-l-1)!}{2n[(n+l)!]^3}\right]^{1/2} e^{-\rho/2}\rho^l L_{n+l}^{2l+1}(\rho), \quad \rho \equiv \left(\frac{2Z}{a_0 n}\right) r \tag{8.49}$$

and the energy eigenfunction of the H-atom is given by

$$u_{nlm}(r,\theta,\varphi) = R_{nl}(r)Y_l^m(\theta,\varphi) \tag{8.50}$$

The typical examples of u_{nlm} are listed in Table 8.2.

8.3.2
Eigenfunction and Atomic Orbital

The eigenfunction $u_{nlm}(r,\theta,\varphi)$ carries a wealth of information and is discussed next. The function u_{nlm} is characterized by three sets of quantum numbers: (i) the principal quantum number, $n = 1, 2, \ldots$ (See Eq. (8.43a)); (ii) the angular momentum quantum number l ranging from 0 to $n-1$ for given n (Eq. (8.43b)); and (iii) the magnetic quantum number m varying from $-l$ to $+l$ in steps of unity Eq. (8.19).The quantum states with l having the values 0, 1, 2, and 3 are designated by s, p, d, and f states in spectroscopy.

Degenerate States

The eigenenergy associated with u_{nlm} depends at this stage solely on the principal quantum number n as clear from Eq. (8.46). This suggests that the multiple quantum states with different quantum numbers l, m for given n all share a common energy E_n. The number of such states can be found by summing over l from 0 to $n-1$, with each l associated with $2l+1$ different m values. Furthermore, for given l, m, there are two possible spin states of the electron, spin-up and spin-down.

Table 8.2 Hydrogenic energy eigenfunctions.

$$a_0 \equiv \frac{\hbar^2}{\mu e_M^2} = \frac{\hbar^2}{e_M^2 m_e}\left(1+\frac{m_e}{m_N}\right) = a_B\left(1+\frac{m_e}{m_N}\right), \quad \rho \equiv \left(\frac{2Z}{a_0 n}\right) r$$

orbital	u_{nlm}
1s	$u_{100} = \frac{(Z/a_0)^{3/2}}{\pi^{1/2}} e^{-Zr/a_0}$
2s	$u_{200} = \frac{(Z/a_0)^{3/2}}{(32\pi)^{1/2}}\left(2-\frac{Zr}{a_0}\right) e^{-Zr/2a_0}$
2p	$u_{210} = \frac{(Z/a_0)^{3/2}}{(32\pi)^{1/2}}\frac{Zr}{a_0} e^{-Zr/2a_0}\cos\theta$
2p	$u_{21\pm1} = \frac{(Z/a_0)^{3/2}}{(64\pi)^{1/2}}\frac{Zr}{a_0} e^{-Zr/2a_0}\sin\theta e^{\pm i\varphi}$
3s	$u_{300} = \frac{(Z/a_0)^{3/2}}{81(3\pi)^{1/2}}\left(27-18\frac{Zr}{a_0}+2\frac{Z^2r^2}{a_0^2}\right) e^{-Zr/3a_0}$
3p	$u_{310} = \frac{2^{1/2}(Z/a_0)^{3/2}}{81(\pi)^{1/2}}\left(6-\frac{Zr}{a_0}\right)\frac{Zr}{a_0} e^{-Zr/3a_0}\cos\theta$
3p	$u_{31\pm1} = \frac{(Z/a_0)^{3/2}}{81(\pi)^{1/2}}\left(6-\frac{Zr}{a_0}\right)\frac{Zr}{a_0} e^{-Zr/3a_0}\sin\theta e^{\pm i\varphi}$
3d	$u_{320} = \frac{(Z/a_0)^{3/2}}{81(6\pi)^{1/2}}\frac{Z^2r^2}{a_0^2} e^{-Zr/3a_0}(3\cos^2\theta-1)$
3d	$u_{32\pm1} = \frac{(Z/a_0)^{3/2}}{81(\pi)^{1/2}}\frac{Z^2r^2}{a_0^2} e^{-Zr/3a_0}\sin\theta\cos\theta e^{\pm i\varphi}$
3d	$u_{31\pm2} = \frac{(Z/a_0)^{3/2}}{162(\pi)^{1/2}}\frac{Z^2r^2}{a_0^2} e^{-Zr/3a_0}\sin^2\theta e^{\pm 2i\varphi}$

Thus, the total number of degenerate states for given n is given by

$$g_n = 2\sum_{l=0}^{n-1}(2l+1) = 2\left\{\left[2\frac{n(n-1)}{2}\right] + n\right\} = 2n^2 \tag{8.51}$$

The energy spectrum is shown in Figure 8.7 together with spectroscopic notations.

Reduced Probability Density

The probability density $u^*_{nlm}u_{nlm}$ depends on three variables r, θ, and φ, and therefore represents the joint probability density of finding the electron in the volume element $\sin\theta r^2 \mathrm{d}r\mathrm{d}\theta\mathrm{d}\varphi$ at r, θ, and φ (Figure 8.2). One can therefore introduce the reduced probability density of finding the electron between r and $r + \mathrm{d}r$ regardless of θ, φ by integrating $u^*_{nlm}u_{nlm}$ over the angular variables:

$$P(r)dr \equiv \int_0^{2\pi} d\varphi \int_0^{\pi} \sin\theta d\theta r^2 dr |Y_l^m|^2 |R_{nl}^2| = r^2 |R_{nl}^2| dr \tag{8.52}$$

where the normalization property of the spherical harmonics has been used. In Figure 8.8 are plotted the reduced radial probability densities $P(r) = r^2 R^*_{nl} R_{nl}$ versus r. Clearly, $P(r)$ vanishes at $r = 0$, as it should, as electrons do not reside in the nucleus. Moreover, the profiles of $P(r)$ exhibit the gross features of the electron clouds around the nucleus, with its peak values roughly corresponding to Bohr's electron orbits. Also, for given n, the value of r at which $P(r)$ is peaked increases with increasing quantum number l, that is, with increasing centrifugal force.

Atomic Orbitals

The s-orbitals for $l = 0$ are spherically symmetric, while others for $l \neq 0$ are non-symmetric and depend sensitively on the orientation. For instance, the p-orbitals

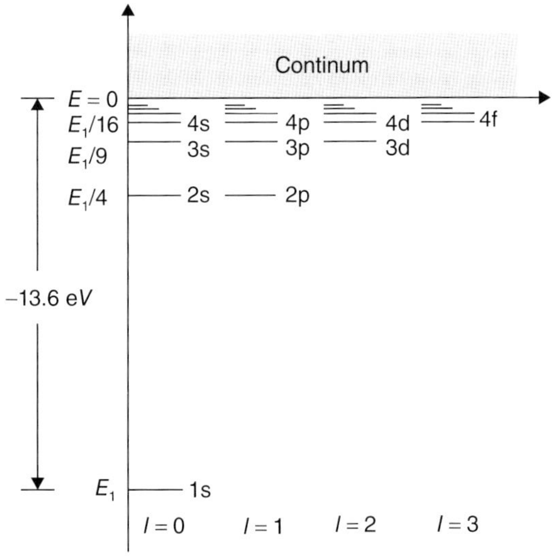

Figure 8.7 The energy spectrum of the H-atom as denoted by the spectroscopic notations.

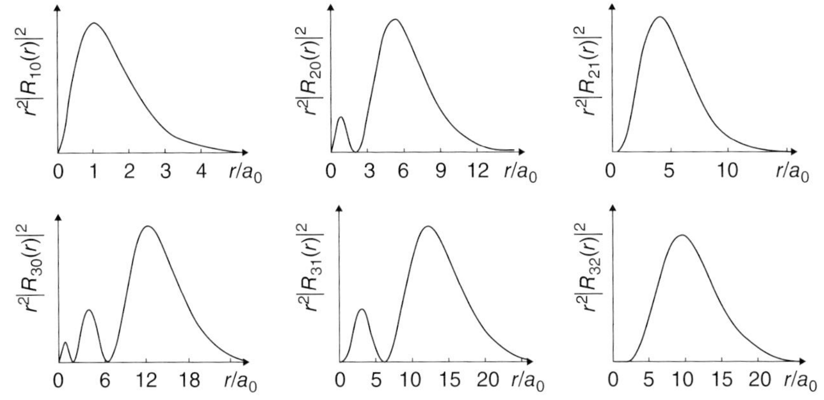

Figure 8.8 Typical reduced radial probability densities versus r for different n, l.

with $l = 1$ for given n are characterized by $m = 0, \pm 1$. The wavefunctions associated are denoted by

$$|np_z\rangle \equiv R_{n1} Y_1^0 = \left(\frac{3}{4\pi}\right)^{1/2} R_{n1} \cos\theta \tag{8.53a}$$

$$|np_+\rangle \equiv R_{n1} Y_1^1 = -\left(\frac{3}{8\pi}\right)^{1/2} R_{n1} \sin\theta e^{i\varphi} \tag{8.53b}$$

$$|np_-\rangle \equiv R_{n1} Y_1^{-1} = \left(\frac{3}{8\pi}\right)^{1/2} R_{n1} \sin\theta e^{-i\varphi} \tag{8.53c}$$

The two complex eigenfunctions (8.53b) and (8.53c) are often combined into two real orthonormal eigenfunctions as

$$|np_x\rangle \equiv \frac{1}{\sqrt{2}}(|p_-\rangle - |p_+\rangle) = \left(\frac{3}{4\pi}\right)^{1/2} R_{n1} \cos\theta \cos\varphi \tag{8.54a}$$

$$|np_y\rangle \equiv \frac{i}{\sqrt{2}}(|p_-\rangle + |p_+\rangle) = \left(\frac{3}{4\pi}\right)^{1/2} R_{n1} \sin\theta \sin\varphi \tag{8.54b}$$

The three p-orbitals Eqs. (8.53a), (8.54a), and (8.54b) are used for describing atomic and molecular structures. The transformation of three p-orbitals Eq. (8.53) into a new set Eqs. (8.53a), (8.54a), and (8.54b) is equivalent to transforming one set of basis vectors into another via rotation. The boundary surfaces of the p-orbitals are shown in Figure 8.9.

8.3.3
Doppler Shift

An atom in an excited state n_i emits radiation when the electron makes the transition to a lower-lying state n_f to conserve energy. The frequency of the emitted radiation undergoes shift due to the motion of the center of mass. Such shift in frequency is known as *Doppler shift*, and the schematic is shown in Figure 8.10.

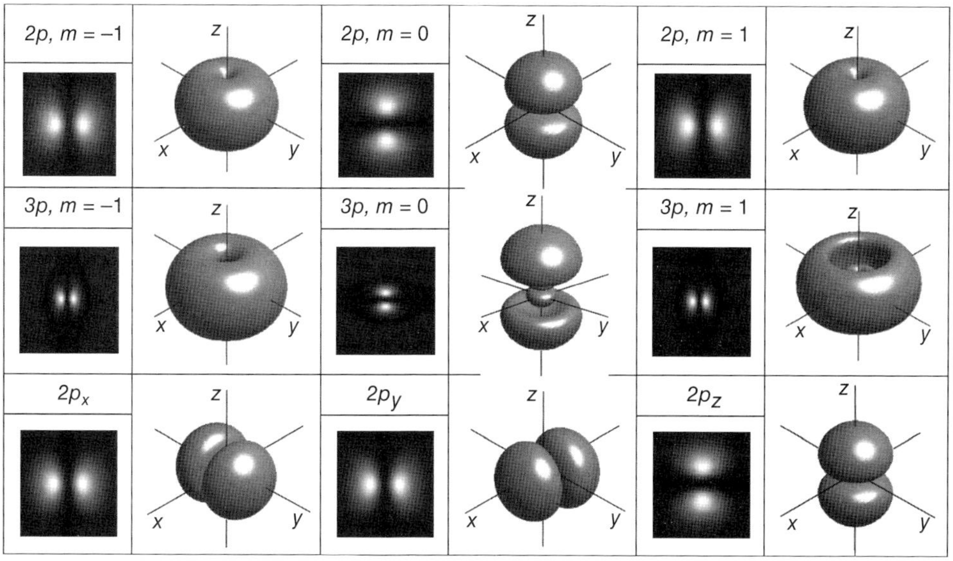

Figure 8.9 The boundary surfaces and side projections of the p-orbitals for $n = 2, 3$ and of p_x, p_y, p_z states.

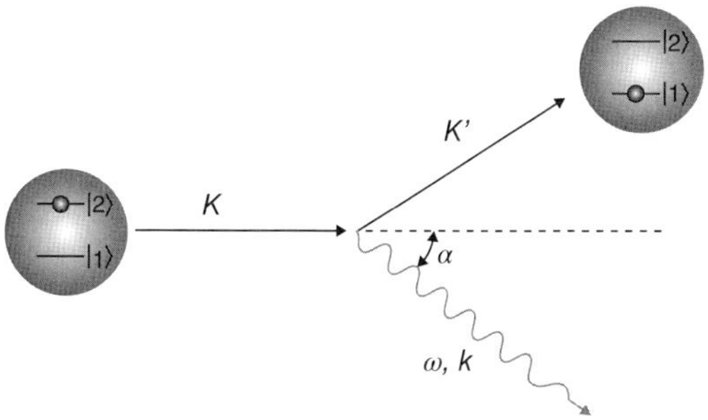

Figure 8.10 The Doppler shift of the radiation emitted from a moving atom. A photon is emitted when the electron makes the transition from upper to lower states.

The Doppler shift can be analyzed based on the conservation principles. During the emission of radiation, the energy and momentum are conserved

$$\frac{\hbar^2 K^2}{2M} + E_n = \frac{\hbar^2 K'^2}{2M} + E_{n'} + \hbar\omega \tag{8.55}$$

$$\hbar \underline{K} = \hbar \underline{K}' + \hbar \underline{k} \tag{8.56}$$

where $\boldsymbol{K}$, $\boldsymbol{K}'$ are the wave vectors of CM representing its kinetic energy, E_n, $E_{n'}$ the internal eigenenergies before and after the emission, and $\boldsymbol{k}$, ω the wave vector and frequency of the emitted photon. The frequency of the photon ω is therefore found from Eq. (8.55) as

$$\omega = \omega_0 + \frac{\hbar}{2M}(K^2 - K'^2); \quad \hbar\omega_0 \equiv E_n - E_{n'} \tag{8.57}$$

and consists of the atomic transition frequency ω_0 and the difference of the kinetic energies before and after the emission. As one can write from Eq. (8.56)

$$(\underline{K} - \underline{k}) \cdot (\underline{K} - \underline{k}) = K'^2$$

the difference between two kinetic energies is given by

$$K^2 - K'^2 = 2Kk\cos\alpha - k^2; \quad \underline{K} \cdot \underline{k} = Kk\cos\alpha \tag{8.58}$$

where α is the angle between $\boldsymbol{K}$ and $\boldsymbol{k}$. Hence, by substituting Eq. (8.58) into Eq. (8.57) and identifying the momentum of the photon $\hbar k = p = \hbar\omega/c$, and the atomic velocity $\hbar K/M = v$, we find

$$\omega - \omega_0 = \omega\left(\frac{v}{c}\cos\alpha - \frac{\hbar\omega}{2Mc^2}\right) \tag{8.59}$$

Since $|\omega - \omega_0| \ll \omega, \omega_0$, we may replace ω on the right-hand side by ω_0 to the first order of approximation and obtain the shift in frequency as

$$\omega - \omega_0 \simeq \omega_0\left(\frac{v}{c}\cos\alpha - \frac{\hbar\omega_0}{2Mc^2}\right) \tag{8.60}$$

The first and second terms on the right-hand side are called the *Doppler shift* of the first and second kinds, respectively. The first term increases or decreases the frequency, depending on whether the atom is moving toward $+v$ or away $-v$ from the detector. The second term accounts for the small shift of frequency caused by atomic recoil.

The atoms in the thermal equilibrium undergo the random thermal motion. Hence, the frequency of radiation emanating from an ensemble of atoms and detected on the y–z plane should exhibit a Gaussian spectral profile due to atoms moving in the x-direction

$$\mathrm{E}(\omega) \propto \exp - \frac{(\omega - \omega_0)^2}{\Delta\omega^2} \tag{8.61}$$

Here the variance $\Delta\omega$ resulting from the thermal motion in the x-direction is obtained by using the Boltzmann distribution function (Eq. (1.11)) as

$$\begin{aligned}(\Delta\omega)^2 &\simeq \left(\frac{M}{2\pi k_B T}\right)^{1/2} \int_{-\infty}^{\infty} dv_x e^{-\beta v_x^2}\left(\omega_0 \frac{v_x}{c}\right)^2; \quad \beta = \frac{M}{2k_B T} \\ &= \omega_0^2\left(\frac{k_B T}{Mc^2}\right)\end{aligned} \tag{8.62}$$

where the atomic recoil term is neglected. This kind of broadening of the emitted radiation is called the *inhomogeneous line broadening*.

Problems

8.1 The angular momentum of a particle is defined as the vector product of $\boldsymbol{r}$ and $\boldsymbol{p}$ **:**

$$\hat{\underline{l}} = \underline{r} \times \underline{\hat{p}} = (\hat{x}x + \hat{y}y + \hat{z}z) \times ((\hat{x}p_x + \hat{y}p_y + \hat{z}p_z)$$

(a) By using the cyclic properties of the vector product $\hat{x} \times \hat{y} = \hat{z}$, $\hat{y} \times \hat{z} = \hat{x}$, $\hat{z} \times \hat{x} = \hat{y}$, derive Eq. (8.2).

(b) By using the commutation relations between (x, p_x), (y, p_y), and (z, p_z), derive the cyclic commutation relations of the angular momentum operators (Eq. (8.7)) and commutation relation (Eq. (8.10b)).

8.2 Consider two particles with the mass m_1, m_2 bound by a central force. Show that the total kinetic energy of the two particle system can be expressed in terms of kinetic energies of center of mass and relative motion

$$\frac{p_1^2}{2m_1} + \frac{p_2^2}{2m_2} = \frac{P^2}{2M} + \frac{p^2}{2\mu}$$

where $\boldsymbol{P}$ is the momentum of the center of mass and $\boldsymbol{p}$ that of a fictitious particle with reduced mass μ (see Eq. (8.26)–(8.28)).

8.3 Find the effective Bohr radius for

(i) the singly ionized helium atom He^+ consisting of two protons and two neutrons at the nucleus and a single electron revolving around it.

(ii) Positronium consisting of a positron and an electron bound together by attractive Coulomb force (the positron has the same mass as electron but a positive charge +e).

8.4 Calculate the average values of the radius r and r^2 and the variance $(\Delta r)^2$ in 1, 2, and 3 s states in H-atom and ionized He^+-atom.

8.5 Show that the average kinetic and potential energies of the ground state of the H-atom are related by

$$\left\langle \frac{p^2}{2\mu} \right\rangle_{100} = -\frac{1}{2}\langle V \rangle_{100}; \quad V = \frac{e_M^2}{r}$$

8.6 (a) Find the wavelengths resulting from the electron making the transition from $n = 2$ to $n = 1$ states in H-atom, deuterium atom (one proton and one neutron and one electron), and ionized He-atom (two protons, two neutrons, and one electron).

(b) If the H-atom is to be optically excited from $n = 1$ to $n = 3$ states, what frequencies will be required?

8.7 The phosphorus atom, when incorporated into the silicon, can be modeled as a hydrogen-like atom, consisting of an outermost electron in $n = 3$ state and bound by a single proton in the nucleus. The dielectric constant of the medium is $\varepsilon_r = 11.9$, and the effective mass of the electron is $m_n \approx 0.2m_0$ with m_0 denoting the rest mass of electron. Calculate the ionization energy, the effective Bohr radius, and the de Broglie wavelength of the electron in the ground state.

Suggested Readings

1. A. Yariv, An Introduction to Theory and Applications of Quantum Mechanics, John Wiley & Sons, 1982.
2. D. M. Kim, Introductory Quantum Mechanics for Semiconductor Nanotechnology, Wiley-VCH, 2010.
3. H. Haken and H. C. Wolf, The Physics of Atoms and Quanta: Introduction to Experiments and Theory, Fifth Edition, Springer, 2004.
4. S. Gasiorowics, Quantum Physics, Third Edition, John Wiley & Sons, 2003.
5. A. I. M. Rae, Quantum Mechanics, Fourth Edition, Taylor & Francis, 2002.
6. M. Karplus and R. N. Porter, Atoms and Molecules: An Introduction for Students of Physical Chemistry, Addison Wesley Publishing Company, 1970.

9
The Perturbation Theory

Rigorous analytical treatments of dynamical systems are generally difficult, and various perturbation or iteration schemes have been devised to deal with such systems. The perturbation theories are capable of shedding an overall insight of the problem at hand and provide the general background information by which to access the accuracy of numerical computations as well. Moreover, the time-dependent perturbation scheme can describe the coupling of dynamical systems such as light and matter. The time-independent and time-dependent perturbation theories are presented and applied for analyzing the Stark effects, atomic polarizability, and Fermi's golden rule.

9.1
Time-Independent Perturbation Theory

Given a system with the Hamiltonian $\hat{H}$, we have to solve the energy eigenequation

$$\hat{H}\varphi = W\varphi \tag{9.1}$$

The crux of the perturbation scheme is to divide $\hat{H}$ into two parts as

$$\hat{H} = \hat{H}_0 + \lambda\hat{H}', \quad |\hat{H}_0| \gg |\hat{H}'| \tag{9.2}$$

where $\hat{H}_0$ can be treated analytically, while $\hat{H}'$ is the remainder to be treated as the perturbing term. Obviously, the accuracy of the scheme depends on relative magnitudes of the two terms. The smallness parameter λ is introduced to keep track of the order of iterations.

Nondegenerate Theory

We first introduce a set of orthonormal eigenfunctions satisfying the eigenequation

$$\hat{H}_0 u_n = E_n u_n \tag{9.3}$$

Introductory Quantum Mechanics for Applied Nanotechnology, First Edition. Dae Mann Kim.

and expand φ and W in Eq. (9.1) in powers of λ:

$$(\hat{H}_0 + \lambda\hat{H}')(\varphi_0 + \lambda\varphi_1 + \lambda^2\varphi_2 + \cdots)$$
$$= (W_0 + \lambda W_1 + \lambda^2 W_2 + \cdots)(\varphi_0 + \lambda\varphi_1 + \lambda^2\varphi_2 + \cdots) \tag{9.4}$$

We can then set up a hierarchy of equations by equating the coefficients of equal powers of λ from both sides of Eq. (9.4). Up to the second order, one can write

$$\hat{H}_0\varphi_0 = W_0\varphi_0 \tag{9.5a}$$

$$\hat{H}'\varphi_0 + \hat{H}_0\varphi_1 = W_0\varphi_1 + W_1\varphi_0 \tag{9.5b}$$

$$\hat{H}'\varphi_1 + \hat{H}_0\varphi_2 = W_0\varphi_2 + W_1\varphi_1 + W_2\varphi_0 \tag{9.5c}$$

Thus, given φ_0 and W_0, the true eigenfunction and eigenvalue φ and W in Eq. (9.1) can be obtained iteratively by incorporating the effect of $\hat{H}'$ to an arbitrary order of accuracy. Let us choose u_m, E_m as φ_0, W_0 and examine the modifications due to $\hat{H}'$:

$$\varphi_0 = u_m, \quad W_0 = E_m \tag{9.6}$$

First-Order Analysis

We can expand φ_1 in terms of $\{u_n\}$ as

$$\varphi_1 = \sum_n a_n^{(1)} u_n \tag{9.7}$$

and insert Eqs. (9.6) and (9.7) into Eq. (9.5b) and rewrite it as

$$\hat{H}'u_m + \hat{H}_0\sum_n a_n^{(1)}u_n = E_m\sum_n a_n^{(1)}u_n + W_1 u_m \tag{9.8}$$

The problem is then reduced to determining $\{a_n^{(1)}\}$ and W_1. To this end, we take the inner product on both sides of Eq. (9.8) with respect to u_k, obtaining

$$\hat{H}'_{km} + \sum_n a_n^{(1)}\langle u_k|\hat{H}_0|u_n\rangle = E_m\sum_n a_n^{(1)}\langle u_k|u_n\rangle + W_1\langle u_k|u_m\rangle \tag{9.9}$$

where

$$\hat{H}'_{km} \equiv \langle u_k|\hat{H}'|u_m\rangle = \int d\underline{r}\, u_k^*(\underline{r})\hat{H}'u_m(\underline{r}) \tag{9.10}$$

is the matrix element of the perturbing Hamiltonian. As $\{u_n\}$ is an orthonormal set, Eq. (9.9) reduces to

$$\hat{H}'_{km} + a_k^{(1)}E_k = E_m a_k^{(1)} + W_1\delta_{km} \tag{9.11}$$

with δ_{km} denoting the Kronecker delta function. It is important to note that k is the dummy quantum number, while m represents a particular state chosen for investigation.

In the nondegenerate system, $E_k \neq E_m$ if $k \neq m$, hence the first-order expansion coefficients are found from Eq. (9.11) as

$$a_k^{(1)} = \frac{\hat{H}'_{km}}{E_m - E_k}, \quad k \neq m \tag{9.12}$$

For $k = m$, the shift in energy of the mth state is also found from Eq. (9.11) as

$$W_{1m} \equiv \hat{H}'_{mm} = \langle u_m|\hat{H}'|u_m\rangle \tag{9.13}$$

Hence u_m, E_m are modified up to the first-order correction as

$$E_m^{(1)} = E_m + \hat{H}'_{mm}; \quad \hat{H}'_{mm} \equiv \langle u_m|\hat{H}'|u_m\rangle \tag{9.14a}$$

$$\varphi_m^{(1)} = u_m(1 + a_m^{(1)}) + \sum_{k\neq m} \frac{\hat{H}'_{km}}{E_m - E_k} u_k; \quad \hat{H}'_{km} \equiv \langle u_k|\hat{H}'|u_m\rangle \tag{9.14b}$$

Although $a_m^{(1)}$ still remains unknown, it can be determined from the normalization condition imposed up to the first order

$$\begin{aligned} 1 = \langle \varphi_m^{(1)}|\varphi_m^{(1)}\rangle &= \langle u_m(1 + \lambda a_m^{(1)})|u_m(1 + \lambda a_m^{(1)})\rangle \\ &= 1 + [a_m^{(1)}]^* + [a_m^{(1)}] + O(\lambda^2) \end{aligned}$$

Since the terms proportional to λ^2 are relegated to the second-order analysis, we can put $a_m^{(1)} = 0$.

Second-Order Analysis

We again expand φ_2 in terms of $\{u_n\}$as

$$\varphi_2 = \sum_n a_n^{(2)} u_n \tag{9.15}$$

By inserting Eqs. (9.15) and (9.14b) into Eq. (9.5c), performing the inner product of both sides with respect to u_k and using the orthonormality of $\{u_n\}$, we find

$$\sum_n a_n^{(1)} \hat{H}'_{kn} + a_k^{(2)} E_k = E_m a_k^{(2)} + W_1 a_k^{(1)} + W_2 \delta_{km} \tag{9.16}$$

Therefore, we can obtain W_2 by putting $k = m$ and using $a_m^{(1)} = 0$:

$$W_2 = \sum_n a_n^{(1)} \hat{H}'_{mn} = \sum_{n\neq m} \frac{|\hat{H}'_{nm}|^2}{E_m - E_n}, \quad \hat{H}'_{nm} = (\hat{H}'_{mn})^* \tag{9.17}$$

where Eq. (9.12) has been used. We can likewise find in the second-order modification of u_m by considering the case $k \neq m$. The energy eigenvalue up to the second order of perturbation analysis is given from Egs. (9.17), (9.14a) by

$$E_m^{(2)} = E_m + \hat{H}'_{mm} + \sum_{n\neq m} \frac{|\hat{H}'_{nm}|^2}{E_m - E_n} \tag{9.18}$$

The Stark Shift in Harmonic Oscillator

Stark shift refers to the shift in energy level when the system is subjected to the electric field. Thus, consider the harmonic oscillator with charge q and placed in an electric field E. The oscillator is then subjected to the force $q\mathrm{E}$ or the potential $-q\mathrm{E}x$. The Hamiltonian is therefore given by

$$\hat{H} = \hat{H}_0 + \hat{H}', \quad \hat{H}_0 = -\frac{\hbar^2}{2m}\frac{\partial^2}{\partial x^2} + \frac{1}{2}kx^2, \quad \hat{H}' = -q\mathrm{E}x \tag{9.19}$$

Let us choose the lth eigenstate $u_l(x)$ of the HO for examination. As the parity of $u_l^*(x)u_l(x)$ is even regardless of whether $u_l(x)$ is even or odd in x while that of $\hat{H}'$ is odd, there is no first-order level shift, and the result applies to all other states as well:

$$W_1 = -q\mathrm{E}\langle u_l|x|u_l\rangle = -q\mathrm{E}\int_{-\infty}^{\infty} u_l^* x u_l = 0 \tag{9.20}$$

Also, the x-matrix elements of the eigenfunctions of the HO connect two nearest neighbor states (see Eq. (7.31)). Hence, for given l, it follows from Eqs. (7.31) and (9.14b) that

$$\begin{aligned}\varphi_l^{(1)} &= u_l + \frac{\hat{H}'_{l+1,l}}{E_l - E_{l+1}}u_{l+1} + \frac{\hat{H}'_{l-1,l}}{E_l - E_{l-1}}u_{l-1}\\ &= u_l + \Lambda\left[(l+1)^{\frac{1}{2}}u_{l+1} - (l)^{\frac{1}{2}}u_{l-1}\right]; \quad \Lambda = \frac{q\mathrm{E}}{\hbar\omega}\sqrt{\frac{\hbar}{2m\omega}}, \quad E_l - E_{l\pm1} = \mp\hbar\omega\end{aligned} \tag{9.21}$$

Similarly, the second-order level shift is contributed by two nearest neighbor states (see Eq. (7.31)); hence, W_2 is obtained from Eqs. (7.31) and (9.18) as

$$W_2 = \frac{|\hat{H}'_{l+1,l}|^2}{E_l - E_{l+1}} + \frac{|\hat{H}'_{l-1,l}|^2}{E_l - E_{l-1}} = \frac{q^2\mathrm{E}^2}{\hbar\omega}\frac{\hbar}{2m\omega}[-(l+1)+l] = -\frac{q^2\mathrm{E}^2}{2m\omega^2} \tag{9.22}$$

and is shown the same for all eigenstates.

The Polarizability of H-Atom

When an electric field E is applied in the z-direction, for instance, the interaction Hamiltonian of the H-atom is given by

$$H' = -\int_0^z (-e\mathrm{E})dz = e\mathrm{E}z \tag{9.23}$$

The dipole moment of the H-atom is then given to the first order of approximation by

$$\langle\mu\rangle \equiv -e\langle z\rangle = -e\langle u_{100} + \Delta u_{100}|z|u_{100} + \Delta u_{100}\rangle \tag{9.24a}$$

with the first-order correction in u_{100} given from Eq. (9.14b) by

$$\Delta u_{100} = e\mathrm{E}\sum_{k\neq100}\frac{\langle u_{100}|z|u_k\rangle}{E_{100} - E_k}u_k \tag{9.24b}$$

Hence, by inserting Eq. (9.24b) into Eq. (9.24a) and retaining the first-order corrections, we obtain the atomic polarizability α as

$$\langle \mu \rangle \equiv \varepsilon_0 \alpha \mathrm{E}, \quad \alpha = 2\frac{e^2}{\varepsilon_0} \sum_{k \neq 100} \frac{|\langle u_{100}|z|u_k\rangle|^2}{E_k - E_{100}} \tag{9.25}$$

where ε_0 is the vacuum permittivity.

The summation over the z-matrix elements in Eq. (9.25) can be carried out with the use of the closure property of $\{u_n\}$as given in Eq. (3.24), that is,

$$\sum_{k \neq 100} |\langle u_{100}|z|u_k\rangle|^2 = \sum_k \langle u_{100}|z|u_k\rangle\langle u_k|z|u_{100}\rangle = \langle u_{100}|z^2|u_{100}\rangle \tag{9.26a}$$

In Eq. (9.26a), the ground state u_{100} has been included in the summation over k, as its z-matrix element is zero. The expectation value of z^2 can be calculated as

$$\langle u_{100}|z^2|u_{100}\rangle = \frac{1}{\pi a_0^2}\int_0^{2\pi} d\varphi \int_0^{\pi} \sin\theta d\theta \int_0^{\infty} r^2 dr e^{-(2r/a_0)}(r^2 \cos^2\theta) = a_0^2,$$
$$z = r\cos\theta \tag{9.26b}$$

where a_0 is the Bohr radius (see Eq. (8.46c)), and Table 8.2 has been used for u_{100}. Hence, upon inserting Eq. (9.26b) into Eq. (9.25) and putting $E_k - E_{100} \approx E_{200} - E_{100}$ for all eigenstates k, we find the upper limit of α as

$$\alpha \leq \frac{64\pi a_0^3}{3} \tag{9.27}$$

The polarizability of the atom is an important parameter affecting its optical and electrical properties.

Degenerate Perturbation Theory

The nondegenerate perturbation theory discussed thus far cannot be applied to the degenerate system. This is because in the degenerate system, some of the denominators $E_k - E_m$ appearing in Eq. (9.14b) are bound to be zero, which disrupts the completion of the first-order corrections. A possible way out of this impasse is to exploit the coupling of degenerate states induced by the perturbing Hamiltonian and to find a new set of eigenfunctions with different eigenvalues. Hence, the degenerate perturbation theory is primarily focused on lifting the degeneracy.

Thus, consider two degenerate states u_i, u_j sharing a common eigenvalue E_m and look for the new eigenfunction in the form

$$v_\kappa = c_i u_i + c_j u_j, \quad E_i = E_j = E_m \tag{9.28}$$

When the state u_m chosen for examination in Eq. (9.8) is replaced by v_κ instead, there results

$$(\hat{H}' - W_1)(c_i u_i + c_j u_j) = \sum_n a_n^{(1)}(E_m - E_n)u_n \tag{9.29}$$

Hence, by performing the inner products with respect to u_i and u_j on both sides of Eq. (9.29) and by using the orthonormality of $\{u_n\}$, we find after a straightforward algebra

$$\begin{pmatrix} H'_{ii} - W_1 & H'_{ij} \\ H'_{ji} & H'_{jj} - W_1 \end{pmatrix} \begin{pmatrix} c_i \\ c_j \end{pmatrix} = 0, \quad H'_{\alpha\beta} \equiv \langle u_\alpha | \widehat{H'} | u_\beta \rangle \tag{9.30}$$

The coupled equation (9.30) is homogeneous. Hence c_i, c_j become trivial, rendering the eigenfunction v_k trivial, unless the secular equation is satisfied (see Eq. (6.16)):

$$\begin{vmatrix} H'_{ii} - W_1 & H'_{ij} \\ H'_{ji} & H'_{jj} - W_1 \end{vmatrix} = 0 \tag{9.31}$$

We can readily solve the quadratic equation for W_1 and obtain

$$W_{\pm 1} = \frac{1}{2}\left(h_+ \pm h_- \Lambda^{\frac{1}{2}}\right)^{\frac{1}{2}}, \quad h_\pm = H'_{ii} \pm H'_{jj}, \quad \Lambda = 1 + \frac{4H'_{ij}H'_{ji}}{h_-^2} \tag{9.32}$$

Clearly, the two branches of W_1 represent the splitting of the degenerate energy level E_m caused by the perturbing Hamiltonian. When $W_{\pm 1}$ are inserted back into Eq. (9.30), the two equations become redundant, as has been discussed already. Consequently, we can specify c_j in terms of c_i and use c_i for normalizing v_κ:

$$c_j = \frac{W_{\pm 1} - H'_{ii}}{H'_{ij}} c_i, \quad c_i = \left\{ 1 + \left[\frac{W_{\pm 1} - H'_{ii}}{H'_{ij}} \right]^2 \right\}^{-1} \tag{9.33}$$

In this manner, the degeneracy of two states u_i, u_j has been lifted, and the two new eigenfunctions $v_{\kappa 1}$, $v_{\kappa 2}$ have been found. The theory can be straightforwardly extended to the general case of the n-fold degeneracy.

9.1.1
Stark Effect in H-Atom

We next apply the degenerate perturbation theory for analyzing the Stark effect in H-atom. When an electric field is applied in the z-direction, there ensues the perturbation Hamiltonian given by

$$\hat{H}' = e\mathrm{E}z = e\mathrm{E}r\cos\theta \tag{9.34}$$

Let us examine the effect of $\hat{H}'$ on the first excited state of the H-atom, which has the fourfold degeneracy, u_{200}, u_{210}, u_{211}, and $u_{21\bar{1}}$. The corresponding secular equation reads as

$$\begin{vmatrix} -W_1 & -3e\mathrm{E}a_0 & 0 & 0 \\ -3e\mathrm{E}a_0 & -W_1 & 0 & 0 \\ 0 & 0 & -W_1 & 0 \\ 0 & 0 & 0 & -W_1 \end{vmatrix} = 0 \tag{9.35}$$

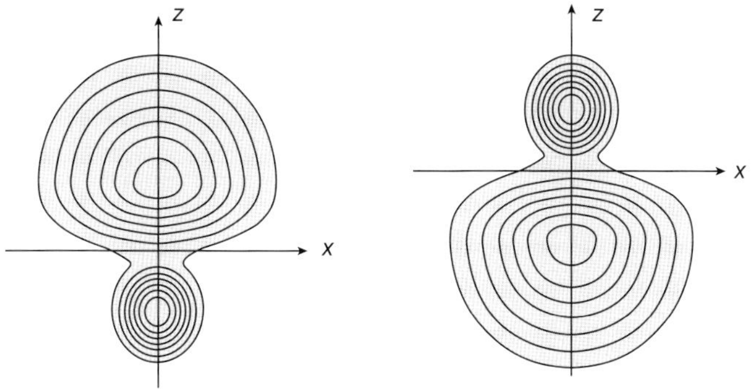

(a) $\nu_{-1} = (u_{200} - u_{210})/\sqrt{2}$ (b) $\nu_{+1} = (u_{200} + u_{210})/\sqrt{2}$

Figure 9.1 Planar view of the probability densities of v_1, v_2 states in the H-atom subjected to the electric field in the z-direction. The probability is commensurate with the number of contours per unit length.

In this case, all off-diagonal matrix elements vanish except for those connecting u_{200} and u_{210}. This means that u_{200} and u_{210} states are coupled via the perturbation, while other states u_{211}, $u_{21\bar{1}}$ still remain uncoupled. Hence, $v_3 = u_{211}$, $v_4 = u_{21\bar{1}}$ but u_{200}, u_{210} are combined to yield new eigenstates v_1, v_2 having different energy eigenvalues.

We can find such shift in the energy level by expanding the determinant Eq. (9.35), obtaining

$$W_1^2 \begin{vmatrix} -W_1 & -3eEa_0 \\ -3eEa_0 & -W_1 \end{vmatrix} = 0 \tag{9.36}$$

Clearly, the reduced secular equation is identical in form to Eq. (9.31). Hence, we can write in strict analogy with Eqs. (9.28), (9.32), and (9.33)

$$v_{\mp 1} = \frac{1}{\sqrt{2}}(u_{200} \mp u_{210}), \quad W_{\mp 1} = \pm 3eEa_0 \tag{9.37}$$

The probability densities of v_+ and v_- are plotted in Figure 9.1. Clearly, the distributions of the electron cloud exhibit the dipole moments of the atom, aligned in parallel and antiparallel directions with respect to the electric field. The dipole moments thus induced in the parallel and antiparallel directions are responsible for the splitting of the degenerate energy level E_2/m.

9.2 Time-Dependent Perturbation Theory

As mentioned, the time-dependent perturbation theory provides the general framework by which to describe the interactions between two dynamical systems. Thus, consider a system with the Hamiltonian

$$\hat{H} = \hat{H}_0 + \lambda \hat{H}'(t) \tag{9.38}$$

where $\hat{H}'(t)$ is the interaction Hamiltonian and λ again denotes the smallness parameter. The Schrödinger equation then reads as

$$i\hbar\frac{\partial}{\partial t}\psi(\underline{r},t) = [\hat{H}_0 + \lambda\hat{H}'(t)]\psi(\underline{r},t) \tag{9.39}$$

We can treat Eq. (9.39) by expanding the wavefunction as usual in terms of the complete set of orthonormal eigenfunctions of the unperturbed Hamiltonian as

$$\psi(\underline{r},t) = \sum_n a_n(t)e^{-i(E_n/\hbar)t}u_n(\underline{r}); \quad \hat{H}_0 u_n = E_n u_n \tag{9.40}$$

Here the expansion coefficient $a_n(t)$ has been taken time dependent to account for the electron making transitions from one state to another driven by the interaction Hamiltonian.

By inserting Eq. (9.40) into Eq. (9.39), there results

$$\begin{aligned} &i\hbar\sum_n \dot{a}_n(t)e^{-i(E_n/\hbar)t}u_n + \sum_n E_n a_n(t)e^{-i(E_n/\hbar)t}u_n \\ &= \sum_n a_n(t)E_n e^{-i(E_n/\hbar)t}u_n + \lambda\hat{H}'(t)\sum_n a_n(t)e^{-i(E_n/\hbar)t}u_n \end{aligned} \tag{9.41}$$

Apparently, the second term on the left-hand side and the first term on the right-hand side of Eq. (9.41) cancel each other out, leaving only two terms to consider. Upon performing the inner product on both sides of Eq. (9.41) with respect to u_k, there results

$$\dot{a}_k = -\frac{i}{\hbar}\lambda\sum_n \hat{H}'_{kn}a_n e^{i\omega_{kn}t}; \quad \omega_{kn} \equiv \frac{E_k - E_n}{\hbar}, \quad H'_{kn} = \langle u_k|H'(t)|u_n\rangle \tag{9.42}$$

where ω_{kn} denotes the transition frequency between u_n and u_k states, and the orthonormality of $\{u_n\}$ has been used.

The problem of solving the Schrödinger equation (9.39) has thus been reduced to obtaining the expansion coefficients as the function of time. For this purpose, let us expand as usual $a_k(t)$ in powers of λ

$$a_k = a_k^{(0)} + \lambda a_k^{(1)} + \lambda^2 a_k^{(2)} + \cdots \tag{9.43}$$

and insert Eq. (9.43) into Eq. (9.42), obtaining

$$\dot{a}_k^{(0)} + \lambda\dot{a}_k^{(1)} + \lambda^2\dot{a}_k^{(2)} + \cdots = \lambda\frac{(-i)}{\hbar}\sum_n \hat{H}'_{kn}e^{i\omega_{kn}t}(a_n^{(0)} + \lambda a_n^{(1)} + \lambda^2 a_n^{(2)} + \cdots) \tag{9.44}$$

Hence, by equating the coefficients of equal powers of λ from both sides of Eq. (9.44), we can write

$$\dot{a}_k^{(j)} = -\frac{i}{\hbar}\sum_n \hat{H}'_{kn}e^{i\omega_{kn}t}a_n^{(j-1)}; \quad \omega_{kn} \equiv \frac{E_k - E_n}{\hbar}, \quad j = 1, 2, 3, \ldots \tag{9.45}$$

In this manner, $a_k(t)$ can be found iteratively to an arbitrary order in λ, given the initial condition $\{a_n^{(0)}\}$. To be specific, let us consider the system initially in the mth eigenstate

$$a_m^{(0)} = 1, \quad a_n^{(0)} = 0, \quad n \neq m \tag{9.46}$$

Then, Eq. (9.45) simply reduces to

$$\dot{a}_k^{(1)} = -\frac{i}{\hbar}\hat{H}'_{km}e^{i\omega_{km}t} \tag{9.47}$$

Harmonic Perturbation

The interaction of the practical interest is the harmonic interaction between two systems. Thus, consider the perturbation oscillating with frequency ω

$$\hat{H}'(t) = \hat{H}'e^{-i\omega t} + \hat{H}'^{*}e^{+i\omega t} \tag{9.48}$$

When Eq. (9.48) is inserted in Eq. (9.47) and the integration in time is carried out, there results

$$\begin{aligned} a_k^{(1)}(t) &= \frac{-i}{\hbar}\int_0^t dt\hat{H}'(t)e^{i\omega_{km}t} \\ &= \frac{-i}{\hbar}\left[\hat{H}'_{km}\frac{e^{i(\omega_{km}-\omega)t}-1}{i(\omega_{km}-\omega)} + \hat{H}'^{*}_{km}\frac{e^{i(\omega_{km}+\omega)t}-1}{i(\omega_{km}+\omega)}\right] \end{aligned} \tag{9.49}$$

We next consider the resonant interaction in which $|\omega_{km}| \approx \omega$. In this case, one of the two terms is dominant and we may disregard the other term and write

$$\begin{aligned} a_k^{(1)}(t) &= \frac{\mp i\hat{H}'_{km}}{\hbar}\frac{e^{\pm i(\omega_a-\omega)t}-1}{i(\omega_a-\omega)} \equiv \frac{\mp i\hat{H}'_{km}e^{\pm i(\omega_a-\omega)t/2}}{\hbar}\frac{\sin[(\omega_a-\omega)t/2]}{(\omega_a-\omega)/2}, \\ \omega_a &= \left|\frac{E_m-E_k}{\hbar}\right| \end{aligned} \tag{9.50}$$

where ω_a is the magnitude of the atomic transition frequency, and the trigonometric identity $\sin x = [\exp(ix) - \exp-(ix)]/2i$ has been used. Hence, the probability of the atomic system being in the state k at time t is given by

$$|a_k^{(1)}(t)|^2 = \frac{|\hat{H}'_{km}|^2}{\hbar^2}\frac{\sin^2[(\omega_a-\omega)t/2]}{[(\omega_a-\omega)/2]^2} \tag{9.51}$$

9.2.1 Fermi's Golden Rule

In the long time limit, the probability can be expressed in terms of the δ-function. One of the representations of the δ-function is given by

$$\delta(\omega_a - \omega) \equiv \lim_{t\to\infty}\frac{2}{\pi}\frac{\sin^2[(\omega_a-\omega)t/2]}{t(\omega_a-\omega)^2} \tag{9.52}$$

so that in the long time limit, Eq. (9.51) can be expressed as

$$|a_k^{(1)}(t)|^2 = \frac{|\hat{H}'_{km}|^2}{\hbar^2}2\pi t\delta(\omega_a-\omega) \tag{9.53}$$

Therefore, the transition rate from the initial m to the final k state is given from Eq. (9.53) by

$$W_{mk} = \frac{d}{dt}|a_k^{(1)}(t)|^2 = \frac{2\pi|\hat{H}'_{km}|^2}{\hbar^2}\delta(\omega_a-\omega) = \frac{2\pi|\hat{H}'_{km}|^2}{\hbar}\delta(|E_k-E_m|-\hbar\omega) \tag{9.54}$$

where use has been made of the identity $\delta(x) = a\delta(ax)$. Equation (9.54) is the celebrated Fermi's golden rule and is extensively utilized for describing the transitions in a variety of physical processes. When the final state consists of quasi-continuous states as sketched in Figure 9.2, the transition rate is given by

$$W_{m\to k} = \sum_k W_{mk} = \frac{2\pi|\hat{H}'_{km}|^2}{\hbar^2}\int_{-\infty}^{\infty}\delta(\omega_a - \omega)\rho(\omega_a)\mathrm{d}\omega_a$$
$$= \frac{2\pi|\hat{H}'_{km}|^2}{\hbar^2}\rho(\omega) \tag{9.55}$$

where the density of states ρ has been introduced for summing over the final states.

Fermi's golden rule Eq. (9.54) has been derived in the long time limit in which

$$\frac{2\pi}{t} \ll \Delta\omega_a \tag{9.56}$$

where $\Delta\omega_a$ is the atomic line width (see Figure 9.2). Otherwise, the probability (Eq. (9.51)) cannot be approximated by the δ-function. In addition, the transition rate has been derived by using the first-order solution of the expansion coefficients. Therefore, in order for the transition rate to be valid, the condition should hold, namely, $|a_k^{(1)}| \ll 1$, which can be specified from Eq. (9.50) by

$$|a_k^{(1)}(t)| \simeq \frac{\hat{H}'_{km}}{\hbar}\frac{\sin[(\omega_a - \omega)t/2]}{(\omega_a - \omega)/2} \approx \frac{\hat{H}'_{km}t}{\hbar} \ll 1 \tag{9.57}$$

Therefore, the limits of validity of Fermi's golden rule is given from Eq. (9.56) and (9.57) by

$$\frac{|\hat{H}'_{km}|}{\hbar} \ll \frac{1}{t} \ll \Delta\nu_a, \quad \Delta\omega_a = 2\pi\Delta\nu_a \tag{9.58}$$

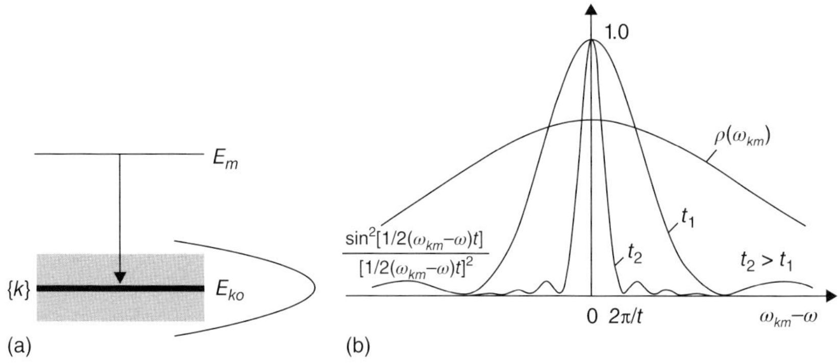

Figure 9.2 The transition from the initial m to final quasi-continuous k states (a). The density of states $\rho(\omega)$ of the final k-states and the frequency profile of the expansion coefficient $|a_k^{(1)}(t)|^2$ at different times (b).

Problems

9.1 The Hamiltonian of an anharmonic oscillator is given by

$$\hat{H} = -\frac{1}{2m}\frac{\partial^2}{\partial x^2} + \frac{1}{2}k_1 x^2 + k_2 x^3 + k_3 x^4; \quad k_1 \gg k_2, k_3$$

Find the first-order corrections in the eigenenergy and eigenfunction.

9.2 Consider an isotropic 2D harmonic oscillator, coupled via a perturbing Hamiltonian:

$$\hat{H} = \left(-\frac{1}{2m}\frac{\partial^2}{\partial x^2} + \frac{1}{2}kx^2\right) + \left(-\frac{1}{2m}\frac{\partial^2}{\partial y^2} + \frac{1}{2}ky^2\right) + \widehat{H'}; \quad \widehat{H'} = Cxy$$

(a) Find the energy eigenfunction and eigenvalue without $\widehat{H'}$.

(b) Find the shift in energy level of the ground and first excited states up to the second-order perturbation analysis.

(c) Introduce new coordinates $\xi = x + y,\ \ \eta = x - y$ and express the total Hamiltonian in terms of ξ, η.

(d) Find energy eigenfunction and eigenvalues in terms of ξ, η and compare the result with those obtained in (b).

9.3 (a) Derive the coupled equation (9.30) from Eq. (9.29) by performing appropriate inner products.

(b) Derive the coupled equation (9.42) from Eq. (9.41) by performing the inner product with respect to u_k.

9.4 Consider an electron in a nanowire in the z-direction with the cross-sectional area on the x–y plane given by $W \times W$. An electric field E is applied in the x-direction.

(a) Find the interaction Hamiltonian and set up the energy eigenequation inside the nanowire.

(b) Find the first-order corrections in the eigenfunction and eigenvalue in the ground state.

9.5 The H-atom is placed in a linearly polarized and circularly polarized electric field varying harmonically in time, so that the perturbing Hamiltonians are given by

$$\hat{H}' = -(-e\underline{E})\cdot\underline{r} = ez\mathrm{E}_0 \cos\omega t; \quad \underline{E} = \hat{z}\mathrm{E}_0 \cos\omega t$$

$$\hat{H}' = e\underline{E}\cdot\underline{r} = ex\mathrm{E}_0 \cos\omega t + ex\mathrm{E}_0 \sin\omega t; \quad \underline{E} = \mathrm{E}_0(\hat{x}\cos\omega t + \hat{y}\sin\omega t)$$

where $\boldsymbol{r}$ is the displacement of the electron from the nucleus.

Given an eigenfunction u_{nl0}, find the final states $u_{n'l'm'}$ to which the electron can make the transition. The condition imposed on n', l', m' is called the *selection rule*.

Hint: Consider the matrix element in Fermi's golden rule.

9.6 Consider a charged 1D harmonic oscillator with the charge-to-mass ratio q/m. The HO is in the nth eigenstate at $t = 0$. A harmonic electric field

$$\mathrm{E}(t) = \mathrm{E}_0 \cos\omega_0 t$$

is applied.

(a) Write down the perturbing Hamiltonian and evaluate the matrix element $\langle u_n|\hat{H}'|u_{n'}\rangle$ and specify the final states to which the transition can occur.

(b) Find the probability that the oscillator makes the transition to those connected final states at $t = \pi/\omega_0$.

Suggested Readings

1. A. Yariv, An Introduction to Theory and Applications of Quantum Mechanics, John Wiley & Sons, 1982.
2. D. M. Kim, Introductory Quantum Mechanics for Semiconductor Nanotechnology, Wiley-VCH, 2010.
3. S. Gasiorowics, Quantum Physics, Third Edition, John Wiley & Sons, 2003.
4. A. I. M. Rae, Quantum Mechanics, Fourth Edition, Taylor & Francis, 2002.

10
System of Identical Particles and Electron Spin

The spin is one of the defining characteristics of the electron and is discussed in conjunction with the system of identical particles. The helium atom is chosen for discussion as a prototypical example of the two spin 1/2 system. Also, the multi-electron atoms are briefly considered by using the H-atom theory, exclusion principle, and the periodic table. Additionally, the fine structure in atomic spectral lines is analyzed in correlation with the spin – orbit coupling, Zeeman splitting, and electron paramagnetic resonance.

10.1
Electron Spin

It has been shown experimentally that the electron possesses two spin states, spin-up and spin-down. The spin critically affects the physical and chemical properties of the atoms and molecules and is also responsible for the fine structures observed in spectral lines. We can treat the spin operators and spin states in analogy with the angular momentum operators and its eigenfunctions. We thus introduce

$$\hat{s}_z \chi_\pm = \left(\pm\frac{1}{2}\hbar\right)\chi_\pm \tag{10.1}$$

$$\hat{s}^2 \chi_\pm = \frac{1}{2}\left(\frac{1}{2}+1\right)\hbar^2\chi_\pm = \frac{3}{4}\hbar^2\chi_\pm \tag{10.2}$$

where $\hat{s}_z$, $\hat{s}^2$, and $\chi_\pm$ correspond to $\hat{l}_z$, $\hat{l}^2$, and Y_l^m (see Eq. (8.24)).

Figure 10.1 shows the two spin states: spin-up and spin-down. We can also introduce the spin flip operators $\hat{s}_+, \hat{s}_-$, which flip the spin-down state to spin-up state and vice versa

$$\hat{s}_+\chi_- = \frac{\hbar}{2}\chi_+, \quad \hat{s}_+ \equiv \frac{1}{2}(\hat{s}_x + i\hat{s}_y), \quad \hat{s}_+\chi_+ = 0 \tag{10.3a}$$

$$\hat{s}_-\chi_+ = \frac{\hbar}{2}\chi_-, \quad \hat{s}_- \equiv \frac{1}{2}(\hat{s}_x - i\hat{s}_y), \quad \hat{s}_-\chi_- = 0 \tag{10.3b}$$

where the operators $\hat{s}_x$ and $\hat{s}_y$ correspond to $\hat{l}_x$ and $\hat{l}_y$, respectively. The spin functions are orthonormal as the spherical harmonics, that is,

$$\langle\chi_\pm|\chi_\pm\rangle = 1, \quad \langle\chi_+|\chi_-\rangle = \langle\chi_-|\chi_+\rangle = 0 \tag{10.4}$$

Introductory Quantum Mechanics for Applied Nanotechnology, First Edition. Dae Mann Kim.

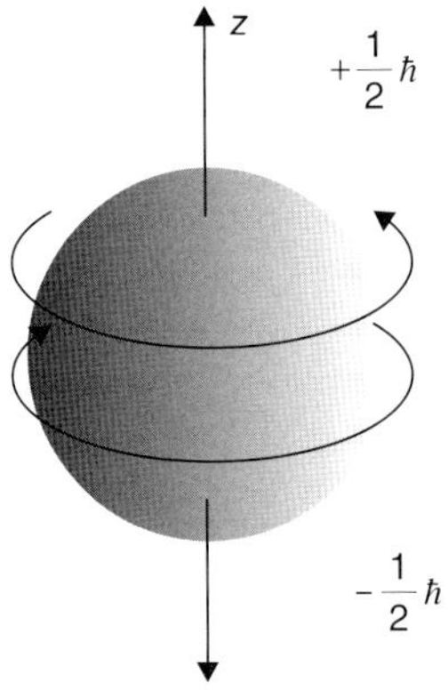

Figure 10.1 The spin-up and spin-down states of the electron.

and the commutation relations involving $\widehat{s}_x, \widehat{s}_y$, and $\widehat{s}_z$ are the same as those involving $\widehat{l}_x, \widehat{l}_y$, and $\widehat{l}_z$ and can thus be summarized with the use of Eq. (8.8) as

$$\widehat{s} \times \widehat{s} = i\frac{\hbar}{2}\widehat{s} \tag{10.5}$$

10.1.1 Pauli Spin Matrices

The basic properties of the spin are conveniently described by the 2×2 Pauli spin matrices. In this matrix representation, the spin-up and spin-down states are denoted by the column vectors

$$\chi_+ = \begin{pmatrix} 1 \\ 0 \end{pmatrix}, \quad \chi_- = \begin{pmatrix} 0 \\ 1 \end{pmatrix} \tag{10.6}$$

and the spin operators are represented by

$$\underline{\widehat{s}} \equiv \frac{\hbar}{2}\underline{\sigma}; \quad \sigma_x = \begin{pmatrix} 0 & 1 \\ 1 & 0 \end{pmatrix}, \quad \sigma_y = \begin{pmatrix} 0 & -i \\ i & 0 \end{pmatrix}, \quad \sigma_z = \begin{pmatrix} 1 & 0 \\ 0 & -1 \end{pmatrix} \tag{10.7}$$

The spin flip operators then read as

$$\widehat{s}_+ = \frac{\hbar}{2}\left[\frac{1}{2}\left(\sigma_x + i\sigma_y\right)\right] = \frac{\hbar}{2}\begin{pmatrix} 0 & 1 \\ 0 & 0 \end{pmatrix}; \quad \widehat{s}_- = \frac{\hbar}{2}\left[\frac{1}{2}\left(\sigma_x - i\sigma_y\right)\right] = \frac{\hbar}{2}\begin{pmatrix} 0 & 0 \\ 1 & 0 \end{pmatrix} \tag{10.8}$$

These spin matrices describe the properties of the spin given in Eqs. (10.1)–(10.5) by the simple matrix algebra.

10.2 Two-Electron System

Let us consider a system of two electrons bound to a common nucleus. Classically, it is possible to distinguish identical particles, but in quantum mechanics, it is not possible to distinguish configurations of identical particles by exchange of particles. This is evident from the fact that an electron is essentially a charge cloud, and it is not possible to disentangle the overlapped charge cloud and to identify each electron.

The Hamiltonian of a system with two noninteracting electrons is given by

$$\hat{H}(1,2) = \hat{H}(1) + \hat{H}(2), \quad j \equiv \underline{r}_j, \quad j = 1,2 \tag{10.9}$$

where each electron is in an eigenstate of the hydrogenic subsystem

$$\hat{H}(j)u_\gamma(j) = E_\gamma u_\gamma(j), \quad \gamma = \alpha, \beta \tag{10.10}$$

with α, β denoting the quantum numbers n, l, m. Then, the product of wavefunctions

$$\varphi_{\alpha\beta}(i,j) = u_\alpha(i)u_\beta(j), \quad i,j = 1,2 \tag{10.11a}$$

satisfy the eigenequation

$$\hat{H}(1,2)\varphi_{\alpha\beta}(i,j) = (E_\alpha + E_\beta)\varphi_{\alpha\beta}(i,j) \tag{10.11b}$$

Also the symmetric and antisymmetric combinations

$$\varphi_\gamma = \frac{1}{\sqrt{2}}[\varphi_{\alpha\beta}(1,2) \pm \varphi_{\alpha\beta}(2,1)], \quad \gamma = s, a \tag{10.12}$$

qualify as the eigenfunctions with the same eigenvalue.

Fermions and Bosons: Electrons belong to the group of particles called *fermions*, having half odd integer spins $\hbar/2, 3\hbar/2, 5\hbar/2$, and so on. Protons and neutrons are also well-known fermions. The fermions are constrained by the Pauli exclusion principle, which prohibits two fermions to occupy simultaneously a common quantum state. Bosons constitute another group of particles with integer spins $\hbar, 2\hbar, 3\hbar$, and so on, and are free of the exclusion principle. Photons, deuterons, and alpha particles are typical examples.

Slater determinant: The fermions are described by antisymmetric wavefunction, which is conveniently represented by the Slater determinant. For the two-electron system, the eigenfunction is represented by 2×2 determinant as

$$\varphi_a(1,2) = \frac{1}{\sqrt{2!}}\begin{vmatrix} u_\alpha(1) & u_\alpha(2) \\ u_\beta(1) & u_\beta(2) \end{vmatrix} = \frac{1}{\sqrt{2!}}[u_\alpha(1)u_\beta(2) - u_\alpha(2)u_\beta(1)] \tag{10.13}$$

Likewise, the wavefunction of N noninteracting fermions is described by $N \times N$ determinant. In this representation, if two quantum numbers are identical, that is, $\alpha = \beta$, the determinant vanishes by definition and is consistent with the exclusion principle.

10.2.1
Helium Atom

The He-atom consists of two electrons bound to the common nucleus with two protons $Z = 2$ (Figure 10.2). The Hamiltonian is thus given by

$$\hat{H} = \sum_{j=1}^{2}\left(-\frac{\hbar^2}{2\mu}\nabla_j^2 - \frac{Ze_M^2}{r_j}\right) + \hat{H}_{12};\quad \hat{H}_{12} = \frac{e_M^2}{r_{12}},\quad Z = 2 \tag{10.14}$$

where the terms in the parenthesis account for two electrons bound to the common nucleus with two protons ($Z = 2$), and the second term represents the repulsive Coulomb interaction between the two electrons.

Singlet and Triplet States

The two electrons as two Fermions should be described by the antisymmetric wavefunction. To construct such wavefunctions, it is convenient to symmetrize and antisymmetrize the two spin states (Figure 10.3):

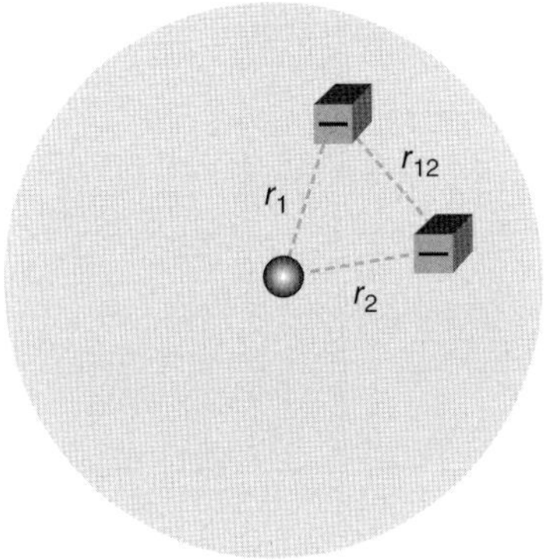

Figure 10.2 The helium atom with two protons in the nucleus ($Z = 2$) and two electrons outside the nucleus. The volume elements of the two-electron charge cloud are also shown.

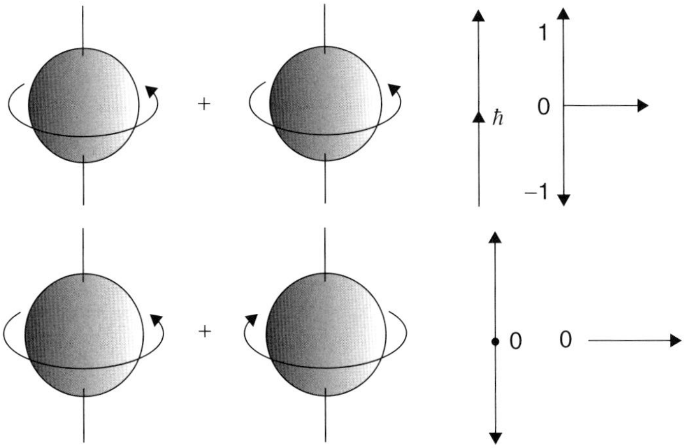

Figure 10.3 The triplet state with $m = -1, 0, 1$ and the singlet state with $m = 0$.

$$\chi_s = \begin{cases} \chi_+(1)\,\chi_+(2) \\ \frac{1}{\sqrt{2}}[\chi_+(1)\chi_-(2) + \chi_-(1)\chi_+(2)] \\ \chi_-(1)\chi_-(2) \end{cases} \tag{10.15a}$$

$$\chi_a = \frac{1}{\sqrt{2}}[\chi_+(1)\chi_-(2) - \chi_-(1)\chi_+(2)] \tag{10.15b}$$

so that

$$s_z\chi_s \equiv (s_{1z} + s_{1z})\chi_s = m\hbar\chi_s, \quad m = 1, 0, -1 \tag{10.16a}$$

$$s_z\chi_a \equiv (s_{1z} + s_{1z})\chi_a = m\hbar\chi_a, \quad m = 0 \tag{10.16b}$$

Clearly, χ_s has three projections onto the z-axis, while χ_a has a single projection, and it is for this reason that χ_s, χ_a are called the *triplet* and *singlet states*, respectively.

Ground State

The ground state of the He-atom consists of two electrons in the respective ground state of two hydrogenic subsystems, and the wavefunction is thus given by

$$\varphi_0(1,2) = u_{100}(\underline{r}_1)u_{100}(\underline{r}_2)\chi_a \tag{10.17}$$

Since $u_{100}(\boldsymbol{r}_1)u_{100}(\boldsymbol{r}_2)$ is symmetric, the singlet state χ_a has to be combined to make the total wavefunction antisymmetric. The ground state energy is then given to the first order of approximation by

$$E_0^{(1)} = \langle\varphi_0(1,2)|\sum_{j=1}^{2}\hat{H}_j + \hat{H}_{12}|\varphi_0(1,2)\rangle = E_0 + \Delta E_0 \tag{10.18a}$$

where the first term

$$E_0 = \sum_{j=1}^{2}\langle u_{100}(j)|\left(-\frac{\hbar^2}{2\mu}\nabla_j^2 - \frac{Ze_M^2}{r_j}\right)|u_{100}(j)\rangle = 2\times\left(-\frac{Z^2e_M^2}{2a_0}\right) \tag{10.18b}$$

is the total energy of two electrons in their respective ground states, while ΔE_0 accounts for the first-order level shift due to $\hat{H}_{12}$ (see Eq. (9.13)).

$$\Delta E_0 = \langle\varphi_0(1,2)|\frac{e_M^2}{r_{12}}|\varphi_0(1,2)\rangle \tag{10.18c}$$

The evaluation of Eq. (10.18c) is facilitated by the fact that $1/r_{12}$ is the generating function of the Legendre polynomial:

$$\frac{1}{r_{12}} \equiv \frac{1}{|\underline{r}_1 - \underline{r}_2|} = \frac{1}{r_>}\left[1 + w\frac{r_<}{r_>} + \frac{1}{2}\left(3w^2 - 1\right)\left(\frac{r_<}{r_>}\right)^2 + \cdots\right], \quad w = \cos\theta \tag{10.19}$$

where $r_>$, $r_<$ denote the greater and lesser of r_1, r_2 and the expansion coefficient of $(r_</r_>)^n$ is the nth order Legendre polynomial $P_n(w)$ and θ the angle between

r_1, r_2. Let us take $\boldsymbol{r}_1$ parallel to the z-axis when carrying out the $\boldsymbol{r}_2$-integration in Eq. (10.18c) by using Eq. (10.19). Then, θ becomes the polar angle, in which case only the first term in Eq. (10.19) $\propto P_0(w)$ contributes to the θ-integration. This is because the Legendre polynomials are orthonormal (see Eq. (8.22)), and $P_0(w)$ is a constant. Therefore, the angular integration yields $(4\pi)^2$, and we can perform the radial integration, obtaining

$$\Delta E_0 = \frac{e_M^2}{\pi^2}\left(\frac{Z}{a_0}\right)^6 (4\pi)^2 \int_0^\infty r_1^2 dr_1 e^{-\frac{2Zr_1}{a_0}} \left[\frac{1}{r_1}\int_o^{r_1} r_2^2 dr_2 e^{-\frac{2Zr_2}{a_0}} + \int_{r_1}^\infty r_2 dr_2 e^{-\frac{2Zr_2}{a_0}}\right]$$
$$= \frac{5Ze_M^2}{8a_0} \tag{10.20}$$

where the $\boldsymbol{r}_2$-integration was carried out in two regions $r_2 \leq r_1$ and $r_2 \geq r_1$.

Ionization Energy

The ground state energy of the He-atom is obtained by combining Eqs. (10.18b) and (10.20):

$$E_0 = 2 \times \left(-\frac{Z^2 e_M^2}{2a_0}\right) + \frac{5Ze_M^2}{8a_0} = -\frac{Ze_M^2}{a_0}\left(Z - \frac{5}{8}\right), \quad Z = 2 \tag{10.21}$$

Without the repulsive interaction between the two electrons, E_0 consists of two ground state energies of the hydrogenic atom. Then, the first and second ionization energies IP_1, IP_2 responsible for He → He^+ + e, He^+ → He^{++} + e should be the same and is given from Eq. (8.46) by

$$\mathrm{IP}_1 = \mathrm{IP}_2 \equiv \left|-\frac{e_M^2}{2a_0} Z^2\right| = 13.6 \times 2^2\,\mathrm{eV} = 54.4\,\mathrm{eV} \tag{10.22}$$

However, the measured data of IP_1 is 24.6 eV while that of IP_2 is 54.4 eV. It is therefore clear that there is a good agreement between theory and experiment with regard to IP_2. This is expected because with one electron left alone after the first ionization, He^+ becomes identical to the H-atom with $Z = 2$, and IP_2 can therefore be precisely quantified by the H-atom theory Eq. (8.46).

The fact that the IP_1-data is smaller than 54.4 eV is explained as follows. The first ionization involves two processes, namely, one electron is removed from the ground state to the vacuum level, while the other forms a hydrogenic atom with $Z = 2$. Therefore, IP_1 is by definition the energy required to boost the ground state energy of the He-atom to that of hydrogenic atom He^+, that is,

$$\mathrm{IP}_1 \equiv -\frac{Z^2 e_M^2}{2a_0} - \left[-\frac{Ze_M^2}{a_0}\left(Z - \frac{5}{8}\right)\right] = \frac{Z_{\mathrm{eff}}^2 e_M^2}{2a_0}, \quad Z_{\mathrm{eff}}^2 = Z\left(Z - \frac{5}{4}\right) \tag{10.23}$$

With $Z = 2$, IP_1 amounts to about 20.46 eV, in better agreement with the data. It is therefore clear that the smaller IP_1 compared with IP_2 is due to the screening of the nuclear charge by one electron to the other, which is being removed for the first ionization. Consequently, Z_{eff} is less than Z. The screening is brought about by the repulsive Coulomb interaction between the two electrons.

The First Excited State

The first excited state of the He-atom consists of one electron in the ground state, while the other in the first excited state of the hydrogenic atom. The wavefunctions are thus given by

$$\varphi_s = \frac{1}{\sqrt{2}}[u_{1s}(1)u_{2s}(2) + u_{1s}(2)u_{2s}(1)]\chi_a \tag{10.24a}$$

$$\varphi_a = \frac{1}{\sqrt{2}}[u_{1s}(1)u_{2s}(2) - u_{1s}(2)u_{2s}(1)]\chi_s \tag{10.24b}$$

The symmetric and antisymmetric energy eigenfunctions have to be combined with the singlet and triplet states Eq. (10.15) to make the total wavefunction antisymmetric.

Overlap and Exchange Integrals

For φ_s, the energy of the first excited state is given to the first order of approximation by

$$E_{0s}^{(1)} = \langle\varphi_s|\hat{H}_1 + \hat{H}_2 + \hat{H}_{12}|\varphi_s\rangle = -\frac{Z^2e_M^2}{2a_0} - \frac{Z^2e_M^2}{2a_0}\frac{1}{4} + \Delta E_s \tag{10.25}$$

where the inner product of χ_a with itself yields unity, and the first two terms correspond to the ground and first excited states of the hydrogenic subsystem. The third term ΔE_s accounts for the repulsive electron–electron interaction and is given by

$$\begin{aligned}\Delta E_s &= \frac{1}{2}\langle u_{1s}(1)u_{2s}(2) + u_{1s}(2)u_{2s}(1)|\hat{H}_{12}|u_{1s}(1)u_{2s}(2) + u_{1s}(2)u_{2s}(1)\rangle \\ &= J + K\end{aligned} \tag{10.26}$$

with J and K denoting the integrals

$$J = \iint d\underline{r}_1 d\underline{r}_2 u_{1s}^2(1)\frac{e_M^2}{r_{12}}u_{2s}^2(2) > 0 \tag{10.27}$$

$$K = \iint d\underline{r}_1 d\underline{r}_2 u_{1s}(1)u_{2s}(2)\frac{e_M^2}{r_{12}}u_{1s}(2)u_{2s}(1) > 0 \tag{10.28}$$

The four integrals in Eq. (10.26) are reduced to J- and K-integrals upon interchanging the variables of integration $\mathbf{r}_1, \mathbf{r}_2$.

The J-integral is known as the *overlap integral* and represents the repulsive Coulomb interaction between the two electrons in 1s and 2s states, respectively. The integrand of the K-integral consists of two products of $u_{1s}\, u_{2s}$ in which the two electrons are exchanged. The integrand results from symmetrizing or antisymmetrizing the wavefunctions, and the K-integral is called the *exchange integral*. We can carry out a similar analysis for the triplet state. Hence, the energy eigenvalues of the singlet and triplet states are given by

$$E_{0s}^{(1)} = E_0 + J + K; \quad E_{0t}^{(1)} = E_0 + J - K \tag{10.29}$$

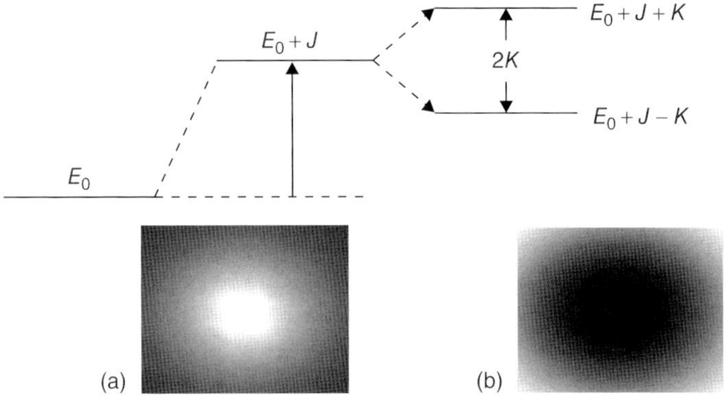

Figure 10.4 The splitting of the energy level of the first excited state of the He-atom. The higher and lower levels correspond to the symmetrized and singlet state and antisymmetrized and triplet state, respectively. Also shown are the electron charge clouds of symmetrized (a) and antisymmetrized (b) states. The probability density is commensurate with the degree of brightness.

It is therefore clear that the energy of the first excited state associated with the triplet state is lower than that of the singlet state by the amount 2 K as shown in Figure 10.4. This is due to the fact that in the singlet state, the symmetric combination of u_{1s}, u_{2s} renders the probability densities high when the two electrons are close to each other (Figure 10.4). But the corresponding probability density is minimal for the triplet state, in which u_{1s}, u_{2s} are combined antisymmetrically. Thus, the repulsive interaction between two electrons is accounted for more effectively in the singlet state, hence higher energy level. In this way, the spin states critically affect the energy level, although the spin functions do not enter in the evaluation of the energy level.

10.2.2
Multi-Electron Atoms and Periodic Table

The Electron Configuration

We next discuss the periodic table based on the H-atom theory and the exclusion principle. The atomic structures are systematically organized in the periodic table. The general features of the table are as follows: (i) it consists of rows, called *periods*, which are comprised of 2, 8, 8, 18, 32, 32 elements from top to down and (ii) the elements in the same column or group exhibit similar properties, including IP_1 and other parameters. The quantum states in the multi-electron atoms are labeled by the quantum numbers n, l, and s. Also, the number of quantum states in one sub-shell for given l is specified by the combination of two spin states and magnetic quantum numbers varying from $-l$ to l in steps of unity.

The energy level of a quantum state is determined primarily by the principal quantum number n, but for given n, the level further splits into the sublevels, depending on l. The sub-shell energy is raised with increasing l or equivalently centrifugal force. The electrons in the multi-electron atom fill the quantum states, one by one according to the exclusion principle, starting from the lowest energy level. For given n electrons fill the sub-shells with l ranging 1, 2, 3 and so on. These states are denoted by s, p, d and so forth. The process goes on until all electrons are assigned to the quantum states available.

The electronic and chemical properties of atoms are mainly determined by valence electrons in the outermost atomic orbital. Also the separation between the highest occupied energy level and lowest unoccupied level on top of it is an important parameter. The specification of electrons with the use of the quantum numbers n, l is called the *electron configuration*. For instance, the electron in the ground state of the H-atom is specified by $n = 1$, $l = 0$, or $1s^1$. Likewise, the two electrons in He-atom are denoted by $1s^2$, which also indicates that the s sub-shell is filled up by two electrons with spin-up and -down states, respectively.

The atoms in the second row of the table starts from Li and ends with Ne and the ground state electron configurations are [He]$2s$, [He]$2s^2$, [He]$2s^22p$, [He]$2s^22p^2$, [He]$2s^22p^3$, [He]$2s^22p^4$, [He]$2s^22p^5$, [He]$2s^22p^6$ for Li, Be, B, C, N, O, F, Ne, respectively. Understandably, the configuration $1s^2$ is often denoted by [He]. A similar electron configuration follows for the third period starting with Na and ending with Ar with Ne serving as the main core (Table 10.1).

First Ionization Energy IP_1 and Electron Affinity

IP_1 is an important parameter of the atom. For H-atom, IP_1 is the energy necessary to release a single electron from the ground state to the vacuum level. For He-atom with two electrons, two ionization energies IP_1, IP_2 are involved as discussed. Given an atom, IP_1 is responsible for the process $A \rightarrow A^+ + e$. The inverse process $A + e \rightarrow A^-$ is associated with the energy called *electron affinity* (*EA*). The EA is the energy released by a free electron at rest when it is captured by a neutral atom into a bound state.

Typical data of IP_1 are shown in Table 10.1. The data clearly indicate that IP_1 increases across a given period, but it drops sharply, as the next period begins. For example, IP_1 of 5.39 eV for Li is much smaller than IP_1 of 24.58 eV for He, although the Li atom has one more proton in the nucleus than the He atom. The behavior

Table 10.1 The first two periods in periodic table, showing the ground state configurations and the first ionization potentials.

H $1s^1$ IP_1 13.595							He $1s^2$ 24.580
Li $2s^1$ 5.390	Be $2s^2$ 9.320	B $2s^2 2p^1$ 8.296	C $2s^2 2p^2$ 11.264	N $2s^2 2p^3$ 14.54	O $2s^2 2p^4$ 13.614	F $2s^2 2p^5$ 17.42	Ne $2s^2 2p^6$ 21.559

of IP_1 can be interpreted, based on the ionization energy of the hydrogenic atom and the screening of the nuclear charge:

$$IP_1 \simeq Z_{\text{eff}}^2 \frac{e_M^2}{2a_0} \frac{1}{n^2}$$

For Li atom, the valence electron in the outermost atomic orbital is in $2s$ state, and the measured IP_1 of 5.39 eV indicates efficient screening of the nuclear charge of three protons by two inner-lying $1s$ electrons. With increasing Z and increasing number of protons in the same period, all electrons added reside in the same sub-shell in the outer orbital. Consequently, the efficiency of screening by these added electrons is low for one of the valence electrons released for the first ionization. Hence, IP_1 increases steadily until the closed shell atom of Ne is reached. With the beginning of new period, starting with Na atom, however, the valence electron is in 3s state alone, while the rest of the electrons fill up the inner lying sub-shells, screening efficiently the nuclear charge. As a result, IP_1 again drops sharply and becomes comparable with that of Li.

10.3 Interaction of Electron Spin with Magnetic Field

It has been found experimentally that in a magnetic field $\boldsymbol{B}$, the frequency of radiation emitted by the H-atom is shifted from the frequency emitted without $\boldsymbol{B}$. Moreover, the magnetic field causes some of the spectral lines to split. The effect is known as *Zeeman splitting*, and these phenomena are discussed semiclassically.

Orbital Magnetic Moments

Thus, consider an electron moving in the circular orbit around the nucleus in the presence of a time-varying magnetic field. The work done on the electron by an electric field E entailed in such a motion is given by

$$W_s \equiv -e \int_C \underline{E} \cdot d\underline{s} = -e \int_S (\nabla \times \underline{E}) \cdot \hat{n} da = e\frac{\partial}{\partial t} \int_S \underline{B}(t) \cdot \hat{n} da \tag{10.30}$$

where W_s is the work done per revolution, $\hat{n}$ the unit vector normal to the orbit plane, and $\boldsymbol{ds}$ the differential line vector along the contour C. The line integral is converted to the surface integral via Stokes theorem in the first equality, and the second one simply reiterates Faraday's law of induction (Eq. (1.21)).

An electron moving with the linear momentum p completes the revolution in the circular orbit of radius r in the time period $T = 2\pi r/(p/m)$. Hence, for a spatially homogeneous $\boldsymbol{B}(t)$, the work dW done on the electron in dt is given by W_s times the number of revolutions made during dt:

$$dW = W_s \times \frac{dt}{T} = \frac{erp}{2m} dB_n(t), \quad W_s = e\pi r^2 \frac{\partial B_n(t)}{\partial t} \tag{10.31}$$

where B_n is the normal component of $\boldsymbol{B}$. Hence, a simple time integration yields

$$W \equiv \hat{H}'_{lm} = \frac{e}{2m}\hat{\underline{l}} \cdot \underline{B}, \quad \hat{\underline{l}} = \underline{r} \times \underline{p} \tag{10.32}$$

In this way, the orbital angular momentum $\boldsymbol{l}$ of the electron naturally enters in the interaction Hamiltonian and singles out B_n by means of the scalar product with $\boldsymbol{B}$. The Hamiltonian can also be expressed in terms of the orbital magnetic moment $\boldsymbol{\mu}_l$ as

$$\hat{H}'_{lm} = -\hat{\underline{\mu}}_l \cdot \underline{B}; \quad \hat{\underline{\mu}}_l \equiv -g_l \frac{e}{2m}\hat{\underline{l}} = -g_l \mu_B \frac{\hat{\underline{l}}}{\hbar}, \quad \mu_B \equiv \frac{e\hbar}{2m}, \quad g_l = 1 \tag{10.33}$$

The magnitude of $\boldsymbol{\mu}_l$ is by definition equal to the product of the orbit area πr^2 and the current $-e/T$, and $\boldsymbol{\mu}_l$ and $\boldsymbol{l}$ are directed in opposite directions due to the negative electron charge. The constant g_l connecting $\boldsymbol{l}$ and $\boldsymbol{\mu}_l$ is called the *gyromagnetic ratio* and is equal to unity for l, and the Bohr magnetron μ_B has the value $9.272 \times 10^{-24}\ (\mathrm{J\,m^2})\,\mathrm{Wb}^{-1}$, with Wb denoting the Weber.

The coupling of the electron spin with $\boldsymbol{B}$ can likewise be expressed in strict analogy with Eq. (10.33) as

$$\hat{H}'_{sm} = -\hat{\underline{\mu}}_s \cdot \underline{B}, \quad \hat{\underline{\mu}}_s \equiv -g_s \mu_B \frac{\hat{\underline{s}}}{\hbar}, \quad g_s = 2 \tag{10.34}$$

where the gyromagnetic ratio g_s is experimentally found twice as large as g_l. The discrepancy between g_l and g_s is referred to as the *magnetic spin anomaly*. Let us take $\boldsymbol{B}$ in the z-direction and express the total interaction Hamiltonian as

$$\hat{H}'_m = B g_l \mu_B \left(\frac{\hat{l}_z}{\hbar} + 2\frac{\hat{s}_z}{\hbar} \right) \tag{10.35}$$

10.3.1 Spin–Orbit Coupling and Fine Structure

An electron moving in a circular orbit around the nucleus generates its own magnetic field $\boldsymbol{B}$ at its site. The $\boldsymbol{B}$-field in turn induces the spin–orbit coupling as follows. Thus, consider an electron moving in circular orbit with a velocity $\boldsymbol{v}$ at $\boldsymbol{r}$ displacement from the nucleus. The electron motion is equivalent to the nucleus moving around the electron at $-\boldsymbol{r}$ with charge Ze in the reference frame in which

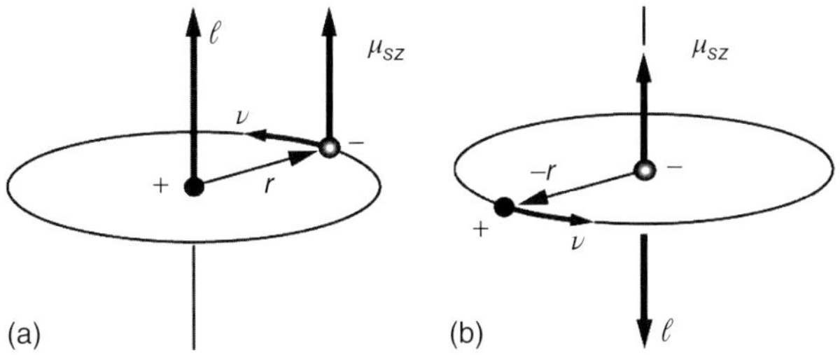

Figure 10.5 The circular motion of an electron around the proton (a). The equivalent circular motion of the proton around the electron (b).

the electron is at rest as shown in Figure 10.5. The magnetic field $\boldsymbol{B}$ generated at the site of the electron can in turn be transformed back to the center of mass frame in which the nucleus is practically at rest:

$$\underline{B}_l = \frac{1}{2}\left\{\frac{Ze\mu_0}{4\pi r^3}\left[\underline{v}\times(-\underline{r})\right]\right\} = \frac{Ze\mu_0}{8\pi r^3 m_e}\underline{l}, \quad \underline{l} \equiv \underline{r}\times m_e\underline{v} \tag{10.36}$$

Equation (10.36) is the well-known Biot–Savart law and μ_0, m_e, $\boldsymbol{l}$ are the vacuum permeability, the rest mass, and angular momentum of the electron, respectively. The factor 1/2 entering in the back transformation is called the *Thomas factor*.

The $\boldsymbol{B}$-field thus generated interacts in turn with the electron spin magnetic moment $\boldsymbol{\mu}_s$ and yields the interaction Hamiltonian given from Eqs. (10.34) and (10.36) by

$$\hat{H}'_{so} \equiv -\underline{\mu}_s \cdot \underline{B}_l = \frac{g_s}{\hbar}\mu_B f(r)(\hat{\underline{s}}\cdot\hat{\underline{l}}), \quad f(r) = \frac{Ze^2\mu_0}{8\pi m_e^2 r^3} \tag{10.37a}$$

Or with the use of the vector identity involving the total angular momentum $\boldsymbol{j}$

$$j^2 \equiv (\underline{l}+\underline{s})\cdot(\underline{l}+\underline{s}) = l^2 + s^2 + 2\underline{s}\cdot\underline{l}$$

the spin orbit coupling (Eq. (10.37a)) can be expressed in terms of $\boldsymbol{j}$ as

$$\hat{H}'_{so} \equiv \frac{g_s}{2\hbar}\mu_B f(r)(\hat{j}^2 - \hat{l}^2 - \hat{s}^2) \tag{10.37b}$$

Naturally, the radius r of the circular orbit has to be treated by its expectation value. The total interaction Hamiltonian of the H-atom under $\boldsymbol{B}$ is given from Eqs. (10.35) and (10.37) by

$$\hat{H}_{\text{int}} = Bg_l\mu_B\left(\frac{\hat{l}_z}{\hbar} + 2\frac{\hat{s}_z}{\hbar}\right) + \frac{g_s}{2\hbar}\mu_B f(r)(\hat{j}^2 - \hat{l}^2 - \hat{s}^2) \tag{10.38}$$

Fine Structure of Spectral Lines

We next consider the effects of the spin–orbit coupling in the absence of $\boldsymbol{B}$. In this case, we can introduce the eigenfunction of $\hat{j}$ in analogy with that of $\hat{l}$ (see Eq. (8.24)):

$$\hat{j}^2|j,m_j\rangle = \hbar^2 j(j+1)|j,m_j\rangle, \quad |j,m_j\rangle = Y_j^{m_j} \tag{10.39a}$$

$$\hat{j}_z|j,m_j\rangle = m_j|j,m_j\rangle, \quad m_j = -j, -j+1, \ldots, j-1, j \tag{10.39b}$$

The detailed derivation of Eq. (10.39) is given in the reference books listed at the end of this chapter. The eigenfunction can also be used as the eigenfunction of $\hat{l}$, $\hat{s}$, as the three angular momentum operators commute.

Therefore, the shift in the energy level due to the spin–orbit coupling can be found by using Eqs. (10.38) and (10.39) as

$$\delta E_{so} \equiv \langle j,m_j|\hat{H}'_{so}|j,m_j\rangle = \frac{g_s\mu_B}{2}\langle f(r)\rangle[j(j+1) - l(l+1) - s(s+1)] \tag{10.40}$$

where the expectation value of $f(r)$ has to be evaluated to the first order of approximation by using the radial part of the wavefunction (Eq. (8.49)). The allowed values

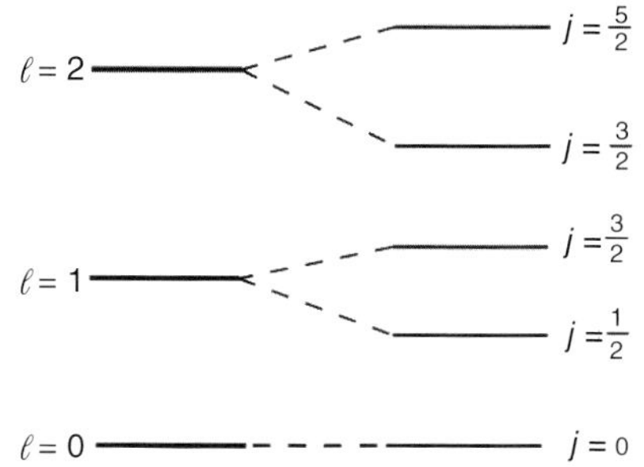

Figure 10.6 The spin–orbit coupling-induced splitting of the energy level of the one-electron atom.

of j for given l and s are $j = l \pm s$, and the associated m_j varies from $-j$ to j in steps of unity. For a single electron, $s = 1/2$, hence $j = l \pm 1/2$. The energy level for given l then splits into two according to Eq. (10.40), one raised while the other lowered by amounts proportional to l and $l+1$, respectively. For $l = 0$, however, $j = s = 1/2$, and there is no splitting. Figure 10.6 shows the energy levels corresponding to $l = 0, 1, 2$.

10.3.2 Zeeman Effect

When a strong magnetic field $\boldsymbol{B}$ is applied, the spin–orbit coupling can be neglected. In this case, the splitting of the energy level due to $\boldsymbol{B}$ is found precisely by using Eq. (10.38) and the unperturbed eigenfunction as

$$\begin{aligned}\Delta E_m &= \frac{eB}{2m_e}\langle u_{nlm}\chi_\pm|\hat{l}_z + 2\hat{s}_z|u_{nlm}\chi_\pm\rangle \\ &= \frac{eB\hbar}{2m_e}(m_l \pm m_s)\end{aligned} \tag{10.41}$$

where m_l, m_s are the magnetic and spin quantum numbers, respectively (see Eqs. (8.50), (10.1)). For $l = 1$, the possible values of m_l are ± 1 and 0, while those of m_s are $\pm 1/2$. Hence, the state with given l and s splits into $(2l+1)(2s+1)$ equally spaced levels. In addition, the spin–orbit coupling can be incorporated as a perturbing term via the perturbation theory. Thus, with the use of Eq. (10.37a), the first-order level shift is given from Eq. (9.13) by

$$\begin{aligned}\delta E_{so} &= \frac{g_s}{\hbar}\mu_B\langle u_{nlm}\chi_\pm|f(r)(\hat{\underline{l}}\cdot\hat{\underline{s}})|u_{nlm}\chi_\pm\rangle \\ &= \frac{g_s}{\hbar}\mu_B\langle f(r)\rangle_{nlm}\hbar^2 m_l m_s\end{aligned} \tag{10.42}$$

where use has been made of $\langle\hat{l}_x\rangle = \langle\hat{l}_y\rangle = \langle\hat{s}_x\rangle = \langle\hat{s}_y\rangle = 0$ (see Eqs. (8.25), (10.7)).

Weak Magnetic Field

The weak-field Zeeman effect refers to the case in which $\hat{H}'_m \ll \hat{H}'_{so}$. We can assume in this case that the eigenfunction $|j, m_j\rangle$ remains unchanged in the presence of a weak magnetic field. The energy level associated with given l

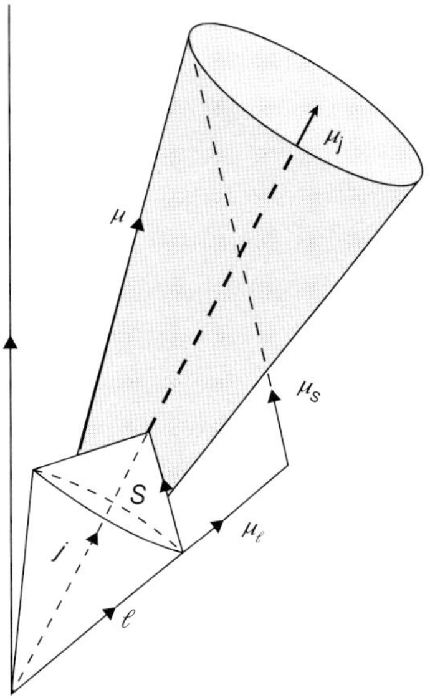

Figure 10.7 The orbital l, spin s, and total j angular momenta and the magnetic moments associated. Due to the magnetic spin anomaly, the sum of μ_l and μ_s is not parallel to μ_j, but its tip stays on the corn surface around μ_j.

and s then splits into $2j+1$ sublevels due to m_j varying from j to $-j$ in steps of unity. However, finding the split energy levels is somewhat complicated, because the magnetic moment $\boldsymbol{\mu}_j$ is not parallel to $\boldsymbol{j}$ due to the magnetic spin anomaly (Figure 10.7). But the tip of $\boldsymbol{\mu}_j$ lies on the surface of the cone with its axis parallel to $\boldsymbol{j}$, and the value of $\boldsymbol{\mu}_j$ projected onto the j-axis is given from Eqs. (10.31), (10.37b) by

$$\mu_j \equiv \underline{\mu}_j \cdot \left(\underline{j}/j\right); \quad \underline{\mu}_j = \frac{e}{2m_e}(\underline{l} + 2\underline{s}) = \frac{e}{2m_e}(\underline{j} + \underline{s})$$
$$= \frac{e}{2m_e j}(\underline{j}^2 + \underline{j} \cdot \underline{s}) = \frac{ej}{2m_e}\left(1 + \frac{j^2 - l^2 + s^2}{2j^2}\right); \quad \underline{j} \cdot \underline{s} = \underline{l} \cdot \underline{s} + s^2 \tag{10.43}$$

Therefore, by taking $\boldsymbol{B}$ in the z-direction without any loss of generality, we can quantify the Zeeman splitting of the energy level in terms of m_j as

$$\delta E_{sm} = \frac{e}{2m_e}\langle j, m_j|\underline{\mu}_j \cdot \hat{z}B|j, m_j\rangle = g_j \mu_B m_j B \tag{10.44}$$

where the gyromagnetic ratio

$$g_j = 1 + \frac{j(j+1) - l(l+1) + s(s+1)}{2j(j+1)} \tag{10.45}$$

is called the *Lande g-factor*. Note that for $s = 0, j = l$, and $g_j = 1$, as it should, since $g_l = 1$. For $l = 0$, $j = s$, and $g_j = 2$ as it should, since $g_s = 2$ (see Eqs. (10.33) and (10.34)).

10.4 Electron Paramagnetic Resonance

The electron paramagnetic resonance is concerned with transitions of the electron between the magnetically split energy levels, and it is used for determining the magnetic moments, structural and dynamical information of liquids and solids, and so on. The transitions are induced by the microwave field $\boldsymbol{B}_1(t)$ applied in the direction perpendicular to the static field $\boldsymbol{B}_0$. Let us consider the electron in the ground state of H-atom subjected to such a magnetic field

$$\underline{B} = \hat{z}B_0 + \hat{x}B_1(t), \quad B_1(t) = B_1\frac{1}{2}(e^{i\omega t} + e^{-i\omega t}) \tag{10.46}$$

The interaction Hamiltonian then reads from Eq. (10.34) as

$$\hat{H}' = -\underline{\mu}_s \cdot \underline{B} = \frac{2\mu_B}{\hbar}(B_0\hat{s}_z + B_1(t)\hat{s}_x) \tag{10.47}$$

In the presence of $\boldsymbol{B}_0$, the ground state energy splits and is given by

$$\langle u_{100}\chi_\pm|\hat{H}_0 + \frac{2\mu_B}{\hbar}B_0\hat{s}_z|u_{100}\chi_\pm\rangle = E_{100} \pm \mu_B B_0 \tag{10.48}$$

When the driving frequency satisfies the condition $\hbar\omega \approx 2\mu_B B_0$, the resonant transition ensues between the two split levels, provided the transition matrix is not zero. The Fermi's golden rule for such a transition is given from Eq. (9.54) by

$$\begin{aligned} W &= \frac{2\pi}{\hbar}\left|\langle\chi_+|\frac{\mu_B}{\hbar}B_1\hat{s}_x|\chi_-\rangle\right|^2 \delta(2\mu_B B_0 - \hbar\omega) \\ &= \frac{\pi\mu_B^2}{2\hbar}B_1^2\delta(2\mu_B B_0 - \hbar\omega) \end{aligned} \tag{10.49a}$$

where the matrix element

$$\langle\chi_+|\hat{s}_x|\chi_-\rangle = \langle\chi_+|(\hat{s}_+ + \hat{s}_-)|\chi_-\rangle = \frac{1}{2}\hbar \tag{10.49b}$$

has been evaluated with the use of Eq. (10.3). Also the harmonic components, $\exp \pm i(\omega + \omega_0)t$, oscillating rapidly in time have been neglected.

In Figure 10.8 are plotted split energy levels versus B_0. The resonant transition is induced in practice by tuning B_0 at a fixed driving frequency ω. Moreover, B_0 is not uniform over the volume of the sample due to imperfections of the magnet or the local variations caused by neighboring atoms with different magnetic

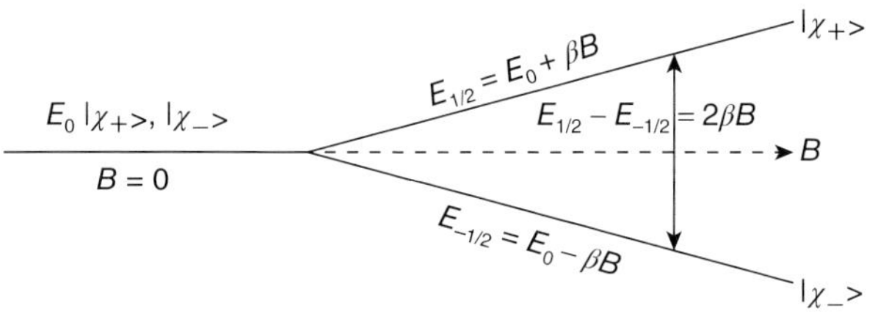

Figure 10.8 The B-field induced splitting of the electron spin up and spin down states: $\beta = e\hbar/2m_e$.

moments. Consequently, atoms in different parts of the sample undergo transitions at different resonant frequencies. The effect is accounted for by introducing the lineshape factor g, which represents the probability of the atom being in magnetic field between B and $B + dB$. The power absorption is thus given by

$$P = (N_- - N_+)\langle W\rangle \hbar\omega V \tag{10.50a}$$

where N_+, N_- are the number of electrons per unit volume in the upper and lower spin states, respectively, $\hbar\omega$ the transition energy, and V the sample volume, and the transition rate is given from Fermi's golden rule by

$$\begin{aligned}\langle W\rangle &= \frac{\pi\mu_B^2}{2\hbar}B_1^2\int_{-\infty}^{\infty} d\varepsilon g(\varepsilon - 2\mu_B B_0)\delta(\varepsilon - \hbar\omega), \quad \varepsilon = 2\mu_B B\\ &= \frac{\pi\mu_B^2}{2\hbar}B_1^2 g(2\mu_B B_0 - \hbar\omega)\end{aligned} \tag{10.50b}$$

The Spin Flip

We next discuss the dynamics of the spin flip between the spin-up and spin-down states. The electron spin state can generally be represented in analogy with Eq. (9.40) by

$$\chi(t) = c_+(t)e^{-i(\omega_0/2)t}\chi_+ + c_-(t)e^{i(\omega_0/2)t}\chi_-; \quad \frac{\hbar\omega_0}{2} = \mu_B B_0 \tag{10.51}$$

where $\pm\hbar\omega_0/2$ are the energy eigenvalues of χ_+, χ_-, and $c_+(t), c_-(t)$ the expansion coefficients. The Schrödinger equation is then given from Eq. (10.47) by

$$\begin{aligned}i\hbar\frac{\partial}{\partial t}\chi(t) &= \hbar\sum_{\pm}\left[\pm\left(\frac{\omega_0}{2}\right)c_\pm(t) + \dot{c}_\pm(t)\right]e^{\mp i(\omega_0 t/2)}\chi_\pm\\ &= \hat{H}'\chi(t) = \sum_{\pm}\left[\pm\mu_B B_0 + \frac{2\mu_B}{\hbar}B_1(t)\hat{s}_x\right]c_\pm(t)e^{\mp(i\omega_0 t/2)}\chi_\pm\end{aligned} \tag{10.52}$$

Obviously, the first terms on both sides are identical and are canceled. Thus, by performing the inner product with respect to $\chi_\pm$ with the remaining terms on both sides and making use of the orthonormality of $\chi_\pm$ Eq. (10.4), we obtain straightforwardly

$$i\hbar\dot{c}_+ = \mu_B B_1 e^{-i\Delta t}c_-, \quad \Delta = \omega - \omega_0 \tag{10.53a}$$

$$i\hbar\dot{c}_- = \mu_B B_1 e^{i\Delta t}c_+ \tag{10.53b}$$

Here $\hbar\omega_0$ is the difference in energy between the spin-up and spin-down states and Δ the frequency detuning between the driving and transition frequencies. In deriving Eq. (10.53), the terms oscillating fast with the frequency $\omega + \omega_0$ have been discarded in rotating wave approximation. Also $\hat{s}_x$ has been replaced by $\hat{s}_\pm$, and the raising and lowering properties of $\hat{s}_\pm$ have been used (see Eq. (10.3)). For the resonant interaction, $\Delta = 0$, and we can decouple c_1, c_2 by differentiating (10.53a) with respect to time and using Eq. (10.53b), obtaining thereby the differential equation of the HO:

$$\ddot{c}_+ + \Omega^2 c_+ = 0, \quad \Omega^2 \equiv \left(\frac{\mu_B B_1}{\hbar}\right)^2 \tag{10.54}$$

Hence, given the initial condition $c_+(0) = 1, c\_(0) = 0$, for example, we obtain from Eqs. (10.54) and (10.53)

$$c_+ = \cos \Omega t, \quad c\_ = -i \sin \Omega t \tag{10.55}$$

and the spin wavefunction (Eq. (10.51)) is then given by

$$\chi(t) = \cos \Omega t e^{-i\omega_0 t/2} \chi_+ - i \sin \Omega t e^{i\omega_0 t/2} \chi_- \tag{10.56}$$

The evolution in time of the electron spin prepared initially at the state χ_+ and driven by the harmonic field is specified by using Eqs. (10.1)–(10.3) as

$$\langle \hat{s}_z \rangle = \langle \chi(t)|\hat{s}_z|\chi(t)\rangle = \frac{\hbar}{2}[\cos^2(\Omega t) - \sin^2(\Omega t)] = \frac{\hbar}{2} \cos 2\Omega t \tag{10.57a}$$

$$\langle \hat{s}_x \rangle = \langle \chi(t)|\hat{s}_x|\chi(t)\rangle = \frac{\hbar}{2} \sin(2\Omega t) \sin(\omega_0 t); \quad \hat{s}_x = (\hat{s}_+ + \hat{s}) \tag{10.57b}$$

$$\langle \hat{s}_y \rangle = \langle \chi(t)|\hat{s}_y|\chi(t)\rangle = -\frac{\hbar}{2} \sin(2\Omega t) \cos(\omega_0 t); \quad \hat{s}_y = \frac{1}{i}(\hat{s}_+ - \hat{s}\_) \tag{10.57c}$$

Indeed, the z-component $s_z(t)$ flips from the spin-up to -down states or vice versa with the frequency 2Ω, as expected. Concomitantly, $s_x(t)$ and $s_y(t)$ precess around the z-axis with the frequency ω_0 called the *Lamar frequency*. Concomitantly, the precessing envelope executes sinusoidal oscillation with the frequency 2Ω (Figure 10.9). This behavior can be viewed in terms of the spinning charge as follows. The initial spin state χ_+ evolves into a linear superposition of χ_+ and χ_- driven by the harmonic field $B_1(t)$, and $s_x(t)$, $s_y(t)$ execute oscillations such that the tip of the spin precesses around the z-axis with the frequency ω_0. Moreover, the radius of the precession is modulated in time in quadrature with $s_z(t)$. An oscillating charge emits or absorbs radiation just as the oscillating atom dipole, flipping thereby the spin. The amplitude of the oscillation of $s_x(t)$ and $s_y(t)$ reaches the maximum level when the tip of the spin lies on the x–y plane.

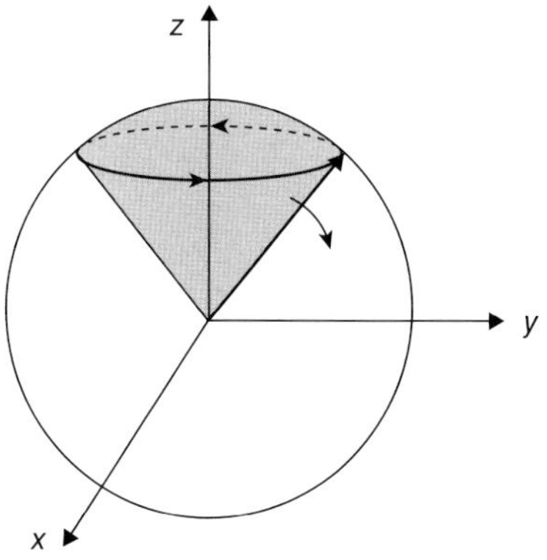

Figure 10.9 The flipping of the electron spin from the spin-up to spin-down states while precessing around the static magnetic field.

$\pi/2$ and π Pulses

The degree of the spin flip of $s_z(t)$ depends on the duration of $B_1(t)$. If the duration τ_p is such that $\tau_p = (\pi/2)/2\Omega \propto 1/B_{10}$, $s_z(t)$ is flipped by $\pi/2$ and lies on the $x-y$ plane. At the same time, the amplitudes of $s_x(t)$, $s_y(t)$ attain the maximum value. The microwave pulse inducing such rotation is called $\pi/2$ pulse. If $\tau_p = \pi/2\Omega$, on the other hand, $s_z(t)$ is flipped by π, completing thereby the flipping of the spin from the spin-up to -down states. Such a pulse is called the π *pulse*.

Problems

10.1 (a) By using the Pauli spin matrices and column vectors given in Eqs. (10.6) and (10.7), verify the basic properties of the spin operators Eqs. (10.1)–(10.4).

(b) Verify the commutation relation (10.5) by showing that

$$[\hat{s}_x, \hat{s}_y] = i\frac{\hbar}{2}\hat{s}_z, \quad [\hat{s}_y, \hat{s}_z] = i\frac{\hbar}{2}\hat{s}_x, \quad [\hat{s}_z, \hat{s}_x] = i\frac{\hbar}{2}\hat{s}_y$$

10.2 By using the ground state wavefunction of the Helium atom given in Eq. (10.17), evaluate the expectation values of total spin operators $\hat{S}^2, \hat{S}_z$ with $\underline{\hat{S}} = \underline{\hat{s}}_1 + \underline{\hat{s}}_2$.

10.3 The first excited singlet and triplet states of the He-atom are given in Eq. (10.24).

(a) Show that the wavefunctions are orthonormal.

(b) Find the expectation values of the total spin operators $\hat{S}^2$ and $\hat{S}_z$ for each state.

10.4 The Li-atom consists of three protons in the nucleus ($Z = 3$) and three electrons revolving around it. The ground state electron configuration is $1s^2 2s^1$, that is, two electrons in u_{100} state with spin-up and -down and the third one in u_{200} state with spin-up or spin-down. Write down the wavefunction using the Slater determinant and find the energy and the total spin in the ground state.

10.5 The sodium atom has 11 protons in the nucleus ($Z = 11$) and 11 electrons revolving around the nucleus.

(a) Assign each electron the quantum numbers, including the spin.

(b) The observed ionization energy and orbital radius of the atom are $5.14\,eV$ and $0.17\,nm$, respectively. Explain the data in terms of the screening of the nuclear charge.

10.6 (a) Find the splitting of the energy levels of the H-atom due to the spin orbit coupling for $n = 1, 2, 3$. Also derive Eq. (10.53) from (10.52).

(b) Find the modification of the largest wavelength of the Balmer series incorporating the fine structure in the energy level.

Suggested Readings

1. A., Yariv, An Introduction to Theory and Applications of Quantum Mechanics, John Wiley & Sons, 1982.
2. D.M., Kim, Introductory Quantum Mechanics for Semiconductor Nanotechnology, Wiley-VCH, 2010.
3. H., Haken and H.C., Wolf, The Physics of Atoms and Quanta: Introduction to Experiments and Theory, Fifth Edition, Springer, 2004.
4. A.I.M., Rae, Quantum Mechanics, Fourth Edition, Taylor & Francis, 2002.
5. M., Karplus and R.N., Porter, Atoms and Molecules: An Introduction for Students of Physical Chemistry, Addison Wesley Publishing Company, 1970.

11
Molecules and Chemical Bonds

Atoms combine to form molecules by means of the chemical bond. The successful elucidation of the chemical bonds is again one of the highlights of the quantum mechanics. There are two kinds of bonds, heteropolar and homopolar. In the former, an electron is transferred from one neutral atom to the other, and the resulting two ions of opposite polarity are bound together via the attractive Coulomb potential. In the latter, two neutral atoms are bound by means of various other mechanisms. The principles of bonding are discussed, using simple examples, together with hybridization, an essential element in the organic molecules.

11.1
Ionized Hydrogen Molecule

Let us consider the chemical bonding in the ionized hydrogen molecule $H_2{}^+$, which consists of one electron interacting with two protons (Figure 11.1). In this structure, the two protons repel each other, while the electron and the two protons attract each other via repulsive and attractive Coulomb forces, respectively. The problem is then to clarify why the two protons do not to break away from each other and form instead a stable molecule.

For simplicity of analysis, let us first take the two protons fixed in space. Then, the Hamiltonian is given by

$$\hat{H}(a,b) = -\frac{\hbar^2}{2m}\nabla^2 - e_M^2\left(\frac{1}{r_a} + \frac{1}{r_b}\right), \quad e_M^2 = \frac{e^2}{4\pi\varepsilon_0} \tag{11.1}$$

and partitions into the Hamiltonian of the hydrogenic subsystem formed by one of two protons, say proton a and the electron with the perturbing term e_M^2/r_b or vice versa. The wavefunctions u_a, u_b of two hydrogenic subsystems represent two identical degenerate states, and we can treat the problem by means of the degenerate perturbation theory. Thus, we look for the solution in the form

$$\varphi(\underline{r}_a, \underline{r}_b) = \sum_{\alpha=a,b} c_\alpha u_\alpha; \quad \left(-\frac{\hbar^2}{2m}\nabla^2 - \frac{e_M^2}{r_\alpha}\right) u_\alpha = E_0 u_\alpha \tag{11.2}$$

Introductory Quantum Mechanics for Applied Nanotechnology, First Edition. Dae Mann Kim.

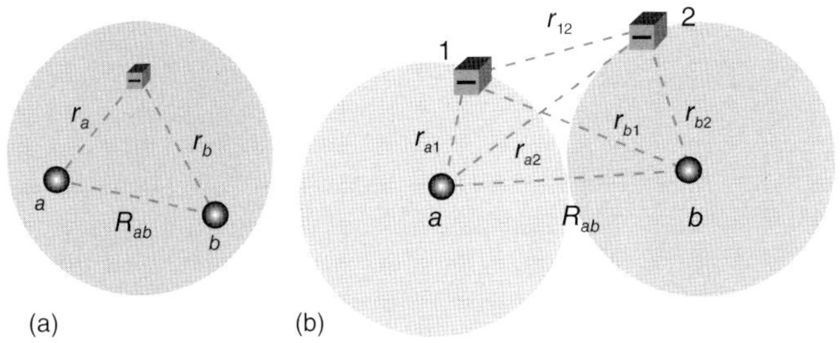

Figure 11.1 The H_2^+ molecule, consisting of two protons and a single electron (a). The H_2 molecule, consisting of two protons and two electrons (b). The electron–proton and electron–electron interactions are distributed over the electron charge cloud.

where E_0 is the ground state energy of the H-atom. In this approach, a single electron is taken to form the hydrogenic subsystems with the two nuclei simultaneously or equivalently to be shared by the two nuclei.

The eigenequation reads as

$$\hat{H}(a,b)\sum_{\alpha=a\cdot b} c_\alpha u_\alpha = E\sum_{\alpha=a\cdot b} c_\alpha u_\alpha \tag{11.3}$$

After rearranging the terms with the use of Eqs. (11.1) and (11.2), we can rewrite Eq. (11.3) as

$$c_a\left(\Delta E - \frac{e_M^2}{r_b}\right)u_a + c_b\left(\Delta E - \frac{e_M^2}{r_a}\right)u_b = 0, \quad \Delta E \equiv E_0 - E \tag{11.4}$$

Hence, finding the wavefunction is reduced to determining c_a and c_b. For this purpose, we can carry out the inner products with respect to u_a, u_b on both sides of Eq. (11.4), using the orthonormality of u_a, u_b and obtain the coupled equation

$$\begin{aligned}(\Delta E + C)c_a + (\Delta ES + D)c_b &= 0\\ (\Delta ES + D)c_a + (\Delta E + C)c_b &= 0\end{aligned} \tag{11.5}$$

where S, C, and D denote the integrals

$$S = \int d\underline{r}u_a^*(\underline{r}_a)u_b(\underline{r}_b) \equiv \langle u_a|u_b\rangle \equiv \langle u_b|u_a\rangle \tag{11.6a}$$

$$C = \langle u_a| - \frac{e_M^2}{r_b}|u_a\rangle = \langle u_b| - \frac{e_M^2}{r_a}|u_b\rangle < 0 \tag{11.6b}$$

$$D = \langle u_a| - \frac{e_M^2}{r_a}|u_b\rangle = \langle u_b| - \frac{e_M^2}{r_b}|u_a\rangle < 0 \tag{11.6c}$$

Overlap, Coulomb, and Exchange Integrals

Three kinds of integrals are involved in the coupled equations (Figure 11.2): S is called the *overlap integral* and specifies the degree of overlap between u_a and u_b at

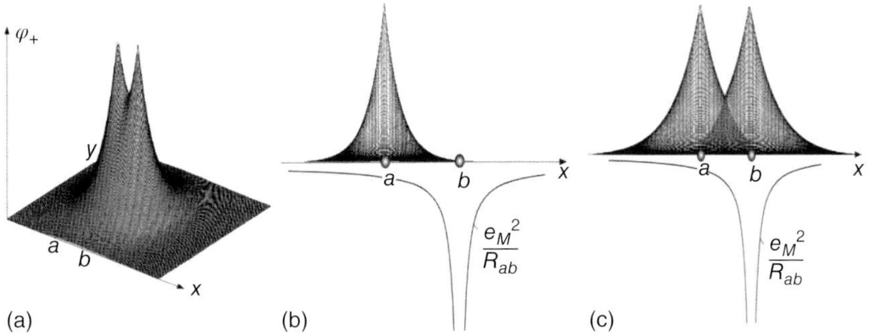

Figure 11.2 The overlapped electron charge cloud with each cloud centered at protons *a* and *b* in the S-integral (a). The Coulomb interaction between the electron cloud centered at proton *a* and proton *b* as a point charge in the C-integral (b). The interaction between the overlapped electron charge cloud with proton *a* or *b* in the D-integral (c).

a given separation of two protons; C is the Coulomb interaction integral, accounting for the interaction between proton b for instance and the electron that forms the hydrogenic subsystem with proton a or vice versa; D is the exchange integral describing the interaction between the exchange probability density $u_a^* u_b$ and proton a or b and represents the interaction between the overlapped electron cloud and proton a or b.

The coupled equation (11.5) is again homogeneous, and therefore c_a, c_b, hence $\varphi(\boldsymbol{r}_a, \boldsymbol{r}_b)$, become trivial, unless the secular equation is satisfied, as discussed:

$$\begin{vmatrix} \Delta E + C & \Delta ES + D \\ \Delta ES + D & \Delta E + C \end{vmatrix} = 0 \tag{11.7}$$

By solving the quadratic equation for ΔE, we obtain

$$\Delta E_\pm \equiv E_0 - E = \frac{\pm D - C}{1 \mp S} \tag{11.8}$$

When Eq. (11.8) is put into Eq. (11.5), the two equations become redundant, and we find

$$c_b = \mp c_a \tag{11.9}$$

Therefore, by expressing c_b in terms of c_a and using c_a to normalize the eigenfunction (Eq. (11.2)), we obtain

$$\varphi_\mp(\underline{r}_a, \underline{r}_b) = \frac{1}{\sqrt{2}}(u_a \mp u_b) \tag{11.10a}$$

with the eigenenergies E given from Eq. (11.8) by

$$E_\mp = E_0 + \frac{C \mp D}{1 \mp S} \tag{11.10b}$$

Thus, the eigenfunction is specified by symmetrical and antisymmetrical combinations of u_a and u_b, and the degeneracy has been lifted.

Bonding and Antibonding

We next consider the binding energy of H_2^+. For this purpose, it is important to note that E_0 is the energy of the electron bound to one proton, while the other is at infinity, and therefore the bonding energy is by definition given by

$$E_b \equiv E_\mp - E_0 + \frac{e_M^2}{R_{ab}} = \frac{C \mp D}{1 \mp S} + \frac{e_M^2}{R_{ab}} \tag{11.11}$$

Let us note at this point that the repulsive Coulomb interaction between the two protons R_{ab} distance apart should be brought in. The distance R_{ab} also critically affects the repulsive Coulomb interaction and the integrals, S, C, and D Eq. (11.6). Naturally, a stable molecule is formed when $\boldsymbol{E}_b < 0$, and such condition has to be examined. As u_a, u_b are normalized eigenfunctions, the overlap integral S Eq. (11.6a) is by definition less than unity if R_{ab} does not collapse to zero. Hence, $1 \mp S$ is positive, leaving C and D as the determining factor for the polarity of E_b.

In the limit $R_{ab} \to 0$, the repulsive Coulomb interaction between the two protons diverges. Concomitantly, $r_a \to r_b$, and C and D simply represent in this case the finite average potential energy of the ground state of the H-atom. Hence, for small R_{ab}, E_b should diverge. On the other hand, in the limit $R_{ab} \to \infty$, $S \to 0$ and C represents the attractive Coulomb interaction between proton b and the electron charge cloud attached to proton a or vice versa. Therefore, C is practically identical to and cancels out e_M^2/R_{ab}. This leaves D as the sole integral dictating the polarity of E_b. Now since $D < 0$ by definition, it is clear from Eqs. (11.10) and (11.8) that $\mathrm{E}_b < 0$ for the symmetric combination of u_a, u_b, namely, for $\varphi_+(\boldsymbol{r}_a,\boldsymbol{r}_b)$.

Figure 11.3 shows E_b versus R_{ab} for both $\varphi_+(\boldsymbol{r}_a,\boldsymbol{r}_b)$ and $\varphi_-(\boldsymbol{r}_a,\boldsymbol{r}_b)$. Indeed, E_b is negative for a range of R_{ab} for $\varphi_+(\boldsymbol{r}_a,\boldsymbol{r}_b)$, and it is in this range of R_{ab} that the ionized H_2^+ is formed. For $\varphi_-(\boldsymbol{r}_a,\boldsymbol{r}_b)$, E_b is positive in the entire range of R_{ab}, so that $\varphi_-(\boldsymbol{r}_a,\boldsymbol{r}_b)$ represents the antibonding mode. The fact that $\varphi_+(\boldsymbol{r}_a,\boldsymbol{r}_b)$ represents the bonding mode can be attributed to a large probability density of electrons in between the two protons as shown in Figure 11.4. In this case, the attractive forces

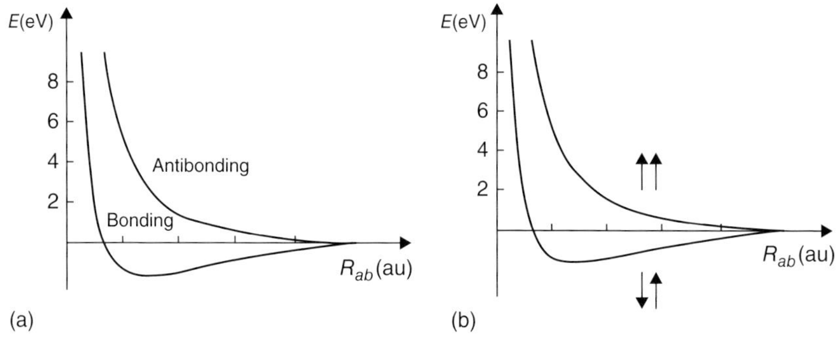

Figure 11.3 The bonding and antibonding curves versus the internuclear distance R in the ionized H_2^+ (a) and neutral H_2 molecules (b).

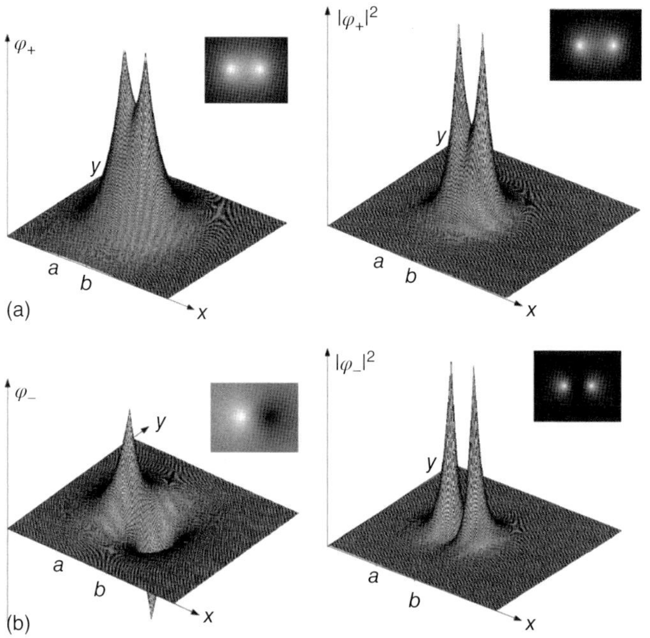

Figure 11.4 The symmetrized (a) and the antisymmetrized (b) eigenfunctions and the probability densities associated. Also shown are the top views and the degree of overlap of the two electron charge clouds.

between the electron cloud and the two protons more than compensate the repulsive force between the two protons.

11.2
H_2 Molecule and Heitler-London Theory

The H_2 molecule consists of two protons and two electrons (Figure 11.1). The Hamiltonian is thus given by

$$\hat{H} = \hat{H}_1 + \hat{H}_2 + \hat{V} \tag{11.12a}$$

where

$$H_1 = -\frac{\hbar^2}{2m}\nabla_1^2 - \frac{e_M^2}{r_{a1}}; \quad \hat{H}_2 = -\frac{\hbar^2}{2m}\nabla_2^2 - \frac{e_M^2}{r_{b2}} \tag{11.12b}$$

are the Hamiltonians of the hydrogenic subsystem formed by proton a with electron 1 and proton b with electron 2, and the potential

$$\hat{V} = -\frac{e_M^2}{r_{b1}} - \frac{e_M^2}{r_{a2}} + \frac{e_M^2}{R_{ab}} + \frac{e_M^2}{r_{12}}, \quad e_M^2 = \frac{e^2}{4\pi\varepsilon_0} \tag{11.12c}$$

lumps together the rest of the interaction terms.

Variational Principle

The H_2 molecule as a four-body central force problem cannot be dealt with analytically, and therefore an approximate treatment is in order. In this context, the variational principle provides a convenient criterion by which to assess the accuracy of the approximate analysis and is discussed first. Given a dynamic system, we have to solve the energy eigenequation

$$\hat{H}\varphi = E\varphi \tag{11.13}$$

and find the average values of the dynamic quantities such as energy

$$E = \frac{\langle\varphi|\hat{H}|\varphi\rangle}{\langle\varphi|\varphi\rangle} = \frac{\int_{-\infty}^{\infty} d\underline{r}\varphi^*\hat{H}\varphi}{\int_{-\infty}^{\infty} d\underline{r}\varphi^*\varphi} \tag{11.14}$$

If the eigenfunction φ is the exact solution of Eq. (11.13), E in Eq. (11.14) represents the true eigenvalue. However, if φ is an approximate solution, E does not represent the true value. In such a case, the variational principle states that the values of E obtained from Eq. (11.14) are always greater than the true value. Therefore, the degree of accuracy of the approximate schemes can be assessed by comparing the resulting E values.

Heitler–London Theory

With this fact in mind, let us consider the Heitler–London theory of the H_2 molecule. The theory introduces the antisymmetrized ground state wavefunctions of the two electrons in the form

$$\varphi(1,2) = \varphi_\pm(1,2)\chi_g, \quad \varphi_\pm(1,2) = [u_a(1)u_b(2) \pm u_b(1)u_a(2)], \quad 1,2 \equiv \underline{r}_1, \underline{r}_2 \tag{11.15}$$

where u_a, u_b are the ground state eigenfunctions of the hydrogenic subsystems with the energy E_0 and χ_g the triplet (s) and singlet (a) spin states, respectively (see(10.15)).

The ground state energy of the H_2 molecule is then given by

$$E_\pm = \frac{\langle\varphi_\pm|\hat{H}|\varphi_\pm\rangle}{\langle\varphi_\pm|\varphi_\pm\rangle} \tag{11.16}$$

The spin functions do not affect the integrals in Eq. (11.16) and have been deleted. However, φ_+, φ_- have to be associated specifically with χ_- and χ_+, respectively. As $u_a(i)$ and $u_b(j)$ are normalized eigenfunctions, the denominator is given in terms of the overlapped integral by

$$\langle\varphi_\pm|\varphi_\pm\rangle = 2(1 \pm S^2); \quad S \equiv \langle u_a(i)|u_b(j)\rangle, \quad i \neq j \tag{11.17}$$

The numerator

$$N = \langle\varphi_\pm|\hat{H}_1 + \hat{H}_2 - \frac{e_M^2}{r_{b1}} - \frac{e_M^2}{r_{a2}} + \frac{e_M^2}{R_{ab}} + \frac{e_M^2}{r_{12}}|\varphi_\pm\rangle \tag{11.18}$$

consists of 24 integrals, 6 Hamiltonian terms paired with 4 different combinations of $u_a(i)u_b(j)$. However, $\hat{H}$ is invariant under the interchange of $\boldsymbol{r}_1$ and $\boldsymbol{r}_2$, so that the four combinations of $u_a(i)u_b(j)$ reduce to two. Thus, N is given by

$$N = 2\langle \hat{H} \rangle_1 \pm 2\langle \hat{H} \rangle_2 \tag{11.19}$$

where the first term

$$\langle \hat{H} \rangle_1 \equiv \langle u_a(1)u_b(2)|\hat{H}_1 + \hat{H}_2 - \frac{e_M^2}{r_{b1}} - \frac{e_M^2}{r_{a2}} + \frac{e_M^2}{R_{ab}} + \frac{e_M^2}{r_{12}}|u_a(1)u_b(2)\rangle$$

$$= 2E_0 + \frac{e_M^2}{R_{ab}} + 2C + E_{RI} \tag{11.20a}$$

is specified in terms of the Coulomb interaction integral C, and the repulsive Coulomb interaction between two electrons E_{RI}:

$$C \equiv \langle u_a(1)|\frac{-e_M^2}{r_{b1}}|u_a(1)\rangle = \langle u_2(2)|\frac{-e_M^2}{r_{a2}}|u_b(2)\rangle \tag{11.20b}$$

$$E_{RI} = \langle u_a(1)u_b(2)|\frac{e_M^2}{r_{12}}|u_a(1)u_b(2)\rangle \tag{11.20c}$$

The second term

$$\langle \hat{H} \rangle_2 = \langle u_b(1)u_a(2)|\hat{H}_1 + \hat{H}_2 - \frac{e_M^2}{r_{b1}} - \frac{e_M^2}{r_{a2}} + \frac{e_M^2}{R_{ab}} + \frac{e_M^2}{r_{12}}|u_a(1)u_b(2)\rangle$$

$$= 2E_0S^2 + \frac{e_M^2}{R_{ab}}S^2 + 2DS + E_{CE} \tag{11.21a}$$

is likewise specified in terms of S, D, and the repulsive interaction computed with the use of exchange densities E_{CE} as

$$D = \langle u_b(1)|\frac{-e_M^2}{r_{b1}}|u_a(1)\rangle = \langle u_a(2)|\frac{-e_M^2}{r_{a2}}|u_b(2)\rangle \tag{11.21b}$$

$$E_{CE} = \langle u_b(1)u_a(2)|\frac{e_M^2}{r_{12}}|u_a(1)u_b(2)\rangle \tag{11.21c}$$

Bonding Energy

By inserting Eqs. (11.17)–(11.21) into Eq. (11.16), we can write

$$E_{\pm} = 2E_0 + \frac{2C + E_{RI}}{1 \pm S^2} \pm \frac{2DS + E_{CE}}{1 \pm S^2} + \frac{e_M^2}{R_{ab}} \tag{11.22}$$

and obtain the bonding energy from Eq. (11.22). As $0 < S < 1$ and $D < 0$ (see Eq. (11.6)), $E_+ < E_-$. Also, when the two protons are taken far apart from each other, with each carrying an electron, the total energy is the sum of the ground state energies of two noninteracting hydrogen atoms, that is, $2E_0$. Therefore, the bonding energy is given by

$$E_b \equiv E_{\pm} - 2E_0 = \frac{2C + E_{RI}}{1 \pm S^2} \pm \frac{2DS + E_{CE}}{1 \pm S^2} + \frac{e_M^2}{R_{ab}} \tag{11.23}$$

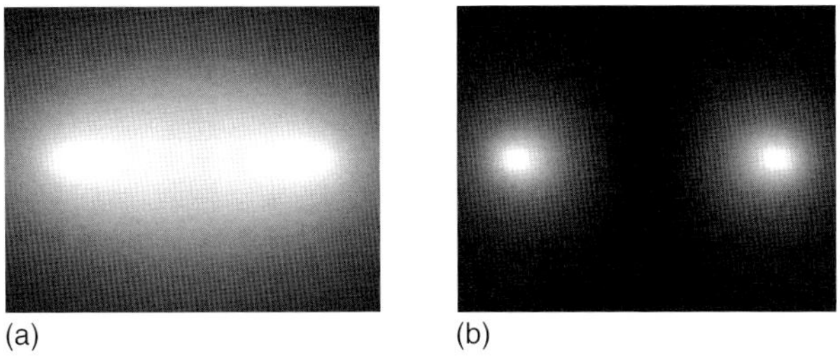

Figure 11.5 The top views of the probability densities of the symmetrized (a) and anti-symmetrized (b) eigenfunctions of the H_2 molecule. The degree of overlap of the electron charge clouds is commensurate with the degree of brightness.

Plotted also in Figure 11.3 is E_b versus R_{ab} curve, which clearly indicates that the symmetric combination of u_a and u_b paired with the singlet spin state constitutes the bonding state. The bonding versus antibonding can again be understood in terms of the degree of the overlap of the electron charge clouds. For φ_+, the overlap is large so that the attractive interaction between the electron charge cloud and the two protons more than compensate the repulsive interaction between the two protons (Figure 11.5). The bonding energy of 3.14 eV as obtained from the minimum value of E_b versus R_{ab} curve is smaller than the measured dissociation energy of 4.48 eV. This indicates that the Heitler–London theory yields the ground state energy higher than the true value, which is consistent with the variational principle. Nevertheless, the theory provides the basis for describing the homopolar bonding.

11.3
Ionic Bond

When atom A transfers an electron to atom B, the resulting two ions A^+ and B^- form a molecule via the ionic bonding. In the sodium chloride, NaCl, for example, the ionization $Na \rightarrow Na^+ + e$ requires IP_1 of 5.14 eV, while the electron capture $Cl + e \rightarrow Cl^-$ releases the energy of 3.65 eV, called the *affinity factor*. Hence, the difference in energy ΔE of 1.49 eV constitutes the bonding energy of NaCl.

When the two ions approach toward each other, they interact via the attractive Coulomb potential. When R is further decreased, the two electron charge clouds overlap. As the two ions have spherically symmetric closed shell configurations, they can be viewed as point charges. Moreover, the exclusion principle requires an additional electron associated with overlapped charge cloud near each ion to behave as though they occupy next higher-lying quantum states. Hence, the energy of the ion pair increases with increasing interpenetration of the electron cloud, adding thereby the repulsive energy term. Thus, ΔE as a function of R for

NaCl is to be modeled as

$$\Delta E(R) = Ae^{-\alpha R} - \frac{e_M^2}{R} + \Delta E(\infty), \quad \Delta E(\infty) = 1.49\,\text{eV} \tag{11.24}$$

where the first term is the empirical representation of the repulsive potential arising from the exclusion principle, while the second term is the attractive Coulomb potential between the two ions.

In Figure 11.6 is plotted ΔE versus R curve. The curve follows the attractive Coulomb potential at large R but is dominated by the repulsive potential for small R. At the minimum point R_e, the attractive and repulsive forces balance each other out, yielding thereby the zero slope of the potential curve. The ionic bond is thus represented by D_e, which denotes the absolute magnitude of the difference between $\Delta \boldsymbol{E}(R_e)$ and $\Delta \boldsymbol{E}(\infty)$. Furthermore, the Taylor expansion of ΔE at R_e yields the expression

$$\Delta \boldsymbol{E}(R) \simeq \Delta \boldsymbol{E}(R_e) + \frac{1}{2}k(R - R_e)^2 + \cdots, \quad k \equiv \frac{\partial^2 \Delta E(R_e)}{\partial R^2} \tag{11.25}$$

where the first expansion term is zero because the Taylor expansion is done at the minimum point of the curve. Then, the ΔE–R curve near R_e is reduced to the potential energy of the harmonic oscillator and therefore indicates that the two nuclei vibrate at the frequency ω $(=(k/\mu)^{1/2})$ with μ denoting the reduced mass $1/\mu = 1/m_{\text{Na}} + 1/m_{\text{Cl}}$.

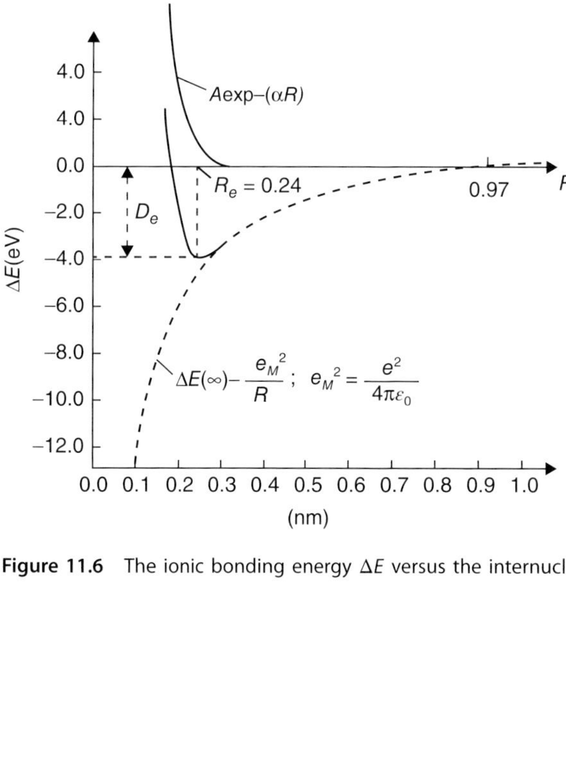

Figure 11.6 The ionic bonding energy ΔE versus the internuclear distance R.

11.4 van der Waals Attraction

The chemical bond underlying the H_2 molecule is called the *covalent bond* and involves the interaction between two open shell neutral atoms, in this case two H-atoms, whose ground states are partially filled by a single electron. The interaction between two closed shell atoms also provides the chemical bond called the *van der Waals attraction*. The resulting attractive interaction occurs in nearly all atoms, and the underlying force is known as the *London dispersion force*. Such chemical bond can be modeled by taking two atoms as two 1D charged harmonic oscillators bound to positive charge centers (Figure 11.7).

The Hamiltonian of two coupled HO is given by

$$\hat{H} = \sum_{j=1}^{2} \hat{H}_{0j} + V(R, x_1, x_2); \quad \hat{H}_{0j} = -\frac{\hbar^2}{2m}\frac{\partial^2}{\partial x_j^2} + \frac{1}{2}kx_j^2 \tag{11.26a}$$

where the potential energy

$$V = e_M^2\left(\frac{1}{R} - \frac{1}{R - x_1} + \frac{1}{R + x_2 - x_1} - \frac{1}{R + x_2}\right), \quad e_M^2 = \frac{e^2}{4\pi\varepsilon_0} \tag{11.26b}$$

represents the Coulomb interactions involving two force centers and two charged oscillators. For $R \gg x_1, x_2$, V can be simplified by expanding the potential terms in powers of x_j/R. For example, we can expand the second term as

$$\frac{1}{R - x_1} = \frac{1}{R}\left(1 - \frac{x_1}{R}\right)^{-1} = \frac{1}{R}\left[1 + \frac{x_1}{R} + \left(\frac{x_1}{R}\right)^2 + \cdots\right]$$

After carrying out similar expansions and adding the terms together, we find

$$V \simeq -\frac{e_M^2 x_1 x_2}{R^3} \tag{11.26c}$$

Therefore, the Hamiltonian is simplified to read as

$$\hat{H} = -\frac{\hbar^2}{2m}\frac{\partial^2}{\partial x_1^2} + \frac{1}{2}kx_1^2 - \frac{\hbar^2}{2m}\frac{\partial^2}{\partial x_2^2} + \frac{1}{2}kx_2^2 - \frac{e_M^2 x_1 x_2}{R^3} \tag{11.27}$$

We can further compact the Hamiltonian by introducing the new variables

$$\xi = x_1 + x_2, \quad \eta = x_2 - x_1$$

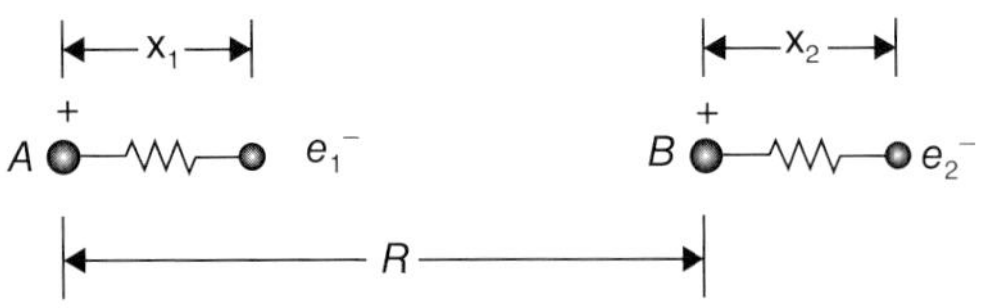

Figure 11.7 Two charged harmonic oscillators coupled via the Coulomb interaction.

and recast Eq. (11.27) into a form

$$H = -\frac{\hbar^2}{2\mu}\frac{\partial^2}{\partial\xi^2} + \frac{1}{2}k_-\xi^2 - \frac{\hbar^2}{2\mu}\frac{\partial^2}{\partial\eta^2} + \frac{1}{2}k_+\eta^2, \quad k_\mp \equiv \left(\frac{k}{2} \mp \frac{e_M^2}{R^3}\right) \tag{11.28}$$

with μ denoting the reduced mass $1/\mu = (1/m) + (1/m) = 2/m$. In this manner, the Hamiltonian of two coupled oscillators is transformed into that of two independent oscillators, oscillating at two different frequencies. The total energy level is therefore given from Eq. (7.12) by

$$E_{n_1,n_2} = \hbar\omega_+\left(n_1 + \frac{1}{2}\right) + \hbar\omega_-\left(n_2 + \frac{1}{2}\right) \tag{11.29a}$$

where n_1, n_2 are the quantum numbers, and oscillation frequencies ω_-, ω_+ are given by

$$\omega_\mp^2 \equiv \frac{k_\mp}{\mu} = \omega_c^2\left(1 \mp \frac{2e_M^2}{kR^3}\right); \quad \omega_c^2 \equiv \frac{k}{m} \tag{11.29b}$$

and are smaller or greater than the characteristic frequency ω_c.

The two frequencies ω_+, ω_- can be shown correlated with the polarizability of the oscillator as follows. When an electric field E is applied, the oscillator charged with $-$e is subjected to the force $-e$E and pushed away from its equilibrium position, while it is simultaneously subjected to the restoring force of the spring $-kx$. These two forces balance at x_e given by $x_e = -e\mathrm{E}/k$. The resulting dipole moment induced is given by

$$\mu_{\mathrm{ind}} \equiv -ex_e = \frac{e^2}{k}\mathrm{E} \equiv 4\pi\varepsilon_0\alpha\mathrm{E}; \quad \alpha = \frac{e^2}{4\pi\varepsilon_0 k} \tag{11.30}$$

where α is the polarizability connecting E and the induced dipole moment. When the spring constant k is replaced by α in Eq. (11.29b), $\omega_\mp$ are expressed in terms of α as

$$\omega_\mp^2 \equiv \frac{k_\mp}{\mu} = \omega_c^2\left(1 \mp \frac{2\alpha}{R^3}\right); \quad \omega_c^2 \equiv \frac{k}{m} \tag{11.31}$$

We can thus expand the dressed frequencies as

$$\omega_\mp = \omega_c\left[1 \mp \frac{1}{2}\left(\frac{2\alpha}{R^3}\right) - \frac{1}{8}\left(\frac{2\alpha}{R^3}\right)^2 + \cdots\right]$$

and obtain the ground state energy as

$$E_{00} = \frac{\hbar}{2}(\omega_- + \omega_+) = \hbar\omega_c - \left(\frac{\hbar\omega_c\alpha^2}{2}\right)\frac{1}{R^6} \tag{11.32}$$

Since the first term is the zero-point energy of the two oscillators in the limit $R \to \infty$, the second term $\propto 1/R^6$ should represent the bonding energy and is known as the *van der Waals attraction*. Obviously, the bonding energy is due to the net potential $V(x)$ providing a net attractive potential Eq. (11.26c). This simple model used for illustrating the dispersion force can also be applied to the two coupled neutral atoms. In this case, the power law dependence of $1/R^6$ is preserved, but the polarizability and binding parameter k of the harmonic oscillator are replaced by the atomic polarizability and the first ionization potential IP_1, respectively.

11.5
Polyatomic Molecules and Hybridized Orbitals

The chemical bonding in polyatomic molecules involves many electrons distributed over several nuclei and is complicated. However, the bonding can be understood based on the bonds operative in diatomic molecules. This is because most of the chemical bonds are localized in which two nuclei are bonded via two electrons as in the case of the H_2 molecule. These bonds are illustrated with the use of a few specific examples.

Methane and sp Hybridization

The methane (CH_4) consists of a carbon atom ($1s^2 2s^2 2p_x 2p_y$) bonded to four H-atoms by four tetrahedral bonds with H–C–H bond angle of 109°28′ as shown in Figure 11.8. In the process, one of the two electrons in $2s$ state is booted to 2p state, forming thereby the valence state ($1s^2 2s 2p_x 2p_y 2p_z$) (see Eqs.(8.53), (8.54) for p_x, p_y, p_z). The bonds are formed by placing the four valence electrons into four hybridized orbitals, given in terms of the single electron eigenfuctions as

$$\begin{aligned}
\chi_1 &= \frac{1}{2}(2s + 2p_x + 2p_y + 2p_z) \\
\chi_2 &= \frac{1}{2}(2s - 2p_x - 2p_y + 2p_z) \\
\chi_3 &= \frac{1}{2}(2s + 2p_x - 2p_y - 2p_z) \\
\chi_4 &= \frac{1}{2}(2s - 2p_x + 2p_y - 2p_z)
\end{aligned} \tag{11.33}$$

These functions are orthonormal, and the probability distribution is maximum along $(1, 1, 1), (1, -1, -1), (-1, -1, 1)$, and $(-1, 1, -1)$ directions, respectively. Thus, the molecular orbitals consisting of the four linear combinations of the atomic orbitals compensate the energy required for an electron to be booted up from 2s to 2p states and stabilize the methane molecule via the attractive Coulomb potential.

Ethane and Directionality of Molecular Orbitals

The spatial directionality of the hybridized molecular wavefunctions is one of the main modes of bonding, as exemplified by the ethane (C_2H_6). As also shown in Figure 11.8, the two carbon atoms are in the configuration similar to that of methane aside from the fact that two of the hybridized molecular orbitals are aligned to the C–C bond. The rest of the orbitals of each carbon atom form the tetrahedral bonds with $2s$ state of three H-atoms.

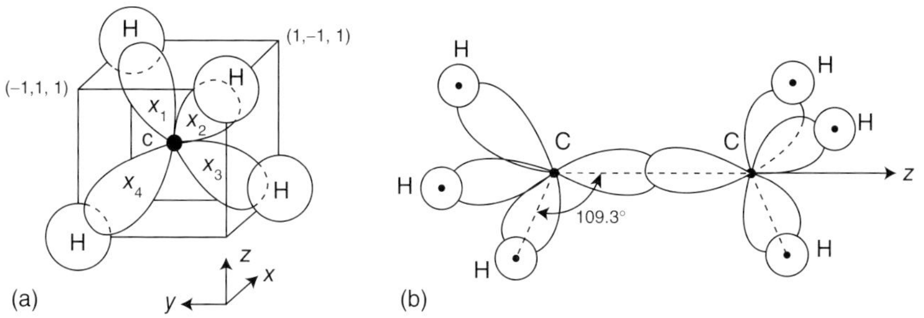

Figure 11.8 The methane molecule consisting of a carbon atom bonded to four H-atoms (a). The ethane molecule formed by two hybridized molecular orbitals aligned to the C–C bond (b).

Problems

11.1 (a) Starting from the eigenequation (11.3), derive the coupled equation (11.5) for the expansion coefficients c_a, c_b in terms of the overlap S-, Coulomb C-, and exchange D-integrals.

(b) Verify the results obtained for the symmetrized and antisymmetrized eigenfunctions and corresponding energy eigenvalues.

11.2 The analysis of the chemical bonding requires the evaluation of various matrix elements, as exemplified by the Heitler–London theory. Starting with the wavefunction given in Eq. (11.15), fill in the detailed algebraic steps and verify the results of the bonding energy Eq. (11.23).

11.3 The interionic distance of NaCl is 0.24 nm, and the vibrational frequency is $\nu_e = \omega_e/2\pi = 1.1 \times 10^{13}\,\text{s}^{-1}$. Determine the parameters A and α in Eq. (11.24) and estimate the bonding energy by using A, α, and $\Delta \boldsymbol{E}(\infty) = 1.49\,\text{eV}$.

11.4 The H-atom is placed in a uniform electric field E in the z-direction. The Hamiltonian is given by

$$H = -\frac{\hbar^2}{2m}\nabla^2 - \frac{e_M^2}{r} + \text{E}er\cos\theta; \quad e_M^2 = \frac{e^2}{4\pi\varepsilon_0}$$

(a) Look for the eigenfunction in the form $\varphi = c_1|u_{100}\rangle + c_2|u_{210}\rangle$ and derive the coupled equation for c_1 and c_2 in analogy with Eq. (11.5).

(b) Find c_a, c_b, and $\varphi_\pm$, $\boldsymbol{E}_\pm$.

(c) Evaluate the atom dipole

$$\mu_\text{ind} = -e\langle \underline{r}\rangle, \langle \underline{r}\rangle = \frac{\langle \varphi_\pm | \underline{r} | \varphi_\pm\rangle}{\langle \varphi_\pm | \varphi_\pm\rangle}$$

and find the atomic polarizability α connecting the input field to the induced atom dipole as $\mu_\text{ind} = -\alpha\text{E}$.

11.5 Starting with the Hamiltonian of two coupled HO Eq. (11.26), fill in the algebra and derive the decoupled Hamiltonian (Eq. (11.28)) in terms of the new variables ξ, η.

Suggested Readings

1. A. Yariv, An Introduction to Theory and Applications of Quantum Mechanics, John Wiley & Sons, 1982.
2. D.M. Kim, Introductory Quantum Mechanics for Semiconductor Nanotechnology, Wiley-VCH, 2010.
3. H. Haken and H.C. Wolf, The Physics of Atoms and Quanta: Introduction to Experiments and Theory, Fifth edn, Springer, 2004.
4. H. Haken and H. Wolf, Molecular Physics and Elements of Quantum Chemistry: Introduction to Experiments and Theory, Second edn, Springer, 2004.
5. M. Karplus and R.N. Porter, Atoms and Molecules: An Introduction for Students of Physical Chemistry, Addison Wesley Publishing Company, 1970.

12
Molecular Spectra

The spectroscopy is an essential element of quantum mechanics. The observed atomic spectra provided a major impetus for developing the quantum theory. Also the molecular spectra constitute a key component of the quantum and molecular chemistry. Moreover, the laser device is the product of an innovative application of atomic and molecular spectroscopy. This chapter is addressed to the molecular spectra. The spectral lines are complicated because of the complexity of electronic structures and the rotational and vibrational motions of the nuclei. But the data carry a wealth of information and are analyzed focused on the diatomic molecules and as a short introduction to the vast area of the molecular chemistry and physics. Included in the discussion are the hyperfine structure, Zeeman splitting, nuclear magnetic resonance, and molecular imaging.

12.1
Theoretical Background

The electromagnetic spectrum encompasses the wavelengths ranging from radiowave to X-ray regions. In the radiowave regime, the wavelength λ spans from 3 km to about 3 m, and the nuclear magnetic resonance (NMR) frequencies are involved therein. In the microwave and far-infrared regions, λ ranges from about 30 cm to 0.03 mm and covers the molecular rotation and the electron spin resonance frequencies. In infrared (IR), visible, and vacuum ultraviolet regions, λ varies from 0.03 mm to 3 nm, and the frequencies of the molecular vibration and the transitions of outer electrons of the atom are involved. Finally, in the X-ray region, λ is shorter than 3 nm and includes the transition frequencies of inner electrons.

Diatomic Molecule

Let us revisit the H_2 molecule as a prototypical example and consider the motion of the two protons. The general features of the spectra are shown in Figure 12.1 in which the potential energy of the ground and first excited states of the electron is plotted versus the internuclear distance R. Also included in the figure are

Introductory Quantum Mechanics for Applied Nanotechnology, First Edition. Dae Mann Kim.

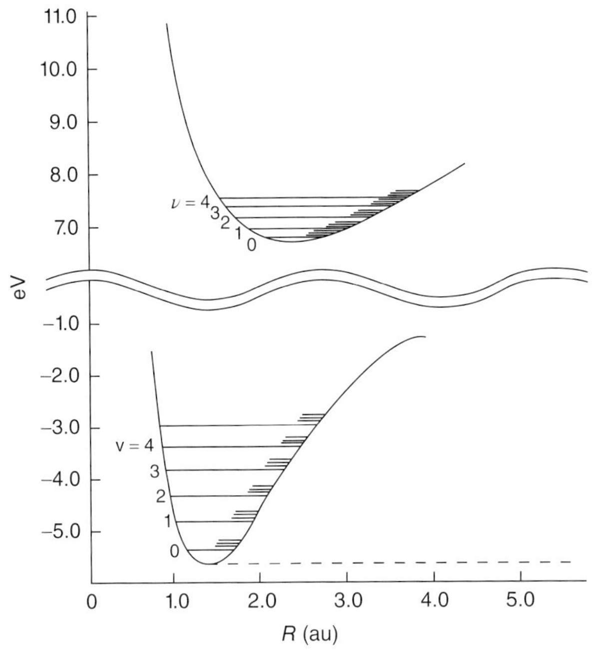

Figure 12.1 The electron potential energy of the ground and first excited states of the H_2 molecule versus the internuclear distance *R*. Shown also are the sublevels resulting from the vibrational and rotational motions of the nucleus.

the nuclear sublevels due to rotational and vibrational motions. The spectral lines are analyzed in two steps. First the electron energy levels are treated by fixing R and then the nuclear sublevels are incorporated by using the time-independent perturbation theory. The procedure is known as the *Born–Oppenheimer approximation.*

The Hamiltonian of the H_2 molecule is given in this case by

$$\left[-\frac{\hbar^2}{2\mu}\nabla_R^2 - \frac{\hbar^2}{2m}\sum_{i=1}^{n}\nabla_i^2 + V\left(\underline{r}, \underline{R}\right)\right]\varphi(\underline{r}, \underline{R}) = E\varphi(\underline{r}, \underline{R}) \tag{12.1}$$

where $\boldsymbol{r}$ represents the set of coordinates $\{\boldsymbol{r}_i\}$ of the two electrons. The motions of the two protons interacting via the Coulomb potential can be partitioned into the motion of the center of mass and the internal motion, as discussed in the H-atom theory (see Eqs. (8.27) and (8.28)). The first term on the left-hand side of Eq. (12.1) is the kinetic energy of the internal motion, and the rest of the terms constitute the usual Hamiltonian of a diatomic molecule Eq. (11.12).

The motion of the center of mass as a free particle has been dealt with already, and only the internal motion is considered. We thus look for the solution of Eq. (12.1) in the form

$$\varphi(\underline{r}, \underline{R}) = \varphi_e(\underline{r}, \underline{R})\chi(\underline{R}) \tag{12.2}$$

where φ_e is the energy eigenfunction of the electron, satisfying the energy eigenequation of the H_2 molecule with the two protons R distance apart

$$\left[-\frac{\hbar^2}{2m}\sum_{i=1}^{n}\nabla_i^2 + V\left(\underline{r}, R\right)\right]\varphi_e(\underline{r}, R) = E(R)\varphi_e(\underline{r}, R) \tag{12.3}$$

Note in Eq. (12.3) that the eigenenergy $E(R)$ should now be taken to depend parametrically on R, and $E(R)$ is also called the *electron potential energy*. By inserting Eqs. (12.2) and (12.3) into Eq. (12.1) and rearranging the terms, there results

$$\left[-\frac{\hbar^2}{2m}\nabla_R^2 + E\left(R\right)\right]\varphi_e(\underline{r}, R)\chi(\underline{R}) = E\varphi_e(\underline{r}, R)\chi(\underline{R}) \tag{12.4}$$

At this point, we resort to the Born–Oppenheimer approximation and put

$$\nabla_R^2[\varphi_e(\underline{r}, R)\chi(\underline{R})] \approx \varphi_e(\underline{r}, R)\nabla_R^2\chi(\underline{R}) \tag{12.5}$$

Upon inserting Eq. (12.5) into Eq. (12.4), dividing both sides by $\varphi_e(\mathbf{r}, R)$ and expressing ∇_R^2 in spherical coordinate frame, we can write

$$-\frac{\hbar^2}{2\mu R^2}\left\{\frac{\partial}{\partial R}\left(R^2\frac{\partial}{\partial R}\right) - \frac{1}{\hbar^2}\hat{L}^2 + E(R)\right\}\chi(\underline{R}) = E\chi(\underline{R}) \tag{12.6}$$

Evidently, Eq. (12.6) is identical in form to the energy eigenequation of the H-atom Eq. (8.35). Therefore, we can carry out a parallel analysis, making use of the results obtained already. Thus, let us look for the solution in the form

$$\chi(R, \theta, \varphi) = \rho(R)Y_L^M(\theta, \varphi) \tag{12.7}$$

where the spherical harmonics Y_L^M is the eigenfunction of $\hat{L}^2$ with the eigenvalue $\hbar^2L(L+1)$ (see Eq. (8.24)). By inserting Eq. (12.7) into Eq. (12.6) and canceling out Y_L^M from both sides, we obtain the radial equation for $\rho(R)$ as

$$-\frac{\hbar^2}{2\mu R^2}\frac{\partial}{\partial R}\left(R^2\frac{\rho\left(R\right)}{\partial R}\right) + \frac{\hbar^2L(L+1)}{2\mu R^2}\rho(R) + E(R)\rho(R) = E\rho(R) \tag{12.8}$$

Note in Eq. (12.8) that E appearing on the right-hand side of Eq. (12.8) consists of the eigenenergy of the electron and the rotational as well as vibrational energies of the two nuclei:

$$E = E_e + E_r + E_v \tag{12.9a}$$

We next expand $E(R)$ at the equilibrium distance R_e between two protons (see Figure 12.1):

$$E(R) = E(R_e) + \varepsilon_v(R), \quad \varepsilon_v(R) = \frac{1}{2}E''(0)\xi^2 + \frac{1}{3!}E'''(0)\xi^3 + \cdots, \quad \xi \equiv R - R_e \tag{12.9b}$$

where the primes denote the differentiation with respect to R. The first expansion term $\propto E'(R_e)$ is zero, as R_e is at the minimum point of the E versus R curve.

Evidently, $E(R_e)$ represents E_e, and therefore E_e is canceled from both sides of Eq. (12.8), and the eigenequation reduces to

$$-\frac{\hbar^2}{2\mu R^2}\frac{\partial}{\partial R}\left(R^2\frac{\partial\rho(R)}{\partial R}\right)+\left[\frac{\hbar^2 L(L+1)}{2\mu R^2}+\varepsilon_v(R)\right]\rho(R)=(E_r+E_v)\rho(R) \quad (12.10)$$

Equation (12.10) provides the starting point of analyzing the molecular spectra.

12.2
Rotational and Vibrational Spectra of Diatomic Molecule

Rotational Spectra

For examining the rotational motion, let us fix R at R_e for simplicity of discussion. In this case, there is no vibrational motion, that is, $E_v = \varepsilon_v(R) = 0$, and the first term on the left-hand side of Eq. (12.10) also vanishes. As a result, the rotational energy naturally follows from Eq. (12.10) as

$$E_r = \frac{\hbar^2 L(L+1)}{2\mu R_e^2}, \quad L = 0, 1, 2, \ldots \quad (12.11a)$$

with L denoting the angular momentum quantum number. Evidently, Eq. (12.11a) is the quantized version of the rotational energy of a classical rigid rotator (Figure 12.2). This can be shown by considering the angular momentum of the rotator

$$|\underline{L}| = \mu R_e v = \mu R_e(R_e\omega) = I_e\omega; \quad I_e \equiv \mu R_e^2 \quad (12.11b)$$

where v is the velocity of the fictitious particle with reduced mass μ at R_e distance from the fixed center and I_e the moment of inertia. Then, the energy of the rotator E_r reads as

$$E_r = \frac{1}{2}\mu v^2 = \frac{1}{2}I_e\omega^2 = \frac{1}{2}\frac{(I_e\omega)^2}{I_e} = \frac{|\underline{L}|^2}{2\mu R_e^2} \quad (12.11c)$$

and is identical to Eq. (12.11a) except for the quantization of the angular momentum.

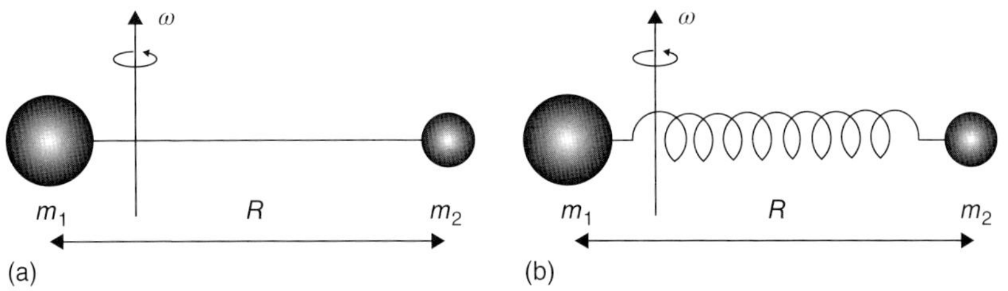

Figure 12.2 The rotational (a) and vibrational (b) motions of the two nuclei in the diatomic molecule.

Selection Rule

The transition from the initial L_i to final L_f rotational states is allowed if the transition matrix element is not zero (see Eqs. (9.42) and (9.55)). The integral involved in the matrix element is not zero, provided the molecule has the permanent dipole and L_i, L_f satisfy the selection rule

$$\Delta L \equiv L_f - L_i = \pm 1 \tag{12.12}$$

This is because the product of two nearest spherical harmonics $Y_L^M(\theta, \varphi)^* \times Y_{L\pm 1}^M(\theta, \varphi)$ and the dipole moment $\propto W(= \cos\theta)$ are the odd functions of w in the interval from -1 to 1. Hence, the matrix element is contributed by the integrand with an even parity in w and does not vanish.

The resulting spectral lines are obtained as follows. For the absorption, $\Delta L = +1$ and the frequency involved is given by $h\nu_r = E_r(L_{i+1}) - E_r(L_i)$. The corresponding frequency and wave number are thus given from Eqs. (12.11a) and (12.12) by

$$\tilde{\nu}_r \equiv \frac{1}{\lambda_r} = \frac{\nu_r}{c} = \tilde{\beta} 2(L_i + 1), \quad \tilde{\beta} \equiv \left(\frac{\hbar^2}{2\mu R_e^2}\right)\left(\frac{1}{ch}\right) \tag{12.13}$$

For emission, on the other hand, $\Delta L = -1$, and the wave number is likewise obtained as

$$\tilde{\nu}_r = \tilde{\beta}|(L_i - 1)L_i - L_i(L_i + 1)| = 2\tilde{\beta} L_i \tag{12.14}$$

Therefore, the absorption and emission spectra are shown to consist of uniformly spaced lines with the spacing $2\tilde{\beta}$. By measuring $\tilde{\beta}$, it is possible to extract the properties of the diatomic molecule, for example, the size, shape, and mass.

Vibrational Spectra

To consider the vibrational motion (see Figure 12.2), the assumption of the fixed R should be relaxed, and R should be taken to vary centered at R_e. Also, as E_r is much smaller than E_v by at least an order of magnitude, we may put $R \simeq R_e$ in the second term on the left-hand side of Eq. (12.10), in which case it is identical to E_r. Under this approximation, E_r is canceled from both sides, and Eq. (12.10) is reduced to describing the vibrational motion only. We further simplify the equation by retaining only the first term in $\varepsilon_v(R)$ in Eq. (12.9b), in which case, Eq. (12.10) reads as

$$-\frac{\hbar^2}{2\mu R^2}\frac{\partial}{\partial R}\left(R^2 \frac{\partial \rho(R)}{\partial R}\right) + \frac{1}{2}k_e(R - R_e)^2 \rho(R) = E_{vib}\rho(R), \quad k_e \equiv \left(\frac{\partial^2 E(R_e)}{\partial R^2}\right) \tag{12.15}$$

We may look for the solution of Eq. (12.15) in the form

$$\rho(R) = \frac{1}{R}\eta(R) \tag{12.16}$$

and reduce Eq. (12.15) to the energy eigenequation of the HO (see Eq. (7.1)):

$$-\frac{\hbar^2}{2\mu}\frac{d^2}{\partial\xi^2}\eta(\xi)+\frac{1}{2}k_e\xi^2\eta(\xi)=E_{vib}\eta(\xi),\quad \xi\equiv R-R_e \tag{12.17}$$

Therefore, we can use all the results obtained in treating HO. The vibrational energy is then given from Eq. (7.12) by

$$E_{vib}=h\nu_{vib}(v+1),\quad v=0,1,2,\ldots;\nu_{vib}=\frac{1}{2\pi}\left(\frac{k_e}{\mu}\right)^{1/2} \tag{12.18}$$

Selection Rule

The transition from the initial v_i to final v_f vibrational states is induced by the oscillating dipole of the molecule. Hence, the transition matrix is proportional to the dipole moment

$$\langle\mu_v\rangle=\int_0^{2\pi}d\varphi\int_0^{\pi}\sin\theta d\theta\int_0^{\infty}R^2dR|Y_L^M|^2\frac{\eta_{vf}(R)}{R}[\mu(R-R_e)]\frac{\eta_{vi}(R)}{R}$$
$$\simeq\int_{-\infty}^{\infty}\eta_{vf}(\xi)\mu(\xi)\eta_{vi}(\xi)d\xi;\quad \xi\equiv R-R_e,\ \mu(\xi)\propto\xi \tag{12.19}$$

where the integration over θ, φ yields unity because of the orthonormality of the spherical harmonics. It then follows from Eq. (12.19) that the dipole moment does not vanish if v_i, v_f satisfy the selection rule

$$\Delta v\equiv v_f-v_i=\pm1 \tag{12.20}$$

This is due to the fact that the product $\eta_v(\xi)\times\eta_{v\pm1}(\xi)$ is odd in ξ (see Eqs. (7.10) and (7.11)), and the dipole moment $\mu\propto\xi$ is also odd in ξ. The moment integral is thus contributed by the integrand having the even parity in ξ and does not vanish. The wave number of absorption or emission is then given from Eq. (12.18) by

$$\tilde{\nu}_{vib}\equiv\frac{1}{\lambda_v}=\frac{\nu_{vib}}{c};\quad \nu_{vib}=\frac{(k_e/\mu)^{1/2}}{2\pi} \tag{12.21}$$

and consists of a single line of frequency ν_{vib}.

Rotation–Vibration

Each vibrational line is accompanied by a number of finely spaced rotational spectral lines as shown in Figure 12.3. This is due to the fact that both transitions occur concurrently. We have analyzed the two transitions, using the rigid rotator and harmonic oscillator models, respectively. The combined energy levels are given from Eqs. (12.11) and (12.18) by

$$E_{v,L}=E_v+E_r=h\nu_{vib}\left(v+\frac{1}{2}\right)+L(L+1)\beta,\quad \beta=\frac{\hbar^2}{2\mu R_e^2} \tag{12.22}$$

with the combined selection rules given from Eqs. (12.12) and (12.20) by

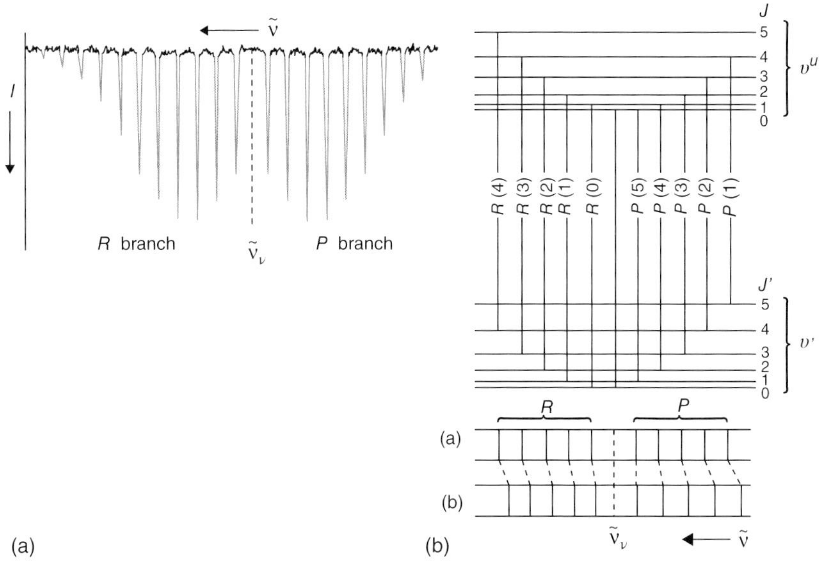

Figure 12.3 The fine structure of a single vibrational spectral line superposed by the *P* and *R* branches of the rotational spectral lines (A). The rotational energy-level diagram and the transitions involved in *P* and *R* branches of a diatomic molecule (B). (Reproduced from Atoms and Molecules, M. Karplus and R. N. Porter, W. A. Benjamin, 1970.)

a) $\Delta \nu = 1, \quad \Delta L = \pm 1$
b) $\Delta \nu = -1, \quad \Delta L = \pm 1$

Clearly, (a) and (b) represent the vibrational absorption and emission with each process accompanied by the rotational absorption and emission.

As noted, the vibrational energy is greater than that of rotation by one or two orders of magnitudes, so that a single vibrational emission or absorption line is accompanied by two groups of lines, called *branches* (Figure 12.3). The *P* branch results from the rotational emission corresponding to $\Delta L = -1$. In this case, the wave number decreases with increasing L_i and is given from Eqs. (12.14) and (12.21) by

$$\frac{1}{\lambda_p} \equiv \tilde{\nu}_P = \tilde{\nu}_{vib} - 2\tilde{\beta} L_i, \quad L_i = 1, 2, 3, \ldots \tag{12.23}$$

The *R* branch results from the rotational absorption corresponding to $\Delta L = +1$. In this case, the wave number increases with L_i and is given from Eqs. (12.14) and (12.22) by

$$\frac{1}{\lambda_R} \equiv \tilde{\nu}_R = \tilde{\nu}_{vib} + 2\tilde{\beta}(L_i + 1), \quad L_i = 1, 2, 3, \ldots \tag{12.24}$$

Figure 12.3 also shows the intensity profiles of *P* and *R* branches, which vary appreciably with L_i. The variation is due to Boltzmann probability factor. The intensity of the spectral line is proportional to the number of molecules present in the initial state, and the number is dictated in equilibrium by the Boltzmann probability factor:

$$N_i \propto g_{vL} \exp - \left(\frac{E_{vL}}{k_B T} \right) \tag{12.25}$$

where $g_{\mathrm{v,L}}$ is the degeneracy factor of the initial state. The vibrational states are free of degeneracy. Hence, $g_{\mathrm{v,L}}$ is due solely to the rotational states, resulting from the quantum number M_i varying from $-L_i$ to $+L_i$ in steps of unity. Thus, N_i is given from Eqs. (1.10) and (12.22) by

$$N_i \propto (2L_i + 1)e^{-\varepsilon_e L_i(L_i+1)/k_B T}, \quad \varepsilon_e = \frac{\hbar^2}{2\mu R_e^2} \tag{12.26a}$$

and varies as a function of L_i. We can therefore find the initial state having the largest number of molecules by differentiating N_i with respect to L_i and putting the result to zero, obtaining

$$L_{i\,\max} = \frac{1}{2}\left[\left(\frac{2k_B T}{\varepsilon_e} \right)^{1/2} - 1 \right] \tag{12.26b}$$

The result given in Eq. (12.26) is in general agreement with the observed intensity profiles of P and R branches.

12.3
Nuclear Spin and Hyperfine Interaction

A nucleus possesses the intrinsic spin angular momentum $\boldsymbol{I}$ and the magnetic moment $\boldsymbol{\mu}_N$ just like the electron. The nuclear spin can therefore be treated in strict analogy with the electron spin. We can thus introduce the eigenfunction and the commutation relations as

$$\hat{I}^2|I, m_I\rangle = I(I+1)\hbar^2|I, m_I\rangle; \quad |I, m_I\rangle = Y_I^{m_I} \tag{12.27a}$$

$$\hat{I}_z|I, m_I\rangle = m_I\hbar|I, m_I\rangle, \quad m_I = -I, -I+1, \ldots, I-1, I \tag{12.27b}$$

$$\hat{I} \times \hat{I} = i\frac{\hbar}{2}\hat{I} \tag{12.28}$$

(see Eqs. (10.1)–(10.5)). The quantum number I has half integer or integer values depending on odd or even atomic number, that is, the number of protons in the nucleus. The proton and neutron have the spin $\hbar/2$.

We can also specify μ_N in analogy with the electron magnetic moment Eqs. (10.33) and (10.34) as

$$\hat{\mu}_N = g_N \left(\frac{e}{2m_N} \right) \hat{I} = g_N \mu_{BN} \frac{\hat{I}}{\hbar}, \quad \mu_{BN} \equiv \frac{e\hbar}{2m_N} \tag{12.29}$$

where m_N is the mass of the nucleus, and the nuclear magnetron $\boldsymbol{\mu}_{BN}$ is parallel to $\boldsymbol{I}$ in this case because the nuclear charge is positive. For proton, $\boldsymbol{\mu}_{BN}$ has the value $5.049 \times 10^{-27}\,\mathrm{J\,m^2\,Wb^{-1}}$, and the gyromagnetic ratio g_N of 2×2.79268 has been determined experimentally.

Hyperfine Interaction

The hyperfine interaction is caused by the nuclear magnetic moment μ_N interacting with the magnetic field, which is induced inherently by the electron circulating the nucleus. The mechanism responsible for the interaction is therefore similar to that of the electron spin–orbit coupling. We can therefore introduce the interaction Hamiltonian in analogy with Eq. (10.37) as

$$\hat{H}'_N \equiv -\underline{\mu}_N \cdot \underline{B}_{\rm J} = \frac{g_N}{\hbar}\mu_{BN} f_N(r)(\underline{\hat{I}} \cdot \underline{\hat{j}}), \quad f_N(r)\frac{e^2\mu_0}{4\pi r^3 m_e} \tag{12.30a}$$

In this case, $\boldsymbol{B}_{\rm J}$ is induced by the electron spinning and circulating the nucleus at the radius r and can thus be specified via the Biot–Savart law as

$$\underline{B}_{\rm J} = \frac{e\mu_0}{4\pi r^3 m_e}\underline{j}, \quad \underline{j} = \underline{l} + \underline{s} \tag{12.30b}$$

Also there is no back-transformation and no Thomas 1/2 factor, as the nucleus can be taken fixed in the laboratory frame. The negative sign in Eq.(12.30a) disappears since j and B_j are anti-parallel. We can introduce the total angular momentum $\boldsymbol{F}$ and express Eq. (12.30a) in analogy with Eq. (10.37b):

$$\hat{H}'_N \equiv \frac{g_N}{2\hbar}\mu_{BN} f_N(r)(\hat{F}^2 - \hat{j}^2 - \hat{I}^2), \quad \hat{F} = \hat{j} + \hat{I} \tag{12.30c}$$

Interaction with Magnetic Field

When the external magnetic field is applied in the z-direction, the interaction Hamiltonian is given again in strict analogy with Eq. (10.38) by

$$\hat{H}'_{Jm} = g_j\frac{\mu_B}{\hbar}B\hat{j}_z - g_N\frac{\mu_N}{\hbar}B\hat{I}_z \tag{12.31}$$

The magnetic moment $\boldsymbol{\mu}_N$ is parallel to $\boldsymbol{I}$, hence the negative sign in the second term, and g_j is the Lande g-factor Eq. (10.45). Therefore, the total hyperfine interaction Hamiltonian is given by the sum of Eqs. (12.30) and (12.31):

$$\hat{H}'_{IJ} = B\left(\frac{g_j\mu_B}{\hbar}\hat{j}_z - \frac{g_N\mu_N}{\hbar}\hat{I}_z\right) + \frac{g_N}{2\hbar}\mu_{BN} f_N(r)(\hat{F}^2 - \hat{j}^2 - \hat{I}^2) \tag{12.32}$$

Hyperfine Splitting of Energy Level

The spectral lines resulting from the hyperfine interaction can be analyzed in parallel with those resulting from the spin–orbit coupling. In the absence of $\boldsymbol{B}$, the eigenfunction of $\boldsymbol{F}$ can be introduced in analogy with the eigenfunction of $\boldsymbol{l}$ given by Eq.(8.34).

$$\hat{F}^2|F, m_F\rangle = \hbar^2 F(F+1)|F, m_F\rangle \tag{12.33a}$$

$$\hat{F}_z|F, m_F\rangle = \hbar m_F|F, m_F\rangle, \quad m_F = -F, -F+1, \ldots, F-1, F \tag{12.33b}$$

The detailed derivation of the eigenfunctions is presented in the first two reference books listed at the end of this chapter. Suffice to say here that the eigenfunction can be used as the common eigenfunction for $\boldsymbol{j}^2$ and $\boldsymbol{I}^2$ as well, since the three operators all commute. The shift in the energy level due to the hyperfine interaction can therefore be evaluated precisely in analogy with Eq. (10.40) and by using Eq.(12.32) without the magnetic field:

$$\delta E_{Ij} \equiv \langle F, m_F|\hat{H}'_{Ij}|F, m_F\rangle = \frac{\hbar g_N \mu_{BN}}{2}\langle f_N(r)\rangle[F(F+1) - j(j+1) - I(I+1)] \quad (12.34)$$

and the allowed values of F for given j and I are specified by the sum rule $F = j \pm I$.

Zeeman Splitting

When a strong magnetic field is applied, we may take the hyperfine interaction term in Eq. (12.32) as a perturbing Hamiltonian, in which case $\boldsymbol{j}$ and $\boldsymbol{I}$ are decoupled. Therefore, the Zeeman splitting can be specified precisely by using the unperturbed eigenfunction as

$$\begin{aligned}\Delta E_m &= \langle I, m_I, j, m_j|\frac{g_j\mu_B}{\hbar}B\hat{j}_z - \frac{g_N\mu_{BN}}{\hbar}B\hat{I}_z|I, m_I, j, m_j\rangle \\ &= g_j\mu_B B m_j - g_N\mu_{BN}Bm_I \end{aligned} \quad (12.35)$$

Additionally, the effect of the hyperfine interaction can be incorporated by means of the time-independent perturbation theory. The resulting shift in energy is then given to the first order of approximation by

$$\begin{aligned}\Delta E_{\text{hf}} &= \frac{g_N}{\hbar}\mu_{BN}\langle I, m_I, j, m_j|f_N(r)(\underline{\hat{I}}\cdot\underline{\hat{j}})|I, m_I, j, m_j\rangle \\ &= \Lambda_N m_I m_j, \quad \Lambda_N = g_N\mu_{BN}\hbar\langle f_N(r)\rangle \end{aligned} \quad (12.36)$$

where use has been made of Eq. (9.13) and $\langle\hat{j}_x\rangle = \langle\hat{j}_y\rangle = \langle\hat{I}_x\rangle = \langle\hat{I}_y\rangle = 0$ (see Eq. (8.25)). Also the function $f_N(r)$ was treated in the same manner as $f(r)$ in the spin–orbit coupling. Hence, the total shift in energy due to the hyperfine interaction in the presence of $\boldsymbol{B}$ is obtained by summing Eqs. (12.35) and (12.36):

$$\Delta E = g_j\mu_B B m_j - g_N\mu_{BN}Bm_I + \Lambda_N m_I m_j \quad (12.37)$$

The hyperfine splitting Eq. (12.37) bears a significant effect on the ground state u_{100} of the H-atom in which $l = 0, j = s = 1/2$. For $B = 0$ F, m_F are good quantum numbers, and the allowed values of F $(= j \pm I)$ are 1 and 0. Hence, the ground state energy E_{100} splits into two according to Eq. (12.34) as

$$E(1, m_F) = \frac{\Lambda}{4}, \quad E(0,0) = -\frac{3\Lambda}{4} \quad (12.38)$$

Also the state with $F = 1$ has threefold degeneracy with m_F ranging from -1 to 1 in steps of unity, while for $F = 0$, there is no degeneracy.

Figure 12.4 shows the ground state energy E_{100} versus B. For $B = 0$, E_{100} splits into two in accordance with Eq. (12.38). In the presence of strong B, on the other hand, m_j and m_I are good quantum numbers (see Eq. (12.35)), and therefore the

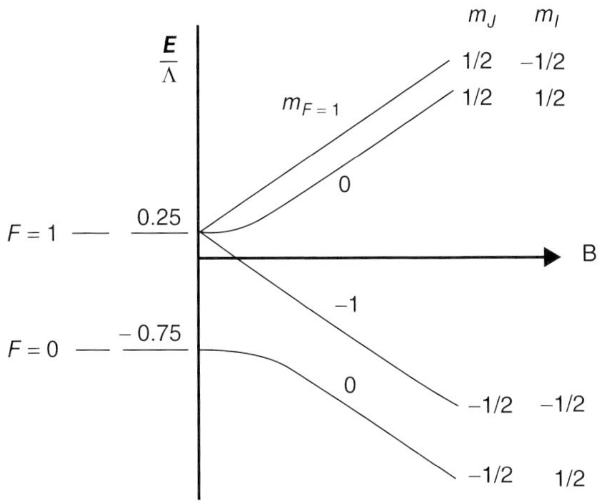

Figure 12.4 The splitting of the ground state energy of the H-atom versus the magnetic field. The splitting is caused by hyperfine interaction and the magnetic field applied. The zero field splitting amounts to ν equaling 1.420405 GHz. (Reproduced from A. Yariv, Theory and Applications of Quantum Mechanics, Wiley & Sons, 1982.)

Zeeman splitting consists of four levels as determined by four combinations of (m_j, m_I), that is, (1/2, −1/2), (1/2, 1/2), (−1/2, −1/2), and (−1/2, 1/2) in Eq. 12.31. These energy levels are raised or lowered by B depending on the polarity of m_j and m_I, but the dependence on B is primarily dictated by m_j since $\mu_B \gg \mu_{BN}$. The four levels in the strong B-field region are joined smoothly by the four levels in the weak-field region, as it should. In the latter region, the upper level for $B=0$ splits into three due to B and provides four levels when combined with the single lower level. In the region of weak magnetic field F, m_F are good quantum numbers (see Eq. (12.33)), and the $E-B$ curves are therefore dictated by m_F as evidenced by the near-zero slope in $E-B$ curves for $m_F = 0$. The wavelength associated with the transition between two split levels due to the hyperfine interaction in the absence of B amounts to

$$\lambda = \left[\frac{c}{(\Lambda/h)}\right] = 21.12\,\text{cm}$$

and is the well-known wavelength emitted by the interstellar hydrogen atom.

12.4 Nuclear Magnetic Resonance (NMR)

NMR is concerned with the transitions between magnetically split nuclear sublevels, driven by the radio-frequency magnetic field. The physical processes involved are essentially same as those operative in the electron paramagnetic resonance and can again be treated in parallel. Thus, consider a nucleus with the magnetic moment μ_N and subjected to a constant static field B in the z-direction.

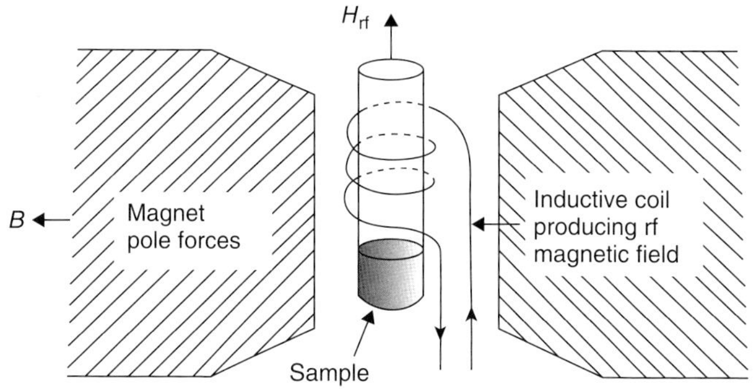

Figure 12.5 The experimental setup of the nuclear magnetic resonance. The static magnetic field is provided by the magnetic pole for inducing the splitting of the energy level. The inductive coil is used to generate the radio-frequency magnetic field, which triggers the transition. (Reproduced from A. Yariv, Theory and Applications of Quantum Mechanics.)

The energy level is then shifted according to Eq. (12.35) as

$$\langle I, m_I| - \frac{g_N \mu_{BN}}{\hbar} B |I, m_I\rangle = -g_N \mu_{BN} B m_I \tag{12.39}$$

and splits into $2I + 1$ sublevels separated by the uniform spacing $g_N \mu_{BN} B$ due to m_I ranging from $-I$ to I in steps of unity. Hence, the resonance transition occurs when the driving frequency satisfies the condition

$$h\nu_R = g_N \mu_{BN} B \tag{12.40}$$

For proton, $\mu_{BN} = 5.049 \times 10^{-27}\,\mathrm{J\,m^2\,Wb^{-1}}$ and $g_N = 2 \times 2.79268$, so that ν_R has the value 42.58 MHz for $B = 1\,\mathrm{Wb\,m^{-2}}$.

The NMR experiment is carried out by placing the sample between two pole faces of electromagnets, subjecting it to a static $\boldsymbol{B}$-field, inducing thereby the Zeeman splitting of the energy level. Concurrently, the radio-frequency field is applied in the direction perpendicular to $\boldsymbol{B}$ to trigger the transition (Figure 12.5). The absorption of power can be treated in a way similar to that of the electron paramagnetic resonance. The NMR has become an important tool in chemistry and condensed matter physics. The high detection sensitivity of the resonant absorption lines enables the sensing of minute variations of the magnetic field as seen by the nucleus. The variations of such fields are caused by the shielding of B by the electrons in the molecule, and the resulting shift of ν_R is called the *chemical shift*. The field also varies depending on the nature and symmetry of environs of the nucleus. This is illustrated in Figure 12.6, in which three resonant lines from ethyl alcohol are shown. The areas under these lines differ because of the varying number of protons participating in the transition. The largest, medium, and smallest lines shown are due to three protons in CH_3, two protons in CH_2, and a single proton in CH, respectively.

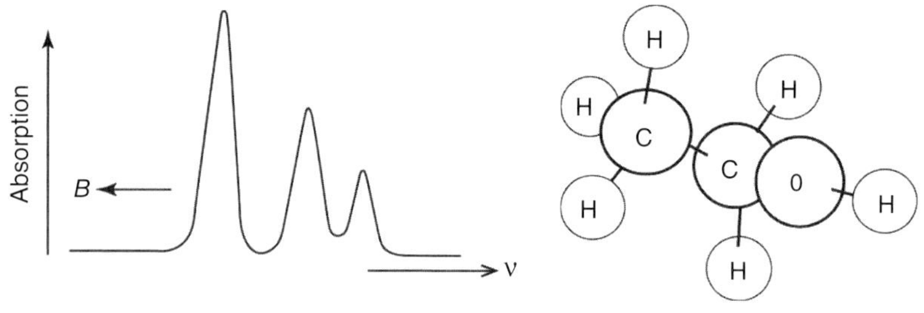

Figure 12.6 Nuclear paramagnetic resonance spectral lines resulting from the H-atoms in ethyl alcohol (CH_3CH_2OH) interacting with the B-field. The peaks are due to three, two, and one protons in CH_3, CH_2, and OH, respectively. (Reproduced from N.F. Ramsey, Nuclear Moments, J. Wiley & Sons, 1953.)

12.4.1 Molecular Imaging

In recent years, NMR has become an efficient tool by which to image the spatial distribution of atomic nuclei inside the body. Furthermore, the capability of measuring the spatial variations of the nuclear spin relaxation times has opened a novel field of the diagnostic medicine and the *in vivo* NMR in biological systems.

The imaging scheme is based on the resonant matching between ν_R and B as given by Eq. (12.40). If B is uniform in space, the measured ν_R is constant and contains no spatial information. However, when the spatial gradient of $\boldsymbol{B}$ is introduced as shown in Figure 12.7, ν_R also varies tracing the gradient. In this manner, NMR signals contain the spatial information in coded form. Thus, by measuring the distribution of ν_R and the NMR signal magnitude, and by transcribing the data, it is possible to image the distributed configuration of the nuclei.

Figure 12.7 specifically illustrates the scheme for such imaging. Given a spatial distribution of protons in the z-direction, for example, a static magnetic field with linear gradient is applied, distributing the B-field in space. Then, with the use of radio-frequency B-field, a series of resonant frequencies ν_R is measured. The measured data of ν_R and the signal strength can be transcribed into the spatial configuration of nuclei. In practice, it is expedient to use the pulsed radio-frequency field. In this case, the resonant frequencies are used concurrently instead of sweeping them, and the flipping of nuclear spin is enhanced as in the case of the $\pi/2$ pulse experiment. The resulting pulsed NMR signal $I(t)$ can be converted to constructing the image via the Fourier transformation

$$I(\omega) = \int_{-\infty}^{\infty} dtI(t)e^{-i\omega t} \tag{12.41}$$

The frequency spectrum $I(\omega)$ extracted from the $I(t)$ data provides the same image information.

Problems

12.1 The radial energy eigenequation (12.10) provides the basis for treating the vibrational and rotational nuclear motions.

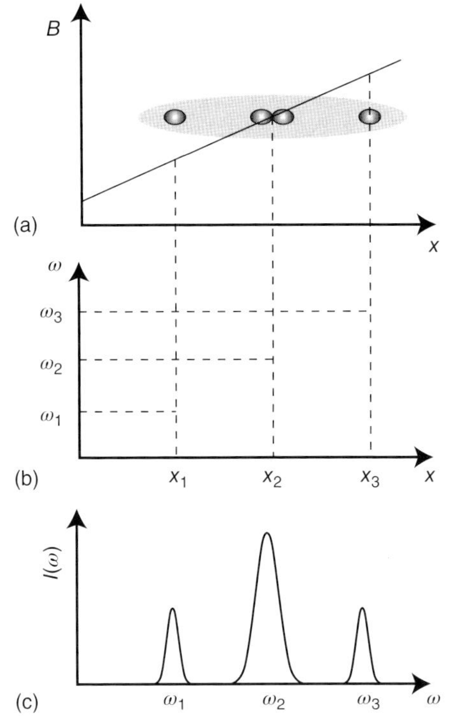

Figure 12.7 The schematics of the molecular imaging via the NMR technique. The spatial distribution of protons to be probed (a). The local resonant frequencies matched with spatially distributed B-field (b). The measured intensity and resonance frequency (c).

(a) Starting from Eq. (12.4), fill in the algebra and derive Eq. (12.6) by using the Born approximation Eq. (12.5).
(b) Starting from Eq. (12.6), fill in the algebra and derive Eq. (12.10) by using Eq. (12.7).

12.2 Consider a diatomic molecule of mass m_1, m_2 and separated by the bond length R.
(a) Show that the moment of inertia of the molecule is given by $I = \mu R^2$ with μ denoting the reduced mass $1/\mu = 1/m_1 + 1/m_2$.
(b) Calculate the moment of inertia of the diatomic molecules H_2, for which $R = 74$ pm, and HCl, for which $R = 126$ pm ($1\ \text{pm} = 10^{-3}$ nm).
(c) Estimate the driving frequencies for inducing the rotational transitions.

12.3 Consider the same diatomic molecules H_2 and HCl.
(a) Find the reduced mass of vibration in each molecule.
(b) If the wave numbers of the vibrational spectrum are 4400.39 and 4138.32 cm^{-1}, respectively, find the effective spring constants and the bonding force.

12.4 Consider a 2D oscillator oscillating in x-, y-directions with the spring constants k_x, k_y, respectively.
Find the energy eigenfunction and eigenvalue for the cases $k_x \neq k_y$, $k_x = k_y$ and discuss the results.

12.5 The observed wave number corresponding to the transition between the vibrational ground and the first excited states of diatomic molecules are $2990.3\,\text{cm}^{-1}$ for D_2, $2143.3\,\text{cm}^{-1}$for CO, $1556.3\,\text{cm}^{-1}$ for O_2, and $378.0\,\text{cm}^{-1}$ for NaCl.
(a) Find the reduced mass of each molecule.
(b) Find the effective spring constant in newton per meter (Nm^{-1}) and binding force.
(c) Find the zero-point energy.

12.6 The diatomic molecule HCl has the following structural data: bond length of 127.5 pm, force constant of the bond $516.3\,\text{N}\,\text{m}^{-1}$, atomic masses of 1.67×10^{-27} for H, and 58.066×10^{-27} kg for Cl, respectively.
(a) Find the vibrational frequency and zero-point energy and
(b) the frequencies of the innermost three P and R lines.

12.7 Describe the flipping of the nuclear spin in the NMR experiment by carrying out a parallel analysis of the electron paramagnetic resonance, that is, by setting up the coupled equation of the two spin states and solving the equation.

Suggested Readings

1. A. Yariv, *An Introduction to Theory and Applications of Quantum Mechanics*, John Wiley & Sons, 1982.
2. A. I. M. Rae, *Quantum Mechanics*, Fourth Edition, Taylor & Francis, 2002.
3. H. Haken and H. C. Wolf, *The Physics of Atoms and Quanta: Introduction to Experiments and Theory*, Fifth Edition, Springer, 2004.
4. H. Haken and H. C. Wolf, *Molecular Physics and Elements of Quantum Chemistry*, Second Edition, Springer, 2004.
5. M. Karplus and R. N. Porter, *Atoms and Molecules: An Introduction for Students of Physical Chemistry*, Addison Wesley Publishing Company, 1970.

13
Atom–Field Interaction

The interaction between the atom and the electromagnetic (EM) field is one of the most important phenomena and is discussed in this chapter. Both semiclassical and quantum mechanical treatments of the interaction are presented in conjunction with the quantized EM field. Also, the stimulated and spontaneous emissions of radiation are highlighted together with the dynamics of a two-level atom driven by the EM field.

13.1
Atom–Field Interaction: Semiclassical Treatment

In the semiclassical description, the field is treated classically, while the atom is treated quantum mechanically. Thus, consider a single atom in resonant interaction with the EM field. If the driving frequency matches closely with the atomic transition frequency between two given levels, the coupling of the field with other levels can be neglected, and we can model the atom as the two-level atom (Figure 13.1).

The atom interacting with the EM field with frequency ω, amplitude E_0, and the polarization vector $\boldsymbol{e}_f$ is described by the dipole interaction Hamiltonian

$$\hat{H}' = -e\underline{\mathrm{E}} \cdot \underline{r} = -\hat{\mu}\frac{\mathrm{E}_0}{2}(e^{i\omega t} + e^{-i\omega t}); \quad \hat{\mu} \equiv e(\hat{e}_f \cdot \underline{r}) \tag{13.1}$$

where $-e\mathbf{E}$ is the force acting on the electron $\boldsymbol{r}$ displacement from the nucleus. The wavelength of the field is much larger than the atomic dimension; hence, the field amplitude E_0 can be taken constant. When the interaction Hamiltonian in Eq. (13.1) is inserted into Fermi's golden rule (Eq. (9.54)), the transition rate between two atomic states u_1, u_2 is given by

$$W_i = \frac{2\pi}{\hbar}|\hat{H}'_{12}|^2\delta(E_2 - E_1 - \hbar\omega), \quad |\hat{H}'_{12}|^2 = \frac{\tilde{\mu}^2\mathrm{E}_0^2}{4}, \quad \tilde{\mu} \equiv e\langle u_1|\hat{e}_f \cdot \underline{r}|u_2\rangle \tag{13.2}$$

where $\tilde{\mu}$ is the atomic dipole moment. In practice, the energy levels E_1, E_2 are not sharply defined but broadened due to the finite lifetime τ of the electron in each level. The level broadening is generally specified via the uncertainty relation $\Delta\mathrm{E} \approx \hbar/\tau$, and τ is generally short because of the collisions the atom encounters.

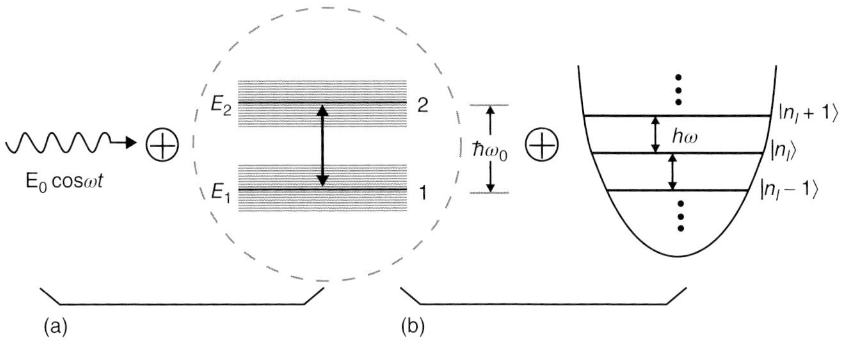

Figure 13.1 The classical (a) and quantum (b) descriptions of the two-level atom driven by classical and quantized fields, respectively. In the quantum treatment, the atom is coupled in essence to a harmonic oscillator.

We can therefore treat $E_2 - E_1$ as a random variable and introduce the lineshape factor $g(E_2 - E_1)$ in integrating Eq. (13.2) over the broadened energy level

$$\begin{aligned} W_i &= \frac{\pi\widetilde{\mu}^2 \mathrm{E}_0^2}{2\hbar}\int_{-\infty}^{\infty} d\varepsilon g(\varepsilon)\delta(\varepsilon - \hbar\omega), \quad \varepsilon = E_2 - E_1 \\ &= \frac{\pi\widetilde{\mu}^2 \mathrm{E}_0^2}{2\hbar} g(\hbar\omega) = \frac{\widetilde{\mu}^2 \mathrm{E}_0^2}{4\hbar^2} g(\nu) \end{aligned} \tag{13.3}$$

where the densities of states in ν- and E-spaces represent the identical number of quantum states, that is, $g(\mathrm{E})d\mathrm{E} = g(\hbar\omega)d(\hbar\omega) = g(\nu)d\nu$.

Stimulated and Spontaneous Transitions

We next consider an ensemble of atoms interacting with the EM field in thermodynamic equilibrium. The number of atoms in each level is then determined by the Boltzmann probability factor as discussed, so that the ratio is given by

$$\frac{N_2}{N_1} = e^{-(E_2 - E_1)/k_B T}; \quad N_j \propto e^{-E_j/k_B T}, \quad j = 1, 2 \tag{13.4}$$

Also every process is balanced by its inverse process in equilibrium, and therefore the number of atoms making the transition from upper to lower level must be equal to that of its inverse transition, that is, $N_2 W_i = N_1 W_i$. But, this equality is in fundamental contradiction with the Boltzmann probability factor, which states that $N_1 > N_2$.

Einstein *A* Coefficient

The apparent inconsistency was resolved by Einstein, who introduced an additional mode of transition from upper to lower level

$$W_{2\to 1} = B\rho(\nu) + A; \quad B\rho(\nu) \propto W_i \propto \widetilde{\mu}^2 \mathrm{E}_0^2 \tag{13.5}$$

where the first term proportional to the field energy density $\rho(\nu)$ (see Eq. (2.3)) represents the stimulated emission. The second term A accounts for additional transition occurring free of the field intensity. As mentioned, the transition rate from the lower to upper level is the same as the first term $W_{1\to 2} = B\rho(\nu)$ in Eq. (13.5). Hence, the balance $N_2 W_{2\to 1} = N_1 W_i$ between two opposing transitions is given by

$$\frac{N_2}{N_1} = \frac{B\rho(\nu)}{[B\rho(\nu) + A]} \tag{13.6}$$

By inserting Eqs. (13.4) and (2.3) into Eq. (13.6), we can write

$$\frac{1}{e^{h\nu/k_B T}} = \frac{1}{1 + (A/B\rho(\nu))} = \frac{1}{1 + (A/B)(c^3/8\pi n^3 h\nu^3)\left(e^{h\nu/k_B T} - 1\right)}; \quad E_2 - E_1 = h\nu \tag{13.7}$$

where ν is the frequency of emission or absorption $h\nu = E_2 - E_1$ and n the index of refraction accounting for the velocity of light in the medium. It is therefore clear from Eq. (13.7) that the detailed balancing holds true, provided

$$\frac{A}{B} = \left(\frac{c^3}{8\pi n^3 h\nu^3}\right)^{-1} \tag{13.8}$$

The constant A is called the *Einstein A coefficient* and represents the spontaneous emission of radiation that occurs irrespective of the presence or absence of the field.

The role of A is best seen by considering an ensemble of atoms prepared in the upper state $u_2(r)$ in the absence of the field. Then, the decay rated of N_2 is governed by

$$\frac{\partial N_2}{\partial t} = -AN_2 \tag{13.9}$$

so that the electron lifetime and the number of atoms in the upper state at t are given by

$$\tau_{sp} \equiv \frac{\int_0^\infty dttN_2(t)}{\int_0^\infty dtN_2(t)} = \frac{1}{A}; \quad N_2(t) = N_2(0)e^{-At} \tag{13.10}$$

Moreover, as A is commensurate with B or the dipole matrix element $\tilde{\mu}^2$ (see Eqs. (13.8) and (13.5)), the spontaneous transition rate is a property inherent in each atomic species.

13.2 Driven Two-Level Atom and Atom Dipole

We next consider the atom – field interaction. There are two regimes of interaction, namely, the collisionless and the collision-dominated regimes. In the former, the field-induced transition time is shorter than the mean collision time, while in the latter, the collision time is much shorter instead.

The Schrödinger equation of the driven two-level atom is given from Eq. (13.1) by

$$i\hbar\frac{\partial}{\partial t}\psi(\underline{r},t) = [\hat{H}_0 + \hat{H}'(t)]\psi(\underline{r},t) \tag{13.11}$$

where $\hat{H}_0$ is the unperturbed Hamiltonian and $\hat{H}'$ the interaction Hamiltonian equation (13.1). The wavefunction of the two-level atom can be generally represented in terms of the two eigenfunctions and is given by

$$\psi(\underline{r},t) = \sum_{j=1}^{2} a_j(t)e^{-i(E_j/\hbar)}|u_j\rangle; \quad \hat{H}_0|u_j\rangle = E_j|u_j\rangle, \quad j = 1,2 \tag{13.12}$$

By inserting Eq. (13.12) into Eq. (13.11) and carrying out the usual inner product with respect to u_1, u_2 on both sides, we obtain the coupled equation in analogy with Eq. (10.53) as

$$\dot{a}_1 = i\frac{\widetilde{\mu}\mathrm{E}(t)}{\hbar}a_2e^{-i\omega_0 t}, \quad \omega_0 \equiv \frac{E_2 - E_1}{\hbar}, \quad \widetilde{\mu} \equiv e\langle 1|\hat{e}_f \cdot \underline{r}|2\rangle \tag{13.13a}$$

$$\dot{a}_2 = i\frac{\widetilde{\mu}\mathrm{E}(t)}{\hbar}a_1e^{i\omega_0 t} \tag{13.13b}$$

where ω_0 is the atomic transition frequency and $\widetilde{\mu}$ the dipole moment. In deriving Eq. (13.13), $u_1(\boldsymbol{r})$ and $u_2(\boldsymbol{r})$ have been taken even and odd in $\boldsymbol{r}$ or vice versa, so that the diagonal matrix element $\propto \langle j|\hat{e}_f \cdot \underline{r}|j\rangle$ is zero, while $\langle u_1|\hat{e}_f \cdot \underline{r}|u_2\rangle \neq 0$.

We next consider the resonant interaction in the collisionless regime in which the driving frequency ω is equal to the transition frequency ω_0. We can then employ the rotating wave approximation and neglect the rapidly oscillating terms $\propto \exp \pm(\omega + \omega_0)$ and obtain straightforwardly from Eq. (13.13)

$$\dot{a}_1 = i\Omega a_2; \quad \Omega \equiv \frac{\widetilde{\mu}\mathrm{E}_0}{2\hbar}; \quad \mathrm{E}(t) = \frac{\mathrm{E}_0}{2}(e^{i\omega t} + e^{-i\omega t}) \tag{13.14a}$$

$$\dot{a}_2 = i\Omega a_1 \tag{13.14b}$$

where Ω is the transition frequency. The coupled equation (13.14) is identical to Eq. (10.53), and we can use the results obtained already. For $a_1(0) = 1$, $a_2(0) = 0$, the solution is given by

$$a_1(t) = \cos\Omega t; \quad a_2(t) = i\sin\Omega t \tag{13.15}$$

and is known as *Rabi flopping formula*, describing the electron swinging between two states with the transition frequency Ω. Also the total probability is conserved, namely, $|a_1(t)|^2 + |a_2(t)|^2 = 1$, as it should (see Figure 13.2).

In the collision-dominated regime, on the other hand, $a_1(t)$, $a_2(t)$ decay rapidly in time, and the oscillation is damped. If the decay time is much shorter than the flipping period, the change in time of $a_1(t)$, $a_2(t)$ is small. In this case, the atom–field interaction simply yields the probability of a photon being absorbed

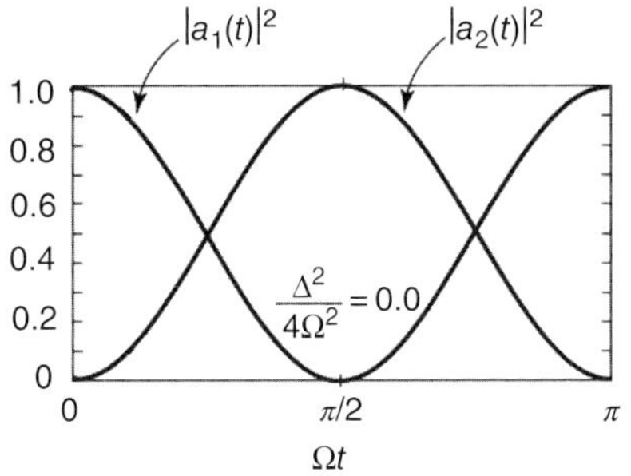

Figure 13.2 The Rabi flopping curve versus time Ωt, describing the evolution in time of the expansion coefficients of the two-level atom, driven by a resonant harmonic field.

or emitted. For example, for the initial condition under consideration, the probability of the photon being absorbed is given by $|a_2(\tau)|^2 \approx \Omega^2\tau^2$ with τ denoting the mean interaction time.

Atom Dipole

We next consider the evolution in time of the atom dipole by using the Rabi flopping formulae Eq. (13.15). When the atom is prepared in upper or lower state, the wavefunction is given by

$$|\psi\rangle = e^{-i\omega_j t}|u_j(\underline{r})\rangle; \quad j = 1 \text{ or } 2 \tag{13.16}$$

and the atom does not possess the dipole moment, that is,

$$\langle u_j|\hat{\mu}|u_j\rangle = 0; \quad \hat{\mu} \equiv e\hat{e}_f \cdot \underline{r}, \quad j = 1 \text{ or } 2 \tag{13.17}$$

regardless of whether u_j is even or odd in $\boldsymbol{r}$. However, when the atom is driven by the field, $a_1(t)$, $a_2(t)$ change in time according to Eq. (13.15). Consequently, the wavefunction evolves into a linear superposition of u_1 and u_2 (see Eq. (13.12)). Moreover, the atom dipole moment

$$\langle\hat{\mu}\rangle = \langle\psi|\hat{\mu}|\psi\rangle = -\tilde{\mu}[a_1^*(t)a_2(t)e^{-i\omega_0 t} + a_1(t)a_2^*(t)e^{i\omega_0 t}] \tag{13.18}$$

oscillates with the atomic transition frequency ω_0 (Figure 13.3). An oscillating electric dipole is well known to emit or absorb the radiation, and therefore a photon is emitted or absorbed by the oscillating atom dipole.

13.3 Atom–Field Interaction: Quantum Treatment

13.3.1 Field Quantization

In the quantum treatment of the interaction, both field and atom are treated quantum mechanically (Figure 13.1). Thus, consider the field quantization, using the resonator, consisting of two parallel metallic plates, L distance apart,

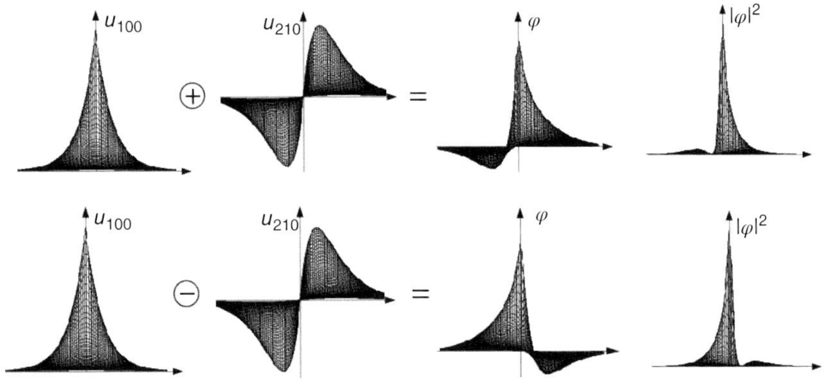

Figure 13.3 The linear superposition of the ground (u_{100}) and first excited (u_{210}) states of the H-atom with equal probability. The resulting atom dipole oscillates in time between two limits ($u_{100} + u_{210}$) and ($u_{100} - u_{210}$).

say in the z-direction. The standing-wave EM fields therein can generally be represented by

$$\underline{E}_l = \hat{y}\sqrt{\frac{2}{V\varepsilon}}p_l(t)\sin k_l z, \quad k_l L = l\pi, \quad l = 1, 2, \ldots \tag{13.19a}$$

$$\underline{H}_l = \hat{x}\sqrt{\frac{2}{V\mu}}q_l(t)\omega_l \cos k_l z \tag{13.19b}$$

where V is the volume of the resonator, and the wave vector k should satisfy the standing-wave boundary condition, so that $\mathbf{E}_l$ vanishes at $z = 0, L$.

Naturally E_l, H_l thus represented should satisfy the wave equation or equivalently Maxwell's equations. Specifically, when Faraday's law of induction (Eq. (1.21)) and Ampere's circuital law (Eq. (1.22)) are applied to Eqs. (13.19a) and (13.19b) in the medium free of charge and current, there results

$$\nabla \times \hat{y}E_l \equiv -\hat{x}\sqrt{\frac{2}{V\varepsilon}}k_l p_l(t)\cos k_l z = -\hat{x}\mu\sqrt{\frac{2}{V\mu}}\omega_l \dot{q}_l \cos k_l z \tag{13.20a}$$

$$\nabla \times \underset{\rightarrow}{H} \equiv \hat{y}(-k_l)\sqrt{\frac{2}{V\mu}}\omega_l q_l(t)\sin k_l z = \hat{y}\varepsilon\sqrt{\frac{2}{V\varepsilon}}\dot{p}_l(t)\sin k_l z \tag{13.20b}$$

As the dispersion relation $k = \omega\sqrt{\mu\varepsilon}$ of the EM wave has to hold, Eqs. (13.20a) and (13.20b) are satisfied, provided

$$p_l(t) = \dot{q}_l(t), \quad \dot{p}_l(t) = -\omega_l^2 q_l(t) \tag{13.21}$$

We can decouple p_l, q_l by differentiating Eq. (13.21) with respect to t, obtaining

$$\ddot{q}_l(t) = \dot{p}_l = -\omega_l^2 q_l(t) \tag{13.22a}$$

$$\ddot{p}_l(t) = -\omega_l^2 \dot{q}_l = -\omega_l^2 p_l(t) \tag{13.22b}$$

Obviously, Eq. (13.22) is precisely the differential equation of the HO. In addition, the energy residing in the lth-mode standing wave can be obtained by integrating field energy density over the cavity volume and is given from Eq. (13.19) by

$$\hat{H}_l = \int_0^L Adz \left(\frac{\varepsilon \underline{E}_l \cdot \underline{E}_l}{2} + \frac{\mu \underline{H}_l \cdot \underline{H}_l}{2} \right) = \frac{1}{2} p_l^2(t) + \frac{1}{2} \omega_l^2 q_l^2(t) \tag{13.23}$$

and is identical to the Hamiltonian of the HO with unit mass and frequency ω_l. In performing the integration, use has been made of the condition $k_l L = l\pi$ and the identities $\cos^2 x = (1 + \cos 2x)/2$, $\sin^2 x = (1 - \cos 2x)/2$, and A is the cross-sectional area. Evidently, $p_l(t)$, $q_l(t)$ are conjugate variables obeying Hamilton's equation of motion (1.6).

The field quantization consists in essence of taking q_l, p_l as operators obeying the commutation relation of the canonically conjugate variables

$$[q_l, p_{l'}] = i\hbar \delta_{ll'} \tag{13.24}$$

(see Eq. (3.27)). With the commutation relation thus imposed, the roles of q_l, p_l in the lth standing wave have become identical to those of x and p_x of the HO. Therefore, the quantum treatment of the EM field can be done in strict analogy with the operator treatment of the HO. Thus, we can introduce the annihilation and creation operators in analogy with Eq. (7.37) as

$$\begin{pmatrix} a_l \\ a_l^+ \end{pmatrix} = \left(\frac{1}{2\hbar\omega_l} \right)^{1/2} (\omega_l q_l \pm i p_l) \tag{13.25}$$

in which case the commutation relation is given from Eq. (13.24) by

$$[a_l, a_{l'}^+] = \delta_{ll'} \tag{13.26}$$

Moreover, by using Eqs. (13.25) and (13.26), the Hamiltonian in Eq. (13.23) can be expressed in strict analogy with Eq. (7.40) as

$$\hat{H}_l = \hbar\omega_l \left(a_l^+ a_l + \frac{1}{2} \right) \tag{13.27}$$

Finally, the eigenfunctions $\{u_n\}$ of the HO can be used as the eigenfunctions of the lth standing wave, and we can write again in strict analogy with Eqs. (7.41)–(7.43)

$$a_l |n_l\rangle = \sqrt{n} |n_l - 1\rangle; \quad |n_l\rangle \equiv |u_l\rangle \tag{13.28a}$$

$$a_l^+ |n_l\rangle = \sqrt{n+1} |n_l + 1\rangle \tag{13.28b}$$

so that

$$a_l^+ a_l |n_l\rangle = \sqrt{n} a_l^+ |n_l - 1\rangle = n |n_l\rangle \tag{13.28c}$$

and

$$\hat{H} |n_l\rangle = \hbar\omega_l \left(a_l{}^+ a_l + \frac{1}{2} \right) |n_l\rangle = \hbar\omega_l \left(n + \frac{1}{2} \right) |n_l\rangle \tag{13.28d}$$

The quantum number n in this case represents the number of photons carrying the quantum of energy $\hbar\omega_l$. Also the operators a_l^+, a_l raise and lower the eigenstate $|n_l\rangle$ by creating and annihilating a photon.

The single-mode treatment can be extended straightforwardly, and the total Hamiltonian in the resonator is given by

$$\hat{H} = \sum_{l=1}^{\infty} \hat{H}_l, \quad \hat{H}_l = \hbar\omega_l \left(a_l^+ a_l + \frac{1}{2}\right) \tag{13.29}$$

with the eigenfunction represented by the product of single-mode eigenfunctions

$$\psi_f = \prod_l u_l = |n_1, n_2, \ldots, n_l, \ldots\rangle \tag{13.30a}$$

and the eigenvalues given by

$$\hat{H}|n_1, n_2, \ldots, n_l, \ldots\rangle = \sum_{l=1}^{\infty} \hbar\omega_l \left(n_l + \frac{1}{2}\right) |n_1, n_2, \ldots, n_l, \ldots\rangle \tag{13.30b}$$

Thus, the field energy in the resonator is represented by the sum of eigenenergies of a denumerable infinite set of harmonic oscillators.

EM Field as Operators

The operator representation of EM field can be made by considering the time rate of change of a_l and a_l^+ with the use of Eqs. (13.22) and (13.25):

$$\begin{pmatrix} \dot{a}_l \\ \dot{a}_l^+ \end{pmatrix} = \left(\frac{1}{2\hbar\omega_l}\right)^{1/2} (\omega_l \dot{q}_l \pm i\dot{p}_l)$$

$$= \mp i\left(\frac{1}{2\hbar\omega_l}\right)^{1/2} \omega_l(\omega_l q_l \pm i p_l) \equiv \mp i\omega_l \begin{pmatrix} a_l \\ a_l^+ \end{pmatrix} \tag{13.31}$$

Hence, a simple time integration of Eq. (13.31) yields

$$a_l(t) = a_l(0)e^{-i\omega_l t}; \quad a_l^+(t) = a_l^+(0)e^{i\omega_l t} \tag{13.32}$$

and upon expressing q_l, p_l in terms of a_l^+, a_l by using Eq. (13.25), the lth standing-wave mode Eq. (13.19) can be represented in terms of $a_l(t)$ and $a_l^+(t)$ as

$$\underline{E}_l = \hat{y} i \sqrt{\frac{\hbar\omega_l}{V\varepsilon}} [a_l^+(t) - a_l(t)] \sin k_l z \tag{13.33a}$$

$$\underline{H}_l = \hat{x} \sqrt{\frac{\hbar\omega_l}{V\mu}} [a_l^+(t) + a_l(t)] \cos k_l z \tag{13.33b}$$

We can also express Eq. (3.33) in terms of the EM field traveling in the $\boldsymbol{k}$-direction by combining Eqs. (13.32) and (13.33). Specifically, the standing-wave modes $\sin k_l z$, $\cos k_l z$ can be transformed into the traveling modes $\exp \pm i(\omega t - \underline{k} \cdot \underline{r})$, and at the same time, **E**, ***H***, and ***k*** can be made mutually perpendicular as required by Maxwell's equations:

$$\underline{E}_k = i\underline{e}_{k\lambda}\sqrt{\frac{\hbar\omega_k}{2V\varepsilon}}[a^+_{k\lambda}(t)e^{-i\underline{k}\cdot\underline{r}} - a(t)_{k\lambda}e^{i\underline{k}\cdot\underline{r}}] \tag{13.34a}$$

$$\underline{H}_{k\lambda} = \left(\underline{e}_{k\lambda} \times \frac{\underline{k}}{k}\right)\sqrt{\frac{\hbar\omega_k}{2V\mu}}[a^+_{k\lambda}(t)e^{-i\underline{k}\cdot\underline{r}} + a_{k\lambda}(y)e^{i\underline{k}\cdot\underline{r}}] \tag{13.34b}$$

where $\boldsymbol{e}_{k\lambda}$ is the polarization vector.

Quantum Treatment of Spontaneous Emission

The interaction Hamiltonian Eq. (13.1) can therefore be expressed in terms of the quantized EM field as

$$\widehat{H}' = -e\underline{E}\cdot\underline{r} = -ie(\widehat{e}_{l\lambda}\cdot\underline{r})\sqrt{\frac{\hbar\omega_l}{2V\varepsilon}}(a^+_{l\lambda}(t)e^{-i\underline{k}\cdot\underline{r}} - a_{l\lambda}(t)e^{i\underline{k}\cdot\underline{r}}) \tag{13.35}$$

and it couples in effect a two-level atom and a harmonic oscillator with frequency ω_l as illustrated in Figure 13.1. The transition rate Eq. (13.2) is then given from Eq. (13.35) by

$$\begin{aligned} W &= \frac{2\pi\omega_l}{2V\varepsilon}\sum_{\lambda=1}^{2}|\langle u_1, n_l+1|e(\widehat{e}_{l\lambda}\cdot\underline{r})(a^+_{k\lambda}(t)e^{-i\underline{k}\cdot\underline{r}} - a_{k\lambda}(t)e^{i\underline{k}\cdot\underline{r}})|u_2, n_l\rangle|^2\delta(E_2 - E_1 - \hbar\omega) \\ &= \frac{2\pi\omega_l}{2V\varepsilon}\sum_{\lambda=1}^{2}\widetilde{\mu}^2_\lambda(n_l+1)\delta(E_2 - E_1 - \hbar\omega) \equiv W^l_{\text{ind}} + W^l_{sp};\quad \widetilde{\mu}_\lambda \equiv e\langle u_1|\widehat{e}_{l\lambda}\cdot\underline{r}|u_2\rangle \end{aligned} \tag{13.36}$$

where the orthonormality of the set of eigenfunctions $\{u_l\}$ has been used and the two polarizations of the wave have been summed over. Thus, W consists of the matrix element connecting the two states, that is, the atom in the lower level and the field in the $(n+1)$ photon state and atom in the upper level with the field in the n photon state. The total energy before and after the transition is then same, namely, $E_2 + n\hbar\omega \approx E_1 + (n+1)\hbar\omega$.

In this manner, the transition rate W is shown to consist of two terms in the quantum treatment of the field. The first term W_{ind} is proportional to the number of photons, n_l, or the light intensity and represents the stimulated emission of radiation. The second term W_{sp} is independent of n_l and should correspond to the spontaneous emission. As the latter term does not rely on n_l, we have to sum over the entire standing-wave modes to obtain the total spontaneous emission rate:

$$W_{sp} \equiv \int_0^\infty d\nu_l W_{sp}^{(l)} \rho_f(\nu_l) V, \quad \rho_f(\nu_l) = \frac{8\pi\nu_l^2 n^3}{c^3}$$
$$= \frac{2\pi\widetilde{\mu}^2}{2V\varepsilon}\int_0^\infty d\nu_l \omega_l \rho_f(\nu_l) V \delta(E_2 - E_1 - h\nu_l)$$
$$= \frac{16\pi^3\widetilde{\mu}^2\nu_0^3 n^3}{\varepsilon h c^3}; \quad h\nu_0 = E_2 - E_1, \quad \delta(h\nu) = \frac{\delta(\nu)}{h} \tag{13.37}$$

where $\rho_f(\nu_l)$ is the density of standing wave modes (Eq. (2.1)).

To identify W_{sp} to Einstein A coefficient, let us formally equate W_{sp} to A in Eq. (13.8) and find $B = \widetilde{\mu}^2/2\varepsilon\hbar^2$. In this case, the induced transition rate in Eq. (13.5) should read as

$$W_{\text{ind}} \equiv B\rho(\nu) = \frac{\widetilde{\mu}^2}{2\varepsilon\hbar^2}\rho(\nu)$$

and should by definition be identical to the transition rate W_i (Eq. (13.3)) as given by Fermi's golden rule. This necessitates the correspondence

$$\frac{\varepsilon \mathrm{E}_0^2}{2} g(\nu) \leftrightarrow \rho(\nu)$$

and obviously the left-hand side is the field energy density at the driving frequency, while the right-hand side denotes the energy density of the radiation field at the same frequency. Obviously, these two quantities are identical. Therefore, the spontaneous emission of radiation is shown an inherent property of the quantized EM field.

Problems

13.1 (a) Starting from the wave equation (13.11), derive the coupled equations involving the expansion coefficients $a_1(t)$, $a_2(t)$ in Eq. (13.13) by using the wavefunction 13.12 and performing the appropriate inner products.

(b) By using the solution given in Eq. (13.15), find the evolution in time of the atom dipole moment (Eq. (13.18)) and interpret the result.

13.2 (a) Given the interaction Hamiltonian of a two-level atom driven by a harmonic field (Eq. (13.1)), use the wavefunction in the Schrödinger picture

$$\psi(\underline{r}, t) = \sum_{j=1}^{2} a_{jS}(t)|u_j\rangle; \quad \widehat{H}_0|u_j\rangle = \mathrm{E}_j|u_j\rangle, \quad j = 1, 2 \tag{A}$$

and show that the coupled equation of $a_{1S}(t)$ and $\mathrm{a}_{2S}(t)$ is given by

$$\dot{a}_{1s} = -i\omega_1 a_{1s} + i\frac{\widetilde{\mu}\mathrm{E}(t)}{\hbar}a_{2s}; \quad \omega_1 = \frac{E_1}{\hbar}$$

$$\dot{a}_{2s} = -i\omega_2 a_{2s} + i\frac{\widetilde{\mu}\mathrm{E}(t)}{\hbar}a_{1s}; \quad \omega_2 = \frac{E_2}{\hbar}$$

(b) Compare the result with Eq. (13.13) in light of $a_{jS}(t) = a_j(t) \exp -(iE_j t/\hbar)$.
(c) Solve the equation by using the initial condition $a_{1S}(0) = 1, a_{2S}(0) = 0$ and the resonant condition $\omega = \omega_0$.
(d) Obtain the expression of the atom dipole moment and compare the result with Eq. (13.18).

13.3 (a) Derive the expression of the field energy in the resonator (Eq. (13.23)) by using $\mathbf{E}_l$, $\boldsymbol{H}_l$ given in Eq. (13.19) and the boundary condition of the standing waves.
(b) By using the commutation relation (13.24), verify the commutation relation of the creation and annihilation operators (Eq. (13.26)).

13.4 (a) By using Eq. (13.25) and the commutation relation (13.26), obtain the Hamiltonian equation (13.27) given in terms of the creation and annihilation operators.
(b) Starting from the standing wave representation $\mathbf{E}_l$, $\boldsymbol{H}_l$ given in Eq. (13.33), obtain the traveling wave representation Eq. (13.34) by combining Eqs. (13.32) and (13.33) and the trigonometric identities existing between $\sin x$, $\cos x$ and $\exp(\pm ix)$.
(c) Using the roles of the creation and annihilation operators given in Eq. (13.28), reproduce the expression of the transition rate W (Eq. (13.36)).

Suggested Readings

1. A. Yariv, An Introduction to Theory and Applications of Quantum Mechanics, John Wiley & Sons, 1982.
2. D. M. Kim, Introductory Quantum Mechanics for Semiconductor Nanotechnology, Wiley-VCH, 2010.
3. M. I. Sargent, M. O. Scully, and W. E. Jr.,, Lamb, Laser Physics, Westview Press, 1978.

14
The Interaction of EM Waves with an Optical Media

The atom–field interaction is extended to the EM waves interacting with an ensemble of atoms in the optical medium. In particular, the absorption, gain, and dispersion of the wave as it propagates in the medium are discussed. Also the operation principles of laser devices are presented in conjunction with the population inversion and controlled emission of radiation.

14.1
Attenuation, Amplification, and Dispersion of Waves

Attenuation and Amplification

In the collision-dominated regime, the atom–field interaction time is much shorter than the transition time period. Hence, the electron simply ends up making a transition from one level to another with a certain probability rather than executing a full Rabi flopping. Thus, given an ensemble of two-level atoms with N_1, N_2 atoms per unit volume, the number of induced transitions is given by

$$N_{1\to 2} = N_1 W_i, \quad N_{2\to 1} = N_2 W_i \tag{14.1a}$$

where the transition rate is given from Eq. (13.3) in terms of the light intensity I_ν and index of refraction n as

$$W_i = \frac{\widetilde{\mu}^2 \mathrm{E}_0^2 g(\nu)}{4\hbar^2} = \frac{\widetilde{\mu}^2 n g(\nu)}{2\hbar^2 c\varepsilon} I_\nu, \quad I_\nu \equiv \frac{\varepsilon \mathrm{E}_0^2}{2}\frac{c}{n} \tag{14.1b}$$

The light incident on a slab at z with unit cross-sectional area and thickness dz (Figure 14.1) is absorbed due to the net upward transition:

$$I_\nu(z+dz) - I_\nu(z) = -(N_1 - N_2) W_i h\nu dz \tag{14.2}$$

We can recast Eq. (14.2) into a differential form by Taylor expanding $I_\nu(z+dz)$ at z as

$$\frac{dI_\nu}{dz} = -\alpha I_\nu, \quad \alpha \equiv (N_1 - N_2)\frac{\widetilde{\mu}^2 n g(\nu)}{2\hbar^2 c\varepsilon} h\nu \tag{14.3}$$

Introductory Quantum Mechanics for Applied Nanotechnology, First Edition. Dae Mann Kim.

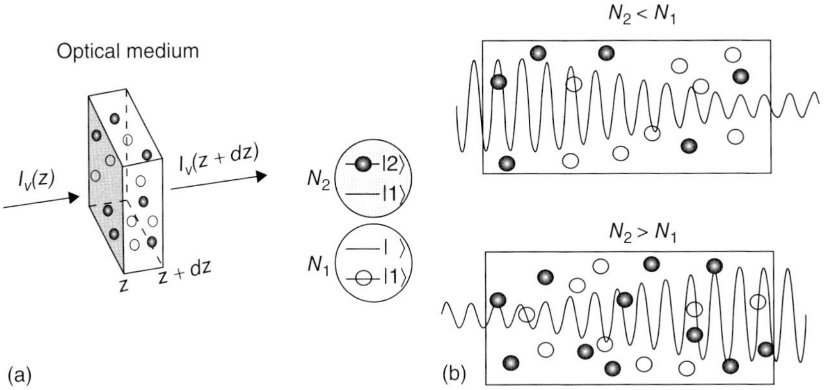

Figure 14.1 The input light intensity at z and the output light intensity at $z + dz$ (a). The light is absorbed or amplified depending on whether $N_2 < N_1$ or $N_2 > N_1$ (b).

The constant α thus introduced with the use of Eq. (14.1b) is called the *linear attenuation coefficient*, and its dependence on the atom dipole moment renders it one of the inherent properties of atomic species. We can easily integrate Eq. (14.3) and obtain

$$I_\nu(z) = I_\nu(0)e^{-\alpha z} \tag{14.4}$$

In an absorbing medium, $N_1 > N_2$ and the light is attenuated, while in the population-inverted medium, $N_1 < N_2$ and light is amplified with the gain coefficient γ given by

$$I_\nu(z) = I_\nu(0)e^{\gamma z}, \quad \gamma = -\alpha \tag{14.5}$$

Dispersion

EM waves also undergo the dispersion while attenuated or amplified. The dispersion comes about because (i) the incident light induces the atom dipole, (ii) an ensemble of such atom dipoles gives rise to the macroscopic polarization vector $\boldsymbol{P}$, and (iii) $\boldsymbol{P}$ in turn acts as the source of the input field:

$$\underline{\mathrm{E}} \to \langle \mu_j \rangle \to \sum_j \langle \mu_j \rangle \to \underline{P} \to \underline{\mathrm{E}}$$

To analyze the feedback process, let us consider P, which is generally represented by

$$\underline{P} = \mathrm{Re}\underline{P}_0 e^{i\omega t} \equiv \mathrm{Re}[\varepsilon_0 \chi_a \underline{\mathrm{E}}_0 e^{i\omega t}] \tag{14.6a}$$

where χ_a connecting $\boldsymbol{E}$ to $\boldsymbol{P}$ is called the *atomic susceptibility* and is a complex quantity

$$\chi_a \equiv \chi_a' - i\chi_a'' \tag{14.6b}$$

Hence, with the use of Eq. (14.6b), we can reexpress Eq. (14.6a) as

$$\underline{P} = \mathrm{Re}\varepsilon_0[(\chi_a' - i\chi_a'')\underline{\mathrm{E}}_0 e^{i\omega t}] = \varepsilon_0\chi_a'\underline{\mathrm{E}}_0 \cos\omega t + \varepsilon_0\chi_a''\underline{\mathrm{E}}_0 \sin\omega t \tag{14.6c}$$

The polarization vector is an integral component of the displacement vector

$$\underline{D} \equiv \varepsilon\underline{\mathrm{E}} \equiv \varepsilon_0\underline{\mathrm{E}} + \underline{P} \tag{14.7}$$

so that $\boldsymbol{D}$ can also be expressed as the output of $\boldsymbol{E}$ as

$$\underline{D} \equiv \varepsilon_0\underline{\mathrm{E}} + \varepsilon_0\chi_a\underline{\mathrm{E}} \equiv \varepsilon\underline{\mathrm{E}} \tag{14.8}$$

The permittivity ε thus defined consists of the background and resonant components, that is, $\varepsilon \equiv \varepsilon_0(1+\chi) = \varepsilon_0(1+\chi_b+\chi_a)$, and we can reexpress ε as

$$\varepsilon \equiv \varepsilon_b\left[1 + \frac{\varepsilon_0\chi_a(\omega)}{\varepsilon_b}\right], \qquad \varepsilon_b \equiv \varepsilon_0(1+\chi_b) \tag{14.9}$$

Therefore, the wave vector of the field is characterized by a complex dispersion relation

$$\begin{aligned} k &\equiv \omega\sqrt{\mu\varepsilon} = \omega\left\{\mu\varepsilon_b\left[1 + \frac{[\chi_a'(\omega) - i\chi_a''(\omega)]}{n^2}\right]\right\}^{\frac{1}{2}} \\ &= k_b\left[1 + \frac{\chi_a'(\omega)}{2n^2}\right] - i\frac{k_b\chi_a''(\omega)}{2n^2}; \quad k_b \equiv \omega\sqrt{\mu\varepsilon_b}, \quad n^2 \equiv \frac{\varepsilon_b}{\varepsilon_0} \end{aligned} \tag{14.10}$$

where an approximation has been made, namely, $(1+\chi_a)^{1/2} \simeq 1 + \chi_a/2$, as $|\chi_a| \ll 1$. Thus, the input field is attenuated or amplified and dispersed at the same time:

$$\begin{aligned} \mathrm{E}(z,t) &= \mathrm{Re}\mathrm{E}_0 e^{i(\omega t - kz)} \\ &= \mathrm{Re}\mathrm{E}_0 e^{i\left\{\omega t - k_b z\left[1+\frac{\chi_a'(\omega)}{2n^2}\right]\right\}} e^{-\frac{k_b z\chi_a''(\omega)}{2n^2}} \end{aligned} \tag{14.11}$$

It follows from Eq. (14.11) that the real part of the susceptibility characterizes the dispersion, while the imaginary part accounts for the attenuation or amplification.

14.2 Atomic Susceptibility

Density Matrix and Ensemble Averaging

We next analyze microscopically the absorption, amplification, and dispersion of the EM waves. Thus consider the two-level atom driven by an external electric field. We can represent the wavefunction of the two-level atom as

$$|\Psi(\underline{r},t)\rangle = a_{1s}(t)|u_1(r)\rangle + a_{2s}(t)|u_2(r)\rangle; \quad a_{js}(t) = a_j(t)e^{-i(\mathrm{E}_j/\hbar)t} \tag{14.12}$$

(see Eq. (13.12)). Note in Eq. (14.12) that the time dependence of the two states is entirely relegated to a_{1s}, a_{2s}, and the representation is known as *Schrödinger picture*. Then, the atomic dipole is given by

$$\langle \hat{\mu} \rangle \equiv \langle \Psi | e(\hat{e}_f \cdot r) | \Psi \rangle = \tilde{\mu}[a_{2s}(t) a_{1s}^*(t) + \text{c.c.}], \quad \tilde{\mu} \equiv \langle u_1 | \hat{e}_f \cdot \underline{r} | u_2 \rangle$$
$$= \tilde{\mu}(\rho_{12} + \rho_{21}); \quad \rho_{ij} \equiv a_{is} a_{js}^*, \quad j = 1, 2 \tag{14.13}$$

where ρ_{ij} thus defined is the off-diagonal element of the density matrix defined as

$$\rho = \begin{pmatrix} \rho_{11} & \rho_{12} \\ \rho_{21} & \rho_{22} \end{pmatrix} \equiv \begin{pmatrix} a_{1s} a_{1s}^* & a_{1s} a_{2s}^* \\ a_{2s} a_{1s}^* & a_{2s} a_{2s}^* \end{pmatrix}; \quad \rho_{21} = \rho_{12}^* \tag{14.14}$$

Obviously, the diagonal element ρ_{jj} represents the probability of finding the atom in the jth state, while the off-diagonal elements describe the atom dipole.

The equations of motion of the density matrix elements can be derived by using the coupled equation of the expansion coefficients Eq. (13.13) but with a_j therein replaced by a_{js} (see Eq. (14.12)). After a straightforward algebra, we obtain

$$\frac{d}{dt}\rho_{21} = -i\omega_0 \rho_{21} + i\frac{\tilde{\mu}\mathrm{E}(t)}{\hbar}(\rho_{11} - \rho_{22}) - \frac{\rho_{21}}{T_2}, \quad \omega_0 = \frac{\mathrm{E}_2 - \mathrm{E}_1}{\hbar} \tag{14.15a}$$

$$\frac{d}{dt}(\rho_{11} - \rho_{22}) = \frac{2i\tilde{\mu}\mathrm{E}(t)}{\hbar}(\rho_{21} - \rho_{21}^*) - \frac{(\rho_{11} - \rho_{22}) - (\rho_{11}^{(0)} - \rho_{22}^{(0)})}{\tau} \tag{14.15b}$$

where ω_0 is the atomic transition frequency. Note in particular that the last term in each equation has been added to incorporate the relaxation processes involved. For instance,

$$\rho_{21}(t) \equiv \frac{1}{N}\sum_{j=1}^{N} a_{2s}^{(j)} \left[a_{1s}^{(j)}\right]^* e^{-i\varphi_j}; \quad \varphi_j = \frac{(\mathrm{E}_{2j} - \mathrm{E}_{1j})t}{\hbar} \tag{14.15c}$$

represents the ensemble-averaged off-diagonal element. When two atoms collide, for example, each atom provides a burst of perturbing Hamiltonian to the other, thereby inducing the shift in energy level or the change in the phase φ_j. The collision is a random process so that φ_j is a random variable. Moreover, expansion coefficients are complex quantities and bring in additional differences in φ_j. As a result, the ensemble-averaged atomic dipole decays in a few T_2 called the *transverse relaxation time*. Likewise, the quantity $\rho_{11}-\rho_{22}$ represents the difference in the number of atoms distributed between the two states, and it also relaxes back to its equilibrium value $\rho_{11}^{(0)} - \rho_{22}^{(0)}$ in time scale τ called the *longitudinal relaxation time*.

The roles of the relaxation terms are best seen by turning off the electric field in Eq. (14.15), in which case we can easily find the solutions of Eq. (14.15) as

$$[\rho_{11}(t) - \rho_{22}(t)] = [\rho_{11}(0) - \rho_{22}(0)]e^{-t/\tau} + [\rho_{11}^{(0)} - \rho_{22}^{(0)}]\left(1 - e^{-t/\tau}\right) \tag{14.16a}$$

$$\rho_{21}(t) = \rho_{21}(0)e^{-i\omega_0 t}e^{-t/T_2} \tag{14.16b}$$

It is therefore clear that that the population difference $\rho_{11}-\rho_{22}$ relaxes back to its equilibrium value in a few τ's irrespective of the initial value. Also ρ_{21} relaxes to its zero equilibrium level in a few T_2's, and T_2 is generally much shorter than τ.

The fast decay of $\rho_{21}(t)$ is due to the rapid dephasing of the atom dipoles caused by collisions or other nonradiative decay (Figure 14.2).

We next consider a harmonic field driving the ensemble of atoms with the resonant frequency

$$\mathrm{E}(t) = \frac{\mathrm{E}_0}{2}(e^{i\omega t} + e^{-i\omega t}), \quad \omega \simeq \omega_0 \tag{14.17}$$

To facilitate the analysis, we decompose the variation in time of ρ_{21} into the component oscillating rapidly with the driving field and the slowly varying matrix element σ_{21} as

$$\rho_{21} = \sigma_{21} \exp -(i\omega t), \quad \sigma_{21} = \sigma_{21}^{(r)} + i\sigma_{21}^{(i)} \tag{14.18}$$

When Eqs. (14.17) and (14.18) are inserted into Eq. (14.15a), it can be reexpressed after rearranging the terms as

$$\left(\frac{d}{dt}\sigma_{21}\right) e^{-i\omega t} = i(\omega - \omega_0)\sigma_{21}e^{-i\omega t} + i\frac{\widetilde{\mu}\mathrm{E}_0}{2\hbar}(e^{i\omega t} + e^{-i\omega t})(\rho_{11} - \rho_{22}) - \frac{\sigma_{21}e^{-i\omega t}}{T_2} \tag{14.19}$$

Also as the variations in time of σ_{ij} and ρ_{jj} are slow compared with the oscillation frequency of the field, we may retain only the synchronous terms from both sides in the rotating-wave approximation. We can thus write

$$\frac{d}{dt}\sigma_{21} = i(\omega - \omega_0)\sigma_{21} + i\frac{\widetilde{\mu}\mathrm{E}_0}{2}(\rho_{11} - \rho_{22}) - \frac{\sigma_{21}}{T_2} \tag{14.20a}$$

We can likewise single out the d.c. components from both sides of Eq. (14.15b), obtaining

$$\frac{d}{dt}(\rho_{11} - \rho_{22}) = \frac{i\widetilde{\mu}\mathrm{E}_0}{\hbar}(\sigma_{21} - \sigma_{21}^*) - \frac{(\rho_{11} - \rho_{22}) - (\rho_{11}^{(0)} - \rho_{22}^{(0)})}{\tau} \tag{14.20b}$$

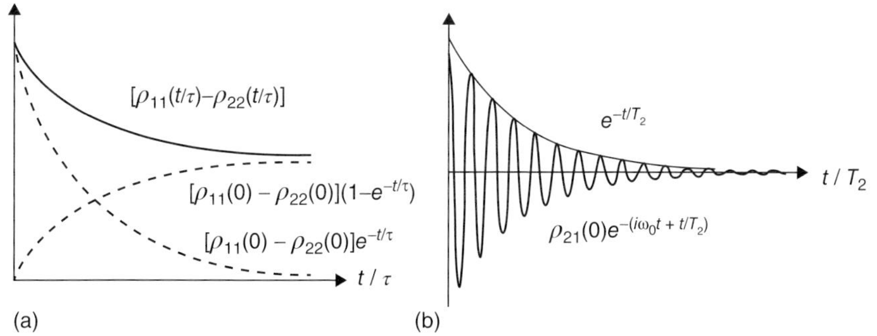

Figure 14.2 The evolution in time of the population difference $\rho_{11} - \rho_{22}$ (a) and the off-diagonal density matrix element ρ_{21} (b). The two quantities relax back to the equilibrium level in the time scale τ, T_2.

Steady-State Analysis

At steady state, the time derivatives in Eq. (14.20) vanish, and therefore by equating the real and imaginary parts from both sides of Eq. (14.20), we obtain three equations involving three unknowns $\sigma_{21}^{(r)}$, $\sigma_{21}^{(i)}$, and $(\rho_{11} - \rho_{22})$. These three unknowns can be readily found in a straightforward manner and are given by

$$\rho_{11} - \rho_{22} = (\rho_{11}^{(0)} - \rho_{22}^{(0)}) \frac{1 + (\omega - \omega_0)^2 T_2^2}{1 + (\omega - \omega_0)^2 T_2^2 + 4\Omega^2 T_2 \tau}; \quad \Omega \equiv \frac{\widetilde{\mu} \mathrm{E}_0}{2\hbar} \tag{14.21a}$$

$$\sigma_{21}^{(r)} = (\rho_{11}^{(0)} - \rho_{22}^{(0)}) \frac{-(\omega - \omega_0) T_2^2 \Omega}{1 + (\omega - \omega_0)^2 T_2^2 + 4\Omega^2 T_2 \tau} \tag{14.21b}$$

$$\sigma_{21}^{(i)} = (\rho_{11}^{(0)} - \rho_{22}^{(0)}) \frac{T_2 \Omega}{1 + (\omega - \omega_0)^2 T_2^2 + 4\Omega^2 T_2 \tau} \tag{14.21c}$$

Atomic Susceptibility

We can now specify the ensemble-averaged atomic dipole by combining Eqs. (14.13), (14.18), and (14.21) as

$$\begin{aligned} \langle \widehat{\mu} \rangle &= \widetilde{\mu}(\rho_{21} + \rho_{12}) = 2\,\mathrm{Re}\, \widetilde{\mu}(\sigma_{21}^{(r)} + i\sigma_{21}^{(i)}) e^{-i\omega t} \\ &= 2\widetilde{\mu}[\sigma_{21}^{(r)} \cos \omega t + \sigma_{21}^{(i)} \sin \omega t] \end{aligned} \tag{14.22}$$

Therefore, the susceptibility is specified by combining Eqs. (14.6c), (14.21), and (14.22) as

$$P(t) \equiv \mathrm{Re} N \langle \mu(t) \rangle = \varepsilon_0 \chi_a' \underline{\mathrm{E}}_0 \cos \omega t + \varepsilon_0 \chi_a'' \underline{\mathrm{E}}_0 \sin \omega t$$

with the identifications

$$\chi_a'(\omega) = \frac{\widetilde{\mu}^2 T_2 (N_1^{(0)} - N_2^{(0)})}{\hbar \varepsilon_0} \frac{-(\omega - \omega_0) T_2}{1 + (\omega_0 - \omega)^2 T_2^2 + 4\Omega^2 T_2 \tau} \tag{14.23a}$$

$$\chi_a''(\omega) = \frac{\widetilde{\mu}^2 T_2 (N_1^{(0)} - N_2^{(0)})}{\hbar \varepsilon_0} \frac{1}{1 + (\omega_0 - \omega)^2 T_2^2 + 4\Omega^2 T_2 \tau} \tag{14.23b}$$

Here $N_j^{(0)} = N\rho_{jj}^{(0)}$ is the density of atoms in each level in equilibrium. Also the phase velocity $v_\mathrm{p} = \omega/k$ and the attenuation coefficient of the light intensity $\alpha(\omega) \propto 2\chi_a''$ can be specified from Eq. (14.11) and (14.23) as

$$v_p(\omega) = \frac{\omega}{k_b(1 + \chi_a'(\omega)/2n^2)} = \frac{c/n}{(1 + \chi_a'(\omega)/2n^2)} \tag{14.24}$$

$$\alpha(\omega) \equiv k_b \frac{\chi_a''(\omega)}{n^2} = \frac{\pi \widetilde{\mu}^2 (N_1^{(0)} - N_2^{(0)})}{\lambda n^2 \hbar \varepsilon_0} g(\nu) \tag{14.25a}$$

where the lineshape function $g(\nu)$ is given in this case by a Lorentzian function

$$g(\nu) = \frac{2T_2}{1 + (\omega - \omega_0)^2 T_2^2 + 4\Omega^2 T_2 \tau} \approx \frac{2T_2}{1 + (\omega - \omega_0)^2 T_2^2} \tag{14.25b}$$

In equilibrium, $N_1 > N_2$ so that $\alpha > 0$ and the light is attenuated, but with the population inversion, that is, $N_1 < N_2$ $\alpha < 0$ and the light is amplified.

14.3
Laser Device

The laser device is based on the Bose Einstein statistics and the feedback mechanism, whereby the input wave is regenerated in cascade via the stimulated emission of radiation. The *cw* operation of the device produces the monochromatic light sources for use in spectroscopic studies. The pulsed-mode operation yields ultra-short light pulses by which to explore fast chemical and physical processes down to attosecond time regimes or shorter. Moreover, the laser diodes are utilized in the optoelectronic applications as well as the fiber communication, and its operation principles are briefly discussed.

Thus, consider a cavity consisting of two parallel mirrors L distance apart (Figure 14.3). The wavelengths of standing-wave modes satisfy therein the boundary condition $(\lambda_l/2)l = L, \quad l = 1, 2, \ldots$, so that the axial frequencies

$$\nu_l^{(0)} = \frac{c/n}{\lambda_l} = l\frac{1}{2L/(c/n)} \tag{14.26a}$$

are separated uniformly by $\Delta\nu$, which corresponds to the inverse cavity round-trip time:

$$\Delta\nu = \nu_{l+1}^{(0)} - \nu_l^{(0)} = \frac{1}{2L/(c/n)} \tag{14.26b}$$

When the population inversion is achieved above the threshold value, the cavity acts as a self-sustaining oscillator. An incident wave is amplified as it traverses the cavity, but it also suffers the loss due to imperfect reflectivity and transmittance of the mirrors and the scattering. The output wave then consists of a string of transmitted beams, with each succeeding beam having undergone one more cavity round-trip and is given by

$$\mathrm{E}_t = t_1 t_2 \mathrm{E}_i e^{-ikL}[1 + s + s^2 + \cdots] = \frac{t_1 t_2 e^{-ikL}}{1 - s}\mathrm{E}_i, \quad s \equiv r_1 r_2 e^{-2ikL} \tag{14.27}$$

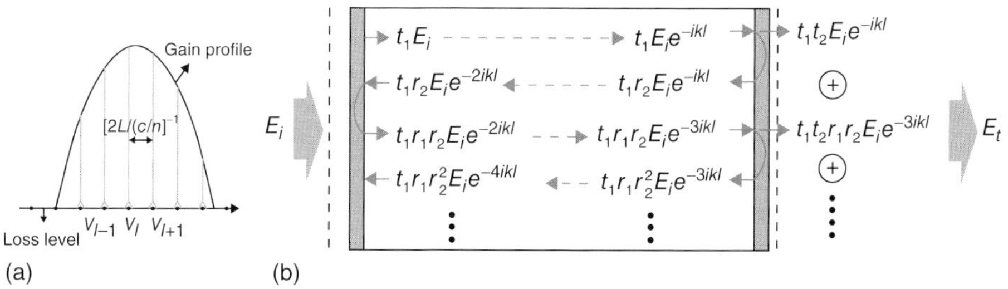

Figure 14.3 The longitudinal standing-wave modes in a Fabry–Perot-type laser cavity (a). The laser oscillator with the input beam yielding a string of output beams with each succeeding one having undergone one more cavity round-trip (b).

where t_j, r_j are the transmission and reflection coefficients of the mirror, and s is the net gain in one cavity round-trip. The infinite geometric series can be summed up if $s < 1$. Also, the amplification and dispersion of the wave are specified by the wave vector as

$$k = k_b + \Delta k + i\frac{1}{2}\gamma; \quad \Delta k = k_b\frac{\chi_a'(\omega)}{2n^2}, \quad \gamma = -k_b\frac{\chi_a''(\omega)}{n^2} \tag{14.28}$$

where k_b is the background term, Δk, γ represent the dispersion and gain, respectively as discussed (see Eqs. (14.10) and (14.23)).

14.3.1 Population Inversion

As mentioned, the gain prevails over the loss when N_2 is greater than N_1, and the population inversion is attained by pumping the atoms. Figure 14.4 shows a two-level atom, pumped, driven, and damped at the same time. The rate equation is given by

$$\dot{\rho}_{22} = \lambda_2 - \left(\frac{1}{\tau_2} + \frac{1}{\tau_{sp}}\right)\rho_{22} - W_i(\rho_{22} - \rho_{11}) \tag{14.29a}$$

$$\dot{\rho}_{11} = \lambda_1 - \frac{1}{\tau_1}\rho_{11} + \frac{1}{\tau_{sp}}\rho_{22} + W_i(\rho_{22} - \rho_{11}) \tag{14.29b}$$

where λ_j is the pumping rate, $1/\tau_j$ the decay rate due to the finite electron lifetime τ_j in each level, and τ_{sp} the spontaneous emission lifetime. As $\tau_1, \tau_2 \ll \tau_{sp}$, the spontaneous emission rate can be neglected. At steady state, the time derivatives vanish, and therefore the population inversion is simply given by

$$N_2 - N_1 \equiv N(\rho_{22} - \rho_{11}) = \frac{N(\lambda_2\tau_2 - \lambda_1\tau_1)}{1 + I_\nu g(\nu)/I_s}, \quad I_s^{-1} = \frac{\tilde{\mu}^2(\tau_1 + \tau_2)}{2\hbar^2 cn\varepsilon_0} \tag{14.30}$$

where N is the density of atoms, and Eq. (14.1b) has been used for W_i. Clearly, the population inversion necessitates a strong pumping rate to the upper level and long lifetime therein in order to meet the condition $\lambda_2\tau_2 > \lambda_1\tau_1$. Note in particular that the population inversion is saturated with light intensity I_ν, and I_s is called the *saturation intensity*.

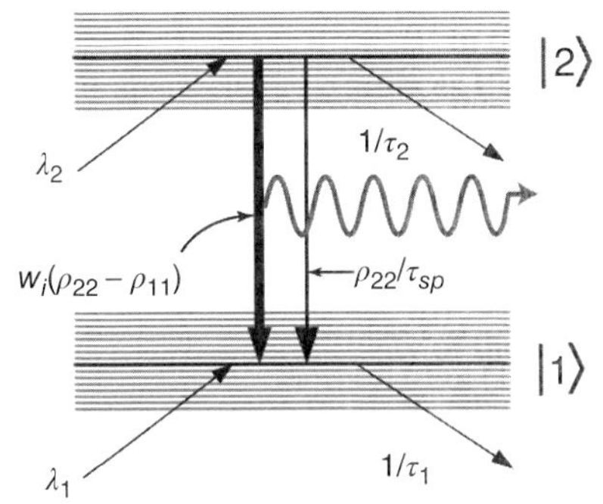

Figure 14.4 A pumped, driven, and damped two-level atom accompanied by stimulated and spontaneous emission of radiation.

Oscillation Condition

The gain coefficient is therefore given by combining Eqs. (14.25) and (14.30) as

$$\gamma \equiv -\alpha = \frac{\gamma_0}{1 + I_\nu g(\nu)/I_s}; \quad \gamma_0 \equiv \frac{\pi \widetilde{\mu}^2 (N_2 - N_1)_0}{\lambda n^2 \hbar \varepsilon_0} g(\nu) \tag{14.31}$$

where γ_0 is the gain factor resulting purely from the pumping rate in the absence of the laser intensity. When γ balances the loss, the factor s appearing in the transfer function Eq. (14.27) becomes unity. Consequently, the denominator of Eq. (14.27) vanishes, and the transmitted field amplitude E_t diverges. The divergence indicates that an infinitesimal input E_i yields finite E_t, that is, the onset of oscillation. The oscillation condition is thus specified explicitly from Eqs. (14.27) and (14.31) as

$$s \equiv r_1 r_2 e^{-2i(k_b+\Delta k)L} e^{(\gamma-\alpha_s)L} = 1 \equiv e^{-2\pi i l}, \quad l = 1, 2, \ldots \tag{14.32}$$

where the loss consists of both the scattering loss α_s and the imperfect mirror reflectivity. Clearly, Eq. (14.32) is satisfied, provided

$$r_1 r_2 e^{(\gamma-\alpha_s)L} = 1 \tag{14.33a}$$

$$2(k_b + \Delta k)L = 2\pi l \tag{14.33b}$$

Threshold Pumping

The amplitude equation (14.33a) determines the threshold pumping level for the onset of oscillation before the laser intensity starts to build up. The level is therefore obtained from Eqs. (14.31) and (14.33a) by putting $I_\nu = 0$ as

$$\frac{\pi \tilde{\mu}^2 (N_2 - N_1)_{TH}}{\lambda n^2 \hbar \varepsilon_0} g(\nu) = \alpha_T; \quad \alpha_T = \alpha_s - \frac{1}{L} \ln(r_1 r_2) \tag{14.34}$$

Once the pumping level exceeds the threshold value, so that $N_2 - N_1 > (N_2 - N_1)_{TH}$, the oscillation sets in. However, the net gain should still be balanced exactly by the total loss α_T so that the steady-state oscillation condition is preserved. Otherwise, the string of transmitted field amplitudes grows without any upper bound.

Laser Intensity

It is at this point that the physical significance of the saturated population inversion Eq. (14.30) becomes apparent. At the onset of oscillation, the laser intensity is at the zero level. However, when the population inversion exceeds the threshold value, the light intensity builds up, so that the gain coefficient γ saturates to balance the total loss, that is,

$$\frac{\pi \tilde{\mu}^2 (N_2 - N_1)}{\lambda n^2 \hbar \varepsilon_0 (1 + I_\nu g(\nu)/I_s)} g(\nu) = \alpha_T \tag{14.35}$$

Clearly, Eq. (14.35) describes the steady-state oscillation condition at a finite laser intensity.

The intensity can therefore be found from Eqs. (14.34), (14.35) as

$$I_\nu = \left[\frac{I_s}{g(\nu)}\right]\left[\frac{N_2 - N_1}{(N_2 - N_1)_{TH}} - 1\right] \tag{14.36}$$

and it increases with increasing pumping level, as it should. Moreover, the intensity depends on the inherent properties of the active medium as entailed in the saturated intensity I_s such as the atomic dipole moment $\widetilde{\mu}$ and the electron lifetimes.

Frequency of Operation

When Eqs. (14.28) and (14.23) are inserted into Eq. (14.33b) for Δk and $\chi'_a(\omega)$, respectively together with the identity $k_b = 2\pi\nu_l/(c/n)$, the phase part of the oscillation condition reads as

$$\frac{2\pi\nu_l L}{c/n}\left[1 + \frac{\chi'_a(\omega)}{2n^2}\right] = l\pi, \quad l = 1, 2, \ldots \tag{14.37}$$

Therefore, the frequency of operation is specified from Eq. (14.37) as

$$\nu_l = \nu_l^{(0)}\frac{1}{1 + \chi'_a(\omega)/2n^2}, \quad \nu_l^{(0)} \equiv \frac{l}{2L/(c/n)} \tag{14.38}$$

where $\nu_l^{(0)}$ is the bare longitudinal cavity mode given in Eq. (14.26). It is therefore clear that the dispersion occurring with the amplification shifts the lasing frequency from its bare standing wave frequency $\nu_l^{(0)}$.

Modes of Operation

The frequency of the laser ranges from the microwave to X-ray regimes, and the lasing medium consists of various kinds of materials such as helium neon, argon, carbon dioxide, dye, ruby, and GaAs. The single-mode *cw* operation generates near ideal monochromatic optical beams. When the longitudinal modes within the broad gain profile are excited simultaneously, powerful free-running light sources are generated. Moreover, when the phases of these waves are locked together, light pulses ensue with the time duration shortened down to attosecond time regime.

Problems

14.1 (a) Show that the coupled equation of the expansion coefficients a_{1s}, a_{2s} introduced in Eq. (14.12) is given in strict analogy with Eq. (13.13) by

$$\dot{a}_{1s} = -i\omega_1 a_{1s} + i\frac{\widetilde{\mu}E(t)}{\hbar}a_{2s}, \quad \dot{a}_{2s} = -i\omega_2 a_{2s} + i\frac{\widetilde{\mu}E(t)}{\hbar}a_{1s} \tag{14.39}$$

(b) Use 14.39 and verify the equation of motion of the density matrix Eq. (14.15) without the two relaxation terms.

14.2 (a) Show that in the absence of the E-field, the differential equation (14.15) can be recast as

$$\frac{d}{dt}\left[\rho_{21}e^{i\omega_0 t+t/T_2}\right]=0,\quad \frac{d}{dt}\left[\left(\rho_{11}-\rho_{22}\right)e^{t/\tau}\right]=\frac{\left(\rho_{11}^{(0)}-\rho_{22}^{(0)}\right)}{\tau}e^{t/\tau} \tag{14.40}$$

(b) Obtain the solution given in Eq. (14.16) by performing the simple integration of (14.40).

14.3 (a) Single out the real and imaginary parts from both sides of Eq. (14.20a) and combine the resulting two equations with Eq. (14.20b) and obtain the solutions given in Eq. (14.21) at steady state.

(b) Using the solution Eq. (14.21) thus obtained, find the real and imaginary parts of the atomic susceptibility Eq. (14.23).

14.4 Find the saturated population inversion Eq. (14.30) from the rate equation (14.29) at steady state.

14.5 Consider a passive Fabry–Perot-type cavity in which $\gamma = \alpha = 0$.

(a) Find the standing-wave modes in frequency and wavelength units for the cavity lengths 1 m, 1 cm, and 100 μm, respectively.

(b) To generate a picosecond optical pulse with the carrier frequency $\nu = c/500\,\text{nm}$, how many standing-wave modes should be locked together?

Suggested Readings

1. A. Yariv, An Introduction to Theory and Applications of Quantum Mechanics, John Wiley & Sons, 1982.
2. M. I. Sargent, M. O. Scully, and W. E. Lamb, Jr, Laser Physics, Westview Press, 1978.
3. D. M. Kim, Introductory Quantum Mechanics for Semiconductor Nanotechnology, Wiley-VCH, 2014.

15
Semiconductor Statistics

The transistors constitute a central element of the digital information technology and are firmly rooted in the concepts inherent in quantum mechanics. The mainstream transistors have thus far been based on the charge control. A factor crucial for the control is the carrier density. The concentrations of electrons and holes are specified in terms of doping level, temperature, and other parameters of the semiconductor material.

15.1
Quantum Statistics

Conductors, Insulators, and Semiconductors

A condensed matter is classified into three groups, namely, conductor, insulator, and semiconductor. The classification results from differing configurations of the valence and conduction bands. In conductors, the valence electrons constitute sea of free electrons, and the valence and conduction bands overlap. Hence, the valence electrons can move up to the conduction band upon acquiring kinetic energy and contribute to the current under bias (Figure 15.1).

In an insulator such as silicon dioxide, the widths of conduction and valence bands are narrow, and the two bands are separated by a large bandgap, typically 10 eV or more. This is because the valence electrons form strong bonds with neighboring atoms, and these bonds are difficult to break. As a consequence, practically no electrons reside in the conduction band to contribute to the current.

The configuration of the conduction and valence bands in semiconductors lies in between those of metals and insulators. The two bands are separated by the bandgap, ranging from about 0.5 to a few electronvolts. The bonds between neighboring atoms are moderately strong and are relatively easy to be broken at room temperature. As a result, an appreciable number of electrons are promoted into the conduction band via the band-to-band thermal excitation to conduct the current. The holes left behind the valence band are also capable of conducting current.

Introductory Quantum Mechanics for Applied Nanotechnology, First Edition. Dae Mann Kim.

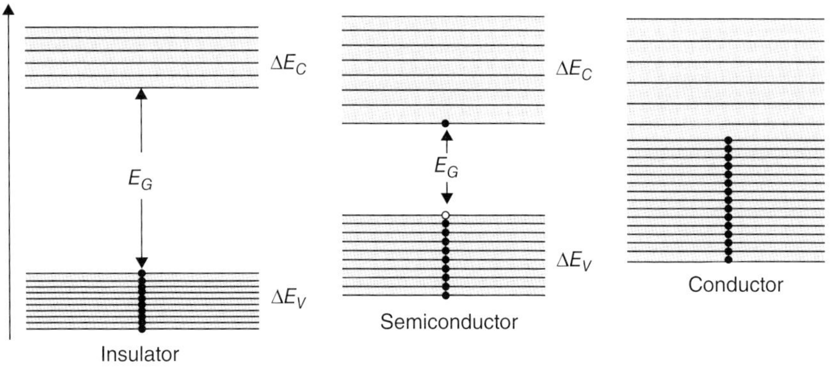

Figure 15.1 The conduction and valence bands and the energy gap in the conductors, insulators, and semiconductors.

15.1.1 Bosons and Fermions

The microscopic world of electrons, atoms, and molecules is manifested in the macroscopic world via the cumulative effects of a large number of such particles. The statistics of the ensemble of such particles are generally different, depending on the kinds of particles. As mentioned, there are three kinds of particles: (i) identical but distinguishable, for example, atoms and molecules; (ii) bosons with integer spins such as photons, phonons, and α particles; and (iii) fermions with half odd integer spins, such as electrons, holes, protons, and neutrons. The fermions are constrained by the Pauli exclusion principle and are prohibited to share a common quantum state between two or more. The distinguishable particles are described by the Boltzmann distribution function in equilibrium, as discussed in Chapter 1.

Bosons

The bosons obey the Bose–Einstein statistics, and the equilibrium distribution function of photons, for example, is given by

$$f(\varepsilon_s) = \frac{1}{e^{h\nu_s/k_BT} - 1}, \quad \varepsilon_s = h\nu_s \tag{15.1}$$

where ε_s is the energy of the photon with frequency ν_s. When $f(\varepsilon_s)$ is multiplied by the number of modes $8\pi n^3\nu_s^2/c^3$ (see Eq. (2.1)), it merges with the celebrated Planck's energy density (Eq. (2.3)), as it should. The distribution function is plotted versus energy in Figure 15.2 at different temperatures, together with the Boltzmann distribution function, for comparison. The photon distribution function clearly exhibits the Bose condensation at low temperature T. As clear from the figure, the Boltzmann distribution function itself decreases exponentially following the power law, $1/T$ but it is practically constant in the temperature range from 0 to 0.001 °K. On the other hand, the photon distribution function decreases in the same temperature range by two orders of magnitude or more.

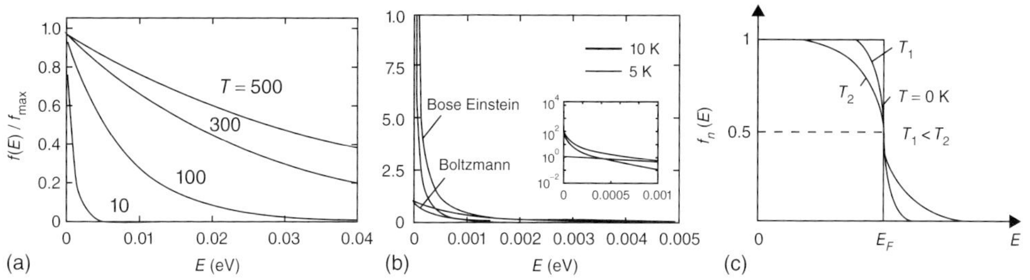

Figure 15.2 The Boltzmann (a), Bose–Einstein (b), and Fermi (c) distribution functions versus energy at different temperatures. Also shown in the inset is the comparison between the Boltzmann and Bose–Einstein distribution functions at the extremely low temperature.

Fermions

The fermions such as electrons and holes are described in equilibrium by the celebrated Fermi function given by

$$f(E) = \frac{1}{1 + e^{(E-E_F)/k_BT}} \tag{15.2}$$

where E_F is the Fermi energy or level. The function is often called the *Fermi occupation factor* and is derived from the fundamental postulate of the exclusion principle. Figure 15.2 also shows the Fermi function versus T. At $T = 0$, $f(E)$ is a step function and is equal to unity for $E < E_F$, representing 100% probability of occupation, while it is zero for $E > E_F$, indicating zero probability. For $T \neq 0$, the shape of $f(E)$ is generally preserved except that the curve is rounded off near E_F. Specifically, $f(E)$ is less than unity a few k_BT below E_F and tails out exponentially a few k_BT above E_F, thereby transferring the occupation probability from below E_F to above E_F. The occupation probability beyond E_F is called the *Boltzmann tail*. With increasing T, the Boltzmann tail is progressively pronounced.

The Fermi distribution function carries far-reaching consequences. For instance, let us consider the 3D concentration of electrons given by

$$n = \int_0^\infty dE \frac{g(E)}{1 + e^{(E-E_F)/k_BT}}; \quad g(E) = \frac{\sqrt{2}m^{3/2}E^{1/2}}{\pi^2\hbar^3} \tag{15.3}$$

where $g(E)$ is the 3D density of states (Eq. (4.15)). The electrons fill up the quantum states one by one in each state from the lowest energy level. For $T \to 0$, the upper limit of the integration is given by $E_F(0)$. We can thus integrate $g(E)$ in the interval $0 \leq E \leq E_F(0)$ by using the step Fermi function (Figure 15.2) and specify $E_F(0)$ in terms of the electron density as

$$E_F(0) = \frac{\hbar^2}{8m}\left(\frac{3}{\pi}n\right)^{2/3} \tag{15.4}$$

It is therefore clear that the Fermi level increases with increasing density. For n equaling Avogadro's number, $n = 10^{29}\ \text{m}^{-3}$, for example, $E_F(0) \approx 7.9$ eV, and the Fermi velocity of the electrons on top of $E_F(0)$

$$\frac{mv_F^2}{2} = E_F(0)$$

is as large as $\simeq 1.7 \times 10^6\ \mathrm{m\,s^{-1}}$ at $T = 0$. This is in drastic contrast with distinguishable particles, which should be completely at rest for $T = 0$.

15.2 Carrier Concentration in Intrinsic Semiconductor

When the electrons in the filled valence band are excited into the conduction band in the intrinsic semiconductor, they leave behind the holes with charge $+e$ as illustrated in Figure 15.1. The holes can be treated as the positive charge carriers with the effective mass m_p just as electrons are the negative charge carriers with the effective mass m_n. In intrinsic semiconductors in which no dopant atoms are present, the concentrations of electrons and holes are the same by definition, that is, $n = p \equiv n_i$.

Thermal Equilibrium

At the outset, we examine the equilibrium from a few different standpoints. The thermodynamic equilibrium is characterized by a few basic facts: (i) the physical quantities are time invariant, as every process is balanced by its inverse process (detailed balancing); (ii) n and p are quantified by a single Fermi level E_F; (iii) E_F is spatially flat and also lines up in composite semiconductors; and (iv) the law of mass action holds, namely $np = n_i^2$, with n_i denoting the intrinsic concentration.

Electron Concentration

The equilibrium concentration of electrons in the conduction band is specified in terms of the Fermi occupation factor $f_n(E)$ given in Eq. (15.2) and 3D density of states $g_n(E)$ as

$$n = \int_{E_C}^{E_C+\Delta E_c} dE g_n(E) f_n(E); \quad g_n(E) = \frac{1}{2\pi^2}\left(\frac{2m_n}{\hbar^2}\right)^{3/2}(E - E_C)^{1/2} \tag{15.5}$$

Here, ΔE_C is the conduction band width and $g_n(E)$ the 3D density of states of the electrons in the conduction band (Eq. (4.15)). The bottom of the conduction band E_C serves as the reference level from which to define the kinetic energy of electrons moving with the effective mass m_n. As discussed $g_n(E)dE$, represents the number of quantum states per unit volume, and when multiplied by $f_n(E)$, it represents the density of state occupied by electrons in the range from E to $E + dE$.

The integration representing n can be reexpressed by introducing a dimensionless variable $\eta \equiv (E - E_C)/k_B T$ and making the approximation $\Delta E_C/k_B T \approx \infty$ as

$$n = \frac{2}{\sqrt{\pi}} N_c F_{1/2}(\eta_{Fn}); \quad \eta_{Fn} \equiv \frac{E_F - E_c}{k_B T}, \quad N_C \equiv 2\left(\frac{2\pi m_n k_B T}{h^2}\right)^{3/2} \tag{15.6a}$$

where N_C is called the *effective density of states* at the conduction band and

$$F_{1/2}(\eta_F) = \int_0^{\infty} \frac{\eta^{1/2} d\eta}{1 + e^{\eta-\eta_F}} \tag{15.6b}$$

is called the *Fermi 1/2-integral*. The approximation $\Delta E_C / k_B T \approx \infty$ is well taken, since ΔE_C is typically few electronvolts, whereas $k_B T \simeq 25$ meV at room temperature. Moreover, the Fermi occupation factor cuts off the contribution from those states a few $k_B T$ above E_C, as illustrated in Figure 15.3. In the nondegenerate regime, E_F ranges in the energy gap below E_C by a few $k_B T$, so that $\exp -\eta_F \gg 1$. In this case, the Fermi integral yields

$$F_{1/2}(\eta_{Fn}) \simeq e^{\eta_{Fn}} \int_0^{\infty} d\eta e^{-\eta} \eta^{1/2} = e^{\eta_{Fn}} \frac{\sqrt{\pi}}{2} \tag{15.7}$$

Hence, by combining Eqs. (15.7) and (15.6a), n can be expressed analytically as

$$n = N_C e^{-(E_C - E_F)/k_B T} \tag{15.8}$$

and Eq. (15.8) indicates explicitly that n increases exponentially with temperature and is equal to N_C when E_F coincides with E_C.

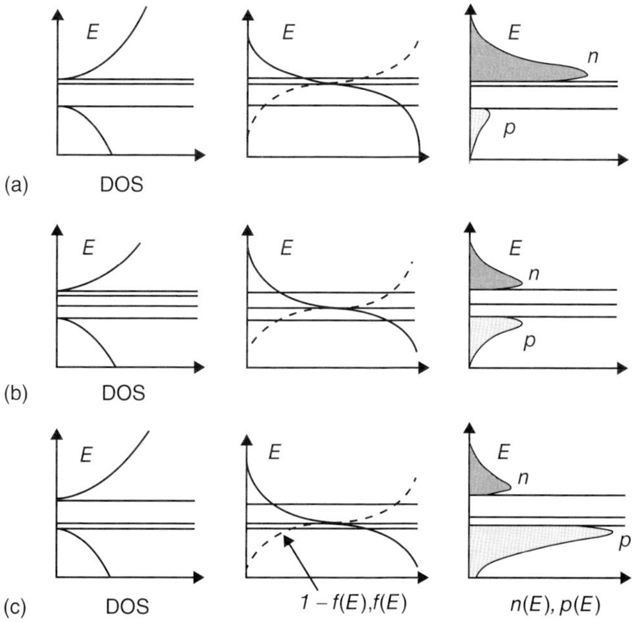

Figure 15.3 The graphical representations of n and p in terms of the 3D density of states, respective occupation factor, and the location of E_F in n-type (a), intrinsic (b), and p-type (c) semiconductors.

Hole Concentration

The hole concentration p is likewise specified by

$$p = \int_{E_V-\Delta E_V}^{E_V} dE g_p(E) f_p(E); \quad g_p(E) = \frac{1}{2\pi^2}\left(\frac{2m_p}{\hbar^2}\right)^{3/2}(E_V - E)^{1/2} \tag{15.9a}$$

where ΔE_V is the valence band width and $g_p(E)$ the hole density of states in the valence band. The top of the valence band E_V serves as the reference level for defining the kinetic energy of holes, moving with the effective mass m_p. With increasing kinetic energy, electrons move up the conduction band from E_C, while holes move down the valence band from E_V, as will become clear in due course. The occupation factor for holes in the quantum state is by definition the probability that the state is not occupied by the electron, that is,

$$f_p(E) \equiv 1 - \frac{1}{1 + e^{(E-E_F)/k_B T}} = \frac{1}{1 + e^{(E_F-E)/k_B T}} \tag{15.9b}$$

and $f_p(E)g_p(E)dE$ represents the total number of states occupied by holes per unit volume between E and $E - dE$.

Again by inserting Eq. (15.9b) into Eq. (15.9a), and introducing the variable of integration $\eta \equiv (E_V - E)/k_B T$ and putting, $\Delta E_V/k_B T \approx \infty$ we can write

$$p = \frac{2}{\sqrt{\pi}} N_V F_{1/2}(\eta_{Fp}); \quad \eta_{Fp} \equiv \frac{E_V - E_F}{k_B T}, \quad N_V \equiv 2\left(\frac{2\pi m_p k_B T}{h^2}\right)^{3/2} \tag{15.10}$$

where N_V is the effective density of states at the valence band. For the nondegenerate case in which E_F stays a few $k_B T$ above E_V in the bandgap, the Fermi 1/2-integral can likewise be evaluated in analogy with Eq. (15.7), and p can be expressed analytically as

$$p = N_V e^{-(E_F-E_V)/k_B T} \tag{15.11}$$

and the hole concentration also increases exponentially with temperature.

Intrinsic Concentration

In intrinsic semiconductors in which there are no impurity atoms present, the electrons excited from valence to conduction bands leave behind the same number of holes in the valence band. Hence, n and p are identical, that is, $n = p = n_i$.The intrinsic concentration is therefore given from Eqs. (15.8) and (15.11) by

$$n_i \equiv \sqrt{np} = \sqrt{N_C N_V} e^{-E_G/2k_B T}; \quad E_G \equiv E_C - E_V \tag{15.12}$$

where E_G is the bandgap. Figure 15.4 shows n_i in silicon, germanium, and gallium arsenide versus the inverse temperature $1/T$. Evidently, n_i varies exponentially with $1/T$, and the variation is accentuated with increasing E_G. Moreover, n_i exponentially increases with decreasing E_G at given T, as more electron–hole pairs are thermally excited across the narrower bandgap. For instance, in Si with E_G of, 1.12 eV $n_i = 1.45 \times 10^{10}$ cm^{-3} at the room temperature, while

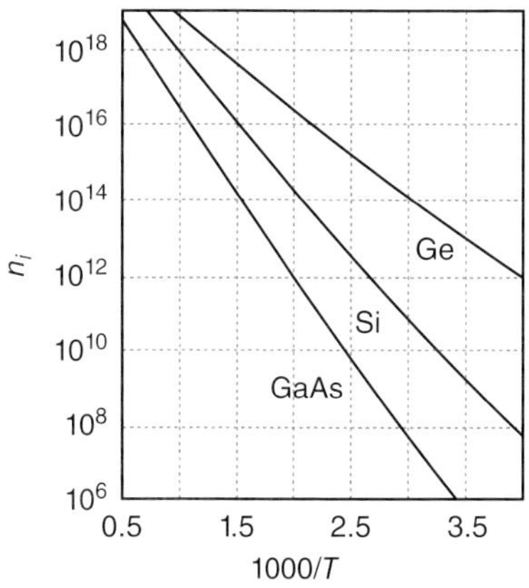

Figure 15.4 The intrinsic carrier concentration versus $1000/T$ in germanium, silicon, and gallium arsenide.

$n_i = 1.79 \times 10^6\ \text{cm}^{-3}$ in GaAs with E_G of 1.424 eV at the same temperature. Thus, the variation of n_i caused by the difference in E_G by 0.3 eV amounts to nearly four orders of magnitude.

Intrinsic Fermi Level

The location of E_F is determined from the charge neutrality condition. In intrinsic semiconductors $n = p$, so that one can write from Eqs. (15.8) and (15.11)

$$N_C e^{-(E_C - E_{Fi})/k_B T} = N_V e^{-(E_{Fi} - E_V)/k_B T} \tag{15.13}$$

and find E_{Fi} as

$$E_{Fi} = \frac{1}{2}(E_C + E_V) + \frac{3k_B T}{4} \ln \frac{m_h}{m_e}; \quad \ln \frac{N_V}{N_C} = \frac{3}{2} \ln \frac{m_p}{m_n} \tag{15.14}$$

Clearly, E_{Fi} is located near the midgap, and the departure from it is due to the difference between m_n and m_p and amounts to a fraction of the thermal energy $k_B T$.

15.3
Carrier Densities in Extrinsic Semiconductors

The control of the carrier concentration is a key factor for charge control, and n and p are controlled primarily by means of the impurity doping. Let us thus consider n and p in extrinsic semiconductors, which are doped with donor or acceptor atoms. To discuss the physics of the impurity doping in silicon, for example, let us revisit the electron configuration of Si [Ne]3$s^2$3p^2. There are four valence electrons outside the closed neon core. The Si atoms are thus covalently bonded with its four neighbors by sharing one valence electron with each other, so that the sub-shell is filled up, as sketched in Figure 15.5. Doping consists of incorporating donors or acceptors at substitutional sites.

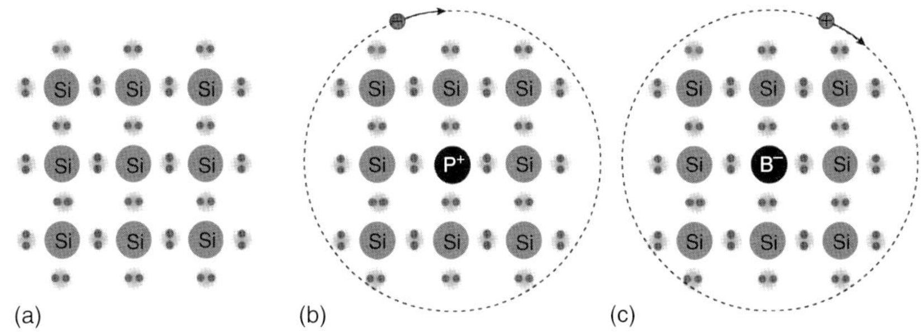

Figure 15.5 The tetrahedrally bonded silicon (a). A donor atom replacing a Si-atom forms a loosely bound hydrogenic atom with a valence electron (b). An acceptor atom in place of Si-atom forms a loosely bound hydrogenic atom with a hole (c).

The donor atoms are from Column V in periodic table, for example, phosphorus ([Ne]$3s^23p^3$) or arsenic ([Ar]$4s^24p^3$), which have five valence electrons. Four of them are used up in the tetrahedral bonding, and the remaining fifth electron forms a hydrogenic atom with P^+ or As^+ ion core (Figure 15.5). On the other hand, the acceptor atoms are from Column III, for instance, boron ([He]$2s^22p$) with three valence electrons, so that it cannot complete the tetrahedral bonding. But it can accept an electron from other Si–Si bonds, to complete bonding. In the process, a hole is generated in the valence band. Thus, the boron as a negative ion and hole again form a hydrogenic atom. The effective radius of hydrogenic donor atom a_0 and the ionization energy E_D of its valence electron can be estimated by using the H-atom theory. In this case, E_D is the energy required to release a valence electron to the conduction band. Specifically, we can specify a_0 and E_D from Eq. (8.46) as

$$a_0 \equiv \frac{\hbar^2 4\pi\varepsilon_S}{m_n e^2} = \left(\frac{\hbar^2 4\pi\varepsilon_0}{m_0 e^2}\right)\left(\frac{m_0}{m_n}\right)\left(\frac{\varepsilon_S}{\varepsilon_0}\right) = 0.05\left(\frac{m_0}{m_n}\right)\left(\frac{\varepsilon_S}{\varepsilon_0}\right)\text{ nm} \quad (15.15a)$$

$$E_D = \frac{e^4 m_n}{2\hbar^2(4\pi\varepsilon_S)^2} = \frac{e^4 m_0}{2\hbar^2(4\pi\varepsilon_0)^2}\left(\frac{m_n}{m_0}\right)\left(\frac{\varepsilon_0}{\varepsilon_S}\right)^2 = 13.64\left(\frac{m_n}{m_0}\right)\left(\frac{\varepsilon_0}{\varepsilon_S}\right)^2\text{ eV} \quad (15.15b)$$

In the estimation, the electron rest mass m_0 and vacuum permittivity ε_0 have been replaced by the effective mass m_n and the permittivity ε_S of Si, respectively. In this manner, a_0 and E_D can be evaluated simply in terms of the Bohr radius 0.05 nm and the ionization energy 13.64 eV of the H-atom.

With the ratio $m_n/m_0 \approx 0.98,\ 0.2$, depending on the crystallographic directions and $\varepsilon_S/\varepsilon_0 \simeq 12$, a_0 ranges from about 0.5 to 2.9 nm, while E_D is in the range 20–100 meV, a few k_BT's at room temperature. Thus, the fifth valence electron in the donor atom is loosely bound to the donor ion and therefore is readily promoted to the conduction band to become a free charge carrier, hence the name the donor (Figure 15.6).The similar estimations can be made for acceptor atoms, and the ionization energy of the hole can also be shown to be about the same as

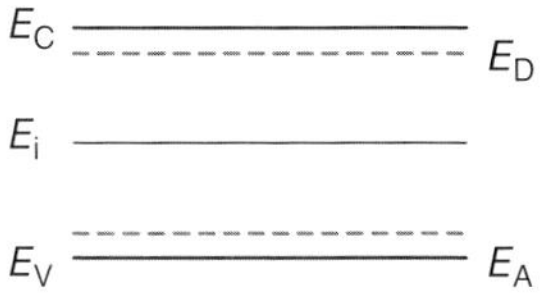

Figure 15.6 The donor and acceptor energy levels. The solid lines represent the extended nature of the conduction and valence bands, while the broken lines denote the localized energy levels of donor and acceptor states.

that of electrons in donor atoms. The ionization energy in this case is the energy required for the acceptor atom to accept an electron from the valence band, exciting a hole therein. As E_A lies above E_V by a few electron volts, the acceptor atoms can readily accept electrons from the valence band, hence the name the acceptor.

15.3.1 Fermi Level in Extrinsic Semiconductors

In the presence of donor and acceptor atoms, E_F is again found from the charge neutrality condition, namely, the electron and ionized acceptor concentrations are equal to the hole and ionized donor concentrations:

$$\frac{2}{\sqrt{\pi}}N_c F_{1/2}(\eta_{Fn}) + \frac{N_A}{1+g_A e^{\frac{(E_A-E_F)}{k_B T}}} = \frac{2}{\sqrt{\pi}}N_V F_{1/2}(\eta_{Fp}) + \frac{N_D}{1+g_D e^{(E_F-E_D)/k_B T}} \quad (15.16)$$

Here Eqs. (15.6) and (15.10) have been used for n and p, and N_D^+ and N_A^- have been specified in terms of E_F, E_D, E_A, and doping levels N_D, N_A and degeneracy factors g_D, g_A of the ground states. For Si, $g_D = 2$ and $g_A = 4$, respectively. Shown in Figure 15.7 is E_F found numerically from Eq. (15.16) as a function of T in n- and p-type silicon for different doping levels. In n-type silicon, in which $N_A = 0$, E_F is monotonously raised above the midgap with increasing N_D, as it should. Also for given N_D, E_F is lowered with increasing T to approach the intrinsic Fermi level E_{Fi} at the midgap. This is because at high T, n is primarily dictated by the thermally excited electrons regardless of the doping level. In p-type silicon in which $N_D = 0$, the behavior of E_F versus N_A and T essentially mirrors that of E_F in n-type silicon.

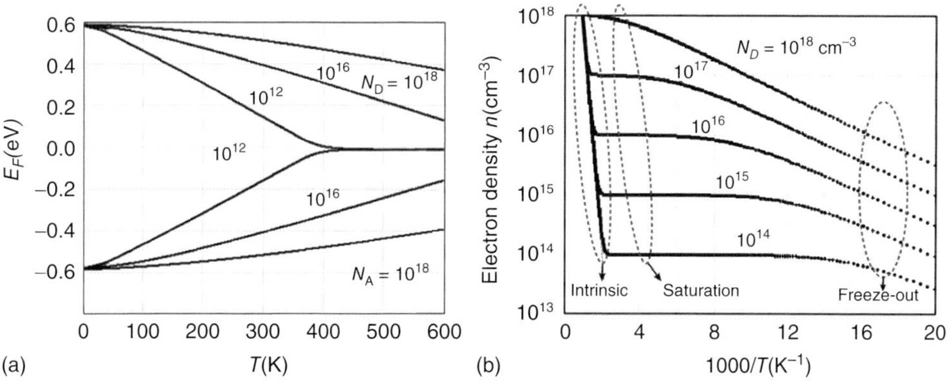

Figure 15.7 The Fermi level versus T in silicon for different N_D, N_A (a), and electron concentration versus $1000/T$ for different doping levels (b).

Also shown in Figure 15.7 are the electron concentrations in the n-type Si versus $1000/T$ for different doping levels. In the region of high T called the *intrinsic regime*, n is mainly determined by T, as more electrons are thermally generated via the band-to-band excitation. With decreasing T, the intrinsic region merges with the saturation regime, in which n is contributed mainly by the donor atoms. In this region, donor atoms are practically all ionized as E_D is still well above E_F. It is in this regime that the charge is controlled via doping N_D. With further decrease in temperature, the freeze-out regime follows in which n is decreased exponentially with decreasing T. This is due to the fact that there is practically no thermal excitation of electrons, and even those electrons donated by the dopant atoms are captured back by the ionized donor atoms as E_F is raised above E_D. The behavior of the hole Fermi level and the hole concentration p versus T and N_A can be interpreted in a similar context.

Fermi Potentials

In the nondegenerate and saturation regime, n is practically equal to N_D and can be specified in terms of the intrinsic carrier concentration n_i via Eq. (15.8) as

$$N_D = n = N_C e^{-(E_C - E_{Fi} + E_{Fi} - E_F)/k_B T} = n_i e^{(E_F - E_{Fi})/k_B T} \tag{15.17}$$

The Fermi potential φ_n of the electron is defined as

$$q\varphi_{Fn} \equiv E_F - E_{Fi} \simeq E_F - E_i \tag{15.18}$$

and is therefore specified in terms of N_D from Eq. (15.17) as

$$\varphi_{Fn} = \left(\frac{k_B T}{q}\right) \ln\left(\frac{N_D}{n_i}\right) \tag{15.19}$$

In the p-type semiconductor, the Fermi potential φ_p of the hole is defined as $q\varphi_{Fp} \equiv E_i - E_F$ and is likewise given in terms of N_A via Eq. (15.11) by

$$\varphi_{Fp} = \left(\frac{k_B T}{q}\right) \ln\left(\frac{N_A}{n_i}\right) \tag{15.20}$$

It is therefore clear that the Fermi level E_F in n-type semiconductor is raised above E_F in p type by the sum of these two Fermi potentials, as can be clearly seen from Figure 15.8.

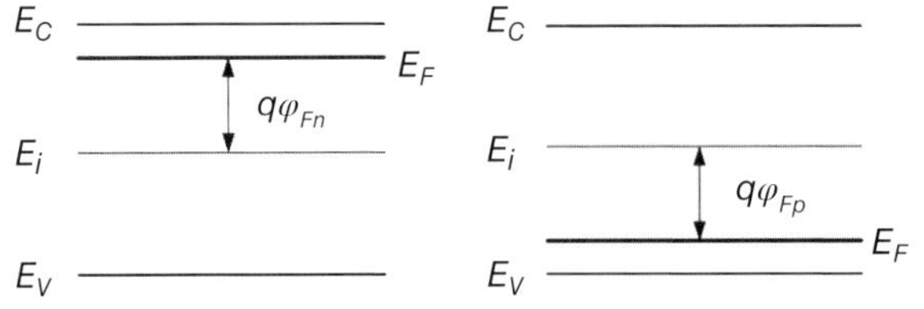

Figure 15.8 The Fermi potentials in the n- and p-type semiconductors.

Problems

15.1 (a) Use the representation of n given in Eqs. (15.6) and (15.8) and plot n versus E_F in the range $-0.15\,\text{eV} \le E_C - E_F \le 0.15\,\text{eV}$ by evaluating the Fermi 1/2-integral numerically.

(b) Use the nondegenerate expression of n given in Eq. (15.8) and plot n versus E_F in the same range as in (a) and compare and discuss the two results.

(c) Starting from the representation of p given in Eq. (15.9), derive the expression of p given in Eqs. (5.10) and (5.11).

15.2 Consider the hole concentrations of 10^2, 10^5, and 10^8 cm^{-3} in Si.

(a) Find the corresponding electron concentrations as the majority carrier at temperatures of 100, 300, and 500 K.

(b) Calculate the corresponding doping level N_D.

(c) Discuss whether or not the analytical expression of n can be taken valid for all concentration ranges considered.

15.3 In the limit $T \to 0\,\text{K}$, E_F is raised above E_D and approaches E_C regardless of the value of N_D in n-type silicon. Likewise, E_F is lowered below E_A and approaches E_V in p-type silicon regardless of N_A (Figure 15.7). Interpret the behavior.

15.4 The fabrication of the laser diode requires heavily doped n- and p-type GaAs.

(a) The E_F in n-type GaAs is located above E_C by 0.1 eV. Find n and the doping level required.

(b) Repeat the analysis as in (a) in the p-type GaAs in which E_F is below E_V by 0.15 eV. The bandgap in GaAs is 1.424 eV at room temperature and $m_n/m_0 = 0.068$ and $m_p/m_0 = 0.54$, respectively.

15.5 (a) Consider a quantum wire consisting of intrinsic silicon of cross-sectional area W^2 for $W = 10\,\text{nm}$. The quantum wire is surrounded by silicon dioxide. Find the subband spectra and specify the 1D electron concentration n_{1D} by using 1D density of states and Fermi occupation factor. For simplicity, approximate the quantum well by an infinite square well potential and take $m_n/m_0 = 0.9$.

(b) Consider a quantum well of width W having the same values as in (a). Find the subband spectra and 2D electron concentration in analogy with (a).

Suggested Readings

1. J. S. Blakemore, Semiconductor Statistics, Dover Publications, 2002.
2. D. M. Kim, Introductory Quantum Mechanics for Semiconductor Nanotechnology, Wiley-VCH, 2010.
3. J. P. McKelvey, Solid State and Semiconductor Physics, Krieger Publishing Company, 1982.
4. R. F. Pierret, Semiconductor Fundamentals, Modular Series on Solid State

Devices, volume **I**, Second Edition, Prentice Hall, 1988.
5. R. S. Muller, T. I. Kamins, and M. Chan, Device Electronics for Integrated Circuits, Third Sub Edition, John Wiley & Sons, 2002.
6. A. Yariv, An Introduction to Theory and Applications of Quantum Mechanics, John Wiley & Sons, 1982.

16
Carrier Transport in Semiconductors

The transport of charge carriers is another key element for operating the charge-based semiconductor devices and is discussed in conjunction with the drift and diffusion currents and the mobility and diffusion coefficient. The former current is driven by the electric field, while the latter by the concentration gradient. Equally important are the generation and recombination currents, and these currents are discussed based on the generation and recombination of electron–hole pairs, band-to-band as well as trap-assisted. Additionally, the thermodynamic equilibrium and nonequilibrium are highlighted in correlation with the Fermi and quasi-Fermi levels.

16.1
Quantum Description of Transport Coefficients

The drift velocity of charge carriers is driven by the electric field and characterized by the mobility, as discussed. The quantum description of the mobility can be done by using the transport equation in strict analogy with the classical theory. Thus, consider an ensemble of free electrons uniformly distributed in space in the presence of the electric field E in the z-direction. In this case, we can again adopt the relaxation approach and specify f in strict analogy with Eq. (1.17) as

$$f = f_0 + \frac{q\mathrm{E}}{m_n}\tau_n\frac{\partial f_0}{\partial v_z};\ f_0 = \frac{1}{1+e^{(E-E_F)/k_BT}},\quad E = \frac{1}{2}m_n v^2 \tag{16.1}$$

where τ_n is the longitudinal relaxation time. The only difference between Eqs. (1.17) and (16.1) is that f_0 in Eq. (1.17) is the Boltzmann distribution function, while f_0 in Eq. (16.1) is the Fermi distribution function.

The differentiation of f_0 in Eq. (16.1) with respect to v_z yields

$$\frac{\partial f_0}{\partial v_z} = \frac{\partial f_0}{\partial E}\frac{\partial \mathrm{E}}{\partial v_z} = -\frac{e^{(E-E_F)/k_BT}}{(1+e^{(E-E_F)/k_BT})^2}\frac{m_n v_z}{k_BT} \equiv -\frac{m_n v_z}{k_BT}f_0(1-f_0)$$

and the distribution function f is obtained to the first order of approximation as

$$f = f_0 - \frac{q\mathrm{E}v_z\tau_n}{k_BT}f_0(1-f_0) \approx f_0 - q\mathrm{E}v_z\tau_n\delta(E-E_F) \tag{16.2a}$$

Introductory Quantum Mechanics for Applied Nanotechnology, First Edition. Dae Mann Kim.

where the product $f_0 \times (1-f_0)$ is peaked sharply at the Fermi level E_F as shown in Figure 16.1 and has been approximated by a delta function

$$f_0(1-f_0) \approx k_B T \delta(E-E_F) \tag{16.2b}$$

16.1.1
Mobility

We can find the average velocity of electrons driven by the E-field in the z-direction by using the distribution function f thus found as

$$\langle v_z \rangle = \frac{\int d\underline{v} v_z f}{\int d\underline{v} f} = \frac{-q\mathrm{E}\int_0^{2\pi} d\varphi \int_{-1}^{1} d\mu \int_0^{\infty} \tau_n(v) v^2 dv v_z^2 \delta(E-E_F)}{\int_0^{2\pi} d\varphi \int_{-1}^{1} d\mu \int_0^{\infty} v^2 dv f_0}, \quad \mu = \cos\theta \tag{16.3}$$

where the first integral in the numerator and the second integral in the denominator vanish due to the odd parity of the integrands involved just as in the case of Eq. (1.18). The remaining $\boldsymbol{v}$-integrals have been expressed in terms of the spherical coordinates, and τ_n has been taken depending on the magnitude of $\boldsymbol{v}$. In this case, the angular integrations in the numerator and the denominator cancel out. Furthermore, the integration in the denominator can be done to a good approximation by taking the Fermi function f_0 as a step function, that is, $f_0 = 1$ for $E \le E_F$, while $f_0 = 0$ for $E > E_F$ (Figure 15.2). Hence, Eq. (16.3) can be expressed in terms of energy as

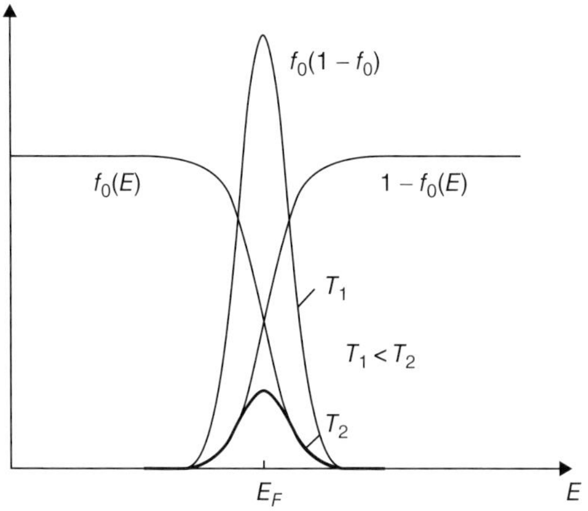

Figure 16.1 The Fermi distribution function $f_0(E)$ and the product $f_0(E) \times (1-f_0(E))$ versus energy for different temperatures.

$$\langle v_z \rangle = -\frac{2q\text{E}}{3m_n} \frac{\left[\int_0^\infty dEE^{3/2}\tau_n(E)\,\delta(E-E_F)\right]}{\int_0^{E_F} dEE^{1/2}}; \quad E = \frac{m_n v^2}{2}$$

$$= -\frac{2q\text{E}}{3m_n}\frac{E_F^{2/3}\tau_n(E_F)}{(2/3)E_F^{2/3}} = -\frac{q\tau_n(E_F)}{\mu_n}\text{E} \tag{16.4}$$

The resulting mobility

$$\mu_n = \frac{q\tau_n(v_F)}{m_n}; \quad \frac{m_n v_F^2}{2} = E_F \tag{16.5}$$

is identical to Eq. (1.19), but the mean collision or relaxation time as a function of v is specified explicitly by the Fermi velocity v_F in the quantum description.

16.1.2
Diffusion Coefficient

We next consider the spatially nonuniform electron concentration and introduce the mean free path of electrons l_n on both sides of x as shown in Figure 16.2. The electrons can then be treated free of scattering in the volume elements from $x - l_n$ to x and from x to $x + l_n$. The net number of electrons crossing x per unit area from left to right is then given by

$$N = \frac{1}{2}n(x-l_n)l_n - \frac{1}{2}n(x+l_n)l_n$$
$$= \frac{l_n}{2}\left\{\left[n(x) - \frac{\partial n(x)}{\partial x}l_n\right] - \left[n(x) + \frac{\partial n(x)}{\partial x}l_n\right]\right\} \simeq -l_n^2\frac{\partial n(x)}{\partial x} \tag{16.6}$$

where 1/2 factor accounts for the fact that due to the random thermal motion in equilibrium, only one half of the electrons are moving from left to right or vice versa. Also, as l_n is generally much less than the spatial range over which n changes appreciably, $n(x \pm l_n)$ has been Taylor expanded and truncated after the first expansion.

The flux of electrons from left to right is thus given by dividing N by the mean collision time τ_n:

$$F_n \equiv \frac{N}{\tau_n} = -D_n\frac{\partial n(x)}{\partial x}; \quad D_n \equiv \frac{l_n^2}{\tau_n} \tag{16.7a}$$

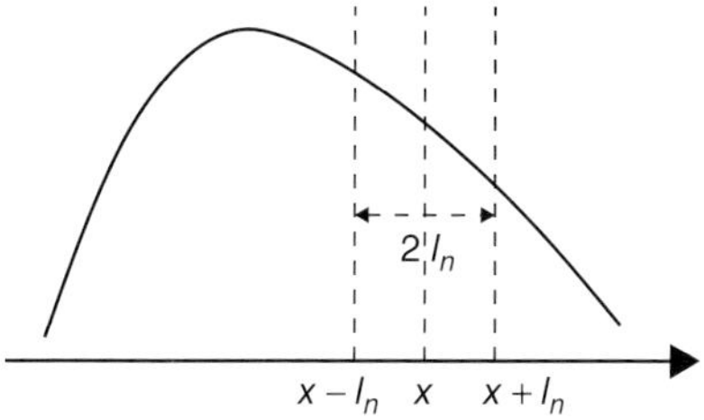

Figure 16.2 The spatially nonuniform concentration profile of electrons and a differential volume element at x, consisting of two parallel planes at $x \pm l_n$ with l_n denoting the electron diffusion length.

The constant D_n is known as the *diffusion coefficient*, and it connects the input concentration gradient to the output electron flux. Now the mean free path l_n is by definition the average distance the electron traverses during the mean collision time τ_n:

$$l_n = v_T \tau_n; \qquad \frac{m_n v_T^2}{2} = \frac{k_B T}{2} \tag{16.7b}$$

where v_T is the thermal speed in the x-direction specified via the equipartition theorem (Eq. (1.14)). It is therefore clear from Eqs. (16.5), (16.7a), and (16.7b) that ratio of two transport coefficients is given by

$$\frac{D_n}{\mu_n} = \frac{l_n^2/\tau_n}{q\tau_n/m_n} = \frac{k_B T}{q} \tag{16.8a}$$

We can carry out a similar analysis for the holes and obtain

$$\frac{D_p}{\mu_p} = \frac{l_p^2/\tau_p}{q\tau_p/m_p} = \frac{k_B T}{q} \tag{16.8b}$$

Equations (16.8a) and (16.8b) are known as the *Einstein relation for electrons and holes*, respectively.

The total current densities of electrons and holes consist of the drift and diffusion components and are given by

$$J_n = qn\mu_n \mathrm{E} + qD_n \frac{dn}{dx}; \quad J_p = qp\mu_p \mathrm{E} - qD_n \frac{dp}{dx} \tag{16.9}$$

16.2 Equilibrium and Nonequilibrium

Single-Semiconductor System

A basic property of the equilibrium is that the carrier densities n, p are quantified by a single Fermi level E_F, as discussed. Also J_n, J_p are inextricably related to E_F, which is shown as follows. Thus, consider the 1D current density of electrons given by Eq. (16.9). The electric field E driving the drift is specified in terms of the electric potential as, $\mathrm{E} = -\partial\varphi/\partial x$ and φ in turn represents the electron potential energy when multiplied by $-q$. Hence, we can treat $-q\varphi$ just like E_C, E_V or the midgap E_i and express E in terms of E_i as

$$\mathrm{E} = -\frac{\partial\varphi}{\partial x} \equiv \frac{(-q\partial\varphi)}{q\partial x} = \frac{1}{q}\frac{\partial E_i}{\partial x} \tag{16.10}$$

where E_C, E_V, and E_i vary in x in parallel with $-q\varphi(x)$.

Additionally, in the nondegenerate regime, n, p are analytically specified in terms of E_i, E_F in equilibrium (see Eqs. (15.8) and (15.11)). Hence, with the use of the Einstein relation (16.8) and the expressions of n and p given in Eqs. (15.8) and

(15.11), J_n can be is expressed as

$$J_n = \mu_n n \left[\frac{dE_i}{dx} + q\left(\frac{k_BT}{q}\right)\frac{d}{dx}\left(\frac{E_F - E_i}{k_BT}\right)\right] = \mu_n n\frac{dE_F}{dx} \tag{16.11a}$$

We can likewise express the hole current density as

$$J_p = \mu_p p \left[\frac{dE_i}{dx} - q\left(\frac{k_BT}{q}\right)\frac{d}{dx}\left(\frac{E_i - E_F}{k_BT}\right)\right] = \mu_p p\frac{dE_F}{dx} \tag{16.11b}$$

In equilibrium, no current flows, and therefore

$$\frac{dE_F}{dx} = 0 \tag{16.12}$$

and E_F in a single-semiconductor system should be flat in equilibrium.

Composite Semiconductor System

We next consider a composite system of two semiconductors in equilibrium contact as shown in Figure 16.3. In equilibrium, the flux of electrons from left to right F_{LR} is balanced by the reverse flux from right to left F_{RL}. These fluxes are dictated by two factors, namely, the density of states g_L, g_R and Fermi occupation factors f_L, f_R on both sides at the energy level E. Specifically, F_{LR} is given by

$$F_{LR} = M[g_L(E)dEf_L(E)] \cdot [g_R(E)dE(1 - f_R(E))] \tag{16.13a}$$

where M is the transfer matrix element. The first bracket is the number of occupied quantum states at E, that is, the electron density on the left-hand side, while the second bracket represents the vacant state density on the right-hand side for electrons to hop in. The flux from right to left is likewise given by

$$F_{RL} = M[g_R(E)dEf_R(E)] \cdot [g_L(E)dE(1 - f_L(E))] \tag{16.13b}$$

As $F_{LR} = F_{RL}$ in equilibrium, it follows from equating (16.13a) and (16.13b)) that, $f_L(E) = f_R(E)$ that is,

$$\frac{1}{1 + e^{(E-E_{FL})/k_BT}} = \frac{1}{1 + e^{(E-E_{FR})/k_BT}} \tag{16.14a}$$

Therefore, the two Fermi levels on both sides should be the same

$$E_{FL} = E_{FR} \tag{16.14b}$$

Equivalently, the Fermi level should line up. This fact together with Eq. (16.12) leads to the general conclusion, namely, that E_F in equilibrium should line up and be flat. The conclusion holds true for any number of semiconductor layers in equilibrium contact.

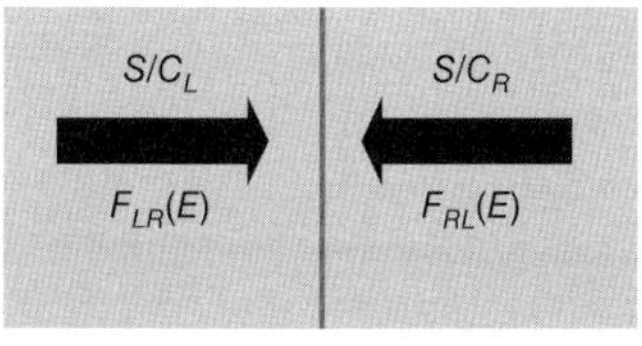

Figure 16.3 A composite semiconductor system, consisting of two semiconductors in equilibrium contact. The electron flux from left to right is balanced by its reverse flux from right to left in equilibrium.

16.2.1 Nonequilibrium and Quasi-Fermi Level

A system, when subjected to irradiation or bias, is driven away from the equilibrium to nonequilibrium conditions. In nonequilibrium, n and p cannot be quantified by a single Fermi level. Rather, two quasi-Fermi levels, one for electrons and the other for holes, are required. This can be seen by considering a slab of the intrinsic semiconductor under irradiation (Figure 16.4). Under illumination, the electron hole pairs are generated via band-to-band excitation and also subjected to recombination. Hence, the rate equations of n and p read as

$$\frac{\partial n}{\partial t} = \frac{\alpha I}{h\nu} - \frac{n}{\tau_n}, \quad \frac{\partial p}{\partial t} = \frac{\alpha I}{h\nu} - \frac{p}{\tau_p} \tag{16.15}$$

where τ_n and τ_p are the recombination times of electrons and holes, respectively, and the generation rate is given in terms of the absorption coefficient α and the flux of photons $I/h\nu$.

At steady state, the derivatives of n and p with respect to time vanish, and the photo-generated n, p are proportional to $I\tau_n/h\nu$, $I\tau_p/h\nu$ respectively. Therefore, the total electron and hole concentrations consist of intrinsic and photo-generated components:

$$n = n_i + n_{\text{ph}} = n_i + \left(\frac{\alpha I \tau_n}{h\nu}\right) \tag{16.16a}$$

$$p = n_i + p_{\text{ph}} = n_i + \left(\frac{\alpha I \tau_p}{h\nu}\right) \tag{16.16b}$$

When the light intensity is high, n_{ph} and p_{ph} can be much greater than n_i, that is, $n_{\text{ph}} \gg n_i$, $p_{\text{ph}} \gg p_i$. The former inequality in Eq. (16.16a) requires that E_F should be higher than the intrinsic Fermi level $E_{Fi} \simeq E_i$ just as in the n-type semiconductor (Figure 15.8). Likewise, the latter inequality in Eq. (16.16b) requires that E_F should be lower than E_i as in the p-type semiconductor. Obviously, these two requirements cannot be met with a single E_F simultaneously. The only way to come out of this inconsistency is to introduce two quasi-Fermi levels, one for electrons E_{Fn} and the other for holes E_{Fp}, and quantify n and p separately

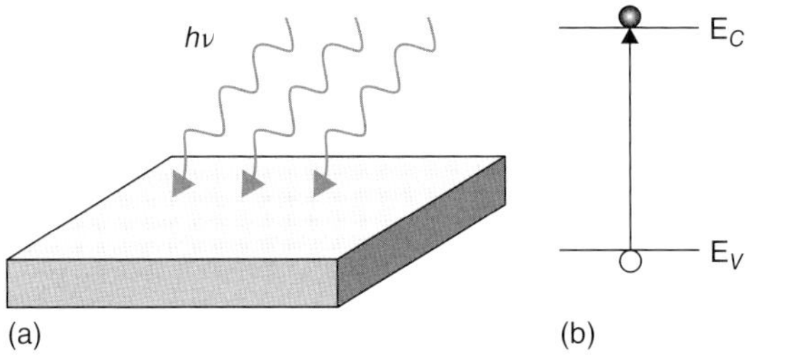

Figure 16.4 A semiconductor sample under uniform irradiation (a) and the photogeneration of the electron–hole pair via band-to-bend excitation (b).

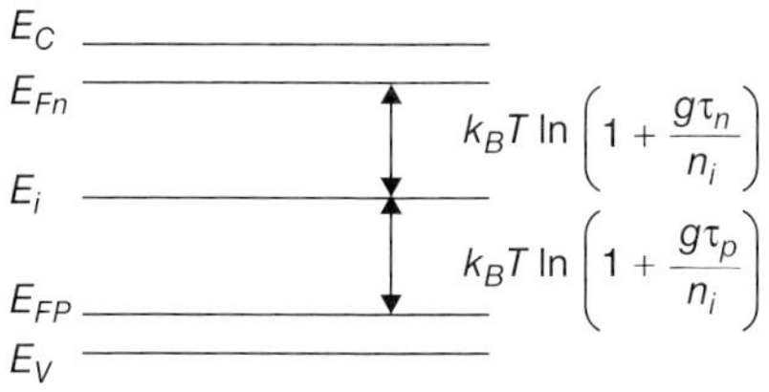

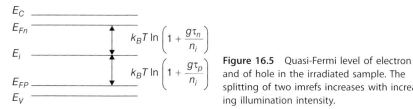

Figure 16.5 Quasi-Fermi level of electron and of hole in the irradiated sample. The splitting of two imrefs increases with increasing illumination intensity.

in strict analogy with E_F in equilibrium. Thus, Eq. (16.16) can be expressed as

$$n = n_i e^{(E_{Fn}-E_i)/k_BT}, \quad p = n_i e^{(E_i-E_{Fp})/k_BT} \tag{16.17}$$

The quasi-Fermi levels are also called *imrefs.*

Moreover, two imrefs split under irradiation by the amount that is given by inserting Eq. (16.17) into Eq. (16.16) and inverting E_{Fn}, E_{Fp}:

$$E_{Fn} - E_{Fp} = k_BT \ln\left[\left(1+\frac{g\tau_n}{n_i}\right)\left(1+\frac{g\tau_p}{n_i}\right)\right], \quad g = \frac{\alpha I}{h\nu} \tag{16.18}$$

Clearly, the splitting increases with increasing light intensity and/or the generation rate, as shown in Figure 16.5. Without illumination, $g = 0$, and the two imrefs collapse into a common Fermi level, that is, $E_{Fn} = E_{Fp} = E_F$ as the system relaxes back to equilibrium. Moreover, just as E_{Fn} and E_{Fp} quantify n and p in exact analogy with E_F in equilibrium, the current densities contributed by drift and diffusion are described in nonequilibrium by the slope of E_{Fn}, E_{Fp} again in strict analogy with Eq. (16.11)

$$J_n = \mu_n n \frac{d}{dx} E_{Fn}, \quad J_p = \mu_p p \frac{d}{dx} E_{Fp} \tag{16.19}$$

Unlike the flat E_F in equilibrium representing the zero current, E_{Fn}, E_{Fp} generally vary as functions of position and account for the current.

16.3 Generation and Recombination Currents

The drift and diffusion currents are due to the motion of electrons and holes in the conduction and valence bands, respectively. There also exist the generation and recombination currents resulting from the law of mass action $np = n_i^2$ in equilibrium being broken. If $np > n_i^2$, the recombination current I_R ensues driven by the reactive force pushing the system back to the equilibrium. By the same token, if $np < n_i^2$, the generation current I_G ensues again to drive the system back to equilibrium.

Band-to-Band Excitation or Recombination

As noted, some of the electrons in the valence band are promoted to the conduction band via the band-to-band thermal excitation, leaving behind the same

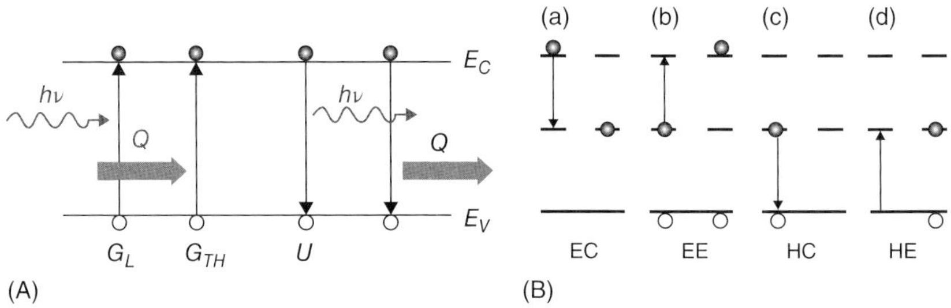

Figure 16.6 The generation of the e–h pairs via thermal and optical band-to-band excitations and the reverse process of radiative and non-radiative recombination of the e–h pairs (A). The single-level trap-assisted emission and capture of electrons and holes (B): (a) electron capture, (b) electron emission, (c) hole capture, and (d) hole emission.

number of holes in the valence band (Figure 16.6). The e–h pairs are also generated by incident photons with energy greater than the bandgap. The inverse process of the recombination of electrons in the conduction band and holes in the valence band also occurs. In carrying out the recombination, the electron has to lose energy amounting to the bandgap. The energy is converted into radiation or consumed via the heat dissipated.

In the n-type semiconductor, for example, the recombination of electrons and holes is proportional to n and p in nonequilibrium and n_{n0} and p_{n0} in equilibrium

$$R = \alpha n_n p_n, \quad \mathrm{R}_e = \alpha n_{n0} p_{n0} \tag{16.20}$$

where α is the proportionality constant, n_{n0} and n_n the equilibrium and nonequilibrium electron concentrations as the majority carrier, and p_{n0} and p_n the respective hole concentrations as the minority carrier. In equilibrium in which the detailed balancing holds true between recombination and generation of e–h pairs, the thermal excitation G_{th} should be equal to R_e. Therefore, in the presence of excess carriers, there ensues a net recombination with the rate given by

$$\begin{aligned} U &\equiv R - G_{th} \\ &= \alpha(n_n p_n - n_{n0} p_{n0}) \simeq \frac{1}{\tau_p}(p_n - p_{n0}); \quad \frac{1}{\tau_p} \equiv \alpha n_{n0} \end{aligned} \tag{16.21}$$

where we have taken $n_n \simeq n_{n0}$ for the low-level injection of charge carriers, and τ_p is the hole recombination lifetime as the minority carrier.

16.3.1
Trap-Assisted Recombination and Generation

The recombination (r) and generation (g) rates are drastically enhanced when trap sites are present in the bandgap, and let us thus consider the trap-assisted r, g processes using the theories of Shockley and Read and also of Hall. For simplicity, a single trap level is considered as sketched in Figure 16.6. There are four possible processes: (i) the capture of an electron from the conduction band by an empty trap site, (ii) the inverse process of electron emission from the trap to the conduction band, or (iii) the trapped electron capturing a hole in the valence band, and (iv) the inverse process of an electron in valence band captured in the trap site, inducing a hole emission.

Evidently, in view of the exclusion principle, the electron capture rate is proportional to n in the conduction band and the empty trap sites, while the electron emission rate is proportional to the filled trap sites. Hence, we can write

$$r_{ec} = (v_{Tn}\sigma_n) \cdot n \cdot [N_t(1-f)]; \quad r_{ee} = e_n \cdot (N_t f) \tag{16.22}$$

where the proportionality constant of the recombination is given by the thermal velocity v_T and the electron capture cross-section, σ_n of the trap, $\approx 10^{-15}\,\text{cm}^2$, and the proportionality constant e_n for emission is called the *electron emission probability*. It is important to note that the distribution function f introduced to account for the filled and empty trap sites away from the equilibrium is not the same as the Fermi distribution function f_0 in equilibrium.

The capture and emission rates of holes can be described in a similar manner. The capture is done by the trapped electron recombining with a hole in the valence band, and the rate is proportional to p and the number of trapped electrons. The hole emission is proportional to the number of empty traps into which electrons in the valence band are captured, hence is proportional to the empty trap density. Thus, we can write

$$r_{hc} = v_{Th}\sigma_p p N_t f; \quad r_{he} = e_p N_t(1-f) \tag{16.23}$$

where the proportionality constant of capture is given by the product of v_{Th} and the hole capture cross-section, while that of emission is called the *hole emission probability*.

Steady State and Equilibrium

The difference between the equilibrium and steady state is best illustrated by considering a semiconductor under uniform irradiation. The rate equations of n and p are given by the photo-generation rate and the net recombination rates of electrons and holes, that is, $r_{ec} - r_{ee}$ and $r_{hc} - r_{he}$:

$$\frac{\partial n}{\partial t} = \frac{\alpha I}{h\nu} - (r_{ec} - r_{ee}), \quad \frac{\partial p}{\partial t} = \frac{\alpha I}{h\nu} - (r_{hc} - r_{he}) \tag{16.24}$$

At steady state in which n, p are independent of time, it follows from Eq. (16.24) that the net recombination of electrons is the same as that of holes. In equilibrium, on the other hand, the respective rates of capture and emission should be balanced. Therefore, we can write

$$r_{ec} - r_{ee} = r_{hc} - r_{he}, \quad r_{ec} = r_{ee}, \quad r_{hc} = r_{he} \tag{16.25}$$

and show that the equilibrium condition is more stringent in that it satisfies automatically the steady-state condition as well.

Steady-State Distribution Function

The equilibrium condition for electrons in Eq. (16.25) is explicitly specified with the use of Eq. (16.22) as

$$v_T\sigma_n nN_t[1-f_0(E_t)] = e_n N_t f_0(E_t); \quad f_0(E_t) = \frac{1}{1+e^{(E_t-E_F)/k_BT}} \tag{16.26}$$

where f in Eq. (16.22) should be replaced by the Fermi function in equilibrium. Therefore, the electron emission probability e_n is specified explicitly from Eq. (16.26) as

$$e_n = v_T\sigma_n n_i e^{(E_t-E_i)/k_BT} \tag{16.27a}$$

Likewise, the hole emission probability is given by

$$e_p = v_T\sigma_p n_i e^{(E_i-E_t)/k_BT} \tag{16.27b}$$

Upon inserting e_n, e_p into Eqs. (16.22) and (16.23), the steady-state condition in Eq. (16.25) reads as

$$v_T\sigma_n nN_t(1-f) - v_T\sigma_n n_i e^{(E_t-E_i)/k_BT}N_t f = v_T\sigma_p pN_t f - v_T\sigma_p n_i e^{(E_i-E_t)/k_BT}N_t(1-f) \tag{16.28}$$

The distribution function f in nonequilibrium is therefore obtained by regrouping the terms in Eq. (16.28). For simplicity, let us take in $\sigma_n = \sigma_p = \sigma$ in which case f is given by

$$f = \frac{n + n_i e^{(E_i-E_t)/k_BT}}{n+p+2n_i\cosh(E_t-E_i)/k_BT} \tag{16.29}$$

In Figure 16.7, f is plotted as a function of $(E_t - E_i)/k_BT$ for the different doping level N_D. Also shown in the figure are the corresponding Fermi functions f_0 for comparison. Although different in mathematical expressions, the two curves of f and f_0 do not depart very much from each other. But the minute departure of f from f_0 accounts for the different recombination and generation rates under bias or irradiation.

Recombination Rate

Now that f has been found, the recombination rate of electron $r_{ec} - r_{ee}$ can be specified by combining Eqs. (16.22), (16.27a), and (16.29) and rearranging the terms as

$$U = \frac{1}{\tau}\frac{(pn-n_i^2)}{n+p+2n_i\cosh(E_t-E_i)/k_BT}; \quad \frac{1}{\tau} \equiv \sigma v_T N_t \tag{16.30}$$

where τ represents the recombination lifetime. We can likewise obtain the same result starting from the net recombination rate of holes. The recombination rate U accounts for the generation rate as well. At equilibrium in which, $np = n_i^2$, $U = 0$ and there is no net recombination, as it should. However for $np > n_i^2$, $U > 0$, and the recombination of e–h pairs ensues. Likewise for, $np < n_i^2$, $U < 0$ and e–h pairs are in this case generated. The r, g processes are mediated by the traps playing the role of stepping sites.

It is clear from Eq. (16.30) that the recombination rate U depends sensitively on $E_t - E_i$. Specifically, U attains the maximum value for $E_i = E_t$, which suggests that the r, g processes are maximally enhanced when $E_C - E_t$, $E_t - E_V$ are approximately the same. Shown in Figure 16.8 are r, g rates for different doping levels. In an n-type semiconductor, where $n_n \gg p_n$, n_i U is mainly dictated by n_n and

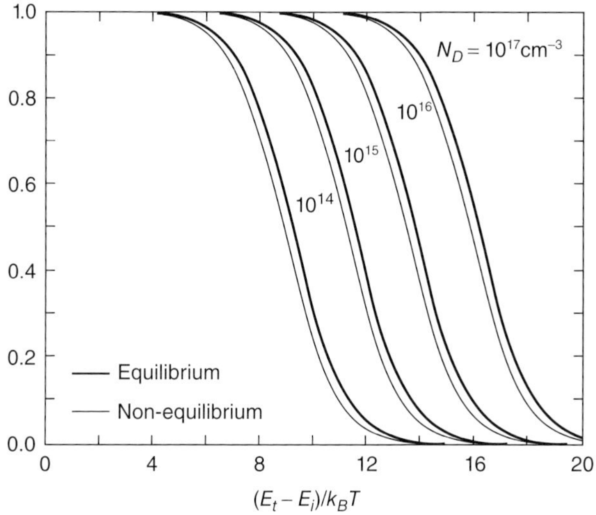

Figure 16.7 The equilibrium Fermi function and the nonequilibrium distribution functions versus $(E_t - E_i)/k_BT$ for different N_D.

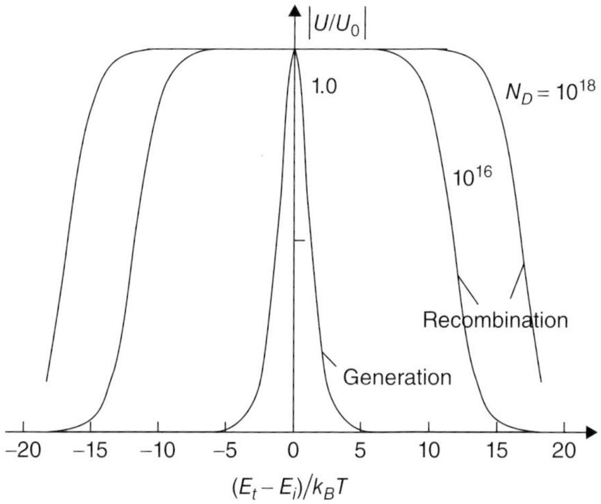

Figure 16.8 The recombination and generation rates versus $(E_t - E_i)/k_BT$ for different N_D.

is pinned at its maximum level over a range in which $n_n \gg n_i \cosh(E_t - E_i)/k_B T$. On the other hand, for $np \ll n_i^2$, the generation rate falls off exponentially as a function of $E_t - E_i$ from its maximum value for $E_t = E_i$. Also in the presence of shallow traps located near E_C, the electrons are easily captured from the conduction band into the trap sites or vice versa. However, the large difference between E_t and E_V slows down the hole emission or capture processes. Consequently, the overall efficiency of the trap-assisted generation or recombination of electron hole pairs is low. Rather, the electron capture is more likely to be accompanied by the inverse process of electron emission. Similarly, for traps near the valence band edge, the hole capture and emission constitute the dominant processes.

Minority Carrier Lifetime

We next consider an n-type semiconductor in which $n_n \gg p_n, \; n_i$. In this case, we may put $n_n \approx n_{n0}$ and simplify the numerator of U as

$$pn - n_i^2 \approx n_{n0}(p_{n0} + p_n - p_{n0}) - n_i^2 = n_{n0}(p_n - p_{n0}); \;\; n_{n0}p_{no} = n_i^2 \quad (16.31)$$

Thus, by inserting Eq. (16.31) into Eq. (16.30) and using the fact that $n_n \gg p_n, \; n_i$, we can write

$$U = \frac{p_n - p_{n0}}{\tau_p}, \quad \frac{1}{\tau_p} = \sigma_p v_T N_t \quad (16.32a)$$

where τ_p is called the *lifetime of hole* as the minority carrier. The recombination rate of electrons in the p-type semiconductor and the lifetime τ_n are likewise given by

$$U = \frac{n_p - n_{p0}}{\tau_n}, \quad \frac{1}{\tau_n} \equiv \sigma_n v_T N_t \quad (16.32b)$$

Problems

16.1 (a) Fill in the algebra and reproduce the Einstein relation for electrons and holes by using the transport coefficients given in Eqs. (16.5) and (16.7).

(b) Starting from the expressions of n, p given in Eq. (16.16), derive Eq. (16.18) specifying the split between E_{Fn} and E_{Fp} under irradiation.

16.2 The conductivity σ and resistivity ρ of the electron are specified in terms of q, n and μ_n as $\sigma \equiv qn\mu_n = 1/\rho$. An n-type Si with resistivity 10 Ωcm is uniformly illuminated, and 10^{21} e–h pairs are generated per cubic centimeter second.

(a) Calculate the dark and photoconductivity.

(b) Calculate the contribution made by electrons and holes to the total conductivity. Use $\mu_n = 800$ cm^2 V^{-1} s^{-1}, $\mu_p = 400$ cm^2 V^{-1} s^{-1}, and take the lifetime of the electron and hole to be 1 μs.

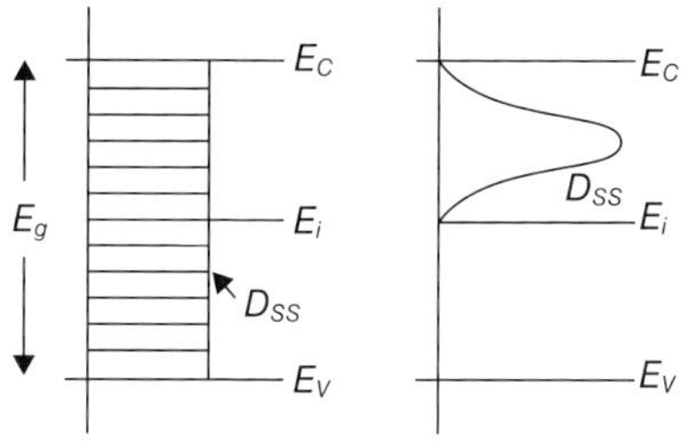

Figure 16.9 The trap levels uniformly and Gaussian distributed.

16.3 An n-type semiconductor is illuminated uniformly with light intensity $10\ \mathrm{W\,cm^{-2}}$. The wavelength and absorption coefficient are given by $\lambda = 500\,\mathrm{nm}$ and $\alpha = 10\ \mathrm{cm^{-1}}$.

(a) Find the photon flux (the number of photons crossing per unit area and per second).

(b) Calculate the number of e–h pairs generated per second.

16.4 The rate equation for p_n as the minority carrier is given by

$$\frac{\partial p_n}{\partial t} = g_L - \frac{p_n - p_{n0}}{\tau_p}; \quad g_L = \frac{\alpha I}{h\nu}$$

Find $p_n(t)$ during the illumination and also after the light is turned off.

16.5 Consider the tap centers uniformly and Gaussian distributed across the energy gap, as sketched in Figure 16.9.

(a) Derive the recombination rates by generalizing the single-level trap model and assuming that $\sigma_n = \sigma_2 = \sigma$, for simplicity.

(b) Derive the expression for minority carrier lifetime.

(c) Repeat the analysis of (a) and (b) for the case of traps Gaussian distributed $N_t \exp -(E_t - E_{tc})^2/2\sigma^2$ centered at E_{tc} half-way between midgap and E_C.

Suggested Readings

1. R. S. Muller, T. I. Kamins, and M. Chan, Device Electronics for Integrated Circuits, Third Sub Edition, John Wiley & Sons, 2002.
2. R. F. Pierret, Advanced Semiconductor Fundamentals, Modular Series on Solid State Devices, volume **VI**, Second Edition, Prentice Hall, 2002.
3. D. M. Kim, Introductory Quantum Mechanics for Semiconductor Nanotechnology, Wiley-VCH, 2010.
4. A. S. Grove, Physics and Technology of Semiconductor Devices, John Wiley & Sons, 1967.
5. A. Yariv, An Introduction to Theory and Applications of Quantum Mechanics, John Wiley & Sons, 1982.

17
P–N Junction Diode: *I–V* Behavior and Device Physics

The p–n junction diode is one of the simplest solid-state switches but is a key hardware element of nanotechnology. The technological platform of the diode is multidisciplinary in nature and covers the central core of the device physics. Hence, the understanding of the diode is essential for comprehending other active devices. Moreover, the diode provides convenient platforms on which to devise the photodiode, light-emitting and laser diodes, solar cells, and so on. Also the p–n junction itself is an essential element of MOSFET. The diode I–V behavior is modeled with an emphasis on the physical principles involved.

17.1
The p–n Junction in Equilibrium

The p–n junction consists of n- and p-type semiconductors in equilibrium contact, as shown in Figure 17.1. When a positive voltage V_F called the *forward voltage* is applied to the p side, a large forward current I_F flows from p to n regions. When a negative voltage V_R called the *reverse voltage* is applied, the minimal level of the reverse current I_R flows from n to p regions. Hence, the diode works as an electrical switch. When the magnitude of V_R is increased beyond a certain value called the *breakdown voltage*, I_R increases exponentially. The explosive growth of I_R is known as the *junction breakdown*.

Junction Band Bending

There are two kinds of junctions: homo and hetero. In the former, the bandgaps in p and n regions are the same, while in the latter, bandgaps are different. We discuss the former junction, but most of the results derived are also applicable in the latter junction. Before contact, the Fermi level E_F in the n bulk region is higher than E_F in the p region by an amount equal to the sum of the Fermi potentials $q\varphi_{Fn} + q\phi_{Fp}$ (Figure 17.1). However, when the n and p regions are brought into the equilibrium contact, E_F should line up and be flat, as detailed. Clearly, the condition necessitates the band bending by the amount $q\varphi_{Fn} + q\phi_{Fp}$ as clear from Figure 17.1. Also, $E_C - E_F$ and $E_F - E_V$ in n and p bulk regions should remain the

Introductory Quantum Mechanics for Applied Nanotechnology, First Edition. Dae Mann Kim.

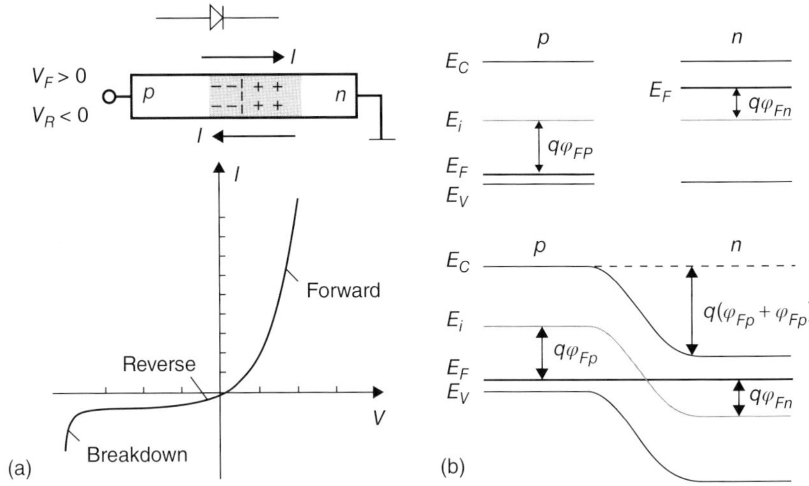

Figure 17.1 The cross-sectional view of the p–n junction diode and the *I–V* curve, consisting of the forward, reverse, and breakdown currents (a). The band diagram of the p–n junction before and after the equilibrium contact (b).

same to preserve the equilibrium carrier concentrations therein before and after the contact.

Space Charge, Field, and Potential

A question then naturally arises as to what physical processes are responsible for such a band bending. The answer to this question lies in the simple electrostatics entailed in the equilibrium contact. Specifically, the electrons diffuse from the higher-concentration n region to the lower-concentration p region. Likewise, holes diffuse from p to n regions. The diffusion of electrons and holes leaves behind the uncompensated donor and acceptor ions near the interface, which induces the dipolar space charge ρ as shown in Figure 17.2. The space charge ρ in turn gives rise to the space charge field **E** and the potential φ and ultimately the potential energy of the electron $-q\varphi$. The potential energy bridges the misaligned E_C, E_V on both sides. In short, the diffusion of electrons and holes triggers the junction band bending.

The space charge ρ induces **E** according to Coulomb's law:

$$\frac{\partial}{\partial x}\mathrm{E} = \frac{\rho}{\varepsilon_S}; \quad \rho = \begin{cases} qN_D; & 0 \le x \le x_n \\ -qN_A; & -x_p \le x \le 0 \end{cases} \tag{17.1}$$

Here, ε_S is the permittivity of the semiconductor, and $\rho(x)$ is taken as a step function, with heights given by the doping levels qN_D and $-qN_A$ in the completely depleted approximation, a nonessential approximation. Also x_n, $-x_p$ demarcate the junction depletion region W from the n and p bulk regions. We can readily

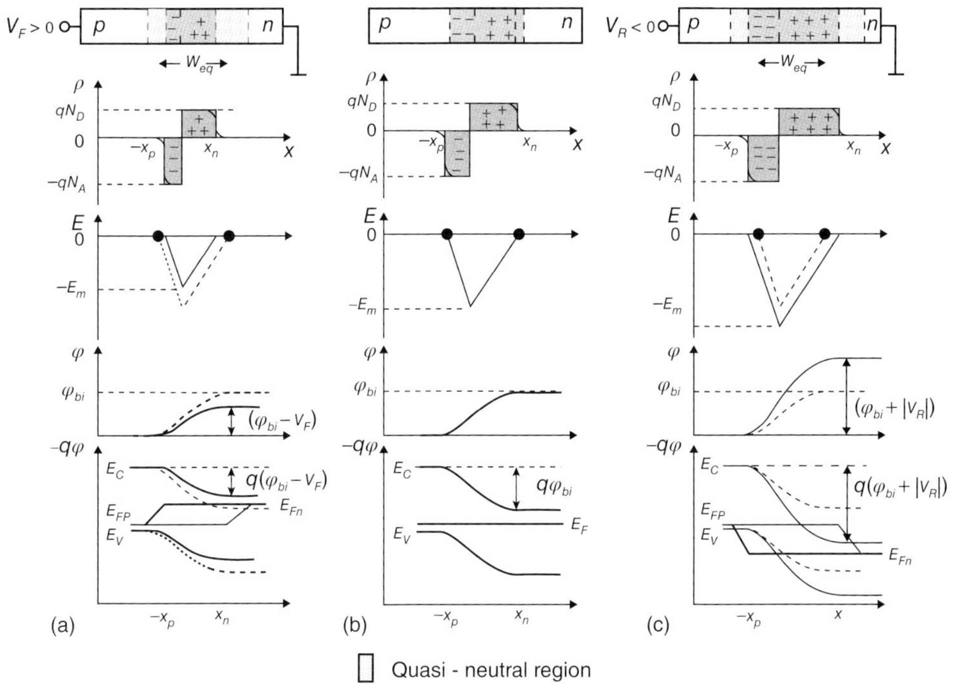

Figure 17.2 The space charge (ρ), field (E), potential (φ), and electron potential energy in equilibrium (b) under forward (a) and reverse (c) biases, in comparison with the equilibrium values.

carry out the integration, obtaining

$$\mathrm{E}(x) = \begin{cases} (qN_D/\varepsilon_S)\,(x - x_n); & 0 \le x \le x_n \\ -(qN_A/\varepsilon_S)(x + x_p); & -x_p \le x \le 0 \end{cases} \tag{17.2}$$

The boundary conditions used are $\mathrm{E}(x_n) = \mathrm{E}(-x_p) = 0$, since $E(x)$ does not penetrate into the neutral bulk regions. Also $\mathrm{E}(x)$ is continuous everywhere; hence, the condition applied at $x = 0$ yields the maximum E and the relationship between x_n and x_p as well

$$|\mathrm{E}_{\max}| = \frac{qN_D x_n}{\varepsilon_S} = \frac{qN_A x_p}{\varepsilon_S}; \quad qN_D x_n = qN_A x_p \tag{17.3}$$

Evidently, Eq. (17.3) shows that the number of electrons spilled over from n to p regions and the number of holes spilled from p to n regions are the same. Once the space charge field, E is set up, electrons and holes are driven by E, electrons to the n region and holes to the p region, compensating thereby the electron and hole diffusion in opposite directions, as it should in equilibrium.

Depletion Depth and Built-In Potential

The depletion depth denotes the total width of the junction region and is given from Eq. (17.3) by

$$\begin{aligned} W &\equiv x_n + x_p \\ &= x_n(1 + N_D/N_A) = x_p(1 + N_A/N_D) \end{aligned} \tag{17.4}$$

Also, the built-in potential φ_{bi} is the difference in $\varphi(x)$ between $-x_p$ and x_n and is obtained simply by finding the triangular area under $\mathrm{E}(x)$ curve

$$\varphi_{bi} \equiv \frac{1}{2}\mathrm{E}_{\max} W = \frac{1}{2}\frac{q}{\varepsilon_S}\frac{N_A N_D}{N_A + N_D} W^2 \tag{17.5a}$$

where $\mathrm{E}_{\max}$ in Eq. (17.3) was expressed in terms of W by using Eq. (17.4). Also φ_{bi} as given by the sum of Fermi potentials is specified in terms of the doping level by using Eqs. (15.19) and (15.20) as

$$\varphi_{bi} \equiv \varphi_{Fn} + \varphi_{Fp} = \frac{k_B T}{q} \ln\left(\frac{N_A N_D}{n_i^2}\right) \tag{17.5b}$$

In this manner, the parameters W, $|\mathrm{E}_{\max}|$, and φ_{bi} are all specified in terms of N_D and N_A.

Equilibrium Carrier Profiles in W

The ratio between n_{n0} and n_{p0} as the majority and minority carrier concentrations in n and p regions, respectively, is given by

$$\frac{n_{p0}}{n_{n0}} = \frac{n_i e^{-q\varphi_{Fp}/k_B T}}{n_i e^{q\varphi_{Fn}/k_B T}} = e^{-q\varphi_{bi}/k_B T}; \quad \varphi_{bi} = \varphi_{Fn} + \varphi_{Fp} \tag{17.6a}$$

By the same token, the ratio between p_{p0} and p_{n0} in p and n regions, respectively, reads as

$$\frac{p_{n0}}{p_{p0}} = e^{-q\varphi_{bi}/k_B T} \tag{17.6b}$$

Also, in the depletion depth W, n and p should depend on x and are given by

$$n(x) = N_C e^{-(E_C(x)-E_F)/k_B T}, \quad p(x) = N_V e^{-(E_F - E_V(x))/k_B T} \tag{17.7a}$$

However, as $E_C(x) - E_V(x) = E_G$ for all x, the law of mass action also holds true in W in equilibrium.

$$n(x)p(x) = N_C N_V e^{-E_G/k_B T} = n_i^2 \tag{17.7b}$$

17.2 The p–n Junction under Bias

Under a bias, the junction is pushed away from the equilibrium to nonequilibrium, and $n(x)$ and $p(x)$ depart from the equilibrium values. When the forward $(+V)$

or reverse ($-V$) voltage is applied to the p side, the band therein is lowered or raised by $-qV$ with respect to the n bulk region. Therefore, the junction band bending decreases or increases from the equilibrium value according to $q(\varphi_{bi} - V)$ (Figure 17.2). However, $E_C - E_F$ and $E_F - E_V$ should remain the same in n and p regions. This is because the equilibrium carrier densities are preserved in n and p bulk regions.

Obviously, the two requirements cannot be met with a single E_F. Instead two quasi-Fermi levels, one for electrons E_{Fn} and the other for holes E_{Fp} are required as clearly shown in Figure 17.2. Furthermore, E_{Fn}, E_{Fp} should split in W by an amount

$$E_{Fn} - E_{Fp} = qV \tag{17.8}$$

In the quasi-equilibrium approximation, E_{Fn} and E_{Fp} are taken flat in W at the level given in Eq. (17.8), but E_{Fn} and E_{Fp} should merge into a single E_F in n and p regions, where the equilibrium bulk properties are preserved. The merging occurs in the quasi-neutral regions. Naturally, the decrease or increase in the band bending under the bias should be accompanied by the concomitant decrease or increase in W and $\mathrm{E}_{\max}$ as dictated by Eq. (17.5a) with φ_{bi} replaced by $\varphi_{bi} - V$. Thus, we can write from Eqs. (17.5a), (17.3), and (17.4)

$$W(V) = \left[\frac{2\varepsilon_S\left(N_A + N_D\right)}{qN_AN_D}(\varphi_{bi} - V)\right]^{1/2} \tag{17.9}$$

$$|\mathrm{E}_{\max}(V)| = \frac{qN_DN_A}{\varepsilon_S(N_A + N_D)}W(V) \tag{17.10}$$

Charge Injection and Extraction

The voltage-controlled n and p are the key to the diode operation. Under the bias and in nonequilibrium, n and p are specified in the usual manner by replacing E_F by E_{Fn} and E_{Fp}, respectively:

$$n(x) = N_C e^{-[E_C(x) - E_{Fn}]/k_BT}, \quad p(x) = N_V e^{-[E_{Fp} - E_V(x)]/k_BT} \tag{17.11}$$

Therefore, the law of mass action is broken in W, and the charge is injected into or extracted out of W under the forward or reverse bias, that is,

$$n(x)p(x) = N_CN_V e^{-[E_C(x) - E_{Fn}(x)]/k_BT} e^{-[E_{Fp}(x) - E_V(x)]/k_BT} = n_i^2 e^{qV/k_BT} \tag{17.12}$$

where Eqs. (17.8) and (15.12) have been used. Also at $x = -x_p$, E_{Fn} lies above or below E_F by qV depending on the polarity of V (see Figure 17.3).

Therefore, as shown in Figure 17.3, n at $-x_p$ is greater or less than its equilibrium value n_{p0} by the amount

$$n_p(-x_p) \equiv N_C e^{-(E_C - E_F - qV)/k_BT} \equiv n_{p0} e^{qV/k_BT} \tag{17.13a}$$

Similarly, p_n at x_n is increased or decreased according to

$$p_n(x_n) = N_V e^{-(E_F - E_V - qV)/k_BT} = p_{n0} e^{qV/k_BT} \tag{17.13b}$$

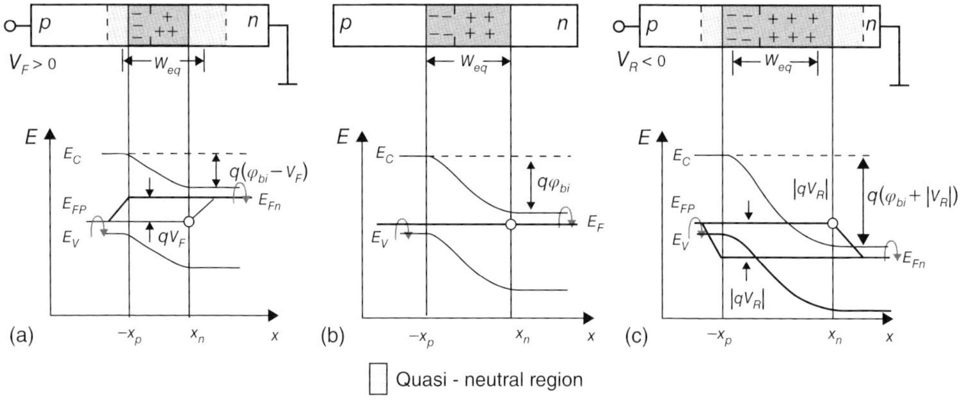

Figure 17.3 The junction band bending under forward (a) and reverse (b) biases and in equilibrium (c). Two quasi-Fermi levels E_{Fn} and E_{Fp} split in the depletion depth and merge in the quasi-neutral regions on both sides of the junction.

The resulting spatial profiles of n and p are shown in Figure 17.4.

The bias-controlled injection or extraction of the minority carriers constitutes the driving force of the diode operation. Once the junction is pushed away from the equilibrium, and the charge is injected or extracted, there ensues the reactive process for bringing the junction back into equilibrium. These reactive processes are responsible for inducing the diode current.

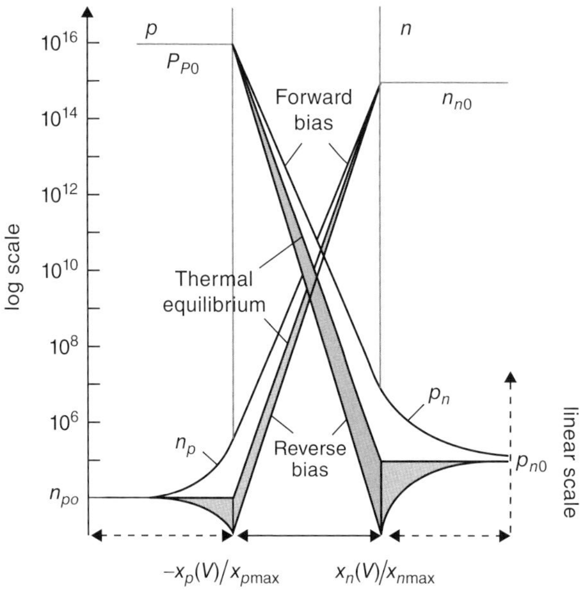

Figure 17.4 The profiles of n and p under the forward and reverse biases and in equilibrium.

17.3 Ideal Diode I–V Behavior

In equilibrium, the diffusion flux of electrons from n to p regions is balanced by the drift flux from p to n regions. For diffusion, the electrons have to overcome the potential barrier $q\varphi_{bi}$ (Figure 17.1). For drift, the electrons simply roll down the potential hill in the opposite direction propelled by the built-in space charge field. Under the forward bias V_F, however, the potential barrier is lowered, and the electric field in the junction is decreased. Hence, the diffusion is enhanced, while the drift is reduced. Similarly, the diffusion of the holes becomes greater than the drift. Consequently, the detailed balancing between the drift and the diffusion is broken.

We next describe the diode current by using the theory of Shockley. In his model, Shockley introduced a few nonessential simplifications: (i) the completely depleted approximation for describing the space charge; (ii) the low-level injection of minority carriers, that is, $n_p(-x_p) \ll n_{n0}$ and $p_n(x_n) \ll p_{p0}$; and finally (iii) the nondegenerate majority carrier concentrations p_{p0} and n_{n0}. In the I–V analysis, the junction diode is divided into three regions, as shown in Figure 17.3: (i) the depletion W, (ii) the quasi-neutral near x_n and $-x_p$, and (iii) the n and p bulk regions.

The Forward Current

As noted, under a forward bias, the diffusion fluxes of electrons and holes become dominant, and excess electrons and holes spill in from n and p regions into the depletion and quasi-neutral regions (Figure 17.4). The change in time of the excess hole concentration p_n injected into the quasi-neutral region on the n side is then governed by

$$\frac{dp_n}{dt} = -\frac{d}{dx}J_p(x) - \frac{p_n - p_{n0}}{\tau_p}; \quad J_p(x) = \left(p_n\mu_p\mathrm{E} - D_p\frac{dp_n}{dx}\right) \tag{17.14}$$

where the recombination term has been added to the usual well-known continuity equation.

In the steady state, p_n is time invariant, that is, $\partial p_n/\partial t = 0$, and the electric field, E in the quasi-neutral region is negligible. Hence, the rate equation reduces to the diffusion equation:

$$\frac{d^2p_n}{dx^2} - \frac{p_n - p_{n0}}{L_p^2} = 0, \quad L_p \equiv (D_p\tau_p)^{1/2} \tag{17.15}$$

where L_p thus defined is the *hole diffusion length*. Obviously, the solution is given by $\exp \pm x/L_p$, but in the region under consideration in which $x > x_n$, the positive branch diverges for large x and should be discarded. Also, the boundary conditions are

$$p_n(x_n) = p_{n0}e^{qV/k_BT}, \quad p_n(x \to \infty) = p_{n0}$$

The first condition accounts for the injection of holes at x_n under bias Eq. (17.13b), while the second condition represents the equilibrium hole concentration in the bulk n region. When the boundary conditions are incorporated, the solution reads as

$$p_n(x) = p_{n0}(e^{qV/k_BT} - 1)e^{-(x-x_n)/L_p} + p_{n0} \tag{17.16a}$$

Likewise, the excess electron concentration in the quasi-neutral region on the p side is given by

$$n_p(x) = n_{p0}(e^{eV/k_BT} - 1)e^{(x+x_p)/L_n} + n_{p0},\quad L_n \equiv (D_n\tau_n)^{1/2} \tag{17.16b}$$

with L_n denoting the electron diffusion length.

The diffusion current is therefore obtained by differentiating $p_n(x)$ and $n_p(x)$ with respect to x:

$$J_p(x) \equiv qD_p\left(-\frac{dp_n}{dx}\right) = \frac{qD_pp_{n0}}{L_p}[e^{qV/k_BT} - 1]e^{-(x-x_n)/L_p},\quad x \geq x_n \tag{17.17a}$$

$$J_n(x) \equiv -qD_n\left(-\frac{dn_p}{dx}\right) = \frac{qD_nn_{p0}}{L_n}[e^{qV/k_BT} - 1]e^{(x+x_p)/L_p},\quad x \leq -x_p \tag{17.17b}$$

The total diffusion current, called the *forward current*, is contributed by the sum of J_n and J_p evaluated at x_n and $-x_p$, respectively:

$$I \equiv I_n(-x_p) + I_p(x_n) = I_S(e^{qV/k_BT} - 1) \tag{17.18a}$$

where the saturation current

$$I_S = \left(\frac{qD_nn_{p0}}{L_n} + \frac{qD_pp_{n0}}{L_p}\right)A_J = qn_i^2\left(\frac{D_n}{L_nN_A} + \frac{D_p}{L_pN_D}\right)A_J \tag{17.18b}$$

is given in terms of the doping level, diffusion length, and the diode cross-section A_J, and use has been made of the identities $n_{p0} = n_i^2/N_A$ and $p_{n0} = n_i^2/N_D$. Clearly, the forward current increases exponentially with V. Also, electrons and holes diffuse in opposite directions, but because of the opposite polarity of the charge, the electron and hole currents add up and flow from p to n regions.

A few comments are due at this point. The two diffusion currents $J_n(x)$ and $J_p(x)$ depend sensitively on x. However, the drift currents contributed by the majority carriers also vary in such a manner that the total current is constant everywhere at a level given by Eq. (17.18), as illustrated in Figure 17.5. If the current is not constant throughout the entire current path, it is not possible to maintain a steady-state current. Also, due to large majority carrier concentrations, the voltage required to induce the respective drift currents takes up a minute fraction of the total junction voltage applied.

The Reverse Current I_R

Under a reverse bias, V_R p_n and n_p are depleted in respective quasi-neutral regions, as shown in Figure 17.4. In this case, electrons diffuse from p to n regions, while

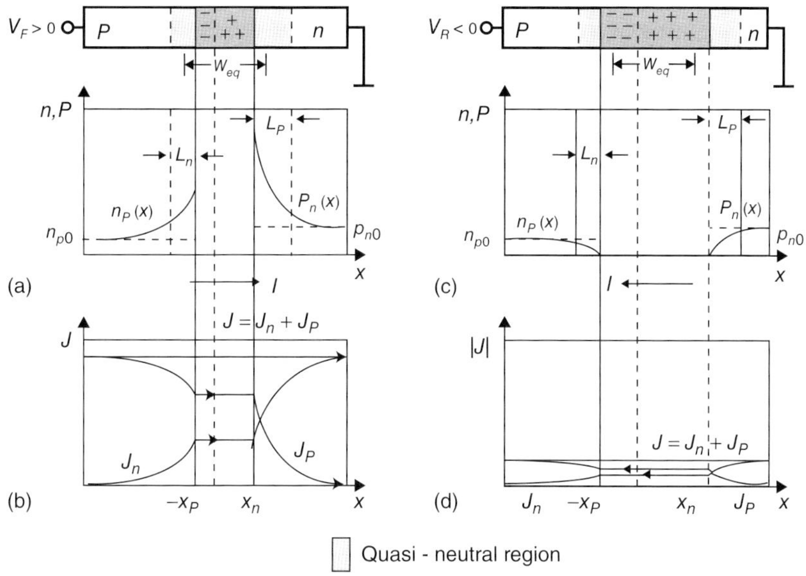

Figure 17.5 The injected minority carrier profiles in the quasi-neutral regions (a) under a forward bias. The accompanying minority carrier diffusion and majority carrier drift currents and the total current (b). The extracted minority carrier profiles in the quasi-neutral regions (c) under a reverse bias. The accompanying minority carrier diffusion currents and the total current (b) and accompanying diffusion current and the total current (d).

holes diffuse from n to p regions. Once diffused into the depletion region W, the electrons and holes are swept across W driven by the strong electric field inherent in the junction and further reinforced by V_R. The resulting reverse current is naturally built into the $I-V$ expression (17.18). For $V > 0$, I_F exponentially increases with V, while $V < 0\, I_R$ flows in the reverse direction and saturates at the level I_S at small V_R, typically a few thermal voltages k_BT/q.

Diffusion Length

The average distance an excess hole, for example, diffuses in the quasi-neutral region on the n side can be found by

$$< x - x_n >= \frac{\int_{x_n}^{\infty} dx p_n(x)(x - x_n)}{\int_{x_n}^{\infty} dx p_n(x)} = L_p, \quad p_n(x) \propto e^{-(x-x_n)/L_p} \tag{17.19}$$

Therefore, it is clear from Eq. (17.19) that the diffusion length L_p represents the average distance an excess minority carrier diffuses before recombination. Hence, electrons and holes within L_n and L_p, respectively, can be taken to diffuse into W and contribute to the reverse current.

17.4 Nonideal I–V Behavior

Generation and Recombination Currents

The nonideal $I-V$ model takes into account the recombination (r) and the generation (g) processes occurring in the depletion region W. The r and g rates are primarily attributed to the trap-assisted r and g processes, and therefore the generation and recombination currents I_G and I_R can be specified by using the recombination rate U derived in Eq. (16.30):

$$U = \frac{1}{\tau}\frac{n_i^2(e^{qV/k_BT} - 1)}{n + p + 2n_i\cosh(E_t - E_i)/k_BT}, \quad np = n_i^2 e^{qV/k_BT} \tag{17.20}$$

For $V > 0$, U attains the maximum level when (i) the trap level lies at the midgap, $E_t = E_i$ so that the third term in the denominator is minimized and (ii) $n = p$, in which case $n = p = n_i \exp(qV/2k_BT)$ and $n + p$ attains the minimum value under the constraint given in (17.11).

Hence, the maximum recombination rate U_R reads as

$$U_R \approx \frac{1}{\tau}\frac{n_i(e^{qV/k_BT} - 1)}{2(e^{qV/2k_BT} + 1)} \approx \frac{1}{2\tau}n_i e^{qV/2k_BT} \tag{17.21}$$

and I_R can be estimated by multiplying U_R with the recombination volume

$$I_R \approx qU_RWA_J \approx \frac{q}{2\tau}n_i e^{qV/2k_BT}WA_J \tag{17.22}$$

where A_J is the cross-sectional area of the diode. Likewise, for $V < 0$, $n,\ p \ll n_i$, $\exp(qV/k_BT) \approx 0$, so that the maximum generation rate is given from Eq. (17.20) by $n_i/2\tau$, and I_G is therefore given by

$$I_G \approx qU_GWA_J; \quad U_G = \frac{n_i}{2\tau} \tag{17.23}$$

Thus, the total nonideal current is obtained by adding I_R and I_G to the ideal diode current:

$$I = \begin{cases} I_{\text{ideal}} + I_R; & V > 0 \\ I_{\text{ideal}} + I_G; & V < 0 \end{cases} \tag{17.24}$$

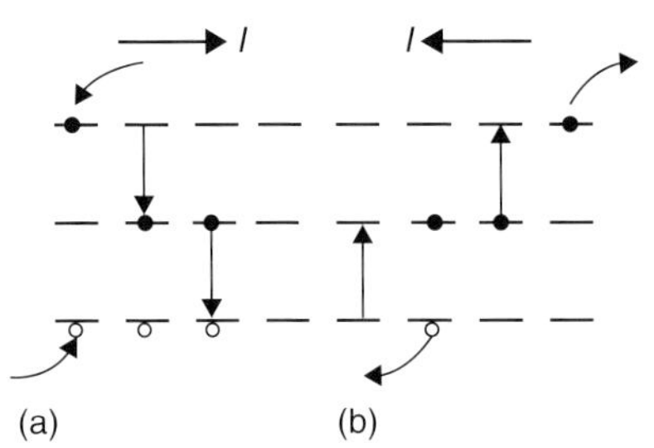

Figure 17.6 The cyclic trap-assisted recombination (a) and generation (b) of the electron–hole pairs in the depletion depth, closing the recombination and generation current loops.

The physical mechanism responsible for I_R is illustrated in Figure 17.6. Under V_F, the excess electrons and holes are constantly injected into W and are recombined in two steps, the electron capture followed by the hole capture, completing thereby the I_R loop. When the process is repeated in cyclic manner, I_R flows in the forward direction. Likewise, under V_R, the electron–hole pairs are generated by alternating emissions of holes and electrons in succession. The electrons and holes thus generated are swept out of W by the space charge field therein and contribute to I_G flowing in the reverse direction. The nonideal I–V expression in Eq. (17.24) can be compacted as

$$I = I_S \left[\exp\left(\frac{qV}{mk_BT}\right) - 1\right], \quad I_S \approx A_J q n_i^2 \left(\frac{D_n}{L_n N_A} + \frac{D_p}{L_p N_D}\right) + A_J \frac{q n_i}{2\tau} W \quad (17.25)$$

In this expression, I_G naturally adds to I_S raising thereby the effective saturation current level. On the other hand, I_R is embedded into the forward current via the ideality factor m, which ranges from 1 to 2. The ideality factor is used as the fitting parameter, and I_F is contributed by both diffusion and recombination currents.

Junction Breakdown

Finally, the reverse current grows explosively beyond V_{BR}, called the *breakdown voltage*. The breakdown is caused by the avalanche multiplication of electrons and holes in W, as illustrated in Figure 17.7. The e–h pairs when generated via the band-to-band or trap-assisted excitations are subjected to the strong electric field in the reverse-biased junction region. In this case, the electrons and holes therein can gain in between collisions kinetic energies sufficient to trigger the impact ionization of the host atoms. The ionization process occurring in cascade gives rise to explosive growth of electron–hole pairs, which are swept out of W, triggering the avalanche breakdown current.

The tunneling is also responsible for the breakdown. For the electrons in the valence band, the energy gap provides the potential barrier, which under a reverse bias typically assumes a triangular shape with height E_G, and the width narrowed by strong electric field as shown in Figure 17.7. In this case, the electrons in

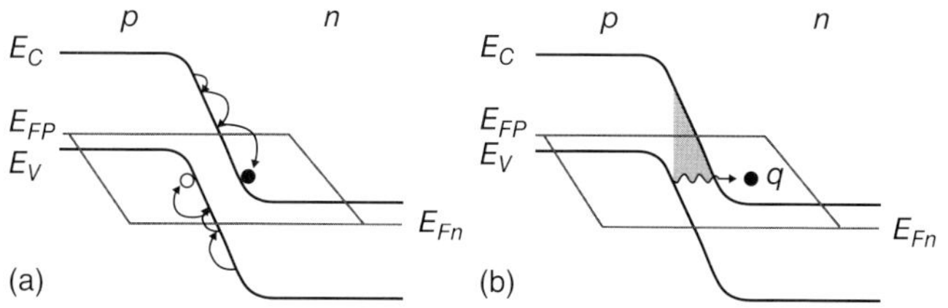

Figure 17.7 The diode breakdown: the avalanche multiplication of the e–h pairs resulting from the ionization occurring in cascade (a) and the Zener breakdown resulting from the F–N tunneling of the electrons from the valence band to the conduction band (b).

the valence band in the p region can tunnel into the conduction band in the n region and contribute to I_R via the FN tunneling, whose probability is given from Eq. (5.26) by

$$T \propto \exp - \left[\frac{4(2m_n)^{1/2}}{3q\text{E}\hbar} E_G^{3/2} \right] \tag{17.26}$$

where $V - E$ in Eq. (5.26) has been replaced by E_G. It is therefore clear that T increases exponentially with the electric field E or the reverse bias V_R, and induces the explosive growth of I_R. The resulting junction breakdown is known as the *Zener breakdown*.

Problems

17.1 Consider the $p^+ - n$ step junctions in silicon in which $N_A = 2 \times 10^{18}\ \text{cm}^{-3}$ and N_D varies from 1×10^{15} to $2 \times 10^{17}\ \text{cm}^{-3}$.
(a) Find the junction parameters x_n, x_p, E_{max}, W, and φ_{bi} as a function of N_D.
(b) At which reverse biases will these $p - n$ junctions undergo breakdown if the maximum field for breakdown is $3 \times 10^5\ \text{V cm}^{-1}$?

17.2 (a) Is it possible to achieve the junction band bending by the amount greater than the junction bandgap?
(b) Estimate N_D and N_A at which $q\varphi_{bi} \simeq E_G$ in silicon with the bandgap 1.12 eV and germanium with the bandgap 0.66 eV.

17.3 The space charge ρ was modeled by Eq. (17.1) in completely depleted approximation. Check the validity of this approximation by estimating the width Δx in which n and p are not vanishingly small near x_n and $-x_p$, respectively, and by comparing Δx with typical x_n and $-x_p$ values.

17.4 (a) Obtain $p_n(x)$ Eq. (17.16a) by solving the diffusion equation (17.15) with the use of the boundary condition under forward bias given in the text.
(b) The maximum recombination rate U in Eq. (17.20) was obtained by minimizing $n + p$ in the denominator. Show that it attains the minimum value if $n = p$ under the forward bias in which $np = n_i^2 \exp[qV/k_BT]$.

17.5 The quasi-Fermi levels E_{Fn} and E_{Fp} in W under forward and reverse biases are taken flat in quasi-equilibrium approximation. Check the accuracy of this approximation by (i) taking $N_D = 10^{16}\,\text{cm}^{-3}$, $N_A = 5 \times 10^{17}\,\text{cm}^{-3}$ and estimating the electron and hole fluxes under forward and reverse biases and (ii) equating these fluxes to the gradient of E_{Fn} and E_{Fp} (Eq. (16.19)).

17.6 In which semiconductor is the Zener breakdown more likely to occur among silicon, germanium. Estimate the reverse biases at which the breakdown occurs in these semiconductors. The bandgap of Si, and Ge, is 1.12, 0.66 eV, respectively.

Suggested Readings

1. B. G. Streetman and S. Banerjee, Solid State Electronic Devices, Sixth Edition, Prentice Hall, 2005.
2. R. S. Muller, T. I. Kamins, and M. Chan, Device Electronics for Integrated Circuits, Third Sub Edition, John Wiley & Sons, 2002.
3. S. M. Sze and K. K. Ng, Physics of Semiconductor Devices, Third Edition, Wiley-Interscience, 2006.
4. A.S. Grove, Physics and Technology of Semiconductor Devices, John Wiley & Sons, 1967.
5. D. M. Kim, Introductory Quantum Mechanics for Semiconductor Nanotechnology, Wiley-VCH, 2010.
6. A. Yariv, An Introduction to Theory and Applications of Quantum Mechanics, John Wiley & Sons, 1982.

18
P–N Junction Diode: Applications

The p – n junction diode has been utilized extensively as the platform on which to devise novel semiconductor devices such as photodiode, solar cell, light-emitting diode (LED), and laser diode (LD). This chapter is focused on discussing these devices. The diode applications are mostly based on its interaction with light, and therefore the light absorption and emission in semiconductors are considered first.

18.1
Optical Absorption

Figure 18.1 shows the conduction and valence bands in a direct bandgap semiconductor. Also shown in the figure are the dispersion curves of electrons and holes in the conduction and valence bands, respectively. In the direct bandgap material, the minimum and maximum points in the two dispersion curves coincide in the k-space. Also, the electrons and holes behave as free particles near the band edges as discussed, and the dispersion relations are thus given by

$$E = \frac{\hbar^2 k^2}{2m_j}, \quad \hbar k = p, \quad j = n, \quad p \tag{18.1}$$

where m_j is the effective mass of the electron or hole.

Absorption Coefficient

Let us consider the band-to-band excitation of an electron by absorbing a photon, as shown in Figure 18.1. The interaction Hamiltonian involving the extended Bloch wavefunction of the electron and the propagating EM wave is given from Eq. (13.1) by

$$\hat{H}' = -\hat{\mu}\frac{\mathrm{E}_0}{2}[e^{i(\omega t - \underline{k}_{\mathrm{opt}} \cdot \underline{r})} + e^{-i(\omega t - \underline{k}_{\mathrm{opt}} \cdot \underline{r})}]; \quad \hat{\mu} \equiv q(\hat{e}_f \cdot \underline{r}) \tag{18.2}$$

where $\hat{e}_f$ is the polarization vector, k_{opt} the optical wave vector, and E_0 the amplitude of the light oscillating with angular frequency ω . Also $\boldsymbol{r}$ is the displacement of the electron from the nucleus and $\hat{\mu}$ the atom dipole moment. The transition

Introductory Quantum Mechanics for Applied Nanotechnology, First Edition. Dae Mann Kim.

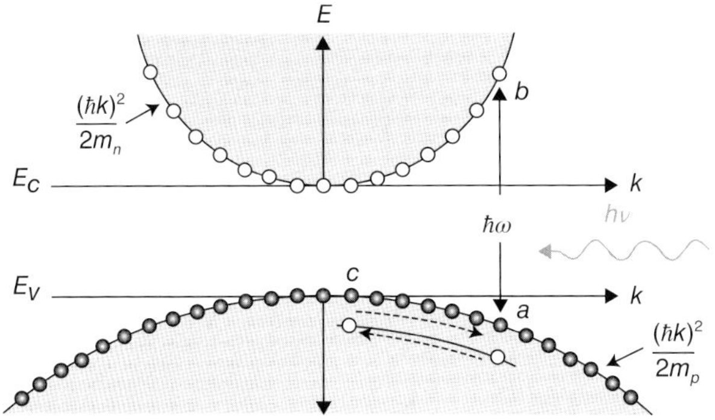

Figure 18.1 The electron and hole dispersion curves in conduction and valence bands, respectively, in a direct bandgap semiconductor. Also shown is the band-to-band excitation of an e–h pair via the absorption of a photon. The electron moving from *c* to *a* to capture a hole and gain the kinetic energy is equivalent to a hole moving from *a* to *c*, gaining the same kinetic energy.

rate of an electron from the valence to conduction bands is given by Fermi's golden rule (Eq. (9.54)) as

$$W_{vc} = \frac{2\pi}{\hbar}|H'|^2\delta(E_b - E_a - \hbar\omega); \quad |H'|^2 = \frac{\tilde{\mu}^2 \mathrm{E}_0^2}{4} \tag{18.3a}$$

where the atom dipole is now specified via the Bloch wavefunction as

$$\tilde{\mu} = q\int d\underline{r}u_c^*(\underline{r})u_v(\underline{r})(\hat{e}_f\cdot\underline{r})e^{-i(\underline{k}_c-\underline{k}_v\pm\underline{k}_{opt})\cdot\underline{r}} \tag{18.3b}$$

In Eq. (18.3b), $u_c(\boldsymbol{r})$ and $u_v(\boldsymbol{r})$ are the modulation functions in conduction and valence bands, respectively, and k_c and k_v the corresponding crystal wave vectors (see Eq. (6.3)).

As the Bloch wavefunction is extended over the entire crystal volume, the integration (Eq. (18.3b)) should be performed over the same extended volume. In this case, the variation of the phase factor in the integrand renders the transition matrix vanishingly small unless the phase-matching condition prevails, that is,

$$\underline{k}_c - \underline{k}_v \pm \underline{k}_{\text{opt}} \approx \underline{k}_c - \underline{k}_v = 0 \tag{18.3c}$$

In the optical wavelength regime, $k \approx 2\pi/\lambda \approx 10^5\ \text{cm}^{-1}$ at $\lambda = 500$ nm, for example, while $k_c \approx k_v \approx 2\pi/d \approx 10^8\ \text{cm}^{-1}$ for the lattice spacing $d \approx 0.5$ nm. Hence, k_{opt} can be neglected, and we may put $\underline{k}_c = \underline{k}_v = \underline{k}$. It thus follows from Eq. (18.3c) that the optical transitions should occur vertically in the k space (Figure 18.1). In this case, the integral of the transition matrix (Eq. (18.3b)) reduces to the expression of dipole moment with respect to $u_c(\boldsymbol{r})$, $u_v(\boldsymbol{r})$. Also, the difference in energy between the initial and final states consists of the energy gap E_G and the kinetic energies of electrons and holes in the conduction and valence bands, $\hbar^2k^2/2m_n$ and $\hbar^2k^2/2m_p$, respectively. We can therefore write

$$E_f - E_i = \frac{\hbar^2k^2}{2\mu_{\text{eff}}} + E_G, \quad \frac{1}{\mu_{\text{eff}}} \equiv \frac{1}{m_n} + \frac{1}{m_p} \tag{18.3d}$$

With the transition rate W_{vc} explicitly specified, the number of transitions N per volume V is obtained by multiplying W_{vc} by the density of states $g(k)$ (Eq. (4.14b)) and carrying out the integration over k:

$$N = \frac{2\pi}{\hbar}\frac{\widetilde{\mu}^2 \mathrm{E}_0^2 V}{4}\int \delta\left(\frac{\hbar^2 k^2}{2\mu_{\text{eff}}} + E_G - \hbar\omega\right) g(k)dk, \quad g(k) = \frac{k^2}{\pi^2} \tag{18.4}$$

An assumption implicitly present in Eq. (18.4) is that the quantum states of electrons in the valence band are all occupied, while empty in conduction band. The condition holds true for $T = 0$ and is also a good approximation unless T is extremely high. We can carry out the integration by introducing a new variable ξ:

$$\frac{N}{V} \propto G\int \delta(\xi)(\xi + \hbar\omega - E_G)^{1/2} d\xi; \ \xi = \frac{\hbar^2 k^2}{2\mu_{\text{eff}}} + E_G - \hbar\omega$$

$$= G(\hbar\omega - E_G)^{1/2}, \ G \equiv \frac{\widetilde{\mu}^2 \mathrm{E}_0^2 \mu_{\text{eff}}^{3/2}}{\sqrt{2}\pi\hbar^4} \tag{18.5}$$

Therefore, the attenuation coefficient $\alpha(\omega)$ of the light intensity is obtained by dividing the optical power absorbed per unit volume $\hbar\omega \times (N/V)$ by the incident power crossing the unit area, that is, the Pointing vector $c\varepsilon \mathrm{E}_0^2$:

$$\alpha(\omega) = A^*(\hbar\omega - E_G)^{1/2}, \ A^* = \frac{\omega\widetilde{\mu}^2 \mu_{\text{eff}}^{3/2}}{\sqrt{2}\pi\hbar^3 c\varepsilon} \tag{18.6}$$

where c and ε are the velocity of light and permittivity of the medium, respectively. It is thus clear from Eq. (18.6) that for the absorption to occur, the photon energy should be larger than the bandgap. Moreover, $\alpha(\omega)$ increases with increasing photon energy as more electron states $\propto k^2$ are available in the absorption process.

18.2 Photodiode

The photodiode is the p–n junction diode used for detecting the optical signal and operates in the reverse bias mode. Thus, consider a p–n junction, reverse biased and irradiated by light, as sketched in Figure 18.2. The e–h pair when generated in W is separated immediately from each other, electrons rolling down the potential hill in the conduction band to the n region, while holes rolling up the hill in the valence band to the p region. An electron in the valence band moving down to capture the hole while gaining the kinetic energy is equivalent to the hole moving up the same trajectory, gaining the same kinetic energy as shown in Figure 18.2.

Photocurrent

The photocurrent is due to the generation and subsequent separation of e–h pairs via the drift in opposite directions. The generation rate of the e–h pairs at the

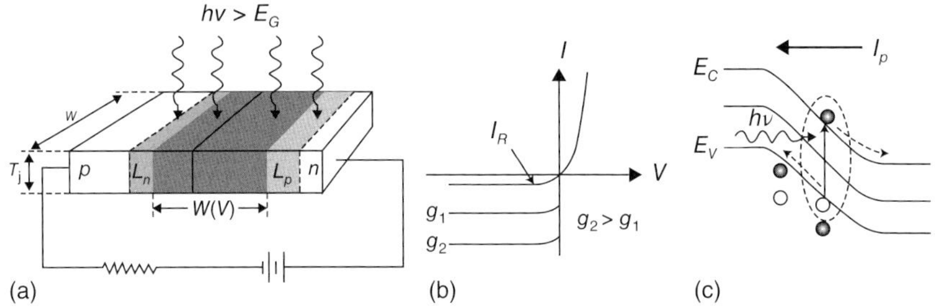

Figure 18.2 The cross-sectional view of the photodiode (a), the photocurrent versus V_R (b), and reverse-biased energy band diagram (c). Electron rolling down the potential hill is equivalent to hole rolling up the same trajectory, gaining the identical kinetic energy.

depth y from the surface is given by

$$g(y) = g_0 e^{-\alpha y}; \quad g_0 = \alpha \left[\frac{I_0 (1 - R)}{h\nu}\right] \tag{18.7}$$

where α is the absorption coefficient (see Eq. (18.6)), R the reflection coefficient, and $I_0/h\nu$ the photon flux given in terms of the light intensity I_0. The photocurrent I_p is then obtained by integrating $g(y)$ over the illuminated volume within the depletion depth W

$$\begin{aligned} I_{p,dr} &= -qWw \int_0^{T_j} dy g(y) \\ &= -qAW\tilde{g}_0, \quad \tilde{g}_0 = g_0 \left[\frac{\left(1 - e^{-\alpha T_j}\right)}{\alpha T_j}\right], \quad A_J = wT_j \end{aligned} \tag{18.8}$$

where T_j, w, and A_J are the thickness, width, and the cross-sectional area of the diode.

The light is also absorbed in the two quasi-neutral regions, and the e–h pairs generated therein also contribute to I_p. Thus, for example, consider the e–h pair generated in the quasi-neutral region on the n side. The electrons drift to the n region driven by the junction field, while the holes diffuse into W. This is because the holes are depleted in W under the reverse bias (see Figure 17.5). Upon reaching the junction edge x_n, holes roll up the junction potential and are swept out of W to the p region, propelled by the electric field in the junction. The resulting I_p is obtained by solving the diffusion equation in the quasi-neutral region, given by

$$p_n'' - \frac{p_n - p_{n0}}{L_p^2} + \frac{\tilde{g}_0}{D_p} = 0 \tag{18.9}$$

In Eq. (18.9), the photo-generation term has been added to the usual diffusion equation (17.15). The boundary conditions involved are (i) $p_n(x = x_n) = 0$ under the reverse bias and (ii) $p_n(x \to \infty) = p_{n0} + \tilde{g}_0\tau_p$ with $\tilde{g}_0\tau_p$ denoting the photo-generated hole concentration in the n bulk. Thus, the solution reads as

$$p_n(x) = (p_{n0} + \tilde{g}_0\tau_p)[1 - e^{-(x-x_n)/L_p}] \tag{18.10}$$

The resulting photocurrent is therefore given by

$$I_{p,diff} \equiv -qAD_p \frac{\partial p_n(x = x_n)}{\partial x}$$
$$= -qAD_p \frac{p_{n0} + \tilde{g}_0 \tau_p}{L_p} \approx -qA\tilde{g}_0 L_p, \quad D_p \tau_p \equiv L_p^2 \tag{18.11a}$$

The first term on the right hand is the reverse current of the diode (Eq. (17.17)), representing in this case the background noise and has been discarded, as it is much less than the photocurrent. Likewise, the photocurrent due to the electron diffusion in the quasi-neutral region on the p side is obtained as

$$I_{n,\mathrm{diff}} = -qA \left(\frac{n_{p0} D_n}{L_n} + \tilde{g}_0 L_n \right) \approx -qA\tilde{g}_0 L_n, \quad D_n \tau_n \equiv L_n^2 \tag{18.11b}$$

Note in Eq. (18.11) that those electrons and holes generated within the respective diffusion lengths from the junction edges reach on the average the depletion region and are swept across W to contribute to the photocurrent. This is consistent with the definition of the diffusion length, as discussed. The total I_p therefore consists of the three components (Eqs. (18.8), (18.11a), and (18.11b)):

$$I_T = -I_l, \quad I_l \equiv qA\tilde{g}_0 (W + L_p + L_n) \tag{18.12}$$

and flows in the reverse direction from n to p regions. The $I_T - V_R$ curves are shown in Figure 18.2. Naturally, the reverse current of the diode constitutes the background noise, and the output photocurrent increases linearly with the input light intensity, as it should. Also the photocurrent is flat and insensitive to the reverse voltage V_R. This is because the electrons and holes are swept across W by the built-in electric field regardless of V_R.

18.3 Solar Cell

Photovoltaic Effect

The solar cell is based on the photovoltaic effect. The effect refers to the physical processes whereby an incident light generates a voltage across a certain portion of the illuminated region of the medium. The p–n junction is a prototypical example exhibiting such effect and carries a most important application, namely, the solar cell.

The solar cell operation is essentially the same as that of the photo-detector, but the bias regime used is different. The photovoltaic effect is again triggered by incident photons, generating e–h pairs in W. The electrons and holes thus generated are separated by the space charge field in the junction. That is, electrons roll down the junction potential hill to the n region, while the holes roll up the hill to the p region. Consequently, the photocurrent flows from n to p regions just as in the photo-detector (Figure 18.3). Simultaneously, the excess electrons and holes pile

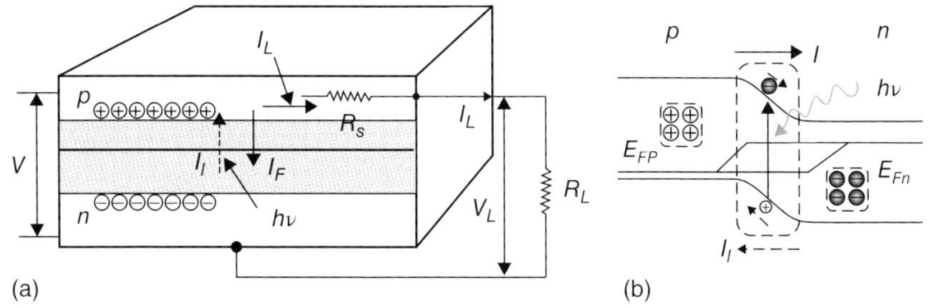

Figure 18.3 The cross-sectional view of the junction solar cell and equivalent circuit (a). The energy band diagram under illumination (b). The photo-generated e–h pairs are separated and contribute to I_l while setting up the forward voltage to induce I_F.

up in the n and p regions, respectively, setting up the forward voltage V and driving the forward current I_F from p to n regions. The total current is therefore given by

$$I = I_F - I_l, \quad I_F = I_S(e^{qV/k_BT} - 1) \tag{18.13}$$

where I_F is taken ideal for simplicity of discussion, and I_l is given by Eq. (18.8) as the physical mechanisms for producing I_l in W are the same in both devices. The open circuit photo-voltage V_{oc} and short circuit current I_{sc} are found by putting $I = 0$ and $V = 0$, respectively, and are given by

$$V_{oc} = \frac{k_BT}{q} \ln\left(\frac{I_l}{I_S} + 1\right) \approx \frac{k_BT}{q} \ln\frac{I_l}{I_S}, \quad I_{sc} = -I_l \tag{18.14}$$

where I_l is in general much greater than I_F, and I_{sc} and V_{oc} are also indicated in Figure 18.4.

When a load resistor is connected to the junction as shown in Figure 18.3, the load voltage V_L is set up by the space charge resulting from excess electrons and

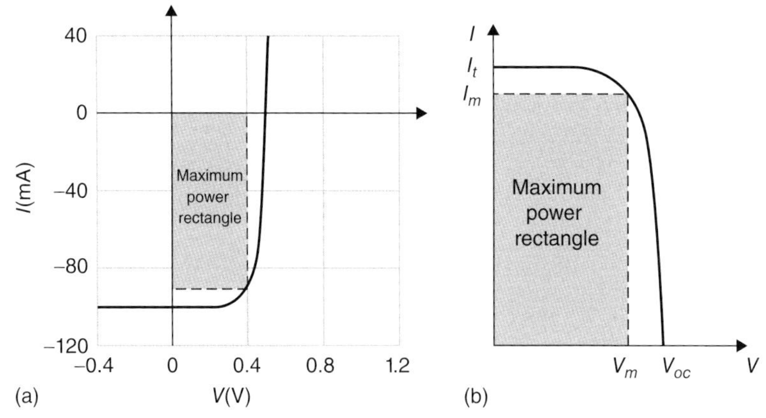

Figure 18.4 The solar cell I–V in the fourth quadrant (a) and I_l versus V with the direction of I_l taken positive (b). Also shown are the short circuit current I_l and open circuit voltage V_{oc}.

holes drifting to n and p regions, respectively. Additionally, the photocurrent I_l flowing against the forward voltage V in the direction from n to p drives the load current I_L across the load resistor. Therefore, $IV < 0$, and the power is extracted. In this manner, the solar cell plays the role of a battery supplying the load current with the voltage charged by the solar radiation. As a consequence, the solar energy is converted into the electrical power.

Clearly, two key processes are involved in the operation of the solar cell: (i) the light absorption and generation of e–h pairs and (ii) the subsequent separation of electrons and holes via the junction band bending. Specifically, electrons roll down the conduction band in W to the n region, while holes roll up the valence band to the p region. The electrons and holes thus separated are recombined through the external circuit to complete the current loop. As shown in Figure 18.4, the $I-V$ curve is located in the fourth quadrant and represents the power extraction $I \times V < 0$. The curve is often plotted by taking I_l positive in which case it intersects with the voltage and current axes at V_{oc} and I_l, respectively. The shaded region represents the maximum power rectangle.

We next consider the equivalent circuit as also shown in Figure 18.3. In the figure, R_S and R_L are the series and load resistances, respectively. In the presence of R_S, V_L is smaller than the junction voltage V as provided by V_{oc}, but R_S is small so that we can put

$$V_L = V - I_L R_S \simeq V \tag{18.15}$$

Also I_L is specified from Eqs. (18.13) and (18.15) as

$$I_L = I_l - I_S(e^{qV_L/k_BT} - 1) \tag{18.16}$$

where the direction of I_l is taken positive. Hence, the power extracted reads as

$$P \equiv V_L I_L = V_L[I_l - I_S(e^{qV_L/k_BT} - 1)] \tag{18.17}$$

We can then find the load voltage V_{Lm} at which the extracted power attains the maximum value by imposing the condition, $\partial P/\partial V_L = 0$, obtaining

$$e^{qV_{Lm}/k_BT} = \frac{(1 + I_l/I_S)}{(1 + qV_{Lm}/k_BT)} \tag{18.18}$$

We can also express V_{Lm} in terms of V_{oc} by using Eq. (18.14) in Eq. (18.18) as

$$V_{Lm} = V_{oc} - \frac{k_BT}{q} \ln\left[1 + \frac{V_{Lm}}{(k_BT/q)}\right] \tag{18.19}$$

Once V_{Lm} is found, I_{Lm} is obtained by combining Eqs. (18.16) and (18.18) as

$$I_{Lm} = I_l - I_S(e^{qV_{Lm}/k_BT} - 1) \approx I_l\left(1 - \frac{k_BT/q}{V_{Lm}}\right) \tag{18.20}$$

where use has been made of $I_l/I_S >> V_{Lm}/(k_BT/q) >> 1$. Clearly, I_{Lm} is commensurate with I_l, which suggests that the cell efficiency depends primarily on the efficient absorption of the solar radiation, as expected.

It is therefore clear that the primary factor for the high-efficiency solar cell is its capability to efficiently absorb the radiation. Ideally, the cell should absorb the entire spectrum of the solar radiation. However, this requires a small bandgap, so that a larger fraction of the solar spectrum is absorbed, as clear from the absorption coefficient (Eq. (18.6)). On the other hand, a larger bandgap induces larger V_{oc} and V_L via reduced I_S as clearly follows from Eqs. (18.18) and (17.18b). Therefore, devising a high-efficiency solar cell requires an innovative engineering by which to attain simultaneously (i) efficient absorption of the solar radiation. (ii) efficient separation of the photo-generated electrons and holes for generating the high photocurrent and (iii) a large V_{oc}.

18.4 LED and LD

The optical fiber communication is endowed with several advantages; (i) the low signal loss, (ii) the wide bandwidth, and (iii) the small diameter of silica fibers. The light source suitable for the fiber communication is the LED and LD. These diode-based light sources are driven by the forward current of the diode and can be incorporated readily into optoelectronic circuits. Moreover, LEDs are fast becoming the mainstream light source with a long lifetime and low power consumption. The p–n junction again provides the platform for LEDs and LDs, and these photonic devices are discussed.

Thus, consider a heavily doped p^+-n^+ junction in a direct bandgap semiconductor such as GaAs. In this case, the carrier concentrations in n and p bulk regions are degenerate, and E_F at equilibrium penetrates deep into the conduction and valence bands as shown in Figure 18.5. Under a forward bias, the junction band bending is reduced, and n and p are specified separately by E_{Fn} and E_{Fp}. Also the splitting of E_{Fn} and E_{Fp}, that is, $E_{Fn} - E_{Fp}$ represents the measure of excess electrons and holes injected into the junction region.

To consider the optical gain or loss at an arbitrary T, the transition rate of an electron between (a) and (b) in Figure 18.1 should include the probability factors associated. This is because the final state should be empty for an electron to enter into it from the initial state, according to the exclusion principle. Thus, the

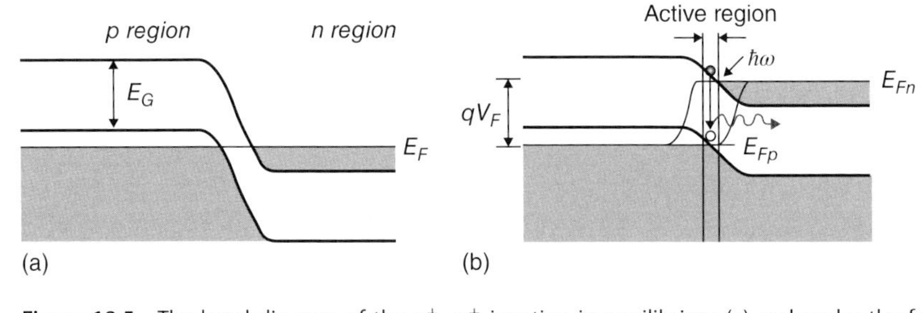

Figure 18.5 The band diagram of the p^+-n^+ junction in equilibrium (a) and under the forward bias (b). The injected electrons and holes recombine and emit the radiation.

transition rate (Eq. (18.4)) should be modified to account for the net absorption from a to b as

$$N_{ab} - N_{ba} = \frac{2\pi}{\hbar} \frac{\widetilde{\mu}^2 E_0^2 V}{4} \int \delta\left(\frac{\hbar^2 k^2}{2\mu_{\text{eff}}} + E_G - \hbar\omega\right) P(E_a, E_b) \frac{k^2 dk}{\pi^2} \tag{18.21a}$$

Here, N_{ab} and N_{ba} represent the number of the upward and downward transitions between two states with the energy E_a and E_b, respectively. The probability factor

$$\begin{aligned} P(E_a, E_b) &= f_v(E_a)(1 - f_c(E_b)) - f_c(E_b)(1 - f_v(E_a)) \\ &= f_v(E_a) - f_c(E_b) \end{aligned} \tag{18.21b}$$

when expressed in terms of the Fermi function

$$f_j(E_\gamma) = \frac{1}{1 + e^{(E_\gamma - E_{Fj})/k_B T}}; \quad \gamma = a, b \;\; j = c, v \tag{18.21c}$$

accounts for the probabilities of upward and downward transitions. Specifically, the first term represents the probability that an electron occupies the state at E_a in the valence band and makes the transition to the empty state at E_b in the conduction band. Similarly, the second term represents the probability for the inverse transition. Under a bias, the distribution function departs from $f_j(E_\gamma)$, but it is still a good approximation to use the Fermi distribution function.

Attenuation and Gain

As the probability factor $P(E_a, E_b)$ does not depend on k, it can be taken out of the integral in Eq. (18.21a), in which case the integral is identical to Eq. (18.4). Hence, we can use the result obtained in Eq. (18.6) and write

$$\widetilde{\alpha}(\omega) = \alpha(\omega)[f_v(E_a) - f_c(E_b)] \tag{18.22}$$

where $\alpha(\omega)$ is the absorption coefficient valid for $T = 0$. It is therefore clear that the absorption or emission occurs if

$$f_v(E_a) > f_c(E_b), \quad or \quad f_v(E_a) < f_c(E_b) \tag{18.23}$$

When the Fermi function is used in Eq. (18.23) with the identification $E_b - E_a = \hbar\omega$, the condition for absorption or emission can also be expressed as

$$\hbar\omega > E_{Fc} - E_{Fv} \quad \text{or} \quad \hbar\omega < E_{Fc} - E_{Fv} \tag{18.24}$$

and the gain coefficient is likewise given by

$$\gamma(\omega) \equiv -\widetilde{\alpha}(\omega) = \alpha(\omega)[f_c(E_b) - f_v(E_a)] \tag{18.25}$$

The first of two inequalities in Eq. (18.23) or (18.24) states that the probability of electrons being at a in the valence band is greater than that of being at b in the conduction band. In this case, there should be more upward transitions, causing the light to be attenuated. By the same token, the light is amplified when more electrons are in the state at b in the conduction band compared with the number of electrons at the state a in the valence band. The gain condition in Eq. (18.24) can be attained by applying a large forward voltage V_F to such an extent that the splitting of two quasi-Fermi levels $E_{Fc} - E_{Fv}(= qV_F)$ exceeds the photon energy $\hbar\omega$ (see Figure 18.5). More specifically, with V_F large enough, the sufficient number of electrons and holes are injected into W for realizing the optical gain. The condition for gain is attained with a modest value of V_F in the $p^+ - n^+$ junction, where there is a considerable overlap between the conduction and valence bands to begin with.

Additionally, a high luminescence efficiency is required for the efficient conversion of the input current to the output light. There are two kinds of recombination processes, namely, radiative and dissipative. The luminescence efficiency is specified by the fraction of the radiative recombination

$$\eta = \frac{1/\tau_r}{1/\tau_r + 1/\tau_{nr}} \tag{18.26}$$

where $1/\tau_r$ and $1/\tau_{nr}$ are the radiative and non-radiative recombination rates with τ_r and τ_{nr} denoting the respective lifetimes. The high efficiency is attained in the direct bandgap material in which the optical transitions are the first-order process. In indirect bandgap material, the optical transitions are the second-order process; hence, the efficiency is low.

Light-Emitting Diode (LED)

LED is a prototype device utilizing the optical conversion of the diode forward current. LEDs have been fabricated by using various kinds of semiconductors. As a result, the emitted radiation spans a wide range of wavelengths from the infrared to visible. Moreover, LEDs are utilized extensively in fiber communications, displays, energy saving lamps, and so on. The junction structure for LED is sketched in Figure 18.6.

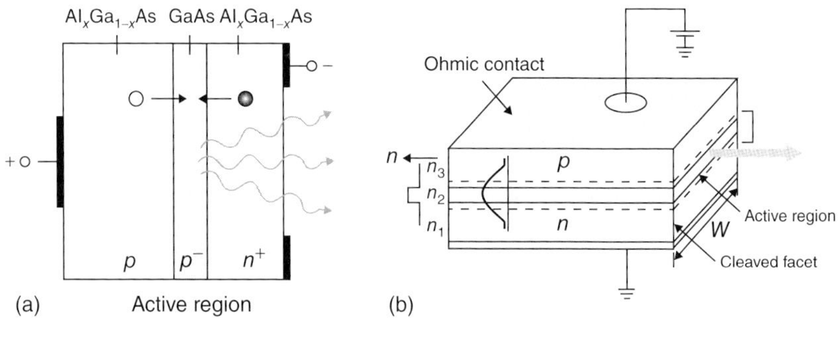

Figure 18.6 The cross-sectional view of LED showing electrons and holes injected into the junction region for the radiative recombination (a). The cross-sectional view of LD showing the lasing layer and optical index profile for confining the laser beam within the active medium in Fabry–Perot-type cavity (b).

Laser Diode (LD)

With further increase in V_F, the gain reaches the critical level to turn the diode into the laser oscillator. Also shown in Figure 18.6 is the cross-sectional view of the LD. The cavity used is the Fabry–Perot type, which consists of a pair of parallel, cleaved planes perpendicular to the lasing layer in the junction region.

Once the gain exceeds the threshold value, the LD operates in strict analogy with laser devices discussed already. A critical factor in this case is the level of the pumping current I_{F0} at which the onset of threshold condition is realized, that is, the gain offsets the loss in the cavity α_T. With I_F increased beyond I_{F0}, the laser intensity starts to grow in the cavity, and the gain is saturated. The steady-state operation requires that the saturated gain be equal to the cavity loss. We can thus write from Eq. (14.35)

$$\frac{\gamma(\omega)}{1 + I/I_S} = \alpha_T, \quad \gamma(\omega) \propto I_F \tag{18.27}$$

where α_T is the total loss in the cavity due to the scattering loss, the imperfect mirror reflectivity, and so on. The condition (18.27) in the absence of the laser intensity represents the threshold condition for the onset of the optical gain at the pumping current I_{F0}. With I_F increased beyond I_{F0}, the laser intensity starts to grow, but the condition (18.27) should still prevail via the saturated gain for the steady-state operation. Hence, I is obtained from the condition as

$$I = I_S\left(\frac{I_F}{I_{F0}} - 1\right); \quad I_{F0} = \frac{I_F}{1 + I/I_S}, \quad I_F > I_{F0} \tag{18.28}$$

Therefore, the operating laser intensity I is determined by the pumping current I_F and the saturated intensity I_S of the lasing medium.

It is thus clear that the lasing process in LD is triggered by the electrical pumping, that is, by injecting electrons and holes into the junction region. Hence, the pumping mechanism is relatively simple and consists of applying the forward voltage to the LD. This in turn points to the fact that the laser intensity can be modulated at high frequencies. A factor essential for lowering α_T and I_{F0} is an optimal wave guiding by which to confine efficiently the laser intensity within the active lasing medium. Otherwise, a substantial fraction of the intensity would tail out of the lasing medium to be dissipated, increasing thereby α_T. For this purpose, a single or double heterostructures are used for implementing the optimal profile of the refractive index (Figure 18.6).

In addition, LDs are often fabricated in the superlattice heterostructures with built-in quantum wells, as shown in Figure 18.7. In this case, electrons and holes are injected into the respective quantum wells in the junction region and reside in the subbands therein. An advantage of this kind of the LDs is the reduced threshold current density. A primary reason for this is that the electron–hole pairs now recombine while residing in respective subbands, well confined in narrow spatial region for relatively long time duration. This is in contrast with conventional LDs in which the injected electrons and holes are swept fast out of the

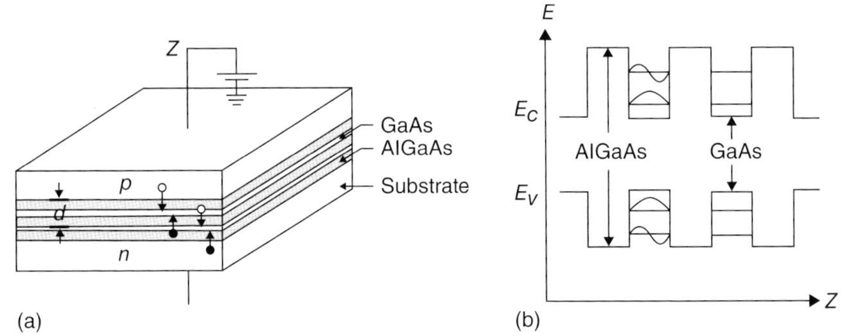

Figure 18.7 The cross-sectional view of a quantum well LD (a). The band diagram of the heterostructure consisting of the multiple quantum wells of electrons and holes (b).

junction region, shortening thereby the time for recombination. Clearly, the superlattice heterostructures are the typical example of the innovative application of the bandgap engineering.

Problems

18.1 The steady-state diffusion of minority carriers under illumination is a key process for the operation of photodiodes and solar cells.

(a) Verify that $p_n(x)$ given in Eq. (18.10) is the solution of the diffusion equation (18.9).

(b) Repeat similar analysis for the diffusion of electrons and find $n(x)$.

18.2 In analyzing the power extraction from the solar cell, the series resistance R_S has been neglected. Examine the effect of R_S in the power extraction either qualitatively or quantitatively.

18.3 (a) What are the key characteristics of the p – n junction that enables the junction to be used as the solar cell?

(b) Describe two key processes involved in the operation of the solar cell.

(c) The efficiency of the solar cell depends on various parameters. Discuss the roles of these parameters and suggest the viable means of enhancing the efficiency of the junction solar cell.

18.4 For fabricating laser diode p and n regions are degenerately doped.

(a) Estimate the donor and acceptor doping levels, for which the conduction and valence bands are overlapped by an amount 0.2 eV in Si and GaAS with the bandgaps 1.12 and 1.424 eV, respectively.

(b) Estimate electron and hole fluxes under forward bias in the two junctions considered in (a).

18.5 The criteria for the attenuation or amplification of light in the LD are given in Eq. (18.23).

(a) Show that the conditions can be recast into the simpler form given in Eq. (18.24).
(b) Discuss the condition (18.25) in specific comparison with the population inversion of laser devices.

Suggested Readings

1. B. G. Streetman and S. Banerjee, Solid State Electronic Devices, Sixth Edition, Prentice Hall, 2005.
2. R. S. Muller, T. I. Kamins, and M. Chan, Device Electronics for Integrated Circuits, Third Sub Edition, John Wiley & Sons, 2002.
3. S. M. Sze and K. K. Ng, Physics of Semiconductor Devices, Third Edition, Wiley-Interscience, 2006.
4. A. S. Grove, Physics and Technology of Semiconductor Devices, John Wiley & Sons, 1967.
5. D. M. Kim, Introductory Quantum Mechanics for Semiconductor Nanotechnology, Wiley-VCH, 2010.
6. A. Yariv, An Introduction to Theory and Applications of Quantum Mechanics, John Wiley & Sons, 1982.

19
Field-Effect Transistors

The idea of field-effect transistors (FETs) was conceived as early as 1930s and successfully implemented in 1960s. The advantages of FET as exemplified by MOSFET consist of the simplicity of structure, low-cost processing, and scalability for use in multifunctional integrated circuits. The well-known theory of MOSFET is first discussed as the general background for modeling the $I-V$ behavior in other kinds of FETs, for example, silicon nanowire (NW), ballistic, and tunneling FETs. An emphasis is placed on highlighting the underlying quantum mechanical concepts.

19.1
The Modeling of MOSFET $I-V$

MOSFET is a three-terminal, normally off, and unipolar device, and its central role consists of electrical switching for the digital logic functions. The device is also extensively utilized as the platform for memory, sensor, and green energy applications and has been downscaled deep into the nanoregime.

I–V Characteristics

Thus, consider NMOS consisting of the n^+ source and drain on p-type substrate and the n^+ polysilicon gate electrode, which is insulated from the substrate via SiO_2 (Figure 19.1). The source and drain electrodes form with the p substrate n^+-p and $p-n^+$ junctions back to back. Hence, with the gate voltage off ($V_G = 0$) and the drain voltage on ($V_D > 0$), the $p-n^+$ junction at the drain end is reverse biased, cutting off the current (off state). But with V_G on greater than the threshold voltage V_T, the junction barrier at the source end is lowered, and electrons are injected from the source into the channel and contribute to the drain current I_D. Also shown in the figure are the transistor $I_D - V_D$ and transfer I_D-V_G curves. Each transistor curve consists of triode and saturation regions. In the former, I_D increases linearly or sublinearly with V_D and saturates at a nearly constant level in the latter. The ON to OFF current ratio typically of 10^6 is a parameter gauging the device as an electrical switch.

Introductory Quantum Mechanics for Applied Nanotechnology, First Edition. Dae Mann Kim.

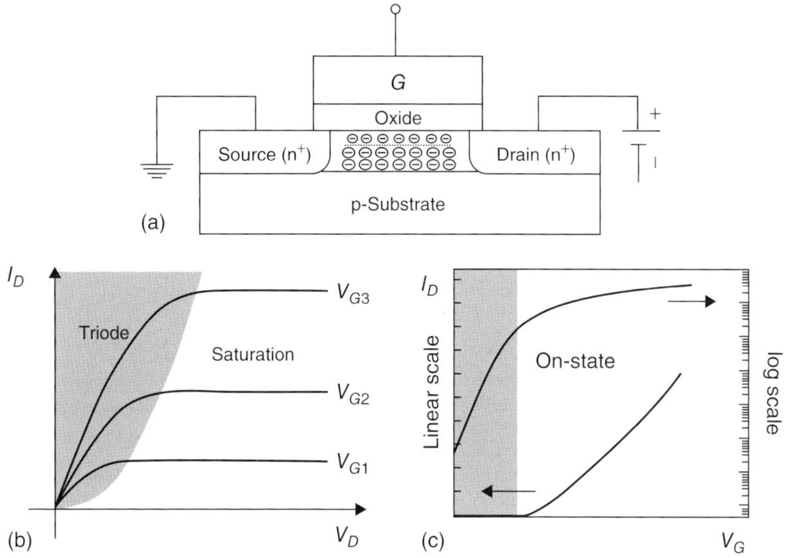

Figure 19.1 The cross-sectional view of NMOS, consisting of the p substrate, n^+ source, drain and gate electrodes (a). The transistor I–V curves (b), and the transfer characteristics (c).

The MOSFET I–V behavior is well summarized by the SPICE model (level 1):

$$I_D = \frac{W}{L} C_{OX} \mu_n \left(V_G - V_T - \frac{1}{2} V_D \right) V_D; \quad V_D \leq V_{DSAT} = V_G - V_T \tag{19.1}$$

Here μ_n is the electron mobility, V_T the threshold voltage, and the ratio between the width and length of the channel W/L is called the *aspect ratio*. The oxide capacitance per unit area is given by the permittivity ε_{OX} and thickness t_{OX} of the oxide as $C_{OX} = \varepsilon_{OX}/t_{OX}$. The I_D increases with V_D until V_{DSAT} is reached, and beyond V_{DSAT}, it remains pinned at the level attained at V_{DSAT}. The $I_D - V_D$ model (Eq. (19.1)) can also be expressed as

$$I_D = Q_L v_D \tag{19.2a}$$

$$Q_L \equiv W C_{OX} \left(V_{GS} - V_T - \frac{1}{2} V_D \right), \quad v_D = \mu_n \left(\frac{V_D}{L} \right) \tag{19.2b}$$

where Q_L is the line charge induced via the capacitive coupling between the gate electrode and substrate, and v_D the drift velocity driven by the longitudinal channel field V_D/L. Thus, I_D is contributed by Q_L, which is constantly injected from the source into the channel and drifts with v_D to the drain to be drained out. The V_T demarcates the channel inversion and the ON state.

19.1.1 Channel Inversion in NMOS

Consider next the n^+ polysilicon gate electrode, SiO_2, and p-type silicon substrate as shown in Figure 19.2. The affinity factors $q\chi$ denote the energy required to

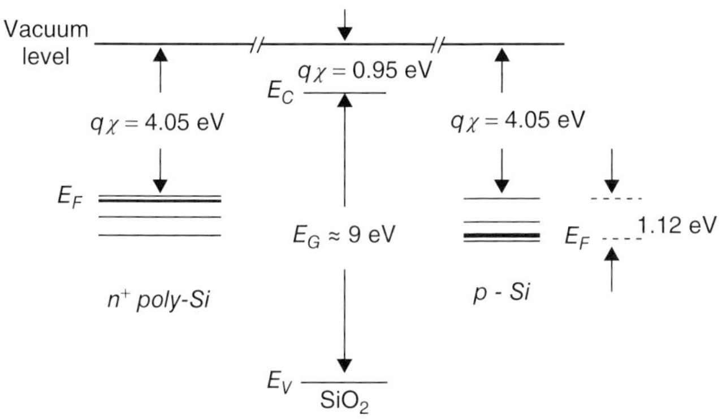

Figure 19.2 The energy bands of the n$^+$ poly-Si, SiO_2, and p-type Si. The Fermi levels E_F and the affinity factors $q\chi$ associated are also shown.

excite an electron from E_C to the vacuum level. When the three elements are brought together into the equilibrium contact, E_F should line up and be flat, as discussed. Obviously, the requirement necessitates the band bending, which occurs via the exchange of electrons between the gate electrode and substrate. As E_F in the n$^+$ gate electrode E_{Fn} is higher than E_{Fp} in the P substrate, electrons are transferred from the gate electrode to the substrate, leaving behind the positive charge sheet at the surface. The charge sheet in turn pushes holes in the substrate away from the interface, exposing thereby acceptor ions uncompensated. Consequently, the space charge is induced, and the band bends downward (Figure 19.3). The total band bending is determined by the difference between Fermi levels $E_{Fn} - E_{Fp}$ and occurs in both the gate oxide and the substrate.

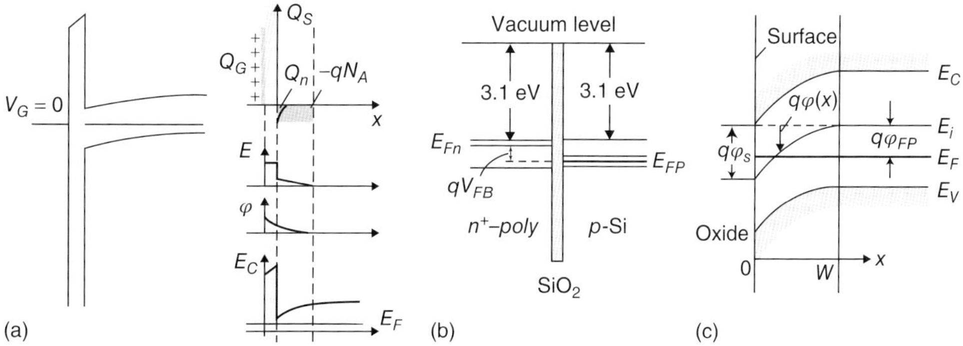

Figure 19.3 The space charge–induced equilibrium band bending of the composite system of the n$^+$ poly-Si, SiO_2, and p-type Si in equilibrium contact (a) and the flattening of the band via the application of the flat band voltage (b). The detailed version of the substrate band bending is with $\varphi(x)$, φ_S, and φ_{FP} denoting the space charge, surface, and Fermi potentials, respectively (c).

Surface Charge and Flat Band Voltage

The band bending is flattened out with the application of V_G given by $qV_{FB} = E_{Fp} - E_{Fn}$, and V_{FB} is called the *flat band voltage*. Since $E_{Fn} > E_{Fp}$, $V_{FB} < 0$, and the positive charge sheet in the gate electrode is canceled, and the space charge disappears and the band bending flattens out (Figure 19.3). Hence, the charging gate voltage

$$V'_G \equiv V_G - V_{FB} = V_{OX} + \varphi_S \tag{19.3}$$

induces the band bending in the oxide V_{OX} and in the substrate φ_S from the zero base. For positive V'_G, the band bends down, and the potential supporting the band bending in the substrate develops in the depletion region W according to the Poisson equation:

$$\frac{d^2\varphi(x)}{dx^2} = -\frac{\rho(x)}{\varepsilon_S}, \quad \rho(x) = q[(p_p(x) - N_A^- - n_p(x)] \tag{19.4a}$$

where the space charge ρ consists of the hole, ionized acceptor, and electron charges. In the bulk p substrate, the charge neutrality prevails, so that $p_{p0} = N_A^- + n_{p0}$. In W, however, n increases while p decreases exponentially as can be clearly seen from Figure 19.3:

$$p_p(x) = p_{p0}e^{-\beta\varphi(x)}, \quad n_p(x) = n_{p0}e^{\beta\varphi(x)}, \beta \equiv \frac{q}{k_BT} \tag{19.4b}$$

Hence, when Eq. (19.4b) is inserted into Eq. (19.4a) with N_A^- replaced by p_{p0} and n_{p0}, there results

$$\frac{d^2\varphi(x)}{dx^2} = -\frac{\rho(x)}{\varepsilon_S}, \quad \rho(x) = q[p_{p0}(e^{-\beta\varphi} - 1) - n_{p0}(e^{\beta\varphi} - 1)] \tag{19.4c}$$

Equation (19.4c) is strongly nonlinear and is difficult to solve. However, it is possible to carry out the first integration by multiplying both sides by $d\varphi$:

$$\int_0^{\varphi} d\varphi\frac{d^2\varphi}{dx^2} \equiv \int_0^{d\varphi/dx} \frac{d\varphi}{dx}d\left(\frac{d\varphi}{dx}\right) = \int_0^{-\mathrm{E}} (\mathrm{E})d(\mathrm{E}) = -\frac{1}{\varepsilon_S}\int_0^{\varphi} \rho(\varphi)d\varphi, \ \mathrm{E} = -\frac{\partial\varphi}{\partial x} \tag{19.5}$$

Because the space field does not penetrate into the bulk substrate $\mathrm{E} = \varphi = 0$, at $x = W$. By performing the integrations in W, we obtain straightforwardly

$$\mathrm{E}_S = \sqrt{2}\frac{k_BT}{q}\frac{1}{L_D}F\left(\beta\varphi_S, \frac{n_{po}}{p_{p0}}\right); \quad L_D = \left(\frac{k_BT\varepsilon_S}{q^2p_{po}}\right)^{1/2}, \quad p_{p0} \simeq N_A \tag{19.6a}$$

where E_S and φ_S are surface field and potential at $x = 0$, and L_D is known as the *Debye length*. Also the F-function is obtained by integrating ρ given in Eq. (19.4c):

$$F(\beta\varphi_S) \equiv [(e^{-\beta\varphi_S} + \beta\varphi_S - 1) + e^{-2\beta\varphi_{Fp}}(e^{\beta\varphi_S} - \beta\varphi_S - 1)]^{1/2}, \quad e^{-2\beta\varphi_{Fp}} = \frac{n_{p0}}{p_{p0}} \tag{19.6b}$$

where φ_{Fp} is the hole Fermi potential in the p substrate (see Eq. (15.20)). Therefore, we can find the surface charge from the well-known boundary condition as

$$Q_S \equiv -\varepsilon_S E_S(\varphi_s) = -\varepsilon_S \sqrt{2}\frac{k_B T}{q}\frac{1}{L_D}F(\beta\varphi_s) \tag{19.7}$$

Figure 19.4 shows Q_S as a function of φ_S together with the profiles of the fixed ionic and mobile electron charges in each φ_s region. At flat band voltage, there is no band bending; hence, $Q_S = 0$. For $\varphi_S < 0$, the band bends up, and the hole concentration p_p is exponentially accumulated near the surface above p_{p0}. The range of φ_S from a to b covers the depletion and weak inversion regimes, and Q_S therein consists mainly of the uncompensated acceptor charge. However, with φ_S attaining the value $2\varphi_{Fp}$, the surface concentration of the electron

$$n_S = n_{p0}\exp(q\varphi_S); \quad \varphi_S = 2\varphi_{Fp} \tag{19.8}$$

reaches the level of the majority carrier concentration p_{p0} in the substrate (see Eq. (19.6b)). Therefore, any further increase in φ_S beyond $2\varphi_{FP}$ increases n_s exponentially above p_{p0}, and the increase in Q_S is then primarily contributed by n_s, that is, the channel is inverted. Also, the electrons thus induced near the oxide reside practically at the surface according to the Boltzmann probability factor, and therefore the band bending is nearly pinned after the channel inversion.

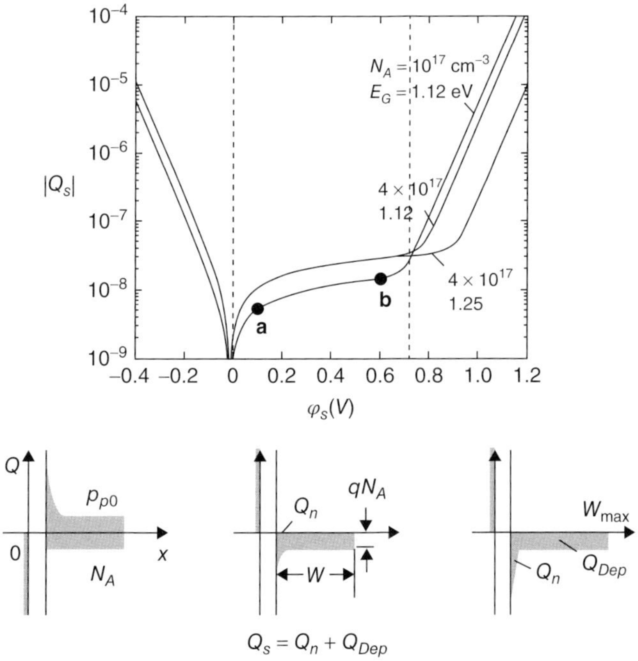

Figure 19.4 The surface charge versus the surface potential in the accumulation, depletion, and inversion regions. Also shown are the fixed ionic and mobile electron charge profiles in each regime.

19.1.2
Threshold Voltage and ON Current

The surface charge Q_S terminating the gate field lines emanating from the gate electrode for $V_G > 0$ consists of the mobile electron (Q_n) and fixed acceptor ionic charges (Q_{DEP}):

$$Q_S \equiv -C_{OX}V_{OX} = Q_n + Q_{DEP} \tag{19.9}$$

where V_{OX} is the fraction of V_G dropped in the gate oxide. The key to modeling I_D is therefore to untangle the mobile charge Q_n from the fixed charge Q_{DEP}. Now the depletion charge can be approximated by

$$Q_{DEP} \equiv -qN_A W = -(2\varepsilon_S qN_A \varphi_S)^{1/2}; \quad \varphi_S = \frac{1}{2\varepsilon_S} qN_A W^2 \tag{19.10}$$

where W has been specified in terms of φ_S in analogy with Eq. (17.5a) for the step junction in which $N_D \gg N_A$. Hence, upon inserting Eq. (19.3) for V_{OX} and Eq. (19.10) for Q_{DEP} in Eq. (19.9), we obtain

$$Q_n = -C_{OX}(V_G - V_{FB} - \varphi_S - \gamma\varphi_S{}^{1/2}); \quad \gamma_n \equiv \frac{(2\varepsilon_S qN_A)^{1/2}}{C_{OX}} \tag{19.11}$$

The constant γ_n is called the *body effect coefficient*. Thus, Q_n beyond the onset of the strong inversion is obtained from Eq. (19.11) by setting $\varphi_S = 2\varphi_{Fp}$ (see Eq. (19.8)):

$$Q_n = -C_{OX}(V_G - V_T); \quad V_T \equiv V_{FB} + 2\varphi_{Fp} + \gamma_n(2\varphi_{Fp})^{1/2} \tag{19.12}$$

In this manner, the channel is inverted, and Q_n is induced by the gate overdrive $V_G - V_T$.

Next, when the drain voltage V_D is turned on, it is distributed in the channel from the source to the drain. The primary effect of the distributed channel voltage V is to reduce the effective gate voltage by V at the channel position y and to decrease Q_n as

$$Q_n(y) = -C_{OX}(V_G - V - V_T) \tag{19.13}$$

The ON current can then be obtained in terms of V_D and V_G by considering the differential voltage drop dV in the channel element from y to $y + dy$, that is,

$$dV \equiv I_D dR; \quad dR \equiv \rho\frac{dy}{Wt_{ch}} = \frac{dy}{\sigma W t_{ch}} = \frac{dy}{W\mu_n |Q_n|}, \quad |Q_n| \equiv qnt_{ch} \tag{19.14}$$

where the resistivity ρ has been expressed in terms of the conductivity $\sigma = q\mu_n n$ (see Eq. (1.20)), and the channel cross-sectional area is given by the product of W and the channel thickness t_{ch}.

Naturally, we can recast Eq. (19.14) into two integrals, involving y and V as

$$\int_0^L I_D dy = \int_0^{V_D} dV \ \mu_n W |Q_n| \tag{19.15}$$

and integrate both sides by using Eq. (19.13) for Q_n and the fact that I_D is constant throughout the channel, obtaining

$$I_D = \frac{W}{L}\mu_n C_{OX}\left(V_{GS} - V_T - \frac{1}{2}V_D\right)V_D \tag{19.16a}$$

Equation (19.16a) is in agreement with Eq. (19.1). Moreover, V_{DSAT} in Eq. (19.1) is shown to originate from the channel pinch-off at the drain end, that is, $V_{DSAT} = V_G - V_T$ as clear from Eq. (19.13). The I_D increases with V_D until $V_G - V_T$, at which point the channel pinches off at the drain. Any further increase in V_D has therefore to be dropped near the drain to keep I_D constant. Consequently, I_D is pinned approximately at the level given by

$$I_{DSAT} = \frac{W}{2L}\mu_n C_{OX}(V_{GS} - V_T)^2, \quad V_{DSAT} \equiv V_G - V_T \tag{19.16b}$$

19.1.3
Subthreshold Current I_{SUB}

The I_{ON} and I_{OFF} are bridged by I_{SUB} in the range $0 < V_G < V_T$ or $0 < \varphi_S < 2\varphi_{Fp}$. In this region, the second term $\beta\varphi_S$ of the F-function Eq. (19.6b) is dominant. Thus, when F is expanded around $\beta\varphi_S$, Q_S is given from Eq. (19.7) by

$$Q_S \equiv Q_{DEP} + Q_n \approx -(2qN_A\varepsilon_S\varphi_S)^{1/2}\left(1 + \frac{1}{2}\frac{e^{\beta(\varphi_S - 2\varphi_{Fp})}}{\beta\varphi_S}\right), \quad \beta = \frac{q}{k_BT} \tag{19.17}$$

Evidently, the two terms on the right-hand side represent Q_{DEP} and Q_n, respectively, and Q_n is indeed shown to increase exponentially with φ_S. When V_D is turned on, φ_S near the drain decreases due to the reduced gate bias, that is, $V_G - V_D$ (see Eq. (19.3)). Consequently, Q_n therein becomes much smaller than the Q_n at the source end. Hence, I_{SUB} is driven by the diffusion, that is,

$$\begin{aligned} |I_{SUB}| &\approx WD_n\frac{Q_n(0) - Q_n(L)}{L} \\ &\approx \frac{W}{L}D_nqN_AL_D\left(\frac{1}{2\beta\varphi_S}\right)^{1/2}e^{\beta(\varphi_S - 2\varphi_{Fp})}, \quad L_D \equiv \left(\frac{k_BT\varepsilon_S}{q^2p_{po}}\right)^{1/2} \end{aligned} \tag{19.18}$$

and increases exponentially with V_G, as φ_S increases with V_G (see Eq. (19.3)).

To sum it up, the MOSFET operation is based on inverting the channel via the capacitive coupling of the gate electrode with the substrate. Above V_T electrons are injected from the source into the channel to sustain Q_n and transported to the drain via drift and diffusion, contributing to I_D. With V_G off, I_{OFF} is limited by the reverse-biased $p-n^+$ junction at the drain. The I_{ON} and I_{OFF} are bridged by I_{SUB}, which increases exponentially with V_G. The PMOS operation can likewise be modeled in strict analogy with NMOS with the roles of electrons replaced by those of holes.

19.2
Silicon Nanowire Field-Effect Transistor

The advantage of MOSFET is its scalability. With the scaling, its performance is improved and, with it, the functionality of ICs. However, the scaling is limited by various physical processes such as the V_T roll-down and roll-up, the punch-through effect, and the leakage current. Most of these adverse effects originate from the extended bulk substrate. Consequently, a variety of novel FET structures has been devised to circumvent the bulk substrate, for example, silicon on oxide, double gate, FinFET, and the gate all around NW FET. The last structure is discussed in this section, focused on the intrinsic silicon NW (see Figure 19.5).

The n-Channel SNWFET

The band diagrams of the n-type NWFET and NMOS before the contact are essentially same. The only difference between the two is the p substrate in NMOS replaced by intrinsic silicon (Figure 19.2). Nevertheless, E_{Fn} is higher than E_{Fi}, so that electrons are also transferred from the gate electrode to the NW. Once transferred into the NW, the electrons reside in the subbands and are not necessarily concentrated near the oxide interface as in NMOS. This is because the electron wavefunction is extended over the entire NW cross-section (Figure 4.5).

Subband Spectra

The electrons in NW move freely in the direction of the wire, say in the x-direction, while confined in the y, z plane as discussed (Figure 4.8). For simplicity of discussion, let us consider the NW with the rectangular cross-section and infinite potential depth. The sublevels are then given from Eq. (4.27) by

$$E_n = \sum_{j=y,\,z} E_j n_j^2, \quad E_j = \frac{\hbar^2 \pi^2}{2m_n W_j^2}, \qquad n_j = 1, 2, \ldots \tag{19.19}$$

where W_j is the width of the rectangle in the y-, z-directions. Figure 19.6 shows the typical subband spectra of the intrinsic silicon NW, obtained numerically by using the finite oxide barrier height of 3.1 eV. We can observe a few general

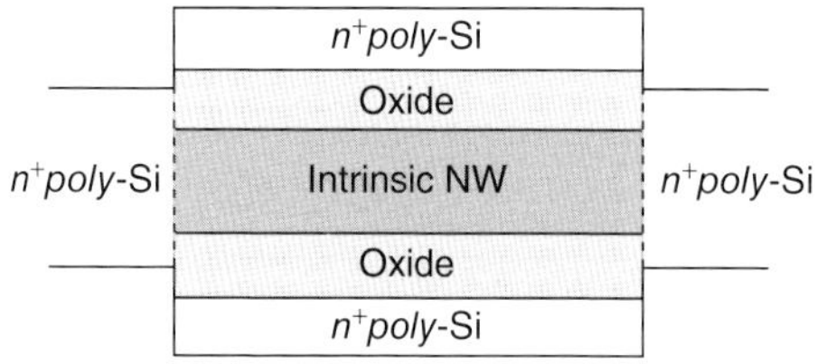

Figure 19.5 The cross-sectional view of the n-type silicon nanowire FET consisting of the intrinsic NW and the n^+ source, drain, and gate electrodes.

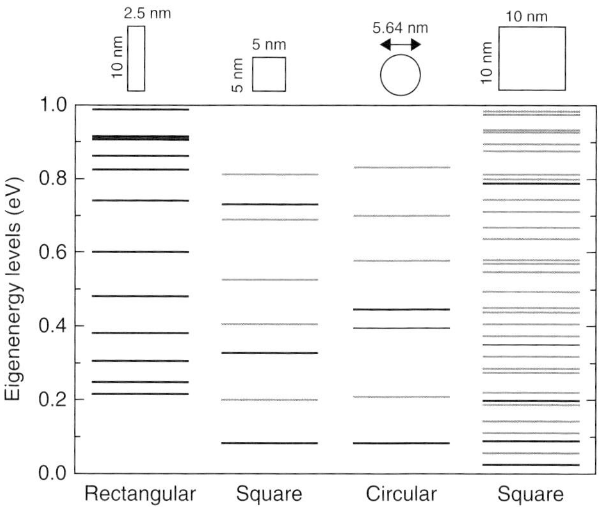

Figure 19.6 The subbands in intrinsic silicon NW surrounded by SiO_2 for different shapes and sizes of the cross-sections. (Figure reproduced from Nanowire Field Effect Transistors; Principles and Applications, Springer 2013.)

features of the spectra. Given the same area, an appreciable difference exists between the rectangular and square cross-sections, but the square and circular cross-sections share the similar spectrum. More important, a noticeable difference exits between the small- and large-area cross-sections. The subbands in the latter are more densely distributed at the levels lower than those in the former. These features are entirely consistent with Eq. (19.19).

Surface Charge

To obtain the surface charge of electron Q_n, it is necessary to find first the 1D density of electrons n_{1D}, which is given by

$$n_{1D}(\varphi) = \sum_{n=1}^{N} \int_{E_C+E_n}^{E_C+\Delta E_C} d\varepsilon \; g_{1D}(\varepsilon)F_n(\varepsilon), \;\; g_{1D}(\varepsilon) = \frac{(\sqrt{2m_n}/\pi\hbar)}{\varepsilon^{1/2}} \tag{19.20a}$$

Here, g_{1D} is the 1D density of states Eq. (4.17), N the total number of subbands in the wire, and ΔE_C the conduction band width. The Fermi occupation factor of electrons in the nth sublevel with eigenenergy E_n reads as

$$F_n(E) = \frac{1}{1+\exp[(E-E_{Fi}-q\varphi)/k_BT]}, \quad E = \varepsilon + E_C + E_n \tag{19.20b}$$

where ε ranges from $E_C + E_n$ to $E_C + \Delta E_C$, and the difference $E - E_{Fi}$ is reduced by the bulk band bending $q\varphi$.

Once $n_{1D}(\varphi)$ is found, the surface charge is obtained in analogy with Eq. (19.7) by

$$Q_n(\varphi) \equiv -\varepsilon_S \mathrm{E}(\varphi); \quad \mathrm{E}(\varphi) = \sqrt{2}\left(\frac{q}{\varepsilon_S}\right)^{1/2} [N(\varphi)]^{1/2} \tag{19.21a}$$

where $N(\varphi)$ is given by

$$N(\varphi) = \int_0^{\varphi} [n_{3\mathrm{D}}(\varphi) - n_{3\mathrm{D}}(0)]d\varphi, \quad n_{3\mathrm{D}}(\varphi) = \frac{n_{1\mathrm{D}}(\varphi)}{A} \tag{19.21b}$$

with A denoting the cross-sectional area of NW. In NMOS, the surface field $\mathbf{E}_S$ was obtained by integrating the space charge density that is induced by the band bending Eq. (19.4). Likewise, in NWFET, $\mathbf{E}_S$ can be found by integrating the 3D space charge density $qn_{3\mathrm{D}}(\varphi)$ induced by $q\varphi$. Thus, by combining Eqs. (19.20) and (19.21), $Q_n(\varphi)$ can be specified as a function of $q\varphi$ and the properties of NW, for example, the shape and size of the cross-section.

Channel Inversion

The channel inversion can be analyzed as in NMOS by using Q_n. Thus, introduce the charging gate voltage in analogy with Eqs. (19.3) and (19.9) as

$$V_G' \equiv V_G - V_{FB} = V_{OX} + \varphi, \quad V_{OX} \equiv \frac{|Q_n(\varphi)|}{C_{OX}} \tag{19.22}$$

where the flat band voltage V_{FB} is given by $qV_{FB} = E_{Fi} - E_{Fn}$. In the intrinsic NW, there is no ionic charge; hence, Q_S consists solely of Q_n, so that Q_n is simply specified by V_G by combining Eqs. (19.21) and (19.22).

Figure 19.7 shows Q_n versus V_G curves for various NW cross-sections. Also shown is a typical $Q_n - V_G$ curve of an NMOS, for comparison. Clearly, the $Q_n - V_G$ curves in the intrinsic NW do not exhibit the distinct transition region as appears in NMOS demarcating the channel inversion. Rather, Q_n in NW exponentially increases for small V_G. In this region, $n_{3\mathrm{D}}$ is still low, so that it requires large $q\varphi$ for inducing enough electrons to terminate the gate field lines as in the case of the subthreshold region of NMOS. However, when V_G exceeds a certain value, n_{3D} has attained such a level that the gate field lines resulting from the increasing V_G can be terminated by electrons that are induced by small changes in $q\varphi$. In this V_G regime, $q\varphi$ is approximately pinned while supplying sufficient excess electrons to terminate the gate field lines. Therefore, Q_n increases in rough proportion with V_G just as in the case of NMOS above V_T. We can thus define V_T as the value of V_G at which a specified level of I_D flows for given V_D, a procedure often used in the I–V characterization. We can thus notice that higher Q_n with smaller V_T is induced in NW with larger cross-sectional area, as expected.

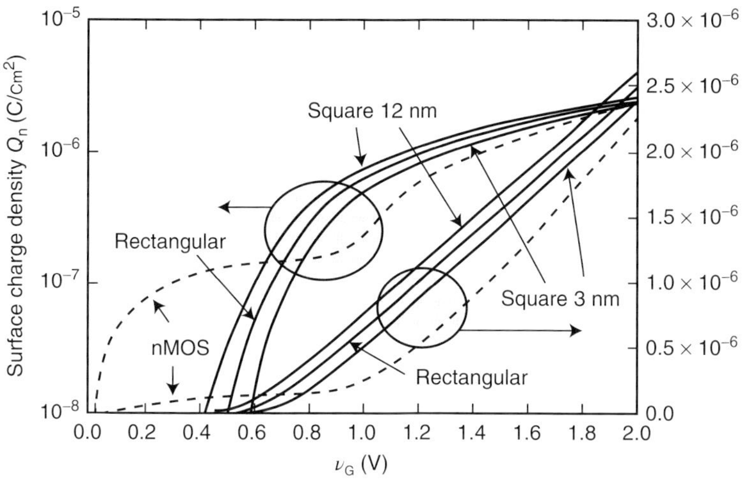

Figure 19.7 The surface charge density of electrons versus V_G in silicon NW with rectangular (3 nm × 12 nm) and square (3 nm × 3 nm, 12 nm × 12 nm) cross-sections. Also plotted for comparison is the electron surface charge in NMOS with the substrate doping of $N_A = 10^{17}$ cm^{-3}. (Figure reproduced from Nanowire Field Effect Transistors; Principles and Applications, Springer 2013.)

Long Channel I–V Behavior

The surface charge Q_n can therefore be expressed in strict analogy with Eq. (19.13) as

$$Q_n(y) = -C_{NW}[V_G - V_{Tn} - V] \tag{19.23a}$$

where V is the channel voltage at the channel position y, and C_{NW} the effective capacitance per unit area. Since Eq. (19.23) is identical to Eq. (19.13), we can use directly the $I-V$ expression derived in Eq. (19.16) for the long channel NWFET. However, there are a few differences existing between the two I_D expressions. For instance, Q_n and V_T in NWFET depend on the shape and size of the cross-section, while in NMOS, the two parameters are determined by the doping level and the substrate bandgap. Moreover, for a cylindrical NW, the gate field lines are uniformly distributed on the oxide surface, and the capacitance per unit area is well known from the basic electromagnetic theory and is given by

$$C_{NW} = \frac{\varepsilon_{OX}}{r\ \ln(1 + t_{OX}/r)} = C_{OX}\gamma; \quad \gamma \equiv \frac{t_{OX}/r}{\ln(1 + t_{OX}/r)}, \quad C_{OX} = \frac{\varepsilon_{OX}}{t_{OX}} \tag{19.23b}$$

where ε_{OX} and t_{OX} are the permittivity and thickness of the gate oxide, respectively, and r the radius of NW. As $\gamma > 1$ for all values of t_{OX}/r, C_{NW} is greater than C_{OX} in NMOS, which indicates the tighter capacitive coupling between the NW and the gate electrode. Moreover, V_T in NWFET is generally less than that of NMOS, further supporting the efficient coupling (Figure 19.7).

19.2.1 Short-Channel I–V Behavior in NWFET

The downscaling of FETs has pushed the channel length L into the mesoscopic regime. In such a short channel, the mean free path of charge carriers cannot be taken much shorter than L, and the ballistic transport has to be taken into account. Figure 19.8 shows the typical band profile of the channel under V_D. Naturally, the maximum point of the profile occurs near the source end, the height of which is determined by the band bending in the n^+-i source junction and is controlled by V_G and V_D.

By way of introducing a convenient background for discussing the $I-V$ behavior in short-channel NWFETs, we first consider I_{DSAT} as derived from the one-flux scattering theory by Lundstrom:

$$I_{DSAT} = Q_{nLS}v_{\text{eff}};\ \ Q_{nLS} = C_{NW}W_{NW}(V_G - V_{Tn}) \tag{19.24a}$$

Here, Q_{nLS} is the line charge induced at the source end at which $V = 0$, and the expression is similar to that of MOSFET I_D Eq. (19.2). However, the effective velocity v_{eff} with which electrons are transported down the channel is given by

$$v_{\text{eff}} = v_{inj}\ \eta;\ \ \eta \equiv \left(\frac{1-r_c}{1+r_c}\right),\ \ v_{inj} \simeq v_T \tag{19.24b}$$

where the injection velocity v_{inj} is approximated by the thermal velocity v_T of the electron, and η denotes the modulation factor. The modulation is specified in terms of the backscattering coefficient r_c, which is in turn given by

$$r_c = \frac{l}{l+\lambda} \tag{19.25}$$

where λ is the mean free path and l the critical length over which the electron gains the kinetic energy equal to the thermal energy k_BT. Naturally, λ and l can be specified as

$$q\text{E}_s l \equiv k_BT,\quad \lambda = v_T\tau_n = v_T\left(\frac{m_n\mu_n}{q}\right) \tag{19.26}$$

where E_s is the longitudinal electric field induced by V_D at the source end and τ_n the mean collision time. When subjected to the force $-q\text{E}_s$, the electron gains

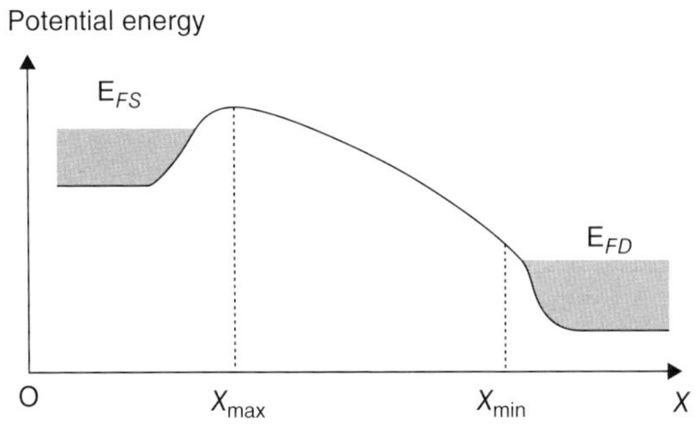

Figure 19.8 A typical band bending in the n-type FETs under the biases of V_G and V_D.

$k_B T$ while traversing the critical length l by definition. Also λ is the distance the electron traverses in the mean collision time τ_n moving with the thermal velocity v_T, and finally τ_n is related to the small signal mobility by $\mu_n = q\tau_n/m_n$ (see Eq. (1.19)).

By combining Eqs. (19.24), (19.25), and (19.26), we can recast the expression of I_{DSAT} as

$$I_{\mathrm{DSAT}} = Q_{nLS}\frac{1}{(1/v_T)+(1/v_D)}; \quad v_D = \mu_n \mathrm{E}_S \tag{19.27}$$

In this representation of I_{DSAT}, the drift–diffusion and ballistic transport naturally fuse in and contribute to I_D with the weighing factors dictated by the channel length L. For instance, in the short channel $v_T \ll \mu_n \mathrm{E}_S$ since $\mathrm{E}_s \propto 1/L$, so that the ballistic transport is prevalent with v_T taken as the saturation velocity. On the other hand, in the long channel, $\mu_n \mathrm{E}_S \ll v_T$, so that the electrons are driven instead by the usual drift velocity in general agreement with Eq. (19.1).

19.2.2
Ballistic NWFET

We next discuss the ballistic NWFET by using the Landauer formulation. In this theory, I_D is specified via the net flux of electrons from the source to the drain given by

$$I_D = \frac{2q}{h}\sum_i \int_{E_c+E_i}^{E_u} dE[F(E, E_{FS}) - F(E, E_{FD})]T_i(E); \; E_u = E_C + \Delta E_C \tag{19.28a}$$

Here the two F-functions are the Fermi occupation factors at the source and drain ends, T_i the transport coefficient of electrons in the ith subband with the energy E_i, and E_u the upper limit of the integration. When a subband is multiplied by F, it represents electrons residing therein. Hence, the two terms in the bracket account for net flux of electrons from the source to the drain in each subband. Also because the band bending in the channel is generally gradual, we may neglect the backscattering and put $T_i(E) \approx 1$.

Now the Fermi functions near the source and drain are given in terms of E_{FS}, E_{FD} as

$$F(E, E_{Fj}) = \frac{1}{1+\exp[E - E_{Fj}/k_B T]}, \quad j = S,\ D \tag{19.28b}$$

Under V_D, the two quasi-Fermi levels split as $E_{FD} - E_{FS} = -qV_D$, as discussed. Hence, for small V_D, we can Taylor expand $F(E,E_{FD})$ and retain only the first expansion term, obtaining

$$F(E,\ E_{FS}) - F(E,\ E_{FS} - qV_D) \approx \frac{\partial F(E,\ E_{FS})}{\partial E} qV_D \approx \delta(E - E_{FS})qV_D \tag{19.29}$$

where the derivative of the F-function near E_F is well approximated by the delta function. Thus, by inserting Eq. (19.29) into Eq. (19.28a), we obtain

$$I_D = G\sum_i g_i V_{DS}, \quad G \equiv \frac{2q^2}{h} \tag{19.30}$$

and specify I_D in terms of the fundamental quantum conductance G and the sum of the contributions from all subbands, including the degeneracy g_i therein.

We can also treat the general case of arbitrary V_D by introducing the variable of integration $\eta = E/k_BT$ and compact the expression of I_D in Eq. (19.28a) as

$$I_D = G\left(\frac{k_BT}{q}\right)\widetilde{M} \tag{19.31a}$$

where the form factor $\widetilde{M}$ reads as

$$\widetilde{M} = \sum_i \int_{\eta_C+\eta_i}^{\eta_u} d\eta \left[\frac{1}{1+e^{(\eta-\eta_{FS})}} - \frac{1}{1+e^{(\eta-\eta_{FS}+qV_{DS}/k_BT)}}\right], \quad \eta_{FS} = \frac{E_{FS}}{k_BT} \tag{19.31b}$$

where E_{FS} is the Fermi level at the source end and $\eta_i = E_i/k_BT$, $\eta_u = (E_C + \Delta E_C)/k_BT$. To evaluate $\widetilde{M}$, $E_C - E_{FS}$ has to be specified as a function of V_G. Figure 19.9 shows the band diagram of the n$^+$ gate, SiO_2, and intrinsic NW both in equilibrium and under the gate bias. In equilibrium, the band bending occurs primarily in SiO_2 by the amount $E_{Fn} - E_{Fi}$ to render E_F flat. But under V_G, the band in the n$^+$-gate electrode is lowered by $-qV_G$ and induces the band bending in both SiO_2 and NW. As a result, $E_C - E_{Fi}$ in NW is reduced by the bulk band bending and is given by

$$E_C - E_{FS} = E_C - E_{Fi} - q\varphi \tag{19.32}$$

and $q\varphi$ is in turn specified in terms of V_G via Eq. (19.22). Therefore, modeling I_D in Landauer formulation consists essentially of solving the coupled equations (19.22), (19.31), and (19.32).

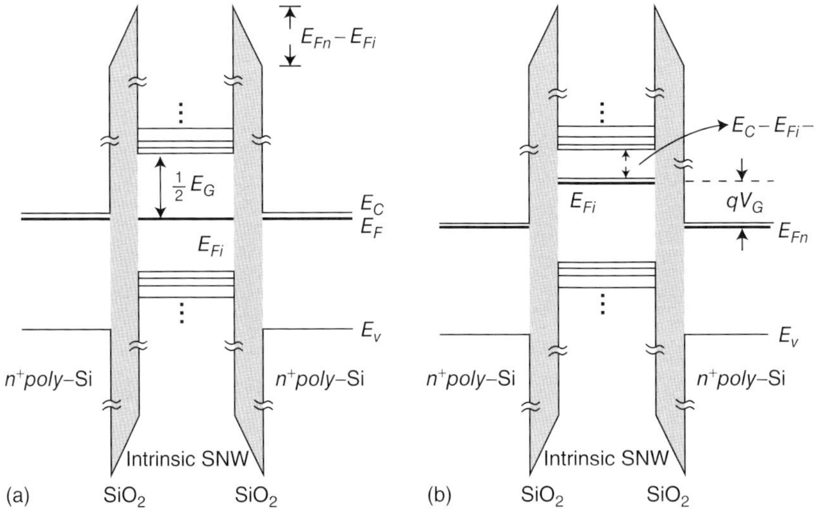

Figure 19.9 The energy band diagram of the n$^+$ poly-gate, SiO_2, and intrinsic silicon NW in equilibrium (a) and under a positive V_G applied at the gate electrode (b). The $q\varphi$ denotes the bulk band bending, and a few subbands of electrons above E_C and of holes below E_V are indicated.

In this manner, I_D is specified as an implicit function of V_G and V_D, but the characteristics of I_D can be seen on a general ground with the aid of Eq. (19.22) and Figure 19.7. For small V_G or φ, most of V_G has to drop in NW and be taken up by φ, so that enough Q_n is induced in the NW to terminate the gate field lines. As a consequence, $E_C - E_{FS}$ shrinks rapidly, and I_D increases exponentially with V_G. Obviously, this range of V_G corresponds to the subthreshold regime of NMOS. Once φ surpasses a certain value, the excess gate field lines originating from further increase in V_G can be terminated with a small increase in φ. At the same time, the Fermi function at the drain end decreases with increasing V_D (see Eq. (19.28b)), decreasing thereby the reverse flux from the drain to the source. In this case, the $I-V$ behavior should be similar to that of NMOS in the triode region. With further increase in V_D for given V_G, the flux of electrons from the drain to source becomes negligible, and I_D becomes free of V_D, and the saturation region sets in.

19.3 Tunneling NWFET as Low-Power Device

The power consumed in the FET operations is a major issue, and the tunneling can offer a viable means of reducing the power. Two kinds of power consumptions are involved, namely, charging and discharging during the switching and the I_{OFF}-induced leakage loss and are specified by

$$P_{\text{switching}} = fCV_{DD}^2, \quad P_{\text{leakage}} = V_{DD}I_{\text{OFF}} \tag{19.33}$$

Here C is the parasitic capacitance of the output node at the drain, V_{DD} the power supply voltage, and f the clock frequency. During the switching from high to low and vice versa, the charging and discharging consume the same amount of power. Also with I_{OFF} not fully eliminated, the power loss due to the leakage is always present during the switching as well as the standby times.

To reduce P, it is therefore desirable to decrease V_{DD}, but it requires the concomitant reduction of V_T, in which case the leakage current increases. This is because the subthreshold V_G regime is narrowed with the V_T reduction, and I_{OFF} at $V_G = 0$ tends to be boosted up. Therefore, it is difficult to reduce the two kinds power consumptions at the same time. A possible way out of this impasse is to decrease the subthreshold swing SS. The SS is defined as the inverse of the slope of $\log(I_D) - V_G$ curve and quantifies ΔV_G over which I_D is increased by one decade. The typical value of the SS in MOSFET at room temperature ranges from 70 to 90 mV per decade.

SS and Thermionic Emission

In conventional FETs, the lowest value of the SS achievable at room temperature is limited to about 60 mV. The limitation originates from the fact that in conventional FETs, the electrons are thermally injected into the channel from the source electrode by overcoming the potential barrier in the gated source junction (Figure 19.8). The barrier therein is lowered with the application of V_G in which case the electrons are injected into the channel and contribute to I_D:

$$I_D \approx K e^{-q(V_{bi}-V_G)/k_B T} \tag{19.34}$$

where V_{bi} is the junction built-in potential for $V_G = 0$. The exponential factor in Eq. (19.34) accounts for the thermionic emission in accordance with the Boltzmann probability factor. Thus, by taking the logarithms on both sides of Eq. (19.34) and performing the differentiation, we obtain

$$\mathrm{SS} = \frac{dV_G}{d(\log\ I_D)} = \frac{k_B T}{q \log e} \tag{19.35}$$

The SS given in Eq. (19.35) represents the lowest limit of 60 mV per decade, since V_G has been taken to drop entirely in the gate oxide.

Tunneling NWFET

The tunneling provides a viable means to improve SS as limited by the thermionic emission. Figure 19.10 shows the band profile of the FET made up of the p^+-n-i-n^+ NW. In equilibrium, E_F is flat, and the band bends in each junction dictated by respective built-in voltages. Thus, the bending in the source junction is more pronounced because φ_{bi} therein is greater than those of other junctions. Nevertheless, the bending is not yet sufficient enough to line up the valence band in the source electrode and the conduction band in the n-type NW. This means that there are no final states the electrons can tunnel into, and therefore the tunneling is prohibited. Likewise, the tunneling in the drain junction is also prohibited.

When the positive V_G is turned on, however, the p^{+-}n junction is reverse-biased, and the conduction band in the channel in the n-region is further lowered. Consequently, the electrons in the valence band of the source electrode can tunnel into the conduction band in the channel. The resulting I_D is dictated by the F–N tunneling probability and is given from Eq. (5.26) by

$$I_D = K\ \exp\left[-\frac{4(2m_n)^{1/2} E_G^{3/2}}{3qE\hbar}\right]; \quad E \approx \sqrt{\frac{2qN_D(\varphi_{bi} + V_G)}{\varepsilon_S}} \tag{19.36}$$

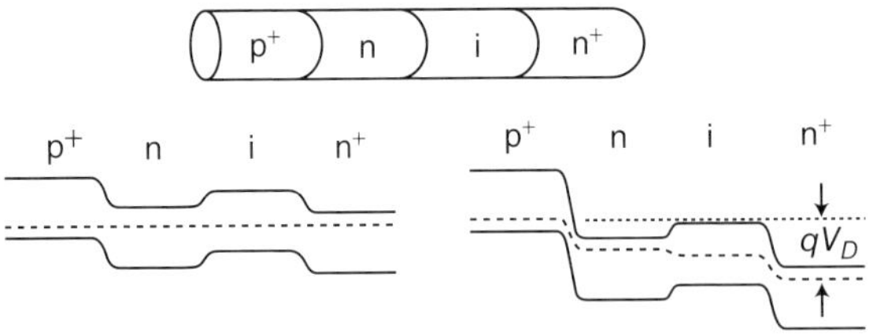

Figure 19.10 The band profile of the p^+-n-i-n^+ tunneling NWFET in equilibrium and under the bias.

where m_n is the effective mass of the electron and ε_S the permittivity of the NW. The barrier potential in this case is the bandgap E_G of the NW, and the space charge field E in the p^+-n step junction has been specified in terms of V_G by combining Eqs. (17.9) and (17.10) for $N_A \gg N_D$. Therefore, it is evident from Eq. (19.36) that the SS can be improved below 60 mV via the gate bias-induced tunneling.

Although the SS can be improved by utilizing the tunneling, I_D in the TNWFET is lower than the typical I_D in MOSFET by more than an order of magnitude. Therefore, it behooves to devise the viable means of improving I_D. The clues for such schemes are clearly contained in Eq. (19.36). Naturally, it is desirable to increase E in the junction, which can be done by increasing the N_D doping level, so that φ_{bi} also becomes greater. Then, the valence band of the source electrode can be readily raised above the conduction band in the channel, making it possible to operate the TFET with the relatively small V_G and to increase I_D efficiently. More important, the use of NW with a small bandgap offers an attractive approach. In this case, the barrier height ΔE_G is reduced, increasing exponentially I_D. However, the narrow bandgap could cause the unwanted F−N tunneling in the drain, giving rise to the high leakage current. Thus, if E_G can be tailored such that E_G is narrow in the source end and gradually widen toward the drain end, it could be an ideal means of enhancing I_D and overcoming the high power consumption. The ultimate aim of this brief discussion is to illustrate the intricate coupling of the process issues and design guidelines offered by the quantum mechanical insights.

Problems

19.1 (a) Sketch the band diagrams of the p^+ polysilicon, SiO_2, and the n substrate in the equilibrium contact.

(b) Find the flat band voltage of the p^+-n junction when the doping level of N_D is 10^{16}, 10^{17} cm^{-3}.

19.2 Carry out the modeling of $I-V$ behavior in PMOS in strict analogy with the NMOS $I-V$ modeling:

(a) Set up the Poisson equation in the n-type substrate and derive the surface charge Q_S of the hole versus the surface potential φ_S for a given N_D doping.

(b) Discuss the accumulation, depletion, weak, and strong inversion regions of the PMOS.

(c) Derive and discuss the $I-V$ behavior, in comparison with that of NMOS.

19.3 The drain current in NMOS is given in linear region by Eq. (19.1). The drain current can be formally expressed in terms of the channel voltage V at y from the source with the replacement $V_D \rightarrow V(y)$ and $L \rightarrow y$

$$I_D = \frac{W}{y} C_{OX} \mu_n \left(V_G - V_T - \frac{1}{2} V(y) \right) V(y) \qquad \text{(A)}$$

(a) Find the profile of the channel voltage by finding $V(y)$ from (A) by taking I_D constant and specified as usual in terms of V_D and V_G Eq. (19.1).

(b) Find the channel field $E(y) = -\partial V(y)/\partial y$.

(c) Using the result of (b), find the transit time of the electron from the source to drain.

$$\tau_{tr} = \int_0^L \frac{dy}{v_d} = \int_0^L \frac{dy}{\mu_n E(y)}$$

19.4 Combine Eqs. (19.24)–(19.26) and obtain I_{DSAT} for the short-channel NWFET given in Eq. (19.27).

19.5 Consider the p^+-n-i-n^+ tunneling NWFET shown in Figure 19.10.

(a) Find the built-in voltages in the three junctions when the donor doping level in the n region ranges from 10^{16} to 10^{17} cm^{-3}.

(b) Estimate V_D at which the electrons can be injected from the p^+ source electrode into the channel via the tunneling for the N_D doping considered and the voltage drop across the p^+-n, n-i, and i-n^+ junctions under the same V_D.

Suggested Readings

1. D. M. Kim and Y. H. Jeong, editors, Nanowire Field Effect Transistors: Principles and Applications, Springer, 2014.
2. R. S. Muller, T. I. Kamins, and M. Chan, Device Electronics for Integrated Circuits, Third Sub Edition, John Wiley & Sons, 2002.
3. S. M. Sze and K. K. Ng, Physics of Semiconductor Devices, Third Edition, Wiley-Interscience, 2006.
4. R. F. Pierret, Field Effect Devices, Modular Series on Solid State Devices, volume **IV**, Second Edition, Prentice Hall, 1990.
5. D. K. Schroder, Advanced MOS Devices, Modular Series on Solid State Devices, volume **VII**, Addison-Wesley Publishing Company, 1987.
6. D. M. Kim, Introductory Quantum Mechanics for Semiconductor Nanotechnology, Wiley-VCH, 2010.
7. M. Lundstrom, Fundamentals of Carrier Transport, Cambridge University Press 2000, Second Edition.

20
The Application and Novel Kinds of FETs

Since the concept of the field-effect transistor (FET) was successfully implemented, the FETs have emerged and remained as the mainstream device for performing the digital logic functions. Additionally, FETs have been extensively utilized as the platform for a variety of innovative applications. Some of the prototypical examples are discussed, including the memory and solar cells, and biosensors. Additionally, an introductory exposition of spintronics is presented in the general context of the novel kinds of FETs, and the quantum computing is briefly discussed.

20.1
Nonvolatile Flash EEPROM Cell

The electrically erasable and programmable read-only memory cell, also called *flash EEPROM cell*, utilizes the MOS structure itself with the floating gate incorporated as the storage site (Figure 20.1). The floating gate electrode lies in between two dielectric layers, thus forming a quantum well. The programming and erase are done by charging and discharging the floating gate. There are two kinds of flash memory cells, namely, NAND and NOR, and the discussion is focused on the former. The dielectric layer deposited on top of the floating gate consists of the oxide/nitride/oxide layers, thick enough to electrically isolate the control and floating gates. Thus, the charging or discharging is carried out through the tunnel oxide layer in between the floating gate and the channel.

Memory Operation

For the programming or erase, electrons are transported across the tunnel oxide from the channel to the floating gate or vice versa via the F–N tunneling. The F–N tunneling is induced in this case by the gate voltage V_{CG} applied at the control gate. The equivalent capacitance of the floating gate is also shown in Figure 20.1. When V_{CG} is applied to the control gate while grounding all other terminals, a part of it is transferred to the floating gate voltage V_{FG} according to the well-known relation in the electrostatics

$$C_{ONO}(V_{CG} - V_{FG}) = (C_{GS} + C_{GB} + C_{GD})V_{FG} \quad (20.1)$$

Introductory Quantum Mechanics for Applied Nanotechnology, First Edition. Dae Mann Kim.

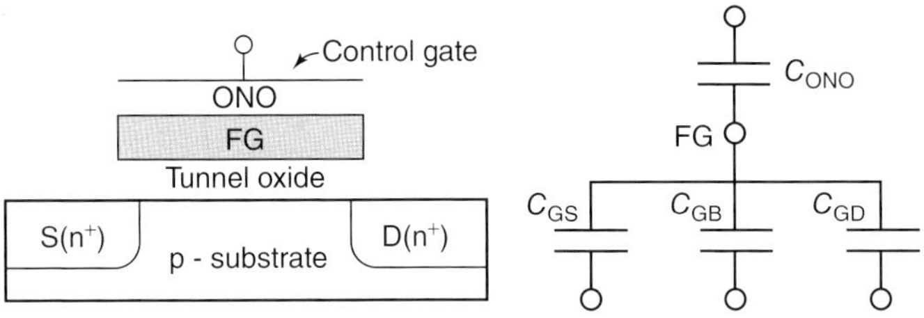

Figure 20.1 The cross-sectional view of the flash EEPROM cell consisting of the MOSFET structure with a floating gate inserted in between the control gate and the tunnel oxide. Also shown is the equivalent capacitance circuit.

Thus, V_{CG} is transferred in part to V_{FG} according to

$$V_{FG} = \alpha_{CG} V_{CG}; \quad \alpha_{CG} = \frac{C_{ONO}}{C_{GS} + C_{GB} + C_{GD} + C_{ONO}} \equiv \frac{C_{ONO}}{C_T} \tag{20.2}$$

where α_{CG} is the coupling coefficient and C_T the total capacitance consisting of all the capacitances connected in parallel. We can likewise introduce other coupling coefficients by applying the bias at each terminal, while grounding the rest. Thus, V_{FG} is generally represented by

$$V_{FG} = \alpha_{CG} V_{CG} + \alpha_S V_S + \alpha_B V_B + \alpha_D V_D + \frac{Q_{FG}}{C_T}, \quad \alpha_j \equiv \frac{C_j}{C_T} \tag{20.3}$$

where j denotes the source, bulk, and the drain terminals, and the last term is the charging voltage of the excess electrons stored in the floating gate.

The programming is done by opening the channel with a positive V_{CG} and by injecting electrons from the channel into the floating gate via F–N tunneling (see Figure 20.2). The triangular potential barrier is formed via V_G during the programming as shown in Figure 20.2 and enhances exponentially the tunneling probability (see Eq. (5.26)). Once injected into the floating gate, electrons reside in the quantum well electrically well isolated. Hence, there is no need to refresh, and the

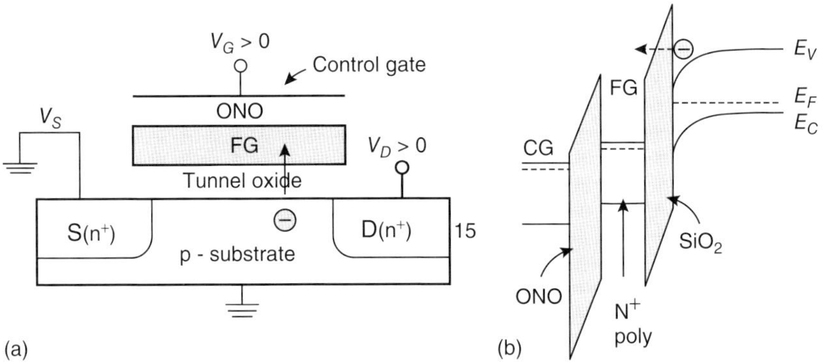

Figure 20.2 The flash EEPROM cell under a positive bias at the control gate for programming (a). The band diagram formed during the programming (b). Electrons are injected into the floating gate from the channel via the F–N tunneling.

device works as the nonvolatile memory cell. The erase is done by applying the negative V_{CG}. In this case, the triangular barrier potential is again formed, and the stored electrons tunnel out of the floating gate into the channel via the F–N tunneling, and the channel remains closed (Figure 20.3).

The reading is carried out by utilizing the different threshold voltages existing between the programmed and erased cells (Figure 20.4). The threshold voltage V_{TCG} at the control gate is taken operationally as the value of V_{CG} at which the given specification of I_D is attained, for instance, 1 μA at V_D of 0.1 V. In the erased cell, there is no excess electron charge, that is, $Q_{FG} = 0$ and the threshold voltage V_{TCGE} therein is the voltage by which to induce V_{FG} according to Eq. (20.2) and invert the channel and satisfy the specified I_D. In the presence of the excess electron charge Q_{FG} in the floating gate of the programmed cell, an additional control gate voltage is required to compensate for Q_{FG}, namely, $\Delta V = |Q_{FG}|/C_{ONO}$ (Figure 20.4). Therefore, the threshold voltage in the programmed cell is greater than that of the erased cell by the amount.

$$V_{\text{TCGP}} = V_{\text{TCGE}} + \frac{|Q_{FG}|}{C_{ONO}} \tag{20.4}$$

Therefore, the reading can be done by probing the cell with the use of V_{CG} in between V_{TCGE} and V_{TCGP} and monitoring I_D. In this case, I_D is equal to or greater than 1 μA in the erased cell while $I_D \approx 0$ in programmed cell. The distributions of the threshold voltages V_{TCGE} and V_{TCGP} should therefore be tight and well separated for the unambiguous reading. To sum it up, the key element of the flash memory cell is the quantum well introduced for the storage site. The electrons are stored therein well isolated electrically, so that the memory cell is nonvolatile. The electrons are injected into or extracted out of the quantum well by means of the F–N tunneling, the transport process unique in quantum mechanics.

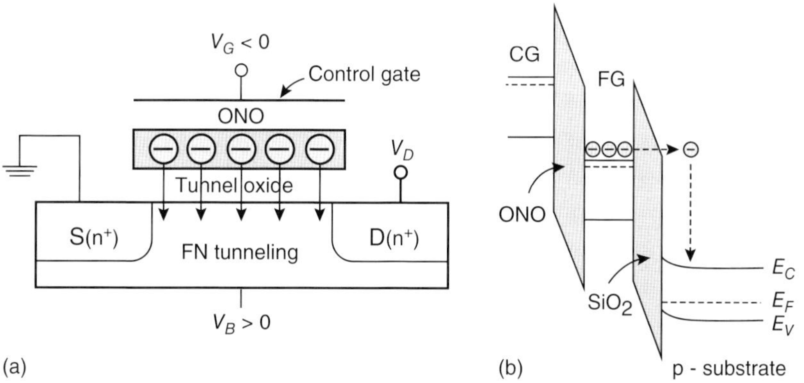

Figure 20.3 The flash EEPROM cell under a negative bias at the control gate for erase (a). The band diagram formed during the erase. The stored electrons are extracted out of the floating gate via the F–N tunneling (b).

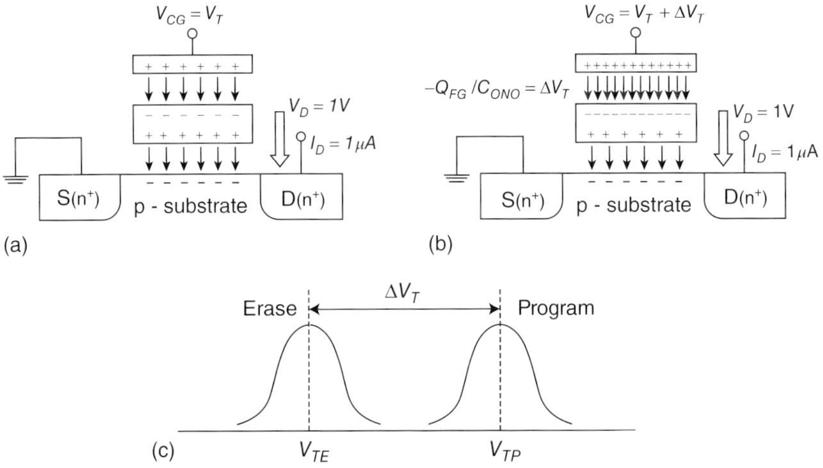

Figure 20.4 The threshold charge configurations in the erased (a) and programmed (b) cells. Also shown are the distributions of V_T in erased and programmed cells (c).

20.2
Semiconductor Solar Cells

The conversion of the solar radiation into the electrical power is a major issue of the nanotechnology. The efficiency of the solar cell is, as discussed, dictated by two factors, namely, the generation and separation of the e–h pairs. A few schemes devised for improving the efficiency are discussed.

Planar Solar Cell

In order to increase the generation of e–h pairs, it is desirable to increase the cell thickness, so that more photons are absorbed therein. However, a thicker absorbing layer is accompanied by the degraded collection efficiency of the e–h pairs generated. Such a tradeoff is illustrated in the n^+-p-p^+ planar solar cell shown in Figure 20.5. Naturally, the e–h pairs generated within the junction depletion

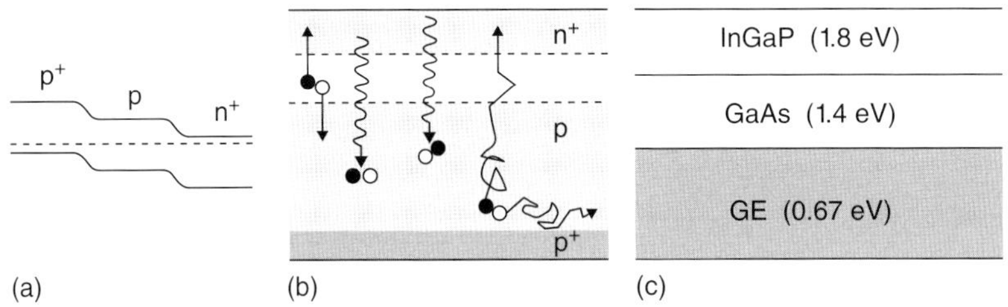

Figure 20.5 The cross-sectional view of the planar solar cell and the band diagram in equilibrium (a). The photo-generated e–h pairs are separated, electrons to the n region and holes to the p region (b). The multi-junction solar cell consisting of the stacked heterojunction semiconductor layers (c).

region are instantly separated and swept out of the region, electrons rolling down the potential hill to the n region, while holes rolling up to the p region just as in the p–n junction solar cell. But those e–h pairs generated outside the depletion region are less likely to reach the destined bulk regions. In this case, the e–h pairs have to traverse the longer distance driven by weaker space charge field as clear from the band profile shown. Moreover, the electrons generated in the p region as the minority carrier are liable to be captured by the holes therein. Also the trap-assisted recombination of e–h pairs further reduces the collection efficiency. Hence, it is generally difficult to attain the efficient generation and collection of e–h pairs at the same time.

Multi-Junction Solar Cell

The multi-junction solar cell also offers a viable means to enhance the cell efficiency as exemplified by the three layers of InGaP, GaAs, and Ge stacked together (Figure 20.5). In this structure, each layer has its own bandgap and the p–n junction built in. Thus, the two-junction solar cells are connected in series, so that the total open circuit voltage V_{oc} consists of the sum of the two V_{oc}'s connected in series. More important, the optical absorption occurs over a wider range of solar spectrum, for example, photons with energy greater than 1.8, 1.4, and 0.67 eV can be absorbed in InGaP, GaAs, and Ge layers, respectively. This is clearly seen from the optical absorption coefficient in the semiconductor (Eq. (18.6)). The collection efficiency is again high for e–h pairs generated within each depletion region, but in between the depletion regions, the efficiency is low for the same reasons as discussed in the planar cell. From the process point of view, the stacked layers should have nearly the same lattice constant in order to reduce the trap density, but the combination of such materials is limited.

Nanowire Solar Cell

The NW (nanowire) solar cell has distinct advantages over traditional wafer-based planar solar cells with regard to the optical absorption and the carrier collection. When a flux of photons is incident on the vertical array of NWs, the photons generally undergo multiple reflections and tend to be trapped therein (Figure 20.6). Consequently, the number of encounters between the photons and NWs is increased prompting more optical absorption. To further increase the absorption, the scattering centers can be inserted in between the NWs to randomize the direction of the photon for more reflections and absorption, irrespective of the incident angle. Moreover, the optical paths of those photons incident along the direction of NW do not lend to the multiple reflections. But the photons can be confined in the NW instead and undergo the resonant interaction with NW, which provides an excellent condition for absorption (Figure 20.6). The confinement of the photons in NW via the resonant interaction is akin to the confinement of light in an optical fiber (see Figure 5.4) and is due to the constructive interference of the waves reflected from the inner surface of the NW.

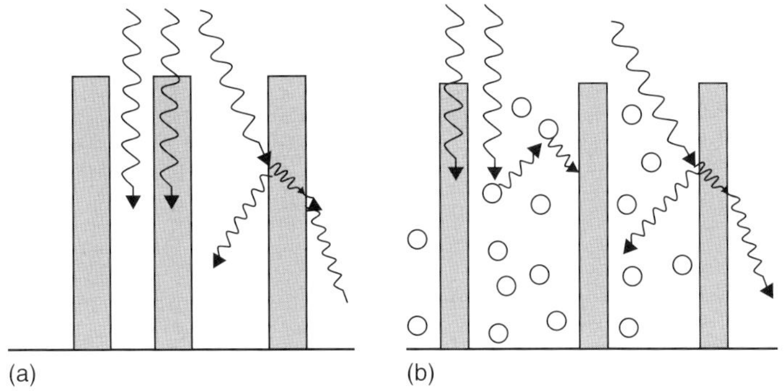

Figure 20.6 The solar radiation incident on the vertically grown NWs: (a) incident in the direction of the NW and undergoing the resonant interaction or incident at an angle and undergoing multiple reflections or transmissions (a). Enhanced multiple reflections aided by the scattering centers (b).

Collection Efficiency of e–h Pairs

In addition to the enhanced light absorption, the efficient collection of e–h pairs can also be attained in NW, as exemplified by the vertical core–shell NW structures (Figure 20.7). In this p-n-p structure, the absorption volume increases with increasing length of NW without the need for the increased footprint. Concomitantly, the entire e–h pairs generated are readily swept out of the narrow junction region in the radial direction regardless of the absorption depth. Therefore, the e–h pairs can be generated and collected simultaneously in an optimal manner. It should be mentioned however that there are various hidden variables hindering the real-life applications of the attractive features and novel ideas, but it behooves to point out such features and concepts.

20.3 Biosensor

The biosensors have become a center piece of nanotechnology by which to carry out the real-time and label-free detection of biochemical species in the sample. The scope of sensing applications is extensive, encompassing the clinical

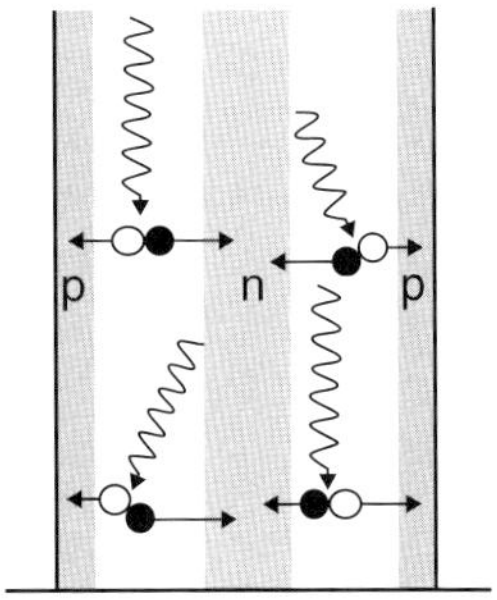

Figure 20.7 The e–h pairs photo-generated in the junction depletion region of the vertically grown core–shell NW and separated efficiently, holes to the p region and electrons to the n region.

diagnostics, molecular medicine, health care, environmental monitoring, and so on. Moreover, the biosensor is the essential element of the lab-on-a-chip, which has been devised for fast and efficient detection and analyses of the biological samples on a chip level. The operation principles of FET-based biosensors are briefly discussed.

The Bio-FET is known as the *ion-sensitive field-effect transistor* called *ISFET* for short, and the device capitalizes on the affinity-based binding of target molecules by probe molecules. For such sensing, the NWFETs are well suited because of the high surface-to-volume ratio and 1D structure of the channel with nanoscale cross-sections. A typical Bio-FET is made up of the usual NWFET as shown is Figure 20.8, but the conventional gate electrode is replaced by the electrolyte and reference gate electrode. In this composite gate structure, the gate dielectric is immersed in the electrolyte and surrounds the channel, providing the sensing surface. The probe or receptor molecules are chosen a priori and attached to the dielectric surface for capturing the target molecules. It is therefore essential that the capture molecules bind the analyte with high affinity and remain stable under varying conditions.

The role of the binding events is to induce the charge exchange between the probe molecules on the sensing surface and the electrolyte containing the sample. The sensing relies on detecting the resulting changes of the channel conductivity. For given V_{GR} at the reference electrode, the gate field lines emanating from it are screened or reinforced by the ionic charge in the electrolyte, depending on its polarity. Moreover, the receptor molecules are protonated or deprotonated on the dielectric surface while capturing the target molecules and form the surface charge sheet. Consequently, the channel conductivity is modified, and the sensing is done by monitoring the changes of the drain current ΔI_D. In this context, there is a parallelism existing between reading in the flash memory cell and sensing in

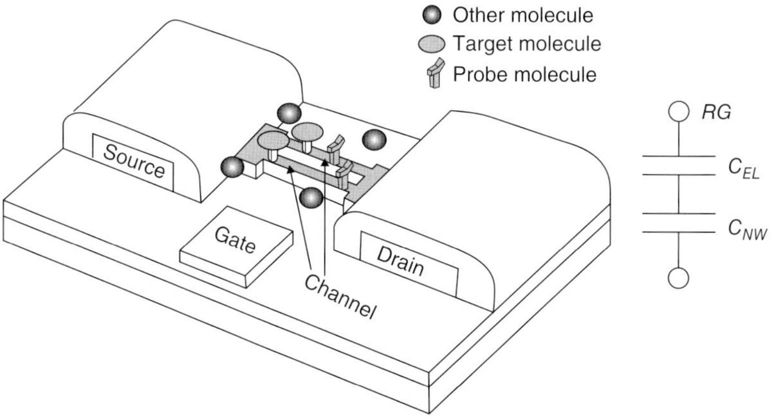

Figure 20.8 The cross-sectional view of the bio-FET consisting of multiple NW channels with the reference gate electrode in the electrolyte. Also shown are the receptor molecules deposited on the gate dielectric for capturing the target molecules and the effective gate capacitance.

the Bio-FET. In both cases, the threshold voltage shift ΔV_T is induced and utilized for reading or sensing.

Thus, consider the total capacitance C_T of the composite gate electrode, which consists of the gate oxide C_{NW} and electrolyte C_{EL} capacitances connected in series (Figure 20.8). The change in I_D caused by the charge exchange ΔQ on the oxide surface is then given from Eq. (19.1) by

$$\Delta I_D = \frac{W_{NW}}{L} C_{NW} \mu_n \Delta V_G V_D; \quad \Delta V_G = V_G - \Delta V_T, \quad 0 < V_D < V_{DSAT} \tag{20.5a}$$

where W_{NW} is the effective channel width of the NW. The change in the gate overdrive ΔV_G is due to the shift in the threshold voltage ΔV_T, which in turn is caused by ΔQ:

$$\Delta V_T = \frac{\Delta Q}{C_T}; \quad \frac{1}{C_T} = \frac{1}{C_{NW}} + \frac{1}{C_{EL}} \tag{20.5b}$$

Evidently, the relative importance of C_{NW} and C_{EL} depends on the geometry of ISFET. Also the expression of I_D in long-channel NWFET has been used for the simplicity of discussion, and ΔV_T was taken much smaller than V_T, a nonessential approximation. In this manner, the presence of the analyte can be quantitatively sensed via ΔI_D.

When the sensing is done in the triode or saturation region of ISFET, the detected signal is proportional to V_D and V_G. But because of the low I_D available in NWFETs and the linear dependence of ΔI_D on ΔV_G, the sensitivity of detection may not be sufficient to sense a minute amount of the sample. To enhance the sensitivity, the detection can be shifted to the subthreshold regime. In this V_G region, the subthreshold current I_{SUB} depends exponentially on V_G (see Eq. (19.18)). Therefore, the effect of the threshold voltage shift ΔV_T caused by ΔQ is exponentially amplified via ΔI_{SUB}. The resulting improvement of the sensitivity can be estimated with the use of the subthreshold slope (see Eq. (19.35)). According to Eq. (19.35), the minimum shift of V_G for inducing the change of I_D by one decade is

$$\Delta V_G = \frac{k_B T}{q \log e} \approx 60\, mV$$

Therefore, the level of the analyte as minute as causing ΔV_G in the range of 60 mV or more can induce the change of I_D by one decade, and the sensitivity is greatly enhanced. Additionally, the drain current level of ISFET can be increased by connecting the multiple NW channels in parallel. In this case, I_D can be increased, but the nonuniformity of each NW channel characteristics gives rise to appreciable variances of I_D and the subthreshold slope, degrading thereby the effective sensitivity.

20.4 Spin Field-Effect Transistor

The "charge" and "spin" are the defining characteristics of the electron together with the "mass." The FETs we have considered thus far are based on the charge control. The binary bits 1 and 0 in such FETs are incorporated via the ON and OFF states, that is, the ON and OFF currents. In the generation of the ON current in the conventional FETs, the electron spins point at random directions and do not play any role. However, the spin-based devices, called the *spintronic devices*, rely exclusively on the electron spin, in particular the difference in transport of the spin-up and spin-down states. The operation principle of the spin FETs is briefly discussed.

A variety of possible schemes for implementing the spintronic devices is under intensive exploration such as the magnetic bipolar diodes and transistors and magnetic tunneling transistors. The device chosen for discussion in this chapter is the Datta–Das spin field-effect transistor (SFET), shown in Figure 20.9. As clear from the figure, the SFET is a three-terminal device, consisting of the source and drain and the gate on top of the channel. Thus, the structure of SFET closely resembles that of the conventional charge-based FETs. Additionally, the role of the gate terminal is also the same in both FETs in that it controls the channel conductivity by means of the gate voltage applied.

However, there also exist the differences between the two FETs. In SFET, the source and drain are made of the ferromagnetic material and possess the parallel magnetic moments. Also the channel consists of a quantum well that is formed by the heterojunction semiconductors in parallel with the gate plate. More important, the operation of SFET is based on an entirely new kind of physical processes, namely, the spin injection and detection by the ferromagnetic source and drain, respectively. In this scheme, only those electrons possessing the spin parallel to the magnetic moment of the source are filtered and injected into the channel. By the same token, only those electrons preserving the input spin while traversing

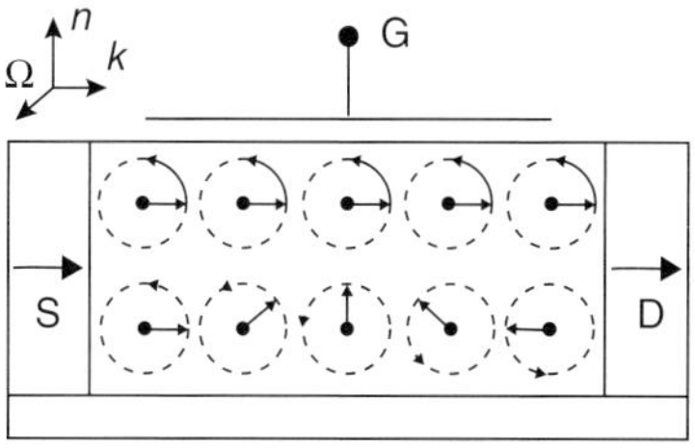

Figure 20.9 The cross-sectional view of Datta–Das spin FET consisting of the source and drain electrodes and the gate on top of the channel. The electron injected into the channel with its spin parallel to the magnetic moment of the source electrode can either exit through the drain terminal by preserving its initial spin free of V_G or is blocked at the drain terminal with its spin flipped via the V_G-driven precession (bottom).

the channel are filtered and transmitted through the drain terminal, contributing to I_D.

Thus, the transistor action in SFET consists of (i) turning on V_G to flip the spin from the spin-up to -down state, blocking the transmission of electrons at the drain terminal and cutting off I_D or (ii) turning off V_G to preserve the input spin and enable the injected electrons to contribute to I_D by passing through the drain terminal. In the conventional FETs, I_{ON} and I_{OFF} are likewise controlled by turning on V_G and opening the channel or turning off V_G to close the channel. In this context, the schemes for the transistor action in both FETs are similar, although the underlying physical processes are different.

Let us next consider an electron that is injected into the channel by passing through the source terminal with its spin parallel to the magnetic moment of the source electrode. With V_G turned on, the spin of the injected electron is driven to precess around the precession vector Ω. The orientation of Ω is specified by the vector product of $\boldsymbol{n}$ and $\boldsymbol{k}$, where $\boldsymbol{k}$ is the propagation vector of the wavefunction of the electron traversing the channel and $\boldsymbol{n}$ is the unit vector normal to the gate plate (Figure 20.9). Naturally, the Datta–Das SFET has an important advantage in that the transistor action is carried out electrically with the use V_G instead of resorting to the external magnetic field. The binary bits 1 and 0 are represented in SFET by the spin-up state passing through the drain and the spin-down state being blocked by the drain.

We next consider the dynamics of the electron spin precession driven by V_G. The spin of the injected electron can be decomposed into two components, parallel s_p and normal s_n with respect to the wave vector $\boldsymbol{k}$. These two components are coupled and evolve in time according to

$$\frac{ds_n}{dt} = 2\alpha_{\mathrm{BR}} k s_p,\ \frac{ds_p}{dt} = -2\alpha_{\mathrm{BR}} k s_n \tag{20.6a}$$

where α_{BR} is the Bychkov–Rashbar structure inversion asymmetry coefficient and is controlled by V_G. We can decouple the Eq. (20.6a) by differentiating with respect to time one of the paired equations in Eq. (20.6a) and using the other, obtaining

$$\frac{d^2 s_j}{dt^2} = -\omega^2 s_j;\quad \omega^2 \equiv (2\alpha_{BR} k)^2,\ j = n,\ p \tag{20.6b}$$

It is therefore clear that the two spin components s_p and s_n process around with the frequency ω, which depends on $\boldsymbol{k}$ and α_{BR}, hence V_G. Let us consider the simple case in which the electrons are injected in the direction parallel with the channel. Then, with V_G turned on the electron executes the precession around the precession axis, and by the time it reaches the drain terminal, the initial parallel component s_{p0} has rotated by an angle θ with respect to $\boldsymbol{k}$ amounting to

$$\theta = \omega\tau_{tr} = \frac{2\alpha_{\mathrm{BR}} mL}{\hbar};\quad \tau_{tr} = \frac{L}{v} = \frac{Lm}{\hbar k} \tag{20.7}$$

Here, τ_{tr} is the transit time of the electron across the channel and m the electron mass. Naturally, the output current I_D is commensurate with the number of electrons passing through the spin filter at the drain terminal. The number is specified by the ensemble-averaged component of s_p projected onto the magnetic moment of the drain, and we can write

$$I_D = I_{\text{in}} \left[1 - \sin^2\left(\frac{\theta}{2}\right)\right] \tag{20.8}$$

It is therefore clear from Eqs. (20.7) and (20.8) that with V_G turned off, there is no precession, that is, $\theta = 0$, so that the injected electrons all pass through the drain, and the ON state ensues. On the other hand, with V_G turned on and the spin precession triggered, the average angle of precession θ can be matched to π by adjusting τ_{tr} and α_{BR} or V_G. In this case, all of the spin-up state can be flipped to the spin-down state. Consequently, the injected electrons are all blocked by the drain electrode, and the OFF state ensues with $I_D \approx 0$.

In summary, there is a strong similarity in the operation of the conventional and spin FETs. In both devices, the ON and OFF states or equivalently the binary bit 1 to 0 are controlled by V_G. In the former FET, the two states are implemented by opening or closing the channel. In SFET, the two states are attained simply by flipping or preserving the input spin in the course of the electron transiting across the channel. In this context, SFET has important advantages in that the flipping of the electron spin requires much less power and can be done much faster. This is in distinct contrast with the conventional charge-based FETs in which the power consumption and the speed of charging or discharging are the major issues.

It is pointed out however that the Datta–Das SFET has yet to be implemented for real-life applications. Some of the major technical difficulties involved are (i) efficient injection of the spin-polarized electrons from the ferromagnetic source into the channel, (ii) tight control of α_{BR} via V_G and uniformity of α_{BR} for the reliability of device performance, and (iii) the ballistic spin-polarized transport rather than the drift–diffusion transport. These technological obstacles are investigated intensively, and for overcoming the technological barriers, the quantum mechanical insights will no doubt be a crucial factor.

20.5 Spin Qubits and Quantum Computing

The spintronics is endowed with an additional advantage in that the two-level nature of the electron spin could possibly be utilized for implementing the quantum computer. Several other approaches have also been proposed for the purpose based on, for example, the ions in magnetic trap, frozen light, the ultracold quantum gases known as *Bose–Einstein condensates*, and the nuclear magnetic resonance of molecules in liquids. However, as an extension of the discussion on spintronics, the two spin states are singled out for highlighting the essential features of the quantum computing.

The basic unit in the quantum computing is the quantum bit called *qubit*, which is the quantum analog of the binary bits 1 or 0 in the conventional digital computers. The qubit is in essence a controllable two-level system such as the spin

1/2 system, two-level atom in a resonant interaction with the EM field. Given an ensemble of n electrons, its Hilbert space dimension of 2^n is the same as the number of configurations of the corresponding classical system. But the advantage of the quantum computer lies in that the computing can be carried out in the superposed state of all basis states. That is, many classical computations can be done simultaneously in parallel via the unitary evolution of the qubits. The unitary evolution is one of the basic premises of the quantum mechanics, namely, that a quantum system evolves in time according to the time-dependent Schrödinger equation.

The Entanglement

The entanglement is closely tied with the quantum computing and is briefly discussed at the outset. It refers to a quantum state involving two or more particles. Given a system of two particles, for example, the essential feature of the entanglement is the fact that the probability of the outcome of the measurement of one particle depends on the state of the other particle although there is no interaction between the two.

Let us consider a specific example of two-spin one-half system. Then, the spin wavefunction of the two Fermions is given from Eq. (10.15) by

$$\varphi(1,\ 2) = 2^{-1/2}[\alpha_z(1)\beta_z(2) - \beta_z(1)\alpha_z(2)], \quad \alpha \equiv \chi_+,\ \beta \equiv \chi_- \tag{20.9}$$

where the spin-up and -down states have been denoted by α and β (see Eq. (10.1)). The information carried by the entangled state Eq. (20.9) is that the spins of the two particles are oriented in the opposite direction, but it does not tell the absolute direction of the spin. In fact, the form of the singlet φ Eq. (20.9) is preserved regardless of the direction of quantization.

To prove it, let us consider, for example, the entangled state in the x-direction. The eigenfunction of the spin-up and spin-down states in that direction can be expressed in terms of the linear combinations of the usual spin-up and -down states along the z-direction as

$$\alpha_x = \frac{1}{\sqrt{2}}(\alpha_z + \beta_z) = \frac{1}{\sqrt{2}}\left[\begin{pmatrix}1\\0\end{pmatrix} + \begin{pmatrix}0\\1\end{pmatrix}\right] = \frac{1}{\sqrt{2}}\begin{pmatrix}1\\1\end{pmatrix} \tag{20.10a}$$

$$\beta_x = \frac{1}{\sqrt{2}}(\alpha_z - \beta_z) = \frac{1}{\sqrt{2}}\left[\begin{pmatrix}1\\0\end{pmatrix} - \begin{pmatrix}0\\1\end{pmatrix}\right] = \frac{1}{\sqrt{2}}\begin{pmatrix}1\\-1\end{pmatrix} \tag{20.10b}$$

where the spin states are represented by the Pauli spin matrices (see Eqs. (10.1) and (10.7). The states α_x and β_x thus combined can indeed be shown to represent the spin-up and -down states by applying the spin operator $\hat{S}_x$:

$$\hat{S}_x\alpha_x = \frac{\hbar}{2}\begin{pmatrix}0 & 1\\1 & 0\end{pmatrix}\frac{1}{\sqrt{2}}\begin{pmatrix}1\\1\end{pmatrix} = \frac{\hbar}{2}\alpha_x, \quad \hat{S}_x\beta_x = \frac{\hbar}{2}\begin{pmatrix}0 & 1\\1 & 0\end{pmatrix}\frac{1}{\sqrt{2}}\begin{pmatrix}1\\-1\end{pmatrix} = -\frac{\hbar}{2}\beta_x \tag{20.10c}$$

Moreover, when the states α_z and β_z are expressed in terms of α_x and β_x from Eqs. (20.10a) and (20.10b) and inserted into Eq. (20.9), the singlet state is transformed into

$$\varphi(1,\ 2) = -2^{-1/2}[\alpha_x(1)\beta_x(2) - \beta_x(1)\alpha_x(2)] \tag{20.11}$$

Clearly, Eq. (20.11) is identical in content and form to Eq. (20.9) aside from the irrelevant negative sign.

Let us next consider the effect of performing a measurement of the z-component of the particle 1, for example, when the system is in the entangled state Eq. (20.9) or (20.11). As clear from Eq. (20.9), the measurement should yield the spin-up or -down state at random with equal probability. But suppose the spin 1 was measured to be in the spin-up state. As a result of this first measurement, the system must disentangle and collapse into the spin eigenstate given by

$$\varphi(1,\ 2) = \alpha_z(1)\beta_z(2) \tag{20.12}$$

This is in accordance with the postulate of the quantum mechanics, namely, that the wavefunction of the system is identical to the eigenfunction associated with the eigenvalue obtained as a result of the measurement. As a consequence, the two particles are in specific eigenstates, and the measurement of the z-component of spin 2 is assured to yield the spin-down state. Thus, the essential feature of the entanglement is that the probabilities of obtaining particular values of the spin of one of the two particles is dependent on what measurement has been carried out on the other.

Quantum Computing

As the first step to understanding the operation of the quantum computing, let us consider the simplest logic gate, NOT. The NOT gate yields an output that is the logic opposite or complement to the input. If the input is a logic 0, the output of the NOT gate is logic 1 or vice versa.

We can devise a unitary operation that will carry out the NOT gate operation. Specifically, we pass the spin through an oscillating magnetic field applied in the z-direction. The wavefunction of the spin 1/2 system is to be represented in general in matrix notation by

$$\psi = c_1(t)\alpha_z + c_2(t)\beta_z = \begin{pmatrix} c_1(t) \\ c_2(t) \end{pmatrix} \tag{20.13a}$$

while the interaction Hamiltonian is given from Eq. (10.33) by

$$\hat{H} = \frac{e}{m_e}B\hat{S}_z = \frac{e\hbar B}{2m_e}\begin{bmatrix} 1 & 0 \\ 0 & -1 \end{bmatrix} \tag{20.13b}$$

where the states α and β and the spin operators have been represented by Pauli spin matrices (see Eqs. (10.6) and (10.7)) and c_1 and c_2 are the expansion coefficients of the spin-up and -down states. The Schrödinger equation then reads in matrix notation as

$$i\hbar\frac{\partial}{\partial t}\begin{pmatrix} c_1(t) \\ c_2(t) \end{pmatrix} = \frac{e\hbar B}{2m_e}\begin{bmatrix} 1 & 0 \\ 0 & -1 \end{bmatrix}\begin{pmatrix} c_1(t) \\ c_2(t) \end{pmatrix} \tag{20.14}$$

By performing the matrix multiplication, we can recast Eq. (20.14) as

$$\frac{\partial}{\partial t}c_1 = -i\Omega c_1; \quad \Omega = \frac{eB}{2m_e}$$
$$\frac{\partial}{\partial t}c_2 = i\Omega c_2 \tag{20.15}$$

and obtain the solution as

$$c_1(t) = c_1(0)\exp(-i\Omega t), \quad c_2(t) = c_2(0)\exp(i\Omega t) \tag{20.16}$$

Thus, if the spin is initially in the α_z state, $c_1(0) = 1$, $c_2(0) = 0$, so that c_2 always remains zero, while the phase factor of c_1 oscillates in time as $c_1(t) = c_1(0)\exp(-i\Omega t)$. By the same token, if the spin is initially in the β_z state, c_1 always remains zero, while the phase factor of c_2 oscillates similarly in time. Thus, the spin state will remain in the same state as initially given.

On the other hand, if the initial state is either in α_x or β_x state, $c_1(0) = 1, c_2(0) = 1$ or $c_1(0) = 1$, $c_2(0) = -1$ (see (Eq. 20.10)). Then, the spin states will undergo the change in time as

$$\frac{1}{\sqrt{2}}\begin{pmatrix} \exp(-i\Omega t) \\ \exp(i\Omega t) \end{pmatrix}, \quad \text{or} \quad \frac{1}{\sqrt{2}}\begin{pmatrix} \exp(-i\Omega t) \\ -\exp(i\Omega t) \end{pmatrix} \tag{20.17}$$

Therefore, the spin-up state is flipped to spin-down state or vice versa at $\Omega t = (2n+1)(\pi/2)$. For $n = 0$, for example, the flipping is done according to

$$\alpha_x \to -i\beta_x \quad \text{or} \quad \beta_x \to -i\alpha_x; \; e^{\pm(i\Omega t)} = \cos(\Omega t) \pm i\sin(\Omega t) \tag{20.18}$$

For other times, each state in Eq. (2.17) always remains as the spin-up or -down state following the direction of the spin matrix given by

$$\hat{S}_\phi = \hat{S}_x\cos\phi + \hat{S}_y\sin\phi = \frac{\hbar}{2}\begin{pmatrix} 0 & \exp(-i\phi) \\ \exp(i\phi) & 0 \end{pmatrix}, \quad \phi = \Omega t \tag{20.19}$$

An important consequence of Eq. (2.17) is that the linearly superposed state can undergo the logic NOT operation as

$$A\alpha + B\beta \to -i(A\beta + B\alpha) \tag{20.20}$$

where A and B are arbitrary constants. Clearly, Eq. (20.20) points to the fact that if we run the program once using the left-hand side wavefunction as the input, the output wavefunction is the linear combination of the outcomes of the logic NOT gate. This carries an enormous advantage and possibility of the parallel quantum computing especially in view of the fact that the number of qubits involved can be readily increased.

It should be pointed out however that in order to have the access to the information, we have to make a measurement, which involves the collapse. In this case, we can determine only one component of the spin, so that there are no practical advantages in quantum computing. However, there exist certain computations that can exploit the advantage via the appropriate interplay of the unitary evolution and collapse. The prime example of such calculation is to determine the period x of a periodic function $f(x)$, which carries a critical bearing in solving the factorization of large numbers into its prime number components. An additional comment is due at this point. As has become clear by now, the quantum computers are inherently associated with the entanglement of a large number of qubits. Such entangled states are extremely sensitive to decoherence and noise, the overcoming of which evidently requires a new technology.

Problems

20.1 (a) The floating gate in the flash memory cell is a quantum well that is formed by the n^+ polysilicon and two dielectric layers. The typical structure of the well is shown in Figure 20.1.
(b) Find the kinetic energy of the electron in the ground state as a function of gate thickness W ranging from 10 to 100 nm. (Use the infinite barrier height for simplicity.)
(c) Find W at which the ground state energy is equal to the thermal energy of the electron.
(d) Find the tunneling probability of the electron across the two barriers.

20.2 Consider the same floating gate as shown in Figure 20.1 with the tunnel oxide thickness of 50 nm.
(a) Estimate the lifetime τ of an electron therein for $W = 10$ nm moving in the well with the thermal velocity at room temperature.
Hint: τ can be defined by $TN = 1$ where T is the tunneling probability and N the total number the electron encounters the barrier during the lifetime.
(b) Calculate the voltage applied at the floating gate at which the lifetime of the electron therein is reduced to 1 μs via inducing the F–N tunneling.

20.3 Consider the ONO dielectric layer with the equivalent thickness of 15 nm of SiO_2. The V_{TP} in the programmed cell is greater than V_{TE} in the erased cell by 3 V. Find the number of electrons stored in the programmed cell. (The dielectric constant of SiO_2 is 11.9.)

20.4 Consider the stacked multi-junction solar cell shown in Figure 20.5.
(a) Draw the equilibrium energy band diagram of the n^+ $InGaP - p\ GaAs - p^+\ Ge$ without the p–n junction in each cell.

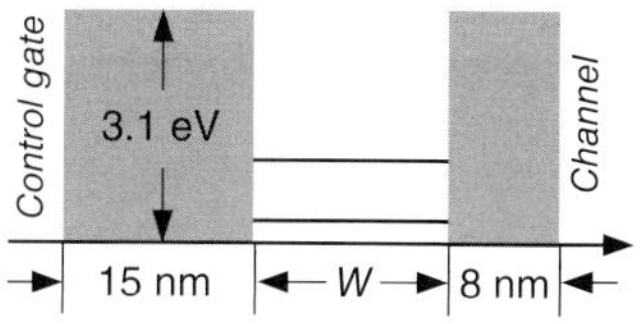

Figure 20.10 Typical quantum well representing the floating gate.

(b) Draw the band diagram under the solar cell operation condition and qualitatively discuss the cell operation.

20.5 (a) Draw the energy band diagram of the vertical core–shell NW across the cross-section of the NW.

(b) Describe the solar cell operation.

Suggested Readings

1. D. M. Kim and Y. H. Jeong, editors, Nanowire Field Effect Transistors: Principles and Applications, Springer, 2014.
2. A. I. M. Rae, Quantum Mechanics, Fourth Edition, Taylor & Francis, 2002.
3. D. M. Kim, Introductory Quantum Mechanics for Semiconductor Nanotechnology, Wiley-VCH, 2010.

Solutions

1.1. (a) The center of mass X and relative x coordinates are defined as

$$(m_1 + m_2)X = m_1x_1 + m_2x_2, \quad x = x_2 - x_1 \tag{1.1}$$

By finding x_1, x_2 in terms of X, x one can write

$$x_1 = X + \frac{m_2}{m_1 + m_2}x, \quad x_2 = X - \frac{m_1}{m_1 + m_2}x \tag{1.2}$$

The Hamiltonian then reads in terms of X, x as

$$H \equiv \frac{1}{2}m_1\dot{x}_1^{\,2} + \frac{1}{2}m_2\dot{x}_2^{\,2} + \frac{1}{2}k(x_1 - x_2)^2 = \frac{1}{2}M\dot{X}^2 + \frac{1}{2}\mu\dot{x}^2 + \frac{1}{2}kx^2 \tag{1.3a}$$

where the center of mass and reduced mass are given by

$$M \equiv m_1 + m_2, \quad \mu \equiv \frac{m_1m_2}{m_1 + m_2} \tag{1.3b}$$

Equivalently H can also be expressed in terms of the momentum as

$$H = \frac{P^2}{2M} + \frac{p^2}{2\mu} + \frac{1}{2}kx^2; \quad P \equiv M\dot{X}, \quad p \equiv \mu\dot{x} \tag{1.4}$$

(b) The Hamilton's equation of motion is then given from Eq. (1.6) by

$$\dot{X} = \frac{\partial H}{\partial P} = \frac{P}{M} \quad \dot{P} = -\frac{\partial H}{\partial X} = 0 \tag{1.5a}$$

so that

$$M\ddot{X} = \dot{P} = 0 \tag{1.5b}$$

Similarly one can write

$$\dot{x} = \frac{\partial H}{\partial p} = \frac{p}{\mu} \quad \dot{p} = -\frac{\partial H}{\partial x} = -kx \tag{1.6a}$$

and

$$\mu\ddot{x} = \dot{p} = -kx \tag{1.6b}$$

Introductory Quantum Mechanics for Applied Nanotechnology, First Edition. Dae Mann Kim.

(c) The equation of motion (1.5b) indicates that the center of mass of the H_2 molecule moves as a free particle, while the relative motion between the two H-atoms is represented by the motion of the harmonic oscillator.

1.2. The thermal velocity of an electron at 300 K is found from (1.15) as

$$v_{T0} = \left(\frac{3k_B T}{m_e}\right)^{1/2} = \left[\frac{3 \times 1.381 \times 10^{-23}\ \text{J K}^{-1} \times 300\ \text{K}}{9.09 \times 10^{-31}\ \text{kg}}\right]^{1/2}$$
$$= 1.17 \times 10^5\ \text{m s}^{-1}$$
$$= 1.17 \times 10^7\ \text{cm s}^{-1}$$

The v_T of the particle with mass m at T is given in terms of v_{T0} as

$$v_T = v_{T0}\left(\frac{m_e}{m}\right)^{1/2}, \quad v_T = v_{T0}\left(\frac{T}{300}\right)^{1/2}$$

	300	10	1000 (cm s^{-1})
electron	1.17×10^7	2.14×10^6	2.24×10^7
proton	2.73×10^5	4.98×10^4	5.10×10^5
H_2	1.93×10^5	3.52×10^4	3.52×10^5
1 g	3.53×10^{-7}	6.44×10^{-8}	6.44×10^{-7}

1.3. (a) Upon inserting the plane wave $\underline{E}(z,t) = \hat{x}E_0 e^{-i(\omega t - kz)}$ into the wave equation there results

$$0 = \left(\nabla^2 - \frac{1}{v^2}\frac{\partial^2}{\partial t^2}\right) E_0 e^{-i(\omega t - kz)} = \left(-k^2 + \frac{1}{v^2}\omega^2\right) E_0 e^{-i(\omega t - kz)}$$

Therefore the plane wave can be made a solution by constraining k and ω to satisfy the dispersion relation, that is,

$$-k^2 + \frac{1}{v^2}\omega^2 = 0 \quad \text{or} \quad \omega^2 = k^2 v^2$$

(b) The E and H fields,

$$\underline{E}(z,t) = \hat{x}E_0 e^{-i(\omega t - kz)}, \quad \underline{H} = \hat{y}\sqrt{\frac{\varepsilon}{\mu}} E_0 e^{-i(\omega t - kz)}$$

satisfy the Faraday's law of induction which is shown as follows. One can insert E on the left hand side of Faraday's law (Eq. (1.21)), obtaining

$$\nabla \times \underline{E} = \left(\hat{x}\frac{\partial}{\partial x} + \hat{y}\frac{\partial}{\partial y} + \hat{z}\frac{\partial}{\partial z}\right) \times \hat{x}E_0 e^{-i(\omega t - kz)}$$
$$= \left(-\hat{z}\frac{\partial}{\partial y} + \hat{y}\frac{\partial}{\partial z}\right) E_0 e^{-i(\omega t - kz)} = \hat{y}ikE_0 e^{-i(\omega t - kz)}$$
$$\hat{x} \times \hat{x} = \hat{y} \times \hat{y} = \hat{z} \times \hat{z} = 0;$$
$$\hat{x} \times \hat{y} = \hat{z}, \quad \hat{y} \times \hat{z} = \hat{x}, \quad \hat{z} \times \hat{x} = \hat{y}$$

Likewise one can also obtain from the right hand side of the Faraday's law

$$-\frac{\partial \underline{B}}{\partial t} = -\mu\frac{\partial \underline{H}}{\partial t} = -\hat{y}\mu\frac{\partial}{\partial t}\left(\sqrt{\frac{\varepsilon}{\mu}}\mathrm{E}_0 e^{-i(\omega t-kz)}\right) = \hat{y}i\omega\sqrt{\varepsilon\mu}\mathrm{E}_0 e^{-i(\omega t-kz)}$$

Hence both sides are equal to each other, provided

$$k = \omega\sqrt{\varepsilon\mu} \quad \text{or} \quad \omega = \frac{k}{(\varepsilon\mu)^{\frac{1}{2}}} = vk, \quad v = \frac{1}{(\varepsilon\mu)^{1/2}}$$

The condition is identical to the dispersion relation and therefore the given fields E, H satisfy the Faraday's law of induction.

One may insert H into the left hand side of the Ampere's law, obtaining

$$\nabla \times \underline{H} = \left(\hat{x}\frac{\partial}{\partial x} + \hat{y}\frac{\partial}{\partial y} + \hat{z}\frac{\partial}{\partial z}\right) \times \hat{y}\sqrt{\frac{\varepsilon}{\mu}}\mathrm{E}_0 e^{-i(\omega t-kz)}$$

$$= \left(\hat{z}\frac{\partial}{\partial x} - \hat{x}\frac{\partial}{\partial z}\right)\sqrt{\frac{\varepsilon}{\mu}}\mathrm{E}_0 e^{-i(\omega t-kz)} = -\hat{x}\sqrt{\frac{\varepsilon}{\mu}}\mathrm{E}_0 ike^{-i(\omega t-kz)}$$

In a medium free of charge $J = 0$ and upon inserting E into the right hand side of the Ampere's law there results

$$\frac{\partial \underline{D}}{\partial t} = \frac{\partial}{\partial t}\varepsilon[\hat{x}\mathrm{E}_0 e^{-i(\omega t-kz)}] = -\hat{x}i\varepsilon\omega\mathrm{E}_0 e^{-i(\omega t-kz)}$$

Again the both hand sides are identical because of the dispersion relation, and fields E, H satisfy the Ampere's circuital law.

(c) For the medium free of the charge the curl operation performed on both sides of Eq. (1.22) yields

$$\nabla \times \nabla \times \underline{H} \equiv [\nabla\nabla\cdot -\nabla^2]\underline{H} = -\nabla^2\underline{H}; \quad \nabla\cdot\underline{B} \propto \nabla\cdot\underline{H} = 0 \tag{1.7}$$

$$\nabla \times \frac{\partial \underline{D}}{\partial t} = \varepsilon\nabla \times \frac{\partial \underline{E}}{\partial t} = -\varepsilon\mu\frac{\partial^2}{\partial t^2}\underline{H} \tag{1.8}$$

In Eq. (1.7) a vector identity was used as in the text and also Eq. (1.24) was used in Eq. (1.8). Therefore by equating Eqs. (1.7) and (1.8) one derives the wave equation

$$\nabla^2\underline{H} - \frac{1}{v^2}\frac{\partial^2}{\partial t^2}\underline{H} = 0, \quad \frac{1}{v^2} \equiv \mu\varepsilon = \mu_0\varepsilon_0\mu_r\varepsilon_r = \frac{1}{(c/n)^2}$$

1.4. To find the variance, the spatial profile of the wave packet given in Eq. (1.35) has to be normalized and one can thus represent the profile as

$$f(z) \propto |E(z,0)|^2 = \frac{\sigma}{\sqrt{\pi}}\exp -(\sigma z)^2 \tag{1.9}$$

and find the associated variance as

$$(\Delta z)^2 = \langle (z - \langle z\rangle)^2\rangle = \langle z^2 - 2z\langle z\rangle + \langle z\rangle^2\rangle; \quad \langle a\rangle \equiv \int_{-\infty}^{\infty} dzaf(z)$$

$$= \langle z^2\rangle - \langle z\rangle^2 = \langle z^2\rangle$$

where $\langle z\rangle = 0$, since $f(z)$ is even in z. Hence the variance can be evaluated as

$$(\Delta z)^2 = \int_{-\infty}^{\infty} dz z^2 f(z) = \frac{3}{2\sigma^2}$$

and is approximately same as Δz as defined by the width of z between 1/e points.

1.5. Given a vector $\boldsymbol{A}$ one may perform the operation

$$\begin{aligned}\nabla\times(\hat{x}A_x+\hat{y}A_y+\hat{z}A_z) &= \left(\hat{x}\frac{\partial}{\partial x}+\hat{y}\frac{\partial}{\partial y}+\hat{z}\frac{\partial}{\partial z}\right)\times(\hat{x}A_x+\hat{y}A_y+\hat{z}A_z)\\ &= \hat{x}\left(\frac{\partial A_z}{\partial y}-\frac{\partial A_y}{\partial z}\right)+\hat{y}\left(\frac{\partial A_x}{\partial z}-\frac{\partial A_z}{\partial x}\right)\\ &\quad+\hat{z}\left(\frac{\partial A_y}{\partial x}-\frac{\partial A_x}{\partial y}\right)\end{aligned}$$

so that

$$\begin{aligned}\nabla\cdot\nabla\times\underline{A} &= \left(\hat{x}\frac{\partial}{\partial x}+\hat{y}\frac{\partial}{\partial y}+\hat{z}\frac{\partial}{\partial z}\right)\cdot\left[\hat{x}\left(\frac{\partial A_z}{\partial y}-\frac{\partial A_y}{\partial z}\right)\right.\\ &\quad\left.+\hat{y}\left(\frac{\partial A_x}{\partial z}-\frac{\partial A_z}{\partial x}\right)+\hat{z}\left(\frac{\partial A_y}{\partial x}-\frac{\partial A_x}{\partial y}\right)\right]\\ &= \frac{\partial}{\partial x}\left(\frac{\partial A_z}{\partial y}-\frac{\partial A_y}{\partial z}\right)+\frac{\partial}{\partial y}\left(\frac{\partial A_x}{\partial z}-\frac{\partial A_z}{\partial x}\right)\\ &\quad+\frac{\partial}{\partial z}\left(\frac{\partial A_y}{\partial x}-\frac{\partial A_x}{\partial y}\right)\\ &= 0\end{aligned}$$

Therefore any vector $\boldsymbol{A}$ is shown solenoidal.

1.6. If the scalar product is performed on both sides of Eq. (1.22) there results with the use of Eqs. (1.23) and (1.25)

$$\nabla\cdot\nabla\times\underline{H} = \nabla\cdot\left(\underline{J}+\frac{\partial\underline{D}}{\partial t}\right) = \nabla\cdot\left(\underline{J}+\varepsilon\frac{\partial\underline{E}}{\partial t}\right) = \nabla\cdot\underline{J}+\frac{\partial\rho}{\partial t} = 0$$

where the well known continuity equation 1.25 has been used.

Chapter 2

2.1. The photon energy is given by

$$h\nu = h\frac{c}{\lambda} = 4.136\times10^{-15}\,(\text{eV s})\frac{3\times10^{8}\,\text{m s}^{-1}}{1\,\text{m}} = 1.24\times10^{-6}\,\text{eV}$$

for the wavelength of 1 m and the wave number is defined as

$$\tilde{\nu} \equiv \frac{1}{\lambda}$$

Thus the energy and wave number of the photons considered are:

λ (nm)	energy (eV)	wave number (cm^{-1})
10^{10}	1.24×10^{-7}	10^{-3}
10^{9}	1.24×10^{-6}	10^{-1}
10^{4}	1.24×10^{-1}	10^{4}
600	2.1	1.67×10^{5}
200	6.2	5×10^{5}
50	25	2×10^{6}
1	1.25×10^{3}	1.0×10^{8}

2.2. (a) The de Broglie wavelength of the electron at room temperature is given in terms of the thermal speed v_{Te} by

$$\lambda_e = \frac{h}{p_e} = \frac{h}{m_e v_{Te}}; \quad \frac{m_e v_{Te}^2}{2} = \frac{3k_B T}{2}$$

$$= \frac{6.626 \times 10^{-34}\ (\mathrm{J\,s})}{9.109 \times 10^{-31}(\mathrm{kg})1.17 \times 10^{5}(\mathrm{m\,s^{-1}})}$$

$$= 6.2 \times 10^{-8}\,\mathrm{m} = 62\,\mathrm{nm}$$

For proton we can find λ in terms of λ_e and the mass ratio

$$\lambda_p = \frac{h}{p_p} = \frac{h}{m_p v_{Tp}} = \frac{h}{m_e v_{Te}}\left(\frac{m_e v_{Te}}{m_p v_{Tp}}\right)$$

Since in equilibrium the thermal kinetic energy same, that is,

$$\frac{m_e v_{Te}^2}{2} = \frac{m_p v_{Tp}^2}{2}$$

λ_p can be evaluated as

$$\lambda_p = \frac{h}{m_e v_{Te}}\left(\frac{m_e}{m_p}\right)^{1/2} = 62 \times \left(\frac{m_e}{m_p}\right)^{1/2} \mathrm{nm} = 1.45\ \mathrm{nm}$$

(b) The de Broglie wavelength of the electron having the energy 1 eV is given by

$$\lambda = \frac{h}{p} = \frac{h}{(2mE)^{1/2}} = \left[\frac{h}{(2m)^{1/2}}\right]\frac{1}{[E(J)]^{1/2}}$$

$$= \frac{6.624 \times 10^{-34}\,\mathrm{J\,s}}{[2 \times 9.109 \times 10^{-31}\mathrm{kg}]^{1/2}[E(\mathrm{eV})/6.2 \times 10^{18}]^{1/2}} = \frac{0.123\,\mathrm{nm}}{\sqrt{E(\mathrm{eV})}}$$

Hence the de Broglie wavelengths of the electron are evaluated as

energy (eV)	1	200	100 K	1 M
λ(nm)	1.23×10^{-1}	8.7×10^{-3}	3.9×10^{-4}	1.23×10^{-4}

(c) The kinetic energy of the electron in the ground state of the H-atom is given from Eqs. 2.16 and 2.17 by

$$K_1 = -\frac{1}{2}V_1, \quad E = K_1 + V_1 = -E_0$$

so that $K_1 = E_0 = 13.6$ eV. Therefore the de Broglie wavelength can be found from (b) as

$$\lambda = 0.123\frac{\text{nm}}{\sqrt{E(\text{eV})}} = 0.123\frac{\text{nm}}{\sqrt{13.6}} = 3.3 \times 10^{-2}\,\text{nm}$$

2.3. (a) Consider the Planck's expression of the blackbody radiation,

$$\rho(\nu) = \frac{8\pi\nu^2}{c^3}\langle\varepsilon\rangle = \frac{8\pi\nu^2}{c^3}\frac{h\nu}{(e^{h\nu/k_BT} - 1)} \tag{2.1}$$

In the limit of the low frequency $h\nu \ll k_BT$ and one can expand the exponential function as

$$e^{h\nu/k_BT} - 1 = 1 + \frac{h\nu}{k_BT} + \frac{1}{2}\left(\frac{h\nu}{k_BT}\right) + \ldots - 1 \approx \frac{h\nu}{k_BT}$$

When this approximation is inserted into Eq. (2.1) the equation is reduced to

$$\rho(\nu) = \frac{8\pi\nu^2}{c^3}k_BT$$

in agreement with Rayleigh–Jean's theory.

(b) We can find the electron momentum from Eqs. 2.6 and 2.7 as

$$p_e^2 = (p_i - p_f + mc)^2 - m^2c^2 = (p_i - p_f)^2 + 2mc(p_i - p_f)$$

$$p_e^2 = \underline{p}_e \cdot \underline{p}_e = (\underline{p}_i - \underline{p}_f)\cdot(\underline{p}_i - \underline{p}_f) = p_i^2 + p_f^2 - 2p_ip_f\cos\theta$$

Hence by equating the right hand sides of the two equations we find

$$2(p_i - p_f)mc = 2p_ip_f(1 - \cos\theta) \equiv 4p_ip_f\sin^2\left(\frac{\theta}{2}\right)$$

where θ is the angle between p_i and p_f, that is, the scattering angle and the well known trigonometric identity has been used. Hence by multiplying both sides with h, while dividing by $p_i\,p_f$ there results.

$$\frac{h}{p_f} - \frac{h}{p_i} \equiv \lambda_f - \lambda_i = \frac{2h}{mc}\sin^2\left(\frac{\theta}{2}\right) \equiv 4\pi\lambda_e\sin^2\left(\frac{\theta}{2}\right); \quad \lambda_e = \frac{\hbar}{mc}, \quad \hbar \equiv \frac{h}{2\pi}$$

(c) From Eq. (2.13) one can express v_n in terms of r_n as

$$v_n = \frac{n\hbar}{(mr_n)}$$

and when inserted into Eq. (2.14), replacing v_n by r_n one obtains Eq. (2.15).

2.4. (a) We can use the results obtained in Eq. (2.1) and write

$$h\nu = h\frac{c}{\lambda} = 1.24 \times 10^{-6}\,\text{eV m}^{-1} = 1.24 \times 10^3\,\text{eV/nm}$$

so that the light of wavelength 300 nm has the energy 4.13 eV. Since the photon energy should be larger than the work function the photoelectric effect can occur only in lithium and beryllium and the stopping powers are given by

$$-(4.13 - 2.3) = -1.83\,\text{V}, \quad -(4.13 - 3.9) = -0.23\,\text{V}$$

(b) The energy conservation equation (2.4) reads in unit of electron volt as

$$\frac{hc}{194 \times 10^{-9}} = e\varphi + 2.3 \tag{2.2}$$

$$\frac{hc}{248 \times 10^{-9}} = e\varphi + 0.9 \tag{2.3}$$

By subtracting Eq. (2.3) from Eq. (2.2) one can write

$$h\left(\frac{c}{194 \times 10^{-9}} - \frac{c}{248 \times 10^{-9}}\right) = 1.4$$

and by inserting $c = 3 \times 10^8\,\text{m}\,\text{s}^{-1}$ h is found as

$$h = 4.17 \times 10^{-15}\ \text{eV}\,\text{s}$$

Also by using the value of h thus found in either Eq. (2.2) or (2.3) one finds the work function of the aluminum as

$$e\varphi = 4.15\,\text{eV}$$

2.5. With two protons the atomic number of the He^+ atom is 2, and the energy spectrum can be found from that of the H-atom, with the modification, $e \to Ze$. We can thus find the ionization energy in terms of the ionization energy of the H-atom as

$$E = \frac{Z^2 e_M^4 m}{2\hbar^2} = 13.6 \times Z^2\,\text{eV} = 54.4\,\text{eV}, \quad e_M^4 \propto (Ze)^2 e^2$$

One can likewise find the radius of the ground state from Eq. (2.15) with the same modifications as given by

$$r_1 = \frac{\hbar^2}{m e_M{}^2 Z} = r_B \cdot \frac{1}{Z} \cong \frac{0.05}{2}\,\text{nm} = 0.025\,\text{nm}, \qquad n = 1$$

The shortest and longest wavelengths λ in Balmer series are given by

$$\frac{hc}{\lambda_s} = 54.4\left(\frac{1}{2^2} - \frac{1}{\infty}\right)\text{eV} = \frac{54.4}{4}\,\text{eV},$$

$$\frac{hc}{\lambda_l} = 54.4\left(\frac{1}{2^2} - \frac{1}{3^2}\right)\text{eV} = 54.4\left(\frac{1}{2^2} - \frac{1}{3^2}\right)\text{eV}$$

Hence

$$\lambda_s = \frac{hc}{(54.4/4)} = \frac{4.136 \times 10^{-15}\,\text{eV}\,\text{s} \times 3 \times 10^8\,\text{m}\,\text{s}^{-1}}{(54.4/4)\,\text{eV}} = 0.91 \times 10^{-7}\,\text{m} = 91\,\text{nm},$$

$$\lambda_l = 91 \times \frac{1}{[1 - (4/9)]}\,\text{nm} = 163.8\,\text{nm}$$

2.6. One can use Eq. (2.9) and find the shift in λ as

$$\Delta\lambda = 4\pi\lambda_e \sin^2\left(\frac{\theta}{2}\right) = 2.5\times10^{-3}\,\text{nm},\quad \lambda_e = 4\times10^{-4}\,\text{nm},\quad \theta = \frac{\pi}{2}$$

The wavelength of X-ray with 200 KeV is given by

$$h\nu = 2\times10^5\,\text{eV} = h\frac{c}{\lambda_i}$$

so that $\lambda_i = 6.2\times10^{-3}$ nm. Hence the wavelength and energy of the scattered radiation are given from Eq. (2.9) by

$$\lambda_f = \lambda_i + \Delta\lambda = 8.7\times10^{-3}\,\text{nm}$$

$$E_f = h\frac{c}{\lambda_f} = 1.43\times10^4\,\text{eV}$$

Also the electron recoil energy is found from the energy conservation by

$$E_{\text{recoil}} = E_i - E_f = (20 - 1.43)\times10^4\,\text{eV} = 1.86\times10^5\,\text{eV}$$

2.7. (a) The radius of the ground state is given from Eq. (2.15) by $r_1 = 0.053$ nm and that of the first excited state is found as $r_2 = r_1 \times 4 = 0.106$ nm. The associated energies of the two states are given from Eq. (2.17) by $E_1 = -13.6$ eV and $E_2 = -(13.6/4)\,\text{eV} = -3.4$ eV.

(b) The transition wavelength between the two levels are then given by

$$\frac{hc}{\lambda} = E_2 - E_1 = 10.2\,\text{eV}$$

so that

$$\lambda = \frac{hc}{10.2\,\text{eV}} = 1.22\times10^{-7}\,\text{m} = 122\,\text{nm}$$

Chapter 3

3.1. The 3D eigenequation of the momentum is given from Eq. (3.11) by

$$-i\hbar\left[\hat{x}\frac{\partial}{\partial x} + \hat{y}\frac{\partial}{\partial y} + \hat{z}\frac{\partial}{\partial z}\right] u(\underline{r}) = (\hat{x}p_x + \hat{y}p_y + \hat{z}p_z)u(\underline{r})$$

One may look for the solution in the form

$$u(r) \propto f(x)g(y)h(z)$$

and insert it into the eigenequation and divide both sides with $u(r)$, obtaining

$$\hat{x}\frac{\partial f(x)/\partial x}{f(x)} + \hat{y}\frac{\partial g(y)/\partial y}{g(y)} + \hat{z}\frac{\partial h(z)/h(z)}{h(z)} = \frac{i}{\hbar}(\hat{x}p_x + \hat{y}p_y + \hat{z}p_z)$$

Hence by singling out the x, y, z components from both sides one obtains three 1D eigenequations,

$$-i\hbar\frac{\partial}{\partial\xi}u(\xi) = p_\xi; \quad \xi = x, y, z, \quad u = f, g, h$$

as given in Eq. (3.8). Therefore we can use the 1D eigenfunction given in Eq. (3.10) and multiply the three to obtain Eq. (3.12).

3.2. Given a wave packet the time duration and frequency bandwidth are constrained by

$$\Delta t \times \Delta \nu \approx 1$$

One may thus take Δt as the pulse duration and estimate the frequency band width as

$$\Delta \nu \approx \frac{1}{\Delta t} = 10^9, 10^{12}, 10^{15}\,\text{Hz}$$

for nano, pico, and femto second pulses.

3.3. Given the diameter of a nucleus D the minimum kinetic energy of the electron therein is to be estimated with the use of the uncertainty principle as

$$\begin{aligned}\Delta E_e &= \frac{\Delta p^2}{2m_e} = \frac{\hbar^2}{2m_e}\frac{1}{D^2}; \quad \Delta p D \simeq \hbar \\ &\approx 6.1 \times 10^{-11}\,\text{J} = 3.8 \times 10^8\,\text{eV}, \quad D = 10^{-14}\,\text{m}\end{aligned}$$

We can likewise estimate the minimum energy of a proton as

$$\Delta E \simeq \frac{\hbar^2}{2m_p}\frac{1}{D^2} = \frac{\hbar^2}{2m_e}\frac{1}{D^2}\left(\frac{m_e}{m_p}\right) = 3.8 \times 10^8\,\text{eV} \cdot \frac{m_e}{m_p} = 2.06 \times 10^5\,\text{eV}$$

Since ΔE_e is greater than the binding energy E_b of a nucleon, while ΔE_p is less than E_b only the proton can reside in the nucleus.

3.4. By using the uncertainty relation,

$$\Delta x \Delta p_x \simeq \hbar, \quad \Delta y \Delta p_y \simeq \hbar, \quad \Delta z \Delta p_z \simeq \hbar$$

the minimum kinetic energy of the electron and proton in a cubic box of length L can be estimated as

$$\Delta E_e \simeq \frac{1}{2m_e}(\Delta p_x^2 + \Delta p_y^2 + \Delta p_z^2) = \frac{\hbar^2}{2m_e}\left(\frac{1}{\Delta x^2} + \frac{1}{\Delta y^2} + \frac{1}{\Delta z^2}\right) = \frac{3\hbar^2}{2m_e}\frac{1}{L^2}$$

$$\Delta E_p \simeq \frac{3\hbar^2}{2m_p}\frac{1}{L^2} = \left(\frac{3\hbar^2}{2m_e}\frac{1}{L^2}\right)\left(\frac{m_e}{m_p}\right)$$

Thus ΔE_e, ΔE_p can be evaluated versus given L as:

L (nm)	ΔE_e (eV)	ΔE_p(eV)
1	1.1×10^{-1}	6.0×10^{-5}
0.1	1.1×10^{1}	6.0×10^{-3}
0.01	1.1×10^{3}	6.0×10^{-1}

3.5. The transition frequency from E_2 to E_1 is given by

$$h\nu = E_2 - E_1 = 13.6\left(1 - \frac{1}{4}\right)\ \text{eV}$$

Hence ν and λ are evaluated as

$$\nu = \frac{10.2\,\text{eV}}{h} = \frac{10.2\,\text{eV}}{4.136 \times 10^{-15}\,\text{eV s}} \simeq 2.47 \times 10^{15}\,\text{s}^{-1},$$

$$\lambda = \frac{c}{\nu} = 1.21 \times 10^{-7}\,\text{m} = 121\,\text{nm}$$

The spread in ν and λ due to the finite lifetime τ can be estimated by using the uncertainty relation as

$$\Delta\nu \approx \frac{h}{\tau \cdot h} \simeq \frac{1}{10^{-8}} = 10^{8}\,\text{Hz}; \quad \Delta E\tau = (h\Delta\nu)\tau \approx h$$

$$\Delta\lambda = \frac{c}{\nu - (\Delta\nu/2)} - \frac{c}{\nu + (\Delta\nu/2)} \approx \lambda\left(\frac{\Delta\nu}{\nu}\right)$$

3.6. The Hamiltonian reads as

$$\hat{H} = -\frac{\hbar^2}{2m}\nabla^2 + V(\underline{r}), \quad V(r) = V^*(r), \quad \nabla^2 = \frac{\partial^2}{\partial x^2} + \frac{\partial^2}{\partial y^2} + \frac{\partial^2}{\partial z^2}$$

where the potential is real. Hence given the two well behaving functions, f, g one can show that V is Hermitian, that is,

$$\int d\underline{r} f^* V g \equiv \int d\underline{r}(Vf)^* g$$

The x-component of the Laplacian can also be shown Hermitian by repeated use of the integration by parts:

$$\int d\underline{r} f^* \frac{\partial^2 g}{\partial x^2} = f^* \frac{\partial g}{\partial x}\bigg|_{-\infty}^{\infty} - \int d\underline{r}\left(\frac{\partial g}{\partial x}\right)\left(\frac{\partial f^*}{\partial x}\right)$$

$$= -g\frac{\partial f^*}{\partial x}\bigg|_{-\infty}^{\infty} + \int d\underline{r} g\frac{\partial^2 f^*}{\partial x^2} = \int dr\left(\frac{\partial^2 f^*}{\partial x^2}\right)g$$

where use has been made of $f(\pm\infty) = g(\pm\infty) = 0$ and the y and z components can likewise be proven hermitian. Therefore the Hamiltonian is Hermitian.

Chapter 4

4.1 (a) The traveling wave representation of the wavefunction,

$$\Psi(\underline{r}, t) = e^{-i\omega t} u(\underline{r}) = \frac{1}{L^{3/2}} e^{-i(\omega t - i\underline{k} \cdot \underline{r})}, \quad \omega = \frac{E}{\hbar}$$

satisfies the Schrödinger equation of a free particle, since

$$i\hbar \frac{\partial}{\partial t} \Psi(\underline{r}, t) = \hbar\omega e^{-i(\omega t - \underline{k} \cdot \underline{r})}, \quad E = \hbar\omega$$

$$-\frac{\hbar^2}{2m} \nabla^2 e^{-i(\omega t - \underline{k} \cdot \underline{r})} = \frac{\hbar^2 k^2}{2m} e^{-i(\omega t - \underline{k} \cdot \underline{r})}, \quad \hat{H} = -\frac{\hbar^2}{2m} \nabla^2$$

and the total energy of a free particle consists solely of the kinetic energy given by

$$E = \frac{p^2}{2m} = \frac{(\hbar k)^2}{2m}$$

(b) The number of particles between k and $k + \mathrm{d}k$ for the cases of 3D, 2D, and 1D is given by

$$g_{3\mathrm{D}}(k) dk = \frac{k^2 dk}{\pi^2}, \quad g_{2\mathrm{D}}(k) dk = \frac{k dk}{\pi}, \quad g_{1\mathrm{D}}(k) dk = \frac{2}{\pi} dk$$

The dispersion relation between E and k of a 3D free particle is given by

$$E = \frac{(\hbar k)^2}{2m}, \quad \text{or} \quad k = \frac{1}{\hbar}(2mE)^{1/2}$$

so that

$$\frac{k^2 dk}{\pi^2} = \frac{1}{\pi^2} \left(\frac{2mE}{\hbar^2}\right) \left[\frac{1}{\hbar}\left(\frac{m}{2E}\right)^{1/2} dE\right]$$

$$\equiv g_{3\mathrm{D}} dE; \quad g_{3\mathrm{D}}(E) = \frac{\sqrt{2} m^{3/2} E^{1/2}}{\pi^2 \hbar^3}$$

Similarly one finds

$$g_{2\mathrm{D}}(k) dk = \frac{k dk}{\pi} = \frac{1}{\pi} \frac{(2mE)^{1/2}}{\hbar} \left[\frac{1}{\hbar}\left(\frac{m}{2E}\right)^{1/2} dE\right]$$

$$\equiv g_{2\mathrm{D}} dE, \quad g_{2\mathrm{D}}(E) = \frac{m}{\pi \hbar^2}$$

$$g_{1\mathrm{D}}(k) dk = \frac{2dk}{\pi} = \frac{2}{\pi} \left[\frac{1}{\hbar}\left(\frac{m}{2E}\right)^{1/2} dE\right]$$

$$\equiv g_{1\mathrm{D}} dE, \quad g_{1\mathrm{D}}(E) = \frac{\sqrt{2} m^{1/2}}{\pi \hbar} \frac{1}{E^{1/2}}$$

4.2. (a) The energy eigenfunction is given from Eq. (4.9) by

$$u_n(x, y, z) = \left(\frac{2}{L}\right)^{3/2} \sin\left(\frac{n_x \pi}{L} x\right) \sin\left(\frac{n_y \pi}{L} y\right) \sin\left(\frac{n_z \pi}{L} z\right); \quad \frac{n_\alpha \pi}{L} = k_\alpha, \quad \alpha = x, y, z$$

The total number of states between k and $k + \mathrm{d}k$ is given by

$$\frac{[(4\pi k^2 dk)/8]}{(\pi/L)^3}$$

where k_x, k_y, k_z values should be confined to positive values to avoid the eigenfunction to be redundant, so that the spherical shell in the k-space should be divided by 8. Hence one can write

$$g_{3D}(k)dk \equiv 2\frac{4\pi k^2 dk}{8(\pi/L)^3}\frac{1}{L^3} = \frac{k^2 dk}{\pi^2}$$

in agreement with the expression of g_{3D} obtained with the use of the traveling wave representation of the eigenfunction.

(b) The wave vector k is related to the frequency by

$$k = \frac{2\pi}{\lambda} = \frac{2\pi}{c}\nu$$

so that the density of states in ν space is given by

$$\frac{k^2 dk}{\pi^2} = \frac{(2\pi\nu/c)^2(2\pi/c)d\nu}{\pi^2} = \left(\frac{8\pi\nu^2}{c^3}\right)d\nu$$

in precise agreement with the density of states used in Eq. (2.1).

4.3. (a) The energy eigenvalue of the electron in a cubic box of length L is given from Eq. (4.9b) by

$$E_n = \frac{\hbar^2\pi^2}{2mL^2}(n_x^2 + n_y^2 + n_z^2), \quad n_x, n_y, n_z = 1, 2, 3 \ldots$$

Now for $L = 1$ m

$$\frac{\hbar^2\pi^2}{2mL^2} = \frac{(1.055\times 10^{-34})^2\pi^2}{2\times 9.106\times 10^{-31}1^2} = 6.0\times 10^{-38}(\mathrm{J/m^2}) = 0.4(\mathrm{eV/nm^2})$$

Hence the lowest three energy levels with lengths 1, 0.1 nm are:

	n_x	n_y	n_z	$E(L = 1\ \mathrm{nm})(\mathrm{eV})$	$E(L = 0.1\ \mathrm{nm})(\mathrm{eV})$
E_1	1	1	1	1.2	120
E_2	2	1	1	2.4	720
	1	2	1	2.4	720
	1	1	2	2.4	720
E_3	2	2	1	3.6	1080
	2	1	2	3.6	1080
	1	2	2	3.6	1080

with the degeneracy of 1, 3, 3 respectively.

(b) The respective ground state energies are larger than the room temperature thermal energy of 25 meV by the factors of 48, 4800.

4.4. (a) The problem can be solved by repeating the analysis discussed in 4.3 and inserting the height and width of the quantum well given

(b) The algorithm can be written based on the graphical method presented in the text.

4.5. (a) The energy eigenequation of a particle in a quantum well is given from Eq. (4.25) by

$$\left[-\frac{\hbar^2}{2m_x}\frac{\partial^2}{\partial x^2}-\frac{\hbar^2}{2m_y}\frac{\partial^2}{\partial y^2}-\frac{\hbar^2}{2m_z}\frac{\partial^2}{\partial z^2}+V(z)\right]u(x,y,z)=Eu(x,y.z)$$

One can look for the solution in the form

$$u(x,y,z)=X(x)Y(y)Z(z)$$

and insert it into the eigen equation and divide both side by $u(x, y, z)$, obtaining

$$\left[-\frac{\hbar^2}{2m_x}\frac{X''}{X}\right]+\left[-\frac{\hbar^2}{2m_y}\frac{Y''}{Y}\right]+\left[-\frac{\hbar^2}{2m_z}\frac{Z''}{Z}+V(z)\right]=E$$

where the double primes denote the second order differentiations with respect to x, y, z.
Since each bracket on the left hand side depends solely on x, y, z respectively, the only way to satisfy the equation is to put each bracket to a constant. In this case there ensues three independent ID equations,

$$\left(\frac{-\hbar^2}{2m_x}\right)\frac{X''}{X}=E_x \text{ or } X''+k_x^2X=0,\ k_x^2\equiv\frac{2m_xE_x}{\hbar^2}=\frac{p_x^2}{\hbar^2} \tag{4.1}$$

$$\left(\frac{-\hbar^2}{2m_y}\right)\frac{Y''}{Y}=E_y \text{ or } Y''+k_y^2Y=0,\ k_y^2\equiv\frac{2m_yE_y}{\hbar^2}=\frac{p_y^2}{\hbar^2} \tag{4.2}$$

$$\left[-\frac{\hbar^2}{2m_z}\frac{Z''}{Z}+V(z)\right]=E_z \text{ or } -\frac{\hbar^2}{2m_z}Z''+V(z)Z=E_zZ \tag{4.3}$$

with the total energy given by the sum of the three kinetic energies,

$$E_x+E_y+E_z=E$$

Evidently Eqs. (4.1) and (4.2) are the eigenequations of a free particle (see Eq. (4.2)), while Eq. (4.3) is that of a particle in a quantum well (see Eq. (4.19)). Therefore Z can be represented by Eq. (4.20) with x replaced by z while X, Y are specified in terms of the trigonometric functions, for example,

$$X(x)\propto\exp(\pm ik_xx),\quad Y(y)\propto\exp(\pm ik_yy)$$

The total energy E then consists of the kinetic energies E_x, E_y and the sublevel E_z in the quantum well of width W:

$$E_n = \frac{\hbar^2 k_x^2}{2m_x} + \frac{\hbar^2 k_y^2}{2m_y} + \frac{\hbar^2 \pi^2}{2m_z W^2} n^2, \quad n = 1, 2, \ldots$$

For simplicity the well depth has been taken infinite and Eq. (4.4) has been used.

(b) For the case of the quantum wire one can likewise write

$$X(x) \propto \exp(\pm i k_x x)$$

and express Y, Z by transcribing Eq. (4.20). The resulting total energy is then given by

$$E_{n,m} = \frac{\hbar^2 k_x^2}{2m_x} + \frac{\hbar^2 \pi^2}{2m_y W_y^2} n^2 + \frac{\hbar^2 \pi^2}{2m_z W_z^2} m^2, \quad n, m = 1, 2, \ldots$$

Chapter 5

5.1. (a) The energy eigenequation reads as

$$\left[-\frac{\hbar^2}{2m} \frac{\partial^2}{\partial x^2} + V(x) \right] u(x) = E u(x)$$

with $V(x)$ given by 0, V_1, V_2 in the interval $x < 0$, $0 \leq x < d_1$, $d_1 < x$ respectively. Since $E > V_2 > V_1$ the energy eigenequations correspond to those of the free particle. Thus one can write

$$u'' + \alpha^2 u = 0;$$

with

$$\alpha = \begin{cases} k_0, & \hbar^2 k_0^2 / 2m = E \\ k_1, & \hbar^2 k_1^2 / 2m = E - V_1 \\ k_2, & \hbar^2 k_2^2 / 2m = E - V_2 \end{cases}$$

The corresponding eigenfunctions are given in analogy with Eq. (5.4) by

$$u = \begin{cases} i_0 e^{i k_0 x} + r e^{-i k_0 x} \\ A e^{i k_1 x} + B e^{-i k_1 x} \\ t e^{i k_2 x} \end{cases}$$

where i_0, r, t represent the incident, reflected and transmitted beams. These constants are used to satisfy the usual boundary conditions applied at 0, d_1:

$$i_0 + r = A + B \tag{5.1a}$$

$$i k_0 (i_0 - r) = i k_1 (A - B) \tag{5.1b}$$

$$Ae^{ik_1d_1} + Be^{-ik_1d_1} = te^{ik_2d_1} \tag{5.1c}$$

$$ik_1(Ae^{ik_1d_1} - Be^{-ik_1d_1}) = ik_2te^{ik_2d_1} \tag{5.1d}$$

The unknown constants r, A, B, t can be found in terms of i_0. For this purpose one can perform the operation, $(5.1a) \pm (5.1b)/ik_1$ and obtain

$$A = \frac{1}{2}e^{-ik_1d_1+ik_2d_1}\left(1+\frac{k_2}{k_1}\right)t \tag{5.2a}$$

$$B = \frac{1}{2}\left[i_0\left(1-\frac{k_0}{k_1}\right)+r\left(1+\frac{k_0}{k_1}\right)\right] \tag{5.2b}$$

The constants A, B can also be found in a similar manner from Eqs. (5.1c) and (5.1d):

$$A = \frac{1}{2}e^{-ik_1d_1+ik_2d_1}\left(1+\frac{k_2}{k_1}\right)t \tag{5.3a}$$

$$B = \frac{1}{2}e^{ik_1d_1+ik_2d_1}\left(1-\frac{k_2}{k_1}\right)t \tag{5.3b}$$

Therefore by equating A, B as obtained in Eqs. (5.2) and (5.3) one can write

$$e^{-ik_1d_1+ik_2d_1}\left(1+\frac{k_2}{k_1}\right)t - \left(1-\frac{k_0}{k_1}\right)r = \left(1+\frac{k_0}{k_1}\right)i_0 \tag{5.4a}$$

$$e^{ik_1d_1+ik_2d_1}\left(1-\frac{k_2}{k_1}\right)t - \left(1+\frac{k_0}{k_1}\right)r = \left(1-\frac{k_0}{k_1}\right)i_0 \tag{5.4b}$$

and find r, t in terms of i_0 as

$$\frac{t}{i_0} = e^{-ik_2d_1}\frac{2k_0/k_1}{D} \tag{5.5a}$$

$$\frac{r}{i_0} = \frac{\frac{k_0-k_2}{k_1}\cos k_1d_1 + i\left(1-\frac{k_0k_2}{{k_1}^2}\right)\sin k_1d_1}{D} \tag{5.5b}$$

$$D = \frac{k_0+k_2}{k_1}\cos k_1d_1 - i\left(1+\frac{k_0k_2}{{k_1}^2}\right)\sin k_1d_1 \tag{5.5c}$$

Hence the transmission and reflection coefficients are given from Eq. (5.10) in the text by

$$R \equiv \frac{\hbar k_0/m}{\hbar k_0/m}\cdot\left|\frac{r}{i_0}\right|^2 = \frac{1}{|D|^2}\left[\left(\frac{k_0-k_2}{k_1}\right)^2 + \left(1-\frac{k_0^2+k_2^2}{{k_1}^2}+\frac{k_0^2k_2^2}{{k_1}^4}\right)\sin^2(k_1d_1)\right] \tag{5.6a}$$

$$T \equiv \frac{\hbar k_2/m}{\hbar k_0/m}\left|\frac{t}{i_0}\right|^2 = \frac{4k_0k_2/k_1^2}{|D|^2} \tag{5.6b}$$

and $R + T = 1$, as it should. In this case no adjustment of k_0, k_1, k_2 can be made for the 100% transmission.

(b) The traveling wave representation of the particle incident on the potential barrier V_2 on the x–z plane is given by

$$\Psi(\underline{r}, t) \sim e^{-(i\omega t - \underline{k} \cdot \underline{r})} = e^{-i\omega t} u(\underline{r})$$

The wave vectors associated with the incident, reflected and transmitted beams are characterized by

$$u_i(x, z) \sim \exp i(k_1 \sin \theta_i x + k_1 \cos \theta_i z), \quad k_1 = \frac{\sqrt{2m(E - V_1)}}{\hbar} \tag{5.7a}$$

$$u_r(x, z) \sim \exp i(k_1 \sin \theta_r x - k_1 \cos \theta_r z), \quad k_1 = \frac{\sqrt{2m(E - V_1)}}{\hbar} \tag{5.7b}$$

$$u_t(x, z) \sim \exp i(k_2 \sin \theta_t x + k_2 \cos \theta_t z), \quad k_2 = \frac{\sqrt{2m(E - V_2)}}{\hbar} \tag{5.7c}$$

where θ_i, θ_r, θ_t are the incident, reflected and transmitted angles, respectively. Since the boundary condition requires that $u_i(x, 0) = u_t(x, 0) = u_r(x, 0)$ it follows from Eq. (5.7) that

$$\theta_i = \theta_r \tag{5.8a}$$

$$k_1 \sin \theta_i = k_2 \sin \theta_2 \tag{5.8b}$$

Evidently Eqs. (5.8a) and (5.8b) represent the law of reflection and reflection of light with the wave vector k_1, k_2 associated with different index of reflection n_1, n_2 appearing in the Snell's law.

5.2. (a) The change in time of the probability density can be accounted for by using the Schrödinger equation as

$$\begin{aligned}
\frac{\partial}{\partial t}\psi^*\psi &= \psi^*\left(\frac{\partial}{\partial t}\psi\right) + \left(\frac{\partial}{\partial t}\psi^*\right)\psi \\
&= \psi^*\frac{1}{i\hbar}\left[\frac{-\hbar^2}{2m}\nabla^2\psi\left(\underline{r}, t\right) + V(\underline{r})\psi(\underline{r}, t)\right] \\
&\quad + \Psi\left(\frac{1}{-i\hbar}\right)\left[\frac{-\hbar^2}{2m}\nabla^2\psi^*\left(\underline{r}, t\right) + V(\underline{r})\psi^*(\underline{r}, t)\right] \\
&= -\left[\psi^*\frac{\hbar}{2mi}\nabla^2\psi\left(\underline{r}, t\right) - \Psi\nabla^2\psi^*(\underline{r}, t)\right]
\end{aligned} \tag{5.9}$$

One can then use the vector identity,

$$\psi^*\nabla^2\psi - \psi\nabla^2\psi^* \equiv \nabla \cdot (\psi^*\nabla\psi - \psi\nabla\psi^*)$$

and recast Eq. (5.9) into a form

$$\frac{\partial}{\partial t}\psi^*\psi = -\nabla \cdot \underline{S}; \quad \underline{S} \equiv \frac{\hbar}{2mi}(\psi^*\nabla\psi - \psi\nabla\psi^*)$$

in which case S represents the probability current density.

(b) With the use of the eigenfunction in Eq. (5.4) one can specify the probability current densities as

$$\underline{S}_I \equiv \frac{\hbar}{2mi}\left[\left(i_0 e^{-ik_0x} + re^{ik_0x}\right)\frac{\partial}{\partial x}(i_0 e^{ik_0x} - r^{-ik_0x}) - c.c\right]$$

$$= \frac{\hbar}{2mi}[(i_0 e^{-ik_0x} + re^{ik_0x})ik_0(i_0 e^{ik_0x} - r^{-ik_0x}) - c.c] = \frac{\hbar k_0}{m}|i_0|^2 - \frac{\hbar k_0}{m}|r|^2$$

$$\underline{S}_{II} \equiv \frac{\hbar}{2mi}\left(te^{ikx}\frac{\partial}{\partial x}te^{ikx} - c.c\right) = \frac{\hbar k}{m}|t|^2$$

5.3. (a) One can find A, B from Eqs. (5.16a) and (5.16b) and also from Eqs. (5.16c) and (5.16d):

$$A = -\frac{e^{-i\alpha}(i_0 e^{-i\alpha_0} + re^{i\alpha_0}) - te^{i(\alpha_0+\alpha)}}{2i\sin 2\alpha}; \quad A = \frac{k_0}{k}\frac{[-e^{-i\alpha}(i_0 e^{-i\alpha_0} - re^{i\alpha_0})] + te^{i(\alpha_0+\alpha)}}{2i\sin 2\alpha}$$

$$B = -\frac{te^{i(\alpha_0-\alpha)} - e^{i\alpha}(i_0 e^{-i\alpha_0} + re^{i\alpha_0})}{2i\sin 2\alpha}; \quad B = \frac{k_0}{k}\frac{te^{i(\alpha_0-\alpha)} - e^{i\alpha}(i_0 e^{-i\alpha_0} + re^{i\alpha_0})}{2i\sin 2\alpha}$$

By equating the two expressions of A and B there result two coupled equations involving r, t and these parameters can be found in terms of i_0 in agreement with Eq. (5.17). Since the velocity of the particle $\propto k_0$ is the same in the regions outside the quantum well, R, T as given by

$$R = \left|\frac{r}{i_0}\right|^2, \quad T = \left|\frac{t}{i_0}\right|^2$$

lead to Eq. (5.18) in the text.

(b) When a particle is incident on a potential barrier of height V and thickness d, the reflection and the transmission coefficients can be found in strict analogy with the corresponding R and T operative for the quantum well. Specifically the eigenfunction is given in this case by

$$u(x) = \begin{cases} i_0 \exp\left(ik_0x\right) + r\exp-(ik_0x) & x < -d/2 \\ A\exp(-\kappa x) + B\exp(\kappa x) & |x| \le 2/d \\ t\exp ik_0x & x \ge d/2 \end{cases}$$

where the only modification required is to replace k by κ defined as

$$\kappa = \sqrt{\frac{2m(V-E)}{\hbar^2}} \quad \text{for} \quad E < V$$

One can then transcribe Eq. (5.17) in the text by replacing k by κ and obtain

$$\frac{t}{i_0} = \exp(-ik_0 d)\frac{2k_0\kappa}{2k_0\kappa\cosh(\kappa d) - i(k_0^2 - \kappa^2)\sinh kd}$$

$$\frac{r}{i_0} = -i\exp(-ik_0 d)\frac{(\kappa^2 + k_0^2)\sinh(2\kappa a)}{2k_0\kappa\cosh(\kappa d) - i(k_0^2 - \kappa^2)\sinh\kappa d}$$

with the use of the trigonometric identities

$$\sin i\kappa = i\sinh\kappa,\quad \cos i\kappa = \cosh\kappa,\quad \cosh^2 x = 1 + \sinh^2 x$$

Since the velocity of the particle outside the potential barrier is the same the tunneling probability T and the reflection coefficient as given by

$$T = \left|\frac{t}{i_0}\right|^2,\quad R = \left|\frac{r}{i_0}\right|^2$$

lead to Eq. (5.22) in the text.

5.4. (a) For infinite barrier height the ground state energy is given from Eq. (4.9) by

$$E_1 = \frac{\hbar^2\pi^2}{2mW^2} = \frac{(1.055\times 10^{-34})^2(\mathrm{J\,s})^2\pi^2}{2\times 9.109\times 10^{-31}\,\mathrm{kg}\,(\mathrm{m}^2)} = 6\times 10^{-37}\,\mathrm{J/(m^2)}$$
$$= 3.7\times 10^{-20}\,\mathrm{eV/(m^2)} = 3.7\times 10^{-2}\,\mathrm{eV/(nm^2)}$$

Hence for $W = 1$, 10 nm the ground state energies are 0.037 and 3.7×10^{-4} eV, respectively.

(b) The lifetime τ can be estimated by

$$\tau = \tau_{rt}\times\tilde{n};\quad T\tilde{n} = 1$$

where τ_{rt} is the round trip time of the electron in the quantum well, T the tunneling probability and $\tilde{n}$ the number of the electron encountering the barrier. Obviously the lifetime is dictated by the thinner barrier, since T is greater. The tunneling probability is given in this case by

$$T = \frac{1}{1+\Lambda};\quad \Lambda = \frac{V^2}{4E(V-E)}\sinh^2 d\sqrt{\frac{2m}{\hbar^2}(V-E)}$$

(see Eq. (5.18)). Also τ_{tr} is specified by the roundtrip distance $2W$ and the thermal velocity of the electron, provided the thermal energy is greater than the ground state energy E_1 of the quantum well.

$$\tau_{rt} = \frac{2W}{v_T},\quad \frac{m}{2}v_T^2 = k_B T$$

If E_1 is greater than $k_B T$ τ_{rt} is determined by

$$\tau_{rt} = \frac{2W}{v}, \quad \frac{m}{2}v^2 = E_1$$

In this manner one can estimate the lifetime by using the parameters given.

5.5. When the electric field is applied, the square potential barrier depends on x as

$$V(x) = V - q|\mathrm{E}|x$$

where $q|\mathrm{E}|$ is the force acting on the electron. Hence the barrier is transformed into either trapezoidal or triangular shape, depending on the kinetic energy of the electron incident on the barrier and thickness of the barrier. The triangular shape ensues if the barrier height at the outer edge is less than the ground state energy E_1, that is,

$$V - q|\mathrm{E}|d \leq E_1, \quad E_1 = \frac{\hbar^2\pi^2}{2mW^2}$$

where d is the thickness of the barrier. The electric field required to transform the square barrier into the triangular shape can be estimated for $V = 3.1$ eV and $d = 10$ nm by

$$|\mathrm{E}| \approx \frac{V}{q \cdot d} \approx \frac{3.1\ \mathrm{eV} \cdot 1.6 \times 10^{-19}\ \mathrm{J eV^{-1}}}{1.6 \times 10^{-19} C \cdot 10 \times 10^{-9}\ \mathrm{m}} = 3.1 \times 10^7\ \mathrm{V m^{-1}} = 31\ \mathrm{mV nm^{-1}}$$

where E_1 has been taken zero. If E is greater than the value thus estimated the F-N tunneling ensues with the probability given from Eq. (5.26) by

$$T \simeq \frac{4\sqrt{2m}}{2q|\mathrm{E}|\hbar}(V - E)^{3/2}, \quad E \approx 0 \tag{5.10}$$

In this expression E denotes either E_1 or $k_B T$ depending on the relative magnitude but both quantities are small and have been discarded. One can find the lifetime based on the F-N tunneling by assuming that E applied is greater than the estimated value. For the well width of 50 nm the ground state energy is less than the thermal energy as can be readily estimated. Hence one can find the lifetime based on

$$\tau = \frac{2W}{v_T} \cdot \frac{1}{T} \tag{5.11}$$

with T given by Eq. (5.10). (b) One can calculate the strength of E which is required to shorten the lifetime to 1 μs combining Eqs. (5.10) and (5.11):

$$10^{-3} = \frac{2W}{v_T} \cdot \frac{1}{T} = \frac{2 \cdot 50 \times 10^{-9}(\mathrm{m})}{1.16 \times 10^5(\mathrm{m\,s^{-1}})} \exp \frac{4\sqrt{2m}}{3q|\mathrm{E}|\hbar} V^{3/2}$$

and obtain $|\mathrm{E}| \simeq 1.78 \times 10^9\,\mathrm{V\,m^{-1}} = 1.78\,\mathrm{V\,nm^{-1}}$.

5.6. (a) The electric field by which to induce the given tunneling probability can be found by putting

$$10^{-4} = \exp -\frac{4\sqrt{2m}}{3q|\mathrm{E}|\hbar} V^{3/2}$$

where V in this case represents the work function, that is, $V = 4.5$ V. Hence upon inserting V one finds

$$|\mathrm{E}| \simeq 7.05 \times 10^9\,\mathrm{V\,m^{-1}} = 7.05\,\mathrm{V\,nm^{-1}}$$

When the field amplitude is multiplied by the distance between the tip and the sample surface one finds

$$|\mathrm{E}| \times 1\,\mathrm{nm} = 7.5\ \mathrm{V}$$

and is greater than the work function forming the barrier. Therefore the use of the F-N tunneling for estimating E is proven correct.

(b) Hence the distance d is given by 50 V/d (nm) = 7.5 V/nm, that is , $d = 6.7$ nm.

To use FN tunneling the voltage between the sample and probe tip should be equal to the work function.

$$\frac{50V}{d} = 7.5\mathrm{V/nm}, \quad d = 6.7\ \mathrm{nm}$$

5.7. The boundary condition

$$u_j(z_j) = u_{j+1}(z_j)$$
$$\frac{\partial}{\partial z}u_j(z_j) = \frac{\partial}{\partial z}u_{j+1}(z_j)$$

when applied to the eigenfunction given in Eq. (5.27) yields

$$\begin{aligned} A_j e^{ikz_j} + B_j e^{-ikz_j} &= A_{j+1}e^{-\kappa z_j} + B_{j+1}e^{\kappa z_j} \\ A_j e^{ikz_j} - B_j e^{-ikz_j} &= \frac{i\kappa}{k}[A_{j+1}e^{-\kappa z_j} - B_{j+1}e^{\kappa z_j}] \end{aligned} \tag{5.12}$$

One can thus find A_j, B_j in terms of A_{j+1}, B_{j+1} as

$$\begin{aligned} A_j &= \frac{e^{-ikz_j}}{2}\left[A_{j+1}e^{-\kappa z_j}\left(1+\frac{i\kappa}{k}\right) + B_{j+1}e^{\kappa z_j}\left(1-\frac{i\kappa}{k}\right)\right] \\ B_j &= \frac{e^{ikz_j}}{2}\left[A_{j+1}e^{-\kappa z_j}\left(1-\frac{i\kappa}{k}\right) + B_{j+1}e^{\kappa z_j}\left(1+\frac{i\kappa}{k}\right)\right] \end{aligned} \tag{5.13}$$

Equation (5.13) can be put into a matrix notation as

$$\begin{pmatrix} A_j \\ B_j \end{pmatrix} = \begin{pmatrix} \alpha & \beta \\ \alpha^* & \beta^* \end{pmatrix} \begin{pmatrix} A_{j+1} \\ B_{j+1} \end{pmatrix} \tag{5.14a}$$

$$\alpha = \frac{1}{2}\left(1 + \frac{i\kappa}{k}\right) e^{i(i\kappa-k)z_j}$$
$$\beta = \frac{1}{2}\left(1 - \frac{i\kappa}{k}\right) e^{i(i\kappa+k)z_j} \tag{5.14b}$$

The constants A_{j+1}, B_{j+1} can be connected to A_{j+2}, B_{j+2} by applying the boundary condition at $z_j + d$ to the eigenfunctions u_{j+1}, u_{j+2}:

$$\begin{pmatrix} A_{j+1} \\ B_{j+1} \end{pmatrix} = \begin{pmatrix} \gamma & \gamma^* \\ \delta & \delta^* \end{pmatrix} \begin{pmatrix} A_{j+2} \\ B_{j+2} \end{pmatrix} \tag{5.15a}$$

$$\gamma = \left(1 + \frac{k}{i\kappa}\right) e^{i(k-i\kappa)(z_j+d)}$$
$$\delta = \left(1 + \frac{k}{(-i\kappa)}\right) e^{i(k+i\kappa)(z_j+d)} \tag{5.15b}$$

Therefore by combining Eqs. (5.14) and (5.15) one obtains

$$\begin{pmatrix} A_j \\ B_j \end{pmatrix} = \begin{pmatrix} \alpha & \beta \\ \alpha^* & \beta^* \end{pmatrix} \begin{pmatrix} \gamma & \gamma^* \\ \delta & \delta^* \end{pmatrix} \begin{pmatrix} A_{j+2} \\ B_{j+2} \end{pmatrix}$$
$$= \begin{pmatrix} m_{11}(d) & m_{12}(z_j, d) \\ m_{21}(z_j, d) & m_{22}(d) \end{pmatrix} \begin{pmatrix} A_{j+2} \\ B_{j+2} \end{pmatrix} \tag{5.16a}$$

with

$$m_{11}(d) = m_{22}^* = \alpha\gamma + \beta\delta$$
$$= e^{ikd}\left(\cosh \kappa d - i\frac{k^2 - \kappa^2}{2k\kappa} \sinh \kappa d\right) \tag{5.16b}$$

$$m_{12}(z_j, d) = m_{21}^* = \alpha\gamma^* + \beta\delta^*$$
$$= ie^{-ik(2z_j+d)}\frac{k^2 + \kappa^2}{2k\kappa} \sinh \kappa d \tag{5.16c}$$

In this manner A_j, B_j are connected to A_{j+2}, B_{j+2}. One can then find T, R for the single barrier by putting B_{j+2} to zero, since there is no reflected component once the particle is transmitted across the single barrier:

$$\begin{pmatrix} A_j \\ B_j \end{pmatrix} = \begin{pmatrix} m_{11}(d) & m_{12}(z_j, d) \\ m_{21}(z_j, d) & m_{22}(d) \end{pmatrix} \begin{pmatrix} A_{j+2} \\ 0 \end{pmatrix}$$

One can therefore specify the tunneling probability as

$$T = \left|\frac{A_{j+2}}{A_j}\right|^2 = \frac{1}{|m_{11}(d)|^2}$$

where

$$|m_{11}(d)|^2 = \cosh^2\kappa d + \left(\frac{k^2-\kappa^2}{2k\kappa}\right)^2 \sinh^2\kappa d$$
$$= 1 + \sinh^2\kappa d + \left(\frac{k^2-\kappa^2}{2k\kappa}\right)^2 \sinh^2\kappa d$$
$$= 1 + \frac{(k^2+\kappa^2)^2}{4k^2\kappa^2}\sinh^2\kappa d = 1 + \frac{V^2}{4(V-E)E}\sinh^2 d\sqrt{\frac{2m}{\hbar^2}(V-E)}$$

in agreement with the result Eq. (5.22).

(b) To analyze the tunneling through the two barriers W distance apart in succession one can likewise connect A_j, B_j to A_{j+4}, B_{j+4} and put B_{j+4} to zero:

$$\begin{pmatrix} A_j \\ B_j \end{pmatrix} = \begin{pmatrix} m_{11}(d) & m_{12}(z_j, d) \\ m_{12}^*(z_j, d) & m_{11}^*(d) \end{pmatrix} \begin{pmatrix} m_{11}(d) & m_{12}(z_{j+2}, d) \\ m_{12}^*(z_{j+2}, d) & m_{11}^*(d) \end{pmatrix} \begin{pmatrix} A_{j+4} \\ 0 \end{pmatrix}$$

where the distance between the input and output plane is given by

$$z_{j+2} = z_j + W + d$$

Then one can find the ratio specifying the tunneling through two barriers T_{2B} as

$$\frac{A_{j+4}}{A_j} = \frac{-e^{-2ikd}\cdot 4k^2\kappa^2}{[D_1 - e^{2ikW}(k^2+\kappa^2)\sinh\kappa d]^2}$$

$$D_1 = [(k^2-\kappa^2)\sinh\kappa d + 2ik\kappa\cosh\kappa d]^2$$
$$= [(k^2-\kappa^2)^2\sinh^2\kappa d + 4k^2\kappa^2\cosh^2\kappa d]e^{2i\theta}$$
$$= [4k^2\kappa^2 + (k^2+\kappa^2)^2\sinh^2\kappa d]e^{2i\theta}$$

where

$$\tan\theta = \frac{2k\kappa\cosh\kappa d}{(k^2-\kappa^2)\sinh\kappa d}$$

and D_1 has been expressed in the phasor notation,

$$x + iy = (x^2+y^2)^{1/2}e^{i\theta}, \ \tan\theta = \frac{y}{x}$$

and a trigonometric relation has been used

$$\cosh^2 x = 1 + \sinh^2 x$$

Since the input and output velocities are the same T_{2B} can be expressed as

$$T_{2B} = \left|\frac{A_{j+4}}{A_j}\right|^2 = \frac{(4k^2\kappa^2)^2}{|D_1|^2\left|1 - e^{2i(kW-\theta)}\cdot\frac{(k^2+\kappa^2)^2\sinh^2\kappa d}{|D_1|}\right|^2}$$
$$= \frac{T_{1B}^2}{|1 - e^{2i(kW-\theta)}\cdot R_{1B}|^2} \tag{5.17}$$

where T_{1B}, R_{1B} represent the corresponding quantities for the single barrier

$$T_{1B} \equiv \frac{4k^2\kappa^2}{|D_1|}, \quad R_{1B} \equiv \frac{(k^2+\kappa^2)^2\sinh^2\kappa d}{|D_1|} \tag{5.18}$$

One can further compact T_{2B} as follows.

$$|1 - e^{i\chi}R_{1B}|^2 = [(1-\cos\chi R_{1B})^2 + \sin^2\chi {R_{1B}}^2]e^{2i\varphi}$$
$$= [1 + {R_{1B}}^2 - 2\cos\chi R_{1B}]e^{2i\varphi} = \left[\left(1-R_{1B}\right)^2 + 4R_{1B}\sin^2\frac{\chi}{2}\right]e^{2i\varphi}$$

with

$$\tan\varphi = \frac{\sin\chi R_{1B}}{1-\cos\chi R_{1B}}, \qquad \chi \equiv 2(kW-\theta)$$

In this expression use has been made of

$$\cos\chi = 1 - 2\sin^2\frac{\chi}{2}$$

Hence by combining Eqs. (5.17) and (5.18) one finds

$$T_{2B} = \frac{1}{1 + 4\frac{R_{1B}}{T_{1B}^2}\sin^2(kW-\theta)}, \qquad T_{1B} = 1 - R_{1B}$$

Chapter 6

6.1. (a) One can solve the coupled equation,

$$a_{11}x + a_{12}y = c_1 \tag{6.1a}$$

$$a_{21}x + a_{22}y = c_2 \tag{6.1b}$$

by performing the operation, $[(6.1a)/a_{12}] - [(6.1b)/a_{22}]$, $[(6.1a)/a_{11}] - [(6.1b)/a_{21}]$, obtaining x,y as

$$[(a_{11}/a_{12}) - (a_{21}/a_{22})]x = (c_1/a_{12}) - (c_2/a_{22})$$

$$[(a_{12}/a_{11}) - (a_{22}/a_{21})]y = (c_1/a_{11}) - (c_2/a_{21})$$

The x, y can also be found in terms of the determinants as

$$x = \frac{\begin{vmatrix} c_1 & a_{12} \\ c_2 & a_{22} \end{vmatrix}}{\begin{vmatrix} a_{11} & a_{12} \\ a_{21} & a_{22} \end{vmatrix}} = \frac{c_1a_{22} - c_2a_{12}}{a_{11}a_{22} - a_{12}a_{21}} \tag{6.2a}$$

$$y = \frac{\begin{vmatrix} a_{11} & c_1 \\ a_{21} & c_2 \end{vmatrix}}{\begin{vmatrix} a_{11} & a_{12} \\ a_{21} & a_{22} \end{vmatrix}} = \frac{c_2a_{11} - c_1a_{21}}{a_{11}a_{22} - a_{12}a_{21}} \tag{6.2b}$$

The method using the determinants is known as the *Kramer's rule.*

(b) Given the characteristic equation,

$$(1-\lambda)x_1 + 2x_2 = 0$$
$$2x_1 + (1-\lambda)x_2 = 0 \tag{6.3}$$

one can find x_1, x_2 by using the Kramer's rule as

$$x_1 = \frac{\begin{vmatrix} 0 & a_{12} \\ 0 & a_{22} \end{vmatrix}}{\begin{vmatrix} 1-\lambda & 2 \\ 2 & 1-\lambda \end{vmatrix}} = \frac{0\times a_{22} - 0\times a_{12}}{(1-\lambda)^2 - 4} = 0 \tag{6.4}$$

Likewise $x_2 = 0$, hence the solutions of the homogeneous coupled equation 6.3 are trivial. The only way to get the non-trivial solution is to put the denominator of Eq. (6.4) to zero, that is

$$\begin{vmatrix} 1-\lambda & 2 \\ 2 & 1-\lambda \end{vmatrix} = 0 \quad \text{or} \quad (1-\lambda)^2 - 4 = 0 \tag{6.5}$$

(c) The resulting two roots of this secular equation are given by $\lambda = -1,\ 3$ and when inserted in either of the two coupled equation (6.3), one finds

$$x_2 = x_1;\ \lambda = 3, \quad x_2 = -x_1;\ \lambda = -1 \tag{6.6}$$

Therefore an infinite number of non-trivial solutions have been found for any value of x_1 as long as x_2 is related to x_1 as given by Eq. (6.6).

(d) With the solution thus found the normalization condition reads as

$$x_1^2 + x_2^2 = x_1^2(1+1) = 2x_1^2 = 1$$

Hence $x_1 = x_2 = \sqrt{2},\ x_1 = -x_2 = \sqrt{2}$ and

$$X_1 = \frac{1}{\sqrt{2}}\begin{pmatrix} 1 \\ 1 \end{pmatrix}, \quad X_2 = \frac{1}{\sqrt{2}}\begin{pmatrix} 1 \\ -1 \end{pmatrix}$$

6.2. (a) One can perform the operations,

$$(6.11a) \pm (6.11b)/ik_1, \quad (6.12a) \pm (6.12b)/ik_1$$

and obtain

$$A = \frac{1}{2}e^{-ikd}(\alpha e^{-\kappa d}C + \alpha^* e^{\kappa d}D), \ B = \frac{1}{2}e^{-ikd}(\alpha^* e^{-\kappa d}C + \alpha e^{\kappa d}D) \tag{6.7}$$

$$A = \frac{1}{2}e^{-ik_1 a}(\alpha e^{-\kappa a}C + \alpha^* e^{\kappa a}D), \quad B = \frac{1}{2}e^{ik_1 a}(\alpha^* e^{-\kappa a}C + \alpha e^{\kappa a}D) \tag{6.8}$$

with

$$\alpha \equiv 1 + \frac{i\kappa}{k_1}$$

By equation A and B appearing in Eqs. (6.7) and (6.8) one readily finds with the use of, say the Kramer's rule the results in Eq. (6.15).

(b) Specifying the resulting secular equation (6.16) into the dispersion relation consists simply of rearranging the terms and is therefore a straightforward algebra. But the algebra is rather lengthy and it is a challenge to derive the dispersion relation.

6.3. (a) The sub-band energy of the infinite potential barrier is given from Eq. (4.4) by

$$E_n = \frac{\hbar^2 \pi^2}{2m_n a^2} n^2$$

so that one can write

$$\Delta E = E_2 - E_1 = \frac{\hbar^2 \pi^2}{2m_n a^2} \times 3 = 40 \text{ meV}$$

and therefore a is found as

$$\begin{aligned} a &= \hbar\pi \left(\frac{3}{2m_n 40 \text{ (meV)}} \right)^{1/2} \\ &= 1.055 \times 10^{-34} (\text{J s}) \\ &\times 3.14 \left\{ \frac{3}{2 \times 9.109 \times 10^{-31} 0.07 \text{ (kg)} \times [(40 \times 10^{-3}) \times 1.60 \times 10^{-19} (\text{J})]} \right\}^{1/2} \\ &= 2 \times 10^{-9} \text{m} = 2\text{nm} \end{aligned}$$

(b) The numerical analysis will yield approximately the same ΔE.

Chapter 7

7.1 (a) The eigenequation,

$$\left[-\frac{\hbar^2}{2m} \frac{\partial^2}{\partial x^2} + \frac{1}{2} k x^2 \right] u(x) = E u(x)$$

can be compacted by introducing the dimensionless variable,

$$\xi = \alpha x, \ \alpha \equiv \left(\frac{m\omega}{\hbar} \right)^{1/2}, \ (\alpha) = \left(\frac{M/T}{ML^2/T} \right)^{1/2} = \left(\frac{1}{\text{length}} \right)$$

Then

$$\frac{\partial}{\partial x} = \frac{\partial}{\partial \xi} \frac{\partial \xi}{\partial x} = \alpha \frac{\partial}{\partial \xi}, \quad \frac{\partial^2}{\partial x^2} = \alpha^2 \frac{\partial^2}{\partial \xi^2}$$

so that

$$-\frac{\hbar^2}{2m}\frac{\partial^2}{\partial x^2} = -\frac{\hbar^2}{2m}\left(\frac{m\omega}{\hbar}\right)\frac{\partial^2}{\partial \xi^2} = -\frac{1}{2}\hbar\omega\frac{\partial^2}{\partial \xi^2},$$

$$\frac{1}{2}kx^2 = \frac{1}{2}m\omega^2\frac{\xi^2}{\alpha^2} = \frac{1}{2}m\omega^2\frac{\hbar}{m\omega}\xi^2 = \frac{1}{2}\hbar\omega\xi^2$$

Therefore the eigenequation reads as

$$u'' + (\lambda - \xi^2)u = 0, \quad \lambda \equiv \frac{2E}{\hbar\omega}, \quad u'' \equiv \frac{\partial^2}{\partial \xi^2} \tag{7.1}$$

(b) When the eigenfunction $u(x)$ is sought in the form,

$$u(\xi) = H(\xi)\exp - \left(\frac{1}{2}\xi^2\right)$$

the derivatives of u read as

$$u' = (H' - \xi H)e^{-\xi^2/2}$$

$$u'' = [H'' - H - \xi H' - \xi(H' - \xi H)]e^{-\xi^2/2} = [H'' - 2\xi H' + (\xi^2 - 1)H]e^{-\xi^2/2}$$

When these derivatives are inserted into Eq. (7.1) it is reduced to Eq. (7.5) in the text.

7.2 (a) Given the 3D eigenequation

$$\left[-\frac{\hbar^2}{2m}\nabla^2 + \frac{1}{2}k_x x^2 + \frac{1}{2}k_y y^2 + \frac{1}{2}k_z z^2\right]u(x, y, z) = E(x, y, z) \tag{7.2}$$

one can look for the solution in the form, $u(x, y, z) = u_x(x)u_y(y)u_z(z)$, insert it into Eq. (7.2) and divide both side with $u(x, y, z)$, obtaining

$$\left[\frac{\left(-\frac{\hbar^2}{2m}\right)\frac{\partial^2 u(x)}{\partial x^2}}{u(x)} + \frac{1}{2}kx^2\right] + \left[\frac{\left(-\frac{\hbar^2}{2m}\right)\frac{\partial^2 u(y)}{\partial y^2}}{u(y)} + \frac{1}{2}ky^2\right] + \left[\frac{\left(-\frac{\hbar^2}{2m}\right)\frac{\partial^2 u(z)}{\partial z^2}}{u(z)} + \frac{1}{2}kz^2\right] = E$$

Since each bracket depends solely on x, y, and z, the only way to satisfy the equation is to put each tem to a constant. In this case three independent 1D harmonic oscillator eigenequations ensue

$$-\frac{\hbar^2}{2m}u''(x) + \frac{1}{2}k_x x^2 u(x) = E_x u(x)$$

$$-\frac{\hbar^2}{2m}u''(x) + \frac{1}{2}k_x x^2 u(x) = E_x u(x)$$

$$-\frac{\hbar^2}{2m}u''(y) + \frac{1}{2}k_y x^2 u(y) = E_y u(y)$$

where the primes denote differentiation with respect to x, y, z and

$$E_x + E_y + E_z = E$$

(b) One can therefore use the eigenfunctions and eigenvalues given in Eqs. (7.12) and (7.25) and write

$$u(x,y,z) = \prod_{j=x,y,z} \left(\frac{\alpha_j}{\sqrt{\pi} 2^{n_j} n_j!} \right)^{1/2} e^{-\frac{\xi_j^2}{2}} H_{nj}(\xi_j), \quad \xi_j = \alpha_j j, \alpha_j = \left(\frac{m\omega_j}{\hbar} \right)^{1/2}$$

$$E_{n_x,n_y,n_z} = \sum_{j=x,y,z} \hbar\omega_j \left(n_j + \frac{1}{2} \right), \quad n_j = 0, 1, 2, \ldots$$

(c) For $k_x = k_y = k_z = k$ the total energy is given by

$$E_{n_x,n_y,n_z} = \hbar\omega \left(n_x + n_y + n_z + \frac{3}{2} \right), \quad \omega = \left(\frac{k}{m} \right)^{1/2}$$

The energy level and degeneracy of the three lowest eigenstates are given as follows:

n_x	n_y	n_z	$E/\hbar\omega$	degeneracy
0	0	0	3/2	none
1	0	0	5/2	
0	1	0	5/2	3
0	0	1	5/2	
1	1	0	7/2	
1	0	1	7/2	
0	1	1	7/2	6
2	0	0	7/2	
0	2	0	7/2	
0	0	2	7/2	

7.3. (a) Differentiating the generating function given in Eq. (7.15) with respect to ξ one obtains

$$\frac{\partial}{\partial \xi} G(\xi, s) = 2se^{-s^2+2s\xi} = 2\sum_{n=0}^{\infty} \frac{H_n(\xi)s^{n+1}}{n!} \equiv \sum_{n=0}^{\infty} \frac{H_n'(\xi)s^n}{n!}$$

One can thus single out the coefficients of equal power of n from both sides and obtain

$$\frac{2H_{n-1}}{(n-1)!} = \frac{H_n'}{n!}, \quad \text{or } H_n' = 2nH_{n-1} \tag{7.3}$$

Also the differentiation of G with respect to s leads to

$$\frac{\partial}{\partial s} G(\xi, s) = (-2s + 2\xi)e^{-s^2+2s\xi} = (-2s + 2\xi)\sum_{n=0}^{\infty} \frac{H_n s^n}{n!}$$

$$\equiv \sum_{n=1}^{\infty} \frac{H_n s^{n-1}}{(n-1)!} = \sum_{n=0}^{\infty} \frac{H_{n+1} s^n}{n!}$$

Hence by singling out the coefficients of equal power of n one finds

$$\xi H_n = \frac{1}{2} H_{n+1} + n H_{n-1} \tag{7.4}$$

(b) The variance Δx is defined as

$$\begin{aligned}(\Delta x)^2 = \langle u_n | (x - < x >)^2 | u_n(x) \rangle &= \langle u_n | x^2 - 2x < x > + < x >^2 | u_n(x) \rangle \\ &= \langle u_n | x^2 - < x >^2 | u_n(x) \rangle = \langle u_n | x^2 | u_n(x) \rangle\end{aligned}$$

Since $u_n(x) \times u_n(x)$ is even in x regardless of n, the average value of x is zero. Therefore one can write

$$(\Delta x)^2 = \int_{-\infty}^{\infty} dx u_n(x) x^2 u_n(x) = \left(\frac{\alpha}{\sqrt{\pi} 2^n n!} \right) \frac{1}{\alpha^3} \int_{-\infty}^{\infty} d\xi e^{-\xi^2} \cdot H_n \xi^2 H_n$$

where Eq. (7.25) was used for $u_n(x)$. Now by using the recurrence relation 7.4 one can write

$$(\xi H_n)^2 = \frac{1}{4} H_{n+1}^2 + n^2 H_{n-1}^2 + H_{n+1} H_{n-1}$$

so that the variance can be evaluated by using the orthonormality of u_n given in Eq. (7.23):

$$\begin{aligned}(\Delta x)^2 &= \frac{1}{\alpha^3} \left(\frac{\alpha}{\sqrt{\pi} 2^n n!} \right) \left[\frac{\sqrt{\pi} 2^{n+1} (n+1)!}{4} + \sqrt{\pi} 2^{n-1} n^2 \right] \\ &= \frac{1}{\alpha^2} \left(\frac{n+1}{2} + \frac{n}{2} \right) = \frac{\hbar}{m\omega} \left(n + \frac{1}{2} \right), \quad \alpha = \left(\frac{m\omega}{\hbar} \right)^{1/2}\end{aligned} \tag{7.5}$$

One can likewise evaluate the variance of p_x by following similar steps. One can write

$$\begin{aligned}(\Delta p_x)^2 = \langle u_n | (p_x - \langle p_x \rangle)^2 | u_n \rangle &= \langle u_n | p_x{}^2 - 2 p_x \langle p_x \rangle + \langle p_x \rangle^2 | u_n \rangle \\ &= \langle u_n | p_x^2 | u_n \rangle\end{aligned}$$

where the average value of $p_x \propto \partial / \partial x$ is zero, since the integrand involved is odd in x. To find the variance one has to carry out the integral

$$\begin{aligned}\langle u_n | p_x^2 | u_n \rangle &= -\hbar^2 \left(\frac{\alpha}{\sqrt{\pi} 2^n n!} \right) \alpha \int_{-\infty}^{\infty} d\xi e^{-\xi^2/2} \cdot H_n \frac{\partial^2}{\partial \xi^2} (H_n e^{-\xi^2/2}) \\ &= -\hbar^2 \frac{\alpha^2}{\sqrt{\pi} 2^n n!} \left\{ e^{-\xi^2/2} H_n \frac{\partial}{\partial \xi} \left(H_n e^{-\xi^2/2} \right) \Big|_{-\infty}^{\infty} - \int_{-\infty}^{\infty} d\xi \right. \\ &\quad \left. \times \left[\frac{\partial}{\partial \xi} \left(e^{-\xi^2/2} H_n \right) \right]^2 \right\} \\ &= \hbar^2 \frac{\alpha^2}{\sqrt{\pi} 2^n n!} \int_{-\infty}^{\infty} d\xi \left[\frac{\partial}{\partial \xi} \left(e^{-\xi^2/2} H_n \right) \right]^2\end{aligned}$$

where the integration in parts has been carried out. Now by using the recurrence relations (7.3) and (7.4) one can write

$$\frac{\partial}{\partial\xi}(e^{-\xi^2/2}H_n) = (-\xi H_n + H_n')e^{-\xi^2/2} = \left(-\frac{1}{2}H_{n+1} - nH_{n-1} + 2nH_{n-1}\right)e^{-\xi^2/2}$$
$$= \left(-\frac{1}{2}H_{n+1} + nH_{n-1}\right)e^{-\xi^2/2}$$

Hence the integrations can be carried out simply with the use of the orthonormality of the eigenfunctions and the variance is to be evaluated as

$$\begin{aligned}(\Delta p_x)^2 &= \hbar^2 \frac{\alpha^2}{\sqrt{\pi}2^n n!}\int_{-\infty}^{\infty} d\xi e^{-\xi^2/2}\cdot\left(-\frac{1}{2}H_{n+1} + nH_{n-1}\right)^2 \\ &= \hbar^2 \frac{\alpha^2}{\sqrt{\pi}2^n n!}\left[\frac{1}{4}\sqrt{\pi}2^{n+1}(n+1)! + n^2\sqrt{\pi}2^{n-1}(n-1)!\right] \\ &= \hbar^2\alpha^2\left(n+\frac{1}{2}\right) = \hbar\omega m\left(n+\frac{1}{2}\right), \quad \alpha^2 = \frac{m\omega}{\hbar} \qquad (7.6)\end{aligned}$$

It therefore follows from Eqs. (7.5) and (7.6) that

$$\Delta x \Delta p_x = \left[\frac{\hbar}{m\omega}\left(n+\frac{1}{2}\right)\cdot \hbar m\omega\left(n+\frac{1}{2}\right)\right]^{1/2} = \hbar\left(n+\frac{1}{2}\right)$$

(c) The matrix element of x can also be evaluated as

$$\begin{aligned}\langle u_l|x|u_{l'}\rangle &= \frac{N_l N_{l'}}{\alpha^2}\int_{-\infty}^{\infty} d\xi e^{-\xi^2/2}\cdot H_l \xi H_{l'} e^{-\xi^2/2}, \qquad \xi = \alpha x \\ &= \frac{N_l N_{l'}}{\alpha^2}\int_{-\infty}^{\infty} d\xi e^{-\xi^2}\cdot H_l\left(\frac{1}{2}H_{l'+1} + l'H_{l'-1}\right) \\ &= \left(\frac{\alpha}{\sqrt{\pi}2^l l!}\right)^{1/2}\left(\frac{\alpha}{\sqrt{\pi}2^{l'} l'!}\right)\frac{1}{\alpha^2}\begin{cases}\frac{1}{2}\sqrt{\pi}2^{l'} l', & l' = l-1 \\ l'\sqrt{\pi}2^{l'} l', & l' = l+1 \\ 0 & \text{otherwise}\end{cases}\end{aligned}$$

where the recurrence relation (7.3) has been used.

7.4. (a) The HO oscillating with the amplitude x_0 is described by

$$x(t) = x_0\cos\omega t, \quad \omega = \sqrt{\frac{k}{m}}$$

The kinetic energy averaged over a period of oscillation is then given by

$$\begin{aligned}< K > &= \frac{1}{T}\int_0^T dx\frac{m}{2}v(t)^2 = \frac{mx_0^2\omega^2}{2T}\int_0^T dx\sin^2(\omega t);\ T = \frac{2\pi}{\omega} \\ &= \frac{mx_0^2\omega^2}{4T}\int_0^T dx(1-\cos(2\omega t)) = \frac{mx_0^2\omega^2}{4} \qquad (7.7)\end{aligned}$$

Also the average potential energy can be evaluated as

$$< V > = \frac{1}{T}\int_0^T dx \frac{k}{2}x(t)^2 = \frac{kx_0^2}{2T}\int_0^T dx\cos^2(\omega t); \quad T = \frac{2\pi}{\omega}$$
$$= \frac{kx_0^2}{4T}\int_0^T dx(1+\cos(2\omega t)) = \frac{mx_0^2\omega^2}{4}; \quad k = m\omega^2 \tag{7.8}$$

The total energy can be specified by the potential energy at $x = x_0$ at which point there is no kinetic energy, that is,

$$E = \frac{kx_0^2}{2} = \frac{m\omega^2 x_0^2}{2}$$

It follows from Eqs. (7.7) and (7.8) that the total energy is equally partitioned into V, K over a period of oscillation:

$$E =< K > + < V >$$

(b) The average kinetic and potential energies in the nth eigenstate are given by

$$< K >= \langle u_n | \frac{p_x^2}{2m} | u_n \rangle = \frac{1}{2m}\hbar\omega m\left(n+\frac{1}{2}\right) = \frac{1}{2}E_n$$
$$< V >= \langle u_n | \frac{k}{2}x^2 | u_n \rangle = \frac{k}{2}\frac{\hbar}{m\omega}\left(n+\frac{1}{2}\right) = \frac{\hbar\omega}{2}\left(n+\frac{1}{2}\right) = \frac{1}{2}E_n, \; k = m\omega^2$$

Here the integrals involved have been carried out in the previous problem. Therefore

$$E_n =< K > + < V >$$

in agreement with the classical results.
(c) The total energy is equally partitioned into $<K>$ and $<V>$ in both descriptions.

7.5. (a) The Hamiltonian of the internal motion of the diatomic molecules is generally modeled by that of the harmonic oscillator

$$\hat{H} = -\frac{\hbar^2}{2\mu}\nabla^2 + \frac{1}{2}kx^2, \quad \frac{1}{\mu} = \frac{1}{m_C} + \frac{1}{m_O}$$

where m_C, m_O represent in this case the mass of the carbon and oxygen molecules. The energy spacing between two nearest vibrational states is therefore given by

$$\hbar\omega = h\nu = hc\tilde{\nu}, \quad \tilde{\nu} \equiv \frac{1}{\lambda}$$

where the inverse wavelength denotes the wave number. Hence ν can be found from the measured wave number as

$$\nu = c\frac{1}{\lambda} = 3 \times 10^8\,(\mathrm{m\,s^{-1}}) \cdot 2170\,\mathrm{cm}^{-1}$$
$$= 3 \times 10^8(\mathrm{m\,s^{-1}}) \cdot 217\,000\,\mathrm{m}^{-1} = 6.51 \times 10^{13}\,\mathrm{s}^{-1}$$

The equivalent spring constant

$$k = \mu\omega^2 = \mu(2\pi\nu)^2$$

specified in terms of the reduced mass

$$\mu = \frac{m_C m_0}{m_C + m_0} = \frac{12 \cdot 16(1.673 \times 10^{-27})^2}{(12+16)(1.673 \times 10^{-27})} = 1.15 \times 10^{-26}\,\text{kg}$$

is thus found as

$$\begin{aligned} k &= 1.15 \times 10^{-26}\,(\text{kg})(2\pi \times 6.51 \times 10^{13})^2(\text{s})^{-2} = 1.92 \times 10^3\,(\text{N m}^{-1}) \\ &= 19.2\,(\text{N cm}^{-1}) \end{aligned}$$

(b) The zero point energy is then given by

$$E_0 = \frac{1}{2}h\nu = \frac{1}{2}4.136 \times 10^{-15}\,\text{eV s}\; 6.51 \times 10^{13}\,\text{s}^{-1} = 1.35 \times 10^{-1}\,\text{eV}$$

7.6 By using Eq. (7.39) the Hamiltonian can be expressed in terms of a and a^+ as

$$\begin{aligned} \hat{H} &= \frac{\hat{p}^2}{2m} + \frac{1}{2}kx^2 \\ &= -\frac{\hbar^2\alpha^2}{2}\frac{1}{2m}(a^+ - a)(a^+ - a) + \frac{k}{2}\frac{1}{2\alpha^2}(a^+ + a)(a^+ + a), \quad \alpha = \left(\frac{m\omega}{\hbar}\right)^{1/2} \end{aligned}$$

Now

$$\frac{\hbar^2\alpha^2}{4m} = \frac{\hbar^2 m\omega}{4m\hbar} = \frac{1}{4}\hbar\omega, \quad \frac{k}{4\alpha^2} = \frac{m\omega^2}{4}\frac{\hbar}{m\omega} = \frac{1}{4}\hbar\omega$$

so that $\hat{H}$ reads as

$$\begin{aligned} \hat{H} &= \frac{\hbar\omega}{4}[-(a^+ - a)(a^+ - a) + (a^+ + a)(a^+ + a)] \\ &= \frac{\hbar\omega}{4} \cdot 2(a^+ a + aa^+) = \frac{\hbar\omega}{2}(a^+ a + 1), \;\; aa^+ \equiv a^+ a + 1 \end{aligned}$$

Chapter 8

8.1. (a) Given the representation of the angular momentum

$$\hat{\underline{l}} = (\hat{x}x + \hat{y}y + \hat{z}z) \times (\hat{x}\hat{p}_x + \hat{y}\hat{p}_y + \hat{z}\hat{p}_z)$$

one can use the cyclic properties of $\hat{x}, \hat{y}, \hat{z}$,

$$\hat{x} \times \hat{y} = \hat{z}, \quad \hat{y} \times \hat{z} = \hat{x}, \quad \hat{z} \times \hat{x} = \hat{y}$$

and single out the x, y, and z components as

$$\hat{l}_x = (y\hat{p}_z - z\hat{p}_y), \quad \hat{l}_y = (z\hat{p}_x - x\hat{p}_z), \quad \hat{l}_z = (x\hat{p}_y - y\hat{p}_x)$$

(b) The commutation relation between the x, y components can be derived as

$$[\hat{l}_x, \hat{l}_y] = [(y\hat{p}_z - z\hat{p}_y), (z\hat{p}_x - x\hat{p}_z)] = [y\hat{p}_z, z\hat{p}_x] + [z\hat{p}_y, x\hat{p}_z]$$
$$= y\hat{p}_x[\hat{p}_z, z] + \hat{p}_y x[z, \hat{p}_z] = i\hbar(x\hat{p}_y - y\hat{p}_x) = \hat{l}_z$$

One can likewise derive

$$[\hat{l}_y, \hat{l}_z] = i\hbar\hat{l}_x, \quad [\hat{l}_z, \hat{l}_x] = i\hbar\hat{l}_y$$

(c) The commutation relation between $\hat{l}^2$ and $\hat{l}_z$ has been proven in Eq. (8.10). Thus, condiser

$$[\hat{l}^2, \hat{l}_x] = [(\hat{l}_x^2 + \hat{l}_y^2 + \hat{l}_z^2), \hat{l}_x] = [(\hat{l}_y^2 + \hat{l}_z^2), \hat{l}_x]$$

Now

$$[\hat{l}_y^2, \hat{l}_x] = \hat{l}_y\hat{l}_y\hat{l}_x - \hat{l}_x\hat{l}_y\hat{l}_y = \hat{l}_y(\hat{l}_x\hat{l}_y - i\hbar\hat{l}_z) - (\hat{l}_y\hat{l}_x + i\hbar\hat{l}_z)\hat{l}_y = -i\hbar(\hat{l}_y\hat{l}_z + \hat{l}_z\hat{l}_y)$$

$$[\hat{l}_z^2, \hat{l}_x] = \hat{l}_z\hat{l}_z\hat{l}_x - \hat{l}_x\hat{l}_z\hat{l}_z = \hat{l}_z(\hat{l}_x\hat{l}_z + i\hbar\hat{l}_y) - (\hat{l}_x\hat{l}_z - i\hbar\hat{l}_y)\hat{l}_z = i\hbar(\hat{l}_y\hat{l}_z + \hat{l}_z\hat{l}_y)$$

where use has been made of commutation relations between l_x, l_y, l_z. Therefore $[\hat{l}^2, \hat{l}_x] = 0$. One can likewise prove $[\hat{l}^2, \hat{l}_y] = 0$.

8.2. The center of mass and relative coordinates are specified in terms of $\boldsymbol{r}_1$, $\boldsymbol{r}_2$ as

$$\underline{R} = \frac{1}{M}(m_1\underline{r_1} + m_2\underline{r_2}), \quad \underline{r} = \underline{r_1} - \underline{r_2}$$

Upon inverting the relation $\boldsymbol{r}_1$, $\boldsymbol{r}_2$ can be expressed as

$$\underline{r_1} = \underline{R} + \frac{m_2}{M}\underline{r}, \quad \underline{r_2} = \underline{R} - \frac{m_1}{M}\underline{r}$$

so that the total kinetic energy is given by

$$\frac{p_1^2}{2m_1} + \frac{p_2^2}{2m_2} = \frac{1}{2}m_1\dot{\underline{r_1}}^2 + \frac{1}{2}m_2\dot{\underline{r_2}}^2$$
$$= \frac{1}{2}m_1\left[\dot{\underline{R}}^2 + \left(\frac{m_2}{M}\right)^2\dot{\underline{r}}^2 + \frac{2m_2}{M}\dot{\underline{R}}\cdot\dot{\underline{r}}\right]$$
$$+ \frac{1}{2}m_2\left[\dot{\underline{R}}^2 + \left(\frac{m_1}{M}\right)^2\dot{\underline{r}}^2 - \frac{2m_1}{M}\dot{\underline{R}}\cdot\dot{\underline{r}}\right]$$
$$= \frac{1}{2}M\dot{\underline{R}}^2 + \frac{1}{2M^2}(m_1m_2^2 + m_1^2m_2)\dot{\underline{r}}^2 = \frac{\underline{P}^2}{2M} + \frac{\underline{p}^2}{2\mu}$$

where

$$\underline{P} \equiv M\underline{R}, \; M = m_1 + m_2$$
$$\underline{p} = \mu\underline{r}, \; \mu = \frac{m_1 \cdot m_2}{m_1 + m_2} \quad \text{or} \quad \frac{1}{\mu} = \frac{1}{m_1} + \frac{1}{m_2}$$

8.3. (a) One can use Eq. (8.46) by incorporating the number of protons in the nucleus, Z:

$$e_M^2 \equiv \frac{e^2}{4\pi\varepsilon_0} \rightarrow Ze_M^2 \equiv \frac{Ze^2}{4\pi\varepsilon_0}$$

Thus,

$$\begin{aligned} a_{0He} &= \frac{\hbar^2}{\mu Z e_M^2} = \frac{1}{Z}\frac{\hbar^2}{e_M^2}\left(\frac{1}{m_e}+\frac{1}{m_N}\right) \\ &= \frac{1}{Z}a_B\left(1+\frac{m_e}{m_N}\right) \simeq \frac{0.049}{2}\,\text{nm} = 0.025\,\text{nm},\ Z=2 \end{aligned}$$

$$\begin{aligned} a_{0po} &= \frac{\hbar^2}{\mu e_M^2} = \frac{\hbar^2}{e_M^2}\left(\frac{1}{m_e}+\frac{1}{m_p}\right) \\ &= a_B\left(1+\frac{m_e}{m_{po}}\right) \simeq 0.049\times 2\,\text{nm} = 0.1\,\text{nm},\ m_\text{e} = m_\text{p} \end{aligned}$$

8.4. The wavefunction of 1s state is given from Table 8.2 as

$$u_{100} = \frac{(Z/a_0)^{3/2}}{\pi^{1/2}}e^{-Zr/a_0}$$

Thus one can find the average values of r and r^2 and the variance as follows:

$$\begin{aligned} <r> = \langle u_{100}|r|u_{100}\rangle &= N^2\int_0^{2\pi}d\varphi\int_0^{\pi}\sin\theta d\theta\int_0^{\infty}r^2drre^{-2Zr/a_0},\ N=\frac{(Z/a_0)^{3/2}}{\sqrt{\pi}} \\ &= N^2\times 4\pi\times\left[\frac{6}{(2Z/a_0)^4}\right] = \frac{1.5a_0}{Z} \end{aligned}$$

$$\begin{aligned} <r^2> &= \langle u_{100}|r^2|u_{100}\rangle \\ &= N^2\int_0^{2\pi}d\varphi\int_0^{\pi}\sin\theta d\theta\int_0^{\infty}r^2drr^2e^{-2Zr/a_0},\quad N=\frac{(Z/a_0)^{3/2}}{\sqrt{\pi}} \\ &= N^2\times 4\pi\times\frac{4!}{(2Z/a_0)^5} = \frac{3a_0^2}{Z^2} \end{aligned}$$

$$\begin{aligned} <(r-<r>)^2> &= <r^2 - 2r<r> + <r>^2> \\ &= <r^2> - <r>^2 = 0.75\left(\frac{a_0^2}{Z^2}\right) \end{aligned}$$

with a_0 denoting the Bohr radius. For H-atom $Z = 1$, while for He$^+$ $Z = 2$. One can likewise find the corresponding quantities u_{200}, u_{300}.

8.5. The average kinetic energy is found as

$$< K >= -\left(\frac{\hbar^2}{2\mu}\right)\langle u_{100}|\nabla^2|u_{100}\rangle, \quad \mu \simeq m_e$$

Now with the use of Eq. (8.35) in the text one can write

$$\nabla^2 u_{100} = \left(\frac{1}{r^2}\frac{\partial}{\partial r}r^2\frac{\partial}{\partial r} - \frac{1}{r^2}\frac{1}{\hbar^2}\hat{l}^2\right)u_{100} = N_{100}\cdot\left[\frac{1}{r^2}\frac{\partial}{\partial r}r^2\frac{\partial}{\partial r}\exp\left(-\frac{r}{a_0}\right)\right]$$
$$= N_{100}\cdot\left(\frac{1}{a_0^2}e^{-r/a_0} - \frac{2}{a_0 r}e^{-r/a_0}\right)$$

where N_{100} is the normalization constant and use has been made of the fact that u_{100} does not carry the angular momentum. Hence one can evaluate the average value of K as

$$\langle K\rangle = \frac{1}{a_0^3\pi}\left(\frac{-\hbar^2}{2\mu}\right)\int_0^{2\pi}d\varphi\int_0^{\pi}\sin\theta\, d\theta\int_0^{\infty}r^2 dr\left(\frac{1}{a_0^2} - \frac{1}{a_0 r}\right)e^{-2r/a_0}$$
$$= \frac{\hbar^2}{2\mu a_0^2} = \frac{\hbar^2}{\mu e_M^2}\frac{e_M^2}{2a_0^2} = \frac{e_M^2}{2a_0}, \quad \frac{\hbar^2}{\mu e_M^2} = a_0, \quad e_M^2 \equiv \frac{e^2}{4\pi\varepsilon_0}$$

We can therefore state that the average kinetic energy is equal to the magnitude of the ground state energy (see (8.46)),

$$< K >= |E_1|$$

One can likewise evaluate the average potential energy as

$$\langle V\rangle_{100} = \frac{1}{a_0^3\pi}\int_0^{2\pi}d\varphi\int_0^{\pi}\sin\theta\, d\theta\int_0^{\infty}r^2 dr\left(-\frac{e_M^2}{r}\right)e^{-2r/a_0}$$
$$= -\frac{4\pi}{a_0^3\pi}\cdot e_M^2\int_0^{\infty}dr\, re^{-2r/a_0} = -\frac{e_M^2}{a_0}$$

Therefore

$$\langle K\rangle_{100} = -\frac{1}{2}\langle V\rangle_{100}$$

One can show in general that

$$\langle K\rangle_{nlm} = -\frac{1}{2}\langle V\rangle_{nlm}$$

8.6. The energy levels are generally given from Eq. (8.46) by

$$E_n \equiv -E_0\frac{1}{n^2}; \quad E_0 = \frac{\mu Z^2 e_M^4}{2\hbar^2} = \frac{Z^2 e_M^2}{2a_B(1 + m_e/m_N)}, \quad n = 1, 2, \ldots;$$
$$a_0 \equiv a_B(1 + m_e/m_N)$$

For the H-atom the nuclear mass is given by the mass of the proton and we can write

$$E_0 = \frac{Z^2 e_M^2}{2a_B(1 + m_e/m_p)}\Lambda = 13.6\,\text{eV} \times \Lambda, \quad \Lambda \equiv \frac{1 + m_e/m_p}{1 + m_e/m_N}$$

The wavelength corresponding to the transition between $n = 2$ and $n = 1$ state

$$h\nu = \frac{hc}{\lambda} = E_2 - E_1$$

can therefore be found in H-atom as

$$\lambda = \frac{hc}{E_2 - E_1} = \frac{4.136 \times 10^{-15}\,(\text{eV s}) \times 3 \times 10^8\,\text{m}}{13.6(1 - 1/4)\,\text{eV}}$$
$$= 12.2 \times 10^{-8}\,\text{m} = 122\,\text{nm}, \ \Lambda = 1$$

One can likewise find λ for the case of deuterium and ionized He atom by evaluating Λ and using the Z values, respectively.

(b) The frequency corresponding to the transition between $n = 1$ and $n = 3$ states is given by

$$\nu = \frac{E_3 - E_1}{h} = \frac{13.6\,\text{eV} \times (1 - 1/9)}{4.136 \times 10^{-15}(\text{eV s})} = 2.92 \times 10^{15}\,\text{s}^{-1}$$

8.7. The ionization energy and atomic radius of the phosphorus atom in Si can be specified in terms of the corresponding values of the H-atom with appropriate scaling of the parameters involved. We can find the parameters with the use of Eq. (8.46):

$$E_{\text{ion}} = 13.6\,\text{eV}\left(\frac{\varepsilon_0}{\varepsilon_0\varepsilon_r}\right)^2\left(\frac{m_0}{m_n}\right) = 13.6 \times \left(\frac{1}{11.9}\right)^2 \times 2\ \text{eV} \simeq 0.1\ \text{eV}$$

$$a_0 = \frac{\hbar^2}{m_n e_M^2} = \frac{\hbar^2 4\pi\varepsilon_0\varepsilon_r}{m_n e^2} \simeq 0.05\varepsilon_\text{r}\left(\frac{m_0}{m_\text{n}}\right)\text{nm} \simeq 1.19\,\text{nm}$$

To find the de Broglie wavelength one can start from the relation, $\lambda = h/p$ and find the linear momentum of the electron in the ground state of P-atom in Si. As discussed in the problem of 8.5 one can put

$$\left\langle \frac{p^2}{2m_n} \right\rangle = |E_1| = 0.2\,\text{eV}$$

and evaluate p as

$$p = (2m_n 0.2\,\text{eV})^{1/2}$$
$$= [2 \times 9.1 \times 10^{-31}(\text{kg}) \cdot 0.2 \times 1.6 \times 10^{-19}\,\text{J}]^{1/2}$$
$$= 2.4 \times 10^{-25}\,\text{kg} \cdot \text{m s}^{-1}$$

Therefore λ is given by

$$\lambda = \frac{h}{p} = \frac{6.626 \times 10^{-34}\,\mathrm{J \cdot s}}{2.4 \times 10^{-25}\,\mathrm{kg \cdot m\,s^{-1}}}$$
$$= 2.8 \times 10^{-9}\mathrm{m} = 2.8\,\mathrm{nm}$$

Chapter 9

9.1. To find the first order corrections in eigenenergy and eigenfuction due to the first term of the perturbating Hamiltonian one has to evaluate the matrix element (see Eq. (9.14)).

$$W_{1m} = \langle u_m | k_2 x^3 | u_m \rangle$$

$$\varphi^{(1)} = u_m + \sum_{k \neq m} \frac{\langle u_k | \hat{H}' | u_m \rangle}{E_m - E_k}$$

For performing the integration the recurrence relation (7.18) is handy to be applied repeatedly:

$$\xi H_m = \frac{1}{2} H_{m+1} + m H_{m-1}$$

$$\begin{aligned}\xi^2 H_m &= \frac{1}{2}\left[\frac{1}{2} H_{m+2} + (m+1) H_m\right] + m\left[\frac{1}{2} H_m + (m-1) H_{m-2}\right] \\ &= \frac{1}{4} H_{m+2} + \left(m + \frac{1}{2}\right) H_m + m(m-1) H_{m-2} \qquad (9.1\mathrm{a})\end{aligned}$$

$$\begin{aligned}\xi^3 H_m &= \frac{1}{4}\left[\frac{1}{2} H_{m+3} + (m+2) H_{m+1}\right] + \left(m + \frac{1}{2}\right)\left[\frac{1}{2} H_{m+1} + m H_{m-1}\right] \\ &\quad + m(m-1)\left[\frac{1}{2} H_{m-1} + (m-2) H_{m-3}\right] \\ &= \frac{1}{8} H_{m+3} + \frac{3m+3}{4} H_{m+1} + \frac{3m^2}{2} H_{m-1} + m(m-1)(m-2) H_{m-3} \qquad (9.1\mathrm{b})\end{aligned}$$

It is clear from Eq. (9.1) that there is no first order level shift caused by k_2 term because the eigfenfunctions are orthogonal and the output of Eq. (9.1b) does not contain the term $\propto u_m$.

$$W_{1m} = \int_{-\infty}^{\infty} dx u_m^* k_2 x^3 u_m = \frac{N_m^2 k_2}{\alpha^4} \int_{-\infty}^{\infty} d\xi e^{-\xi^2} H_m \xi^3 H_m \quad \alpha \equiv \left(\frac{m\omega}{\hbar}\right)^{1/2}$$
$$= 0$$

To find the first order correction in eigenfunction one has to examine the matrix element

$$\langle u_k | k_2 x^3 | u_m \rangle$$

It is clear from Eq. (9.1b) that there are two non-zero matrix elements corresponding to

$$k = m \pm 1, m \pm 2$$

The evaluation of the matrix element for such k can be carried out with the use of Eq. (7.31).
One can treat the k_4 term in a similar manner.

9.2. (a) Without the perturbation term one can look for the solution in the form

$$\varphi(x, y) = u_x(x)u_y(y)$$

and put it into the energy eigenequation and divide both sides with $\varphi(x, y)$, obtaining

$$\frac{\left[-\frac{\hbar^2}{2m}\frac{\partial^2}{\partial x^2} + \frac{1}{2}k_x x^2\right] u_x(x)}{u_x(x)} + \frac{\left[\frac{\hbar^2}{2m}\frac{\partial^2}{\partial y^2} + \frac{1}{2}k_y y^2\right] u_y(y)}{u_y(y)} = E$$

Since each term appearing on the left hand side depends solely on x and y respectively, the only way to satisfy the equation is to put each term to a constant. As a result two 1D eigenequations of HO ensue and one can write from Eqs. (7.12), (7.25) as

$$u_{nx}(\xi_x) = N_{nx}e^{-1/2\xi_x{}^2}H_{nx}(\xi_x), \quad \xi_x = \alpha_x x,$$
$$u_{ny}(\xi_y) = N_{ny}e^{-1/2\xi_y{}^2}H_{ny}(\xi_y), \quad \xi_y = \alpha_y y$$
$$\alpha_j = \left(\frac{m\omega_j}{\hbar}\right)^{1/2}, \quad \omega_j^2 = \frac{k_j}{m}, \quad j = x, y$$
$$E_{n_x,n_y} = \hbar\omega_x\left(n_x + \frac{1}{2}\right) + \hbar\omega_y\left(n_y + \frac{1}{2}\right), \quad n_x, n_y = 0, 1, 2, \ldots \tag{9.2a}$$

(b) The first order level shift

$$W_1 = \langle u_{nx}(\xi_x)u_{ny}(\xi_y)|\hat{H}'|u_{nx}(\xi_x)u_{ny}(\xi_y)\rangle, \quad \hat{H}' = Cxy$$

is zero, that is, $W_1 = 0$, since the integrand of the matrix element is odd. However there is the second order level shift contributed by two non-vanishing matrix elements as clear from Eq. (7.31):

$$\langle u_{nx}|x|u_{nx'}\rangle = \frac{\sqrt{n_x + 1}}{(2m\omega_x/\hbar)^{1/2}}, \quad n_x' = n_x + 1; \quad \frac{\sqrt{n_x}}{(2m\omega_x/\hbar)^{1/2}}, \quad n_x' = n_x - 1$$
$$\langle u_{ny}|y|u_{ny'}\rangle = \frac{\sqrt{n_y + 1}}{(2m\omega_y/\hbar)^{1/2}}, \quad n_y' = n_y + 1; \quad \frac{\sqrt{n_y}}{(2m\omega_y/\hbar)^{1/2}}, \quad n_y' = n_y - 1$$

Thus one can write the second order level shift of the ground state as

$$\Delta E^{(2)} = \sum_{\substack{nx' \neq nx \\ ny' \neq ny}} \frac{|\langle u_{nx'}u_{ny'}|Cxy|u_{nx,ny}\rangle|^2}{[E_{nx,ny} - E_{nx',ny'}]^2}; \quad n_x' = n_x \pm 1, \quad n_y' = n_y \pm 1, \quad n_x = n_y = 0$$
$$= \frac{C^2\hbar}{8m^2\omega^3}, \quad \omega_x = \omega_y = \omega \tag{9.2b}$$

where use has been made of Eq. (9.18) and the recurrence relation,

$$\xi_x H_{nx} = \frac{1}{2} H_{nx+1} + n_x H_{nx-1}$$

(c) Given two new variables

$$\xi = (x + y), \quad \eta = (x - y)$$

one can invert it and obtain

$$x = \frac{1}{2}(\xi + \eta), \quad y = \frac{1}{2}(\xi - \eta)$$

Hence one can express the potential energy as

$$\begin{aligned}\frac{1}{2}k_x{}^2 + \frac{1}{2}k_y y^2 + Cxy &= \frac{1}{2}k_x[(\xi + \eta)/2]^2 + k_y[(\xi - \eta)/2]^2 + C\frac{1}{4}(\xi^2 - \eta^2) \\ &\simeq \frac{1}{8}(\xi^2 + \eta^2)(k_x + k_y) + \frac{1}{4}C(\xi^2 - \eta^2)\end{aligned}$$

where an assumption was made, namely $k_x \approx k_y$ so that $2\xi\eta(k_x - k_y) \approx 0$. The Hamiltonian then reads in terms of η, ξ as

$$\begin{aligned}\hat{H} &= -\frac{\hbar^2}{2m}\frac{\partial^2}{\partial x^2} + \frac{1}{2}k_x x^2 - \frac{\hbar^2}{2m}\frac{\partial^2}{\partial y^2} + \frac{1}{2}k_y y^2 + Cxy \\ &= -\frac{\hbar^2}{2\mu}\frac{\partial^2}{\partial \xi^2} + \frac{1}{2}k_+\xi^2 - \frac{\hbar^2}{2\mu}\frac{\partial^2}{\partial \eta^2} + \frac{1}{2}k_-\eta^2, \quad \frac{1}{\mu} = \frac{1}{m} + \frac{1}{m}\end{aligned}$$

with

$$k_\pm = \frac{1}{2}\left[\frac{1}{2}\left(k_x + k_y\right) \pm C\right]$$

(d) The coupled Hamiltonian has thus been decoupled with the use of new variables and the energy eigenvalues are given from Eq. (7.12) by

$$E_{n\xi,n\eta} = \hbar\omega_+\left(n_\xi + \frac{1}{2}\right) + \hbar\omega_-\left(n_\eta + \frac{1}{2}\right)$$

where

$$\begin{aligned}\omega_\pm &= \left(\frac{k_\pm}{\mu}\right)^{1/2} = \left\{\frac{1}{2}\left[\left(k_x + k_y\right)/2 \pm C\right]/\mu\right\}^{1/2} = \left[\frac{\left(k_x + k_y\right)/2}{m} \pm \frac{C}{m}\right]^{1/2} \\ &= \left[\frac{1}{2}\left(\omega_x^2 + \omega_y^2\right) \pm \frac{C}{m}\right]^{1/2}, \quad \mu = \frac{m}{2}\end{aligned}$$

To compare the result one may expand $\omega_\pm$ as

$$\begin{aligned}\omega_\pm &= \omega\left[1 \pm \frac{C}{m\omega^2}\right]^{1/2}, \quad \omega^2 \equiv \frac{\omega_x^2 + \omega_y^2}{2} \\ &= \omega\left[1 \pm \frac{1}{2}\frac{C}{m\omega^2} - \frac{1}{4}\frac{C^2}{m^2\omega^4} + \dots\right]\end{aligned}$$

Therefore the ground state for instance is given by

$$E_{00} = \hbar\omega - \frac{\hbar C^2}{4m^2}\frac{1}{\omega^3} \tag{9.3}$$

9.3. (a) Given the eigenequation,

$$(\hat{H}' - W_1)(c_i u_i + c_j u_j) = \sum_n a_n^{(1)}(E_m - E_n)u_n$$

one can perform the inner product with respect to u_i by multiplying both sides with u_i^* and carrying out the integrations, obtaining

$$c_i(\hat{H}_{ii} - W_1) + c_j\hat{H}_{ij} = 0, \quad \hat{H}'_{\alpha\beta} \equiv \int_{-\infty}^{\infty} d\underline{r} u_\alpha^* \hat{H}' u_\beta \tag{9.4}$$

where the orthonormality of the eigenfunctions have been used and $E_i = E_j = E_m$ in this case.
One can likewise perform the inner product with respect to u_j, obtaining

$$c_i\hat{H}_{ji} + c_j(\hat{H}_{jj} - W_1) = 0 \tag{9.5}$$

Equations 9.4 and 9.5 when put into the matrix notation, are in agreement with Eq. (9.30).

(b) Given the Schrödinger equation

$$i\hbar\sum_n \dot{a}_n(t)e^{-i(E_n/\hbar)t}u_n + \sum_n E_n a_n(t)e^{-i(E_n/\hbar)t}u_n$$
$$= \sum_n a_n(t)E_n e^{-i(E_n/\hbar)t}u_n + \lambda\hat{H}'(t)\sum_n a_n(t)e^{-i(E_n/\hbar)t}u_n$$

one can carry out the inner product on both sides with respect to u_k, obtaining

$$i\hbar\dot{a}_k e^{-i\omega_k t} + E_k a_k e^{-i\omega_k t} = E_k a_k e^{-i\omega_k t} + \lambda\sum_n \hat{H}'_{kn} a_n e^{-i\omega_n t}, \quad \frac{E}{\hbar} = \omega$$

Obviously the second term on the left hand side and the first term on the right hand side cancel each other out so that one can write

$$\dot{a}_k e^{-i\omega_k t} = -\left(\frac{i\lambda}{\hbar}\right)\sum_n \hat{H}'_{kn} a_n e^{-i\omega_{kn} t}, \quad \omega_{kn} = \omega_k - \omega_n$$

9.4. (a) The interaction Hamiltonian is given by

$$\hat{H}' = -(-e\mathrm{E})x = e\mathrm{E}x$$

where $-e\mathrm{E}$ represents the force acting on the electron.

(b) The unperturbed eigenfunction of the nanowire is given from Eq. (4.9) by

$$u(x,y,z) \propto e^{ikz}\left[\left(\frac{2}{W}\right)^{1/2}\sin\left(\frac{n_x\pi}{W}x\right)\right]\left[\left(\frac{2}{W}\right)^{1/2}\sin\left(\frac{n_y\pi}{W}y\right)\right]$$

and represents the particle traveling along the z-direction as a free particle, while confined in the x, y directions. The ground and first excited states therefore read as

$$u(x,y,z) \propto e^{ikz}\left[\left(\frac{2}{W}\right)^{1/2}\sin\left(\frac{\pi}{W}x\right)\right]\left[\left(\frac{2}{W}\right)^{1/2}\sin\left(\frac{\pi}{W}y\right)\right], \quad n_x = n_y = 1$$

$$u(x,y,z) \propto e^{ikz}\left[\left(\frac{2}{W}\right)^{1/2}\sin\left(\frac{2\pi}{W}x\right)\right]\left[\left(\frac{2}{W}\right)^{1/2}\sin\left(\frac{\pi}{W}y\right)\right],$$

$$n_x = 1, n_y = 2 \text{ or } n_x = 2, n_y = 1$$

Hence the first order level shift of the ground state is given by

$$\langle u(x,y,z)|e\mathrm{E}x|u(x,y,z)\rangle = e\mathrm{E}\frac{2}{W}\int_0^W dxx\ \sin^2\left(\frac{\pi x}{W}\right) = \frac{e\mathrm{E}W}{2}$$

Since the eigenfunction is normalized the y, z integrations automatically yield unity.

9.5. Given the eigenfunction u_{nlm} the transition to other state occurs, provided the matrix element is not zero, that is,

$$\langle u_{nlm}|\hat{H}'|u_{n'l'm'}\rangle \neq 0, \quad \hat{H}' = ez\mathrm{E}_0\cos\omega t, \quad z = r\cos\theta$$

The evaluation of this matrix element requires both the angular and radial integrations but the angular integration dictates whether or not it becomes zero. Thus one has to consider

$$\int_0^{2\pi} d\varphi \int_0^{\pi} \sin\theta d\theta Y_{lm}^* \cos\theta Y_{l'm'} = \int_0^{2\pi} d\varphi \int_{-1}^{1} dwY_{lm}^* wY_{l'm'}, \quad w = \cos\theta$$

Since $Y_{lm} \propto \exp \pm im\varphi$, m, m' should be constrained by $m = m'$ for the φ-integral not to yield zero Also since w is proportional to the first order Legendre polynomial P_1 the product $P_lP_{l'}$ should be odd, so that the total product $P_lP_{l'}P_1$ is even in w to make the w-integral not to vanish. This requires $l - l' = \pm 1$ and therefore the conditions imposed on $l'm'$ called the selection rule are given by

$$\Delta m = m - m' = 0$$
$$\Delta l = l - l' = \pm 1$$

For the perturbing Hamiltonian

$$\hat{H}' = ex\mathrm{E}_0\cos\omega t$$
$$= er\sin\theta\cos\varphi\mathrm{E}_0\cos\omega t$$

the transition matrix reads as

$$\langle u_{nlm}|\hat{H}'|u_{n'l'm'}\rangle \propto \int_0^{2\pi} d\varphi \int_{-1}^{1} dwY_{lm}^* \sin\theta\cos\varphi Y_{l'm'}$$

Now the integrand of the φ-integral is given by

$$e^{i(m'-m)\varphi}\cdot(e^{i\varphi}+e^{-i\varphi})/2 \propto e^{i(m'-m+1)\varphi}+e^{i(m'-m-1)\varphi}$$

so that in order of the φ-integration not to vanish m, m' should be constrained by

$$\Delta m = m' - m = \pm 1$$

Also since $\sin\theta = \sqrt{1-\cos^2\theta} = \sqrt{1-w^2}$ the w-integral does not vanish when the product $P_l^m P_l^{m'}$ is even in w, so that the constraints on l are given by

$$\Delta l = l - l' = 0, 2, \ldots$$

The selection rule for the Hamiltonian

$$\begin{aligned}\hat{H}' &= ey\mathrm{E}_0 \cos\omega t\\ &= er\sin\theta\sin\varphi\mathrm{E}_0\cos\omega t\end{aligned}$$

can likewise be analyzed and the selection rule is the same as considered.

9.6. (a) Given the harmonic field

$$\mathrm{E} = \mathrm{E}_0\cos\omega t$$

the interaction Hamiltonian reads as

$$\hat{H}' = -\int_0^x dx(q\mathrm{E}) = -q\mathrm{E}x = -\frac{q\mathrm{E}_0 x}{2}(e^{i\omega t}+e^{-i\omega t})$$

The HO initially prepared in the state u_n can make the transitions to other state n' via $\hat{H}'$, the rate of which is from Eq. (9.42) by

$$\dot{a}_{n'} = -\frac{i}{\hbar}\hat{H}'_{n'n}e^{i\omega_{n'n}t};\quad \omega_{n'n} = \frac{1}{\hbar}(E_{n'}-E_n),\quad \hat{H}'_{n'n} = \langle u_{n'}|\hat{H}'|u_n\rangle \tag{9.6}$$

Now the transition matrix element can be evaluated with the use of Eq. (7.31) as

$$\begin{aligned}\langle u_{n'}|\hat{H}'|u_n\rangle &= -\frac{q\mathrm{E}_0}{2}(e^{i\omega t}+e^{-i\omega t})\int_{-\infty}^{\infty}dx u_n x u_{n'}\\ &= -\frac{q\mathrm{E}_0}{2}(e^{i\omega t}+e^{-i\omega t})\begin{cases}(n+1)^{1/2}/(2m\omega_c/\hbar)^{1/2}, & n'=n+1\\ n^{1/2}/(2m\omega_c/\hbar)^{1/2}, & n'=n-1\\ 0, & \text{otherwise}\end{cases}\end{aligned} \tag{9.7}$$

Thus by inserting Eq. (9.7) into Eq. (9.6) one can write

$$a_{n'}(t) = -\frac{qE_0}{2\hbar}\left(\frac{n'}{2m\omega_c/\hbar}\right)^{1/2}\cdot\int_0^t dt'[e^{i(\omega+\omega_{n'n})t}+e^{-i(\omega-\omega_{n'n})t}],\quad n'=n\pm 1$$

where

$$\begin{aligned}\omega_{n'n} &\equiv \frac{1}{\hbar}(E_{n'}-E_n) = \frac{1}{\hbar}\left[\hbar\omega_c\left(n'+\frac{1}{2}\right)) - \hbar\omega_c\left(n+\frac{1}{2}\right)\right] = \omega_c(n'-n)\\ &= \pm\omega_c,\quad \omega_c^2 \equiv \frac{k}{m}\end{aligned}$$

depending on $n' = n \pm 1$. Hence the transition occurs between two nearest eigenstates.

(b) Consider the resonant interaction in which $\omega \approx \omega_c$ and $n=0$. The integrand consists of a fast oscillating term $\exp \pm[i(\omega+\omega_c)]$ and the resonant term, $\exp \pm[i(\omega-\omega_c)]$. Because the fast oscillating term averages out to be zero one may disregard it and evaluate the time-integration, obtaining

$$a_{n'}(t) = -\frac{q\mathrm{E}_0}{2}\left(\frac{n'}{2m\omega_c/\hbar}\right)^{1/2} \cdot \frac{1}{i(\omega-\omega_c)}[e^{i(\omega-\omega_c)t} - 1]$$

$$= -\frac{q\mathrm{E}_0}{2}\left(\frac{n'}{2m\omega_c/\hbar}\right)^{1/2} 2e^{\pm[i(\omega-\omega_c)/2]t}\frac{\sin(\omega-\omega_c)t}{(\omega-\omega_c)}; \quad n' = 1$$

The probability of the HO making the resonant transition to n' state at $t = \pi/\omega$ is therefore given by

$$|a_{n'}(t)|^2 = \frac{q^2\mathrm{E}_0^2(\pi/\omega)^2}{2m\omega/\hbar}, \quad n' = 1, \quad t = \frac{\pi}{\omega}$$

where use has been made of

$$\frac{\sin[(\omega-\omega_c)](\pi/\omega)}{(\omega-\omega_c)} = \left\{\frac{\sin\left[(\omega-\omega_c)\right](\pi/\omega)}{(\omega-\omega_c)(\pi/\omega)}\right\} \times (\pi/\omega), \omega \to \omega_c,$$

$$\frac{\sin x}{x} = 1 \quad \text{for} \quad x \to 0$$

Chapter 10

10.1. (a) With the matrix representation of the spin operators and spin states

$$\hat{\underline{s}} \equiv \frac{\hbar}{2}\underline{\sigma}; \quad \sigma_x = \begin{pmatrix} 0 & 1 \\ 1 & 0 \end{pmatrix}, \quad \sigma_y = \begin{pmatrix} 0 & -i \\ i & 0 \end{pmatrix}, \quad \sigma_z = \begin{pmatrix} 1 & 0 \\ 0 & -1 \end{pmatrix};$$

$$\chi_+ = \begin{pmatrix} 1 \\ 0 \end{pmatrix}, \quad \chi_- = \begin{pmatrix} 0 \\ 1 \end{pmatrix}$$

one can write

$$\hat{s}_z\chi_+ = \frac{\hbar}{2}\begin{pmatrix} 1 & 0 \\ 0 & -1 \end{pmatrix}\begin{pmatrix} 1 \\ 0 \end{pmatrix} = \frac{\hbar}{2}\begin{pmatrix} 1 \\ 0 \end{pmatrix},$$

$$\hat{s}_z\chi_- = \frac{\hbar}{2}\begin{pmatrix} 1 & 0 \\ 0 & -1 \end{pmatrix}\begin{pmatrix} 0 \\ 1 \end{pmatrix} = -\frac{\hbar}{2}\begin{pmatrix} 0 \\ 1 \end{pmatrix}$$

$$\hat{s}^2\chi_+ = \frac{\hbar^2}{4}\left[\begin{pmatrix} 0 & 1 \\ 1 & 0 \end{pmatrix}\begin{pmatrix} 0 & 1 \\ 1 & 0 \end{pmatrix} + \begin{pmatrix} 0 & -i \\ i & 0 \end{pmatrix}\begin{pmatrix} 0 & -i \\ i & 0 \end{pmatrix} + \begin{pmatrix} 1 & 0 \\ 0 & -1 \end{pmatrix}\begin{pmatrix} 1 & 0 \\ 0 & -1 \end{pmatrix}\right] \times \begin{pmatrix} 1 \\ 0 \end{pmatrix}$$

$$= \frac{\hbar^2}{4}\left[\begin{pmatrix} 1 & 0 \\ 0 & 1 \end{pmatrix} + \begin{pmatrix} 1 & 0 \\ 0 & 1 \end{pmatrix} + \begin{pmatrix} 1 & 0 \\ 0 & 1 \end{pmatrix}\right]\begin{pmatrix} 1 \\ 0 \end{pmatrix} = \frac{3\hbar^2}{4}\begin{pmatrix} 1 \\ 0 \end{pmatrix}$$

Similarly

$$\hat{s}^2\chi_- = \frac{3\hbar^2}{4}\begin{pmatrix}0\\1\end{pmatrix}$$

Also the raising and lowering operators flip the spin states as

$$\begin{aligned}\hat{s}_+\chi_- &= \frac{1}{2}(\hat{s}_x + i\hat{s}_y)\begin{pmatrix}0\\1\end{pmatrix} = \frac{1}{2}\frac{\hbar}{2}\left[\begin{pmatrix}0 & 1\\1 & 0\end{pmatrix} + i\begin{pmatrix}0 & -i\\i & 0\end{pmatrix}\right]\begin{pmatrix}0\\1\end{pmatrix}\\ &= \frac{1}{2}\frac{\hbar}{2}\left[2\begin{pmatrix}0 & 1\\0 & 0\end{pmatrix}\right]\begin{pmatrix}0\\1\end{pmatrix} = \frac{\hbar}{2}\begin{pmatrix}1\\0\end{pmatrix}\end{aligned}$$

$$\begin{aligned}\hat{s}_-\chi_+ &= \frac{1}{2}(\hat{s}_x - i\hat{s}_y)\begin{pmatrix}1\\0\end{pmatrix} = \frac{1}{2}\frac{\hbar}{2}\left[\begin{pmatrix}0 & 1\\1 & 0\end{pmatrix} - i\begin{pmatrix}0 & -i\\i & 0\end{pmatrix}\right]\begin{pmatrix}1\\0\end{pmatrix}\\ &= \frac{1}{2}\frac{\hbar}{2}\left[2\begin{pmatrix}0 & 0\\1 & 0\end{pmatrix}\right]\begin{pmatrix}1\\0\end{pmatrix} = \frac{\hbar}{2}\begin{pmatrix}0\\1\end{pmatrix}\end{aligned}$$

The orthonormality of spin states can also be shown as

$$\langle\chi_+|\chi_+\rangle = \begin{pmatrix}1 & 0\end{pmatrix}\begin{pmatrix}1\\0\end{pmatrix} = 1,\ \langle\chi_-|\chi_-\rangle = \begin{pmatrix}0 & 1\end{pmatrix}\begin{pmatrix}0\\1\end{pmatrix} = 1$$

$$\langle\chi_-|\chi_+\rangle = \begin{pmatrix}0 & 1\end{pmatrix}\begin{pmatrix}1\\0\end{pmatrix} = 0,\ \langle\chi_+|\chi_-\rangle = \begin{pmatrix}1 & 0\end{pmatrix}\begin{pmatrix}0\\1\end{pmatrix} = 0$$

Thus the spin matrices are capable of describing the properties of the spin operators.

(b) One can prove the commutation relation between the x and y components as follows.

$$\begin{aligned}[\hat{s}_x, \hat{s}_y] &= \frac{\hbar^2}{4}\left(\begin{pmatrix}0 & 1\\1 & 0\end{pmatrix}\begin{pmatrix}0 & -i\\i & 0\end{pmatrix} - \begin{pmatrix}0 & -i\\i & 0\end{pmatrix}\begin{pmatrix}0 & 1\\1 & 0\end{pmatrix}\right)\\ &= \frac{\hbar^2}{4}\left(\begin{pmatrix}i & 0\\0 & -i\end{pmatrix} - \begin{pmatrix}-i & 0\\0 & i\end{pmatrix}\right)\\ &= \frac{\hbar^2}{4}\begin{pmatrix}2i & 0\\0 & -2i\end{pmatrix} = \frac{i\hbar}{2}\frac{\hbar}{2}\begin{pmatrix}2 & 0\\0 & -2\end{pmatrix} = \frac{i\hbar}{2}\begin{pmatrix}1 & 0\\0 & -1\end{pmatrix} = \frac{i\hbar}{2}\hat{s}_z\end{aligned}$$

The remaining cyclic commutation relations can be similarly proven.

10.2. The ground state eigenfunction is given from Eqs. (10.15) and (10.17) by

$$\varphi_0(1,2) = u_{100}(\underline{r}_1)u_{100}(\underline{r}_2)\chi_a = u_{100}(\underline{r}_1)u_{100}(\underline{r}_2)\frac{1}{\sqrt{2}}[\chi_+(1)\chi_-(2) - \chi_-(1)\chi_+(2)]$$

In evaluating the expectation values of spin operators one needs to use χ_a, χ_s only, since the inner product of u_{100} automatically yields unity. Thus one can write

$$\begin{aligned}\hat{S}_z|\chi_a\rangle &= \frac{1}{\sqrt{2}}(\hat{s}_{1z} + \hat{s}_2)|\chi_+(1)\chi_-(2) - \chi_-(1)\chi_+(2)\rangle\\ &= \frac{1}{\sqrt{2}}\left[\left(\frac{\hbar}{2} - \frac{\hbar}{2}\right)|\chi_+(1)\chi_-(2)\rangle - \left(-\frac{\hbar}{2} + \frac{\hbar}{2}\right)|\chi_-(1)\chi_+(2)\rangle\right] = 0|\chi_a\rangle\end{aligned}$$

$$\widehat{S}^2|\chi_a\rangle = \frac{1}{\sqrt{2}}(\widehat{s}_1 + \widehat{s}_2)\cdot(\widehat{s}_1 + \widehat{s}_2)|\chi_+(1)\chi_-(2) - \chi_-(1)\chi_+(2)\rangle$$
$$= \frac{1}{\sqrt{2}}(\widehat{s}_1^2 + \widehat{s}_2^2 + 2\cdot\widehat{s}_1\widehat{s}_2|\chi_+(1)\chi_-(2) - \chi_-(1)\chi_+(2)\rangle$$
$$= \frac{1}{\sqrt{2}}\left[\hbar^2\left(\frac{1}{2}\frac{3}{2} - \frac{1}{2}\frac{3}{2}\right) - 2\hbar^2\left(\frac{1}{4} - \frac{1}{4}\right)\right]|\chi_+(1)\chi_-(2) - \chi_-(1)\chi_+(2)\rangle$$
$$= \hbar^2 0|\chi_a\rangle$$

and therefore

$$\langle\chi_a|\widehat{S}_z|\chi_a\rangle = 0, \quad \langle\chi_a|\widehat{S}^2|\chi_a\rangle = 0$$

10.3. (a) Since both u_{nlm} and $\chi_\pm$ constitute the orthonormal set of eigenfunctions, one can write

$$\langle\chi_+(1)\chi_+(2)|\chi_+(1)\chi_+(2)\rangle = 1, \quad \langle\chi_-(1)\chi_-(2)|\chi_-(1)\chi_-(2)\rangle = 1,$$
$$\langle\chi_+(1)\chi_-(2)|\chi_-(1)\chi_+(2)\rangle = 0$$

so that

$$\langle\chi_s|\chi_s\rangle = \langle\chi_a|\chi_a\rangle = 1, \quad \langle\chi_s|\chi_a\rangle = 0$$

Also,

$$\frac{1}{2}\langle u_{100}(1)u_{200}(2) \pm u_{100}(2)u_{200}(1)|u_{100}(1)u_{200}(2) \pm u_{100}(2)u_{200}(1)\rangle = 1$$
$$\frac{1}{2}\langle u_{100}(1)u_{200}(2) + u_{100}(2)u_{200}(1)|u_{100}(1)u_{200}(2) - u_{100}(2)u_{200}(1)\rangle = 0$$

Therefore

$$\langle\varphi_s|\varphi_s\rangle = \langle\varphi_a|\varphi_a\rangle = 1, \quad \langle\varphi_s|\varphi_a\rangle = 0$$

(b) The expectation values of $\widehat{S}^2$, $\widehat{S}_z$ of the singlet state have already been evaluated in the previous problem. Thus one needs to evaluate the expectation values of the triplet states. The triplet state consists of the three symmetrized states:

$$\widehat{S}_z|\chi_s\rangle = (\widehat{s}_{1z} + \widehat{s}_{2z})|\chi_+(1)\chi_+(2)\rangle = \frac{\hbar}{2}(1+1)|\chi_+(1)\chi_+(2)\rangle = \hbar|\chi_+(1)\chi_+(2)\rangle$$

$$\widehat{S}_z|\chi_s\rangle = (\widehat{s}_{1z} + \widehat{s}_{2z})\frac{1}{\sqrt{2}}|\chi_+(1)\chi_-(2) + \chi_-(1)\chi_+(2)\rangle$$
$$= \frac{\hbar}{2}[(1-1) + (-1+1)]\frac{1}{\sqrt{2}}|\chi_+(1)\chi_-(2) + \chi_-(1)\chi_+(2)\rangle$$
$$= \frac{\hbar}{2}0|\chi_+(1)\chi_-(2) + \chi_-(1)\chi_+(2)\rangle$$

$$\widehat{S}_z|\chi_s\rangle = (\widehat{s}_{1z} + \widehat{s}_{2z})|\chi_-(1)\chi_-(2)\rangle = 2\left(-\frac{\hbar}{2}\right)|\chi_-(1)\chi_-(2)\rangle = -\hbar|\chi_-(1)\chi_-(2)\rangle$$

Also

$$\langle \chi_s | \hat{S}^2 | \chi_s \rangle = \langle \chi_s | (\hat{s}_1 + \hat{s}_2) \cdot (\hat{s}_1 + \hat{s}_2) | \chi_s \rangle = \langle \chi_s | (\hat{s}_1^2 + \hat{s}_2^2) + 2\hat{s}_1 \cdot \hat{s}_2 | \chi_s \rangle$$
$$= \langle \chi_s | (\hat{s}_1^2 + \hat{s}_2^2) + 2(\hat{s}_{+1} \cdot \hat{s}_{-2} + \hat{s}_{-1} \cdot \hat{s}_{+2} | \chi_s \rangle$$
$$= \langle \chi_s | \left[\left(\frac{3\hbar^2}{4} + \frac{3\hbar^2}{4} \right) + 2 \left(\frac{\hbar^2}{4} + \frac{\hbar^2}{4} \right) \right] | \chi_s \rangle = 2\hbar^2$$

where use has been made of

$$\hat{s}_x = \hat{s}_+ + \hat{s}_-, \quad \hat{s}_y = -i(\hat{s}_+ - \hat{s}_-)$$

and the orthonormality of the spin states (see Eq. (10.8)).

10.4. The quantum numbers associated with the ground and first excited states denoted by

$$u_{\alpha 1} = u_{100}\chi_+, \quad u_{\alpha 2} = u_{100}\chi_-, \quad u_{\alpha 3} = u_{200}\chi_+ \text{ or } u_{200}\chi_-$$

so that the three electron wave function is described by

$$\varphi(1,2,3) = \frac{1}{\sqrt{3!}} \begin{vmatrix} u_{\alpha 1}(1) & u_{\alpha 1}(2) & u_{\alpha 1}(3) \\ u_{\alpha 2}(1) & u_{\alpha 2}(2) & u_{\alpha 2}(3) \\ u_{\alpha 3}(1) & u_{\alpha 3}(2) & u_{\alpha 3}(3) \end{vmatrix}$$

The Hamiltonian of the Li-atom reads as

$$\hat{H} = \sum_{j=1}^{3} \hat{H}_{0j} + \hat{H}';$$

$$\hat{H}_{0j} = -\frac{\hbar^2}{2m}\nabla_j^2 + \frac{e_M^2}{r_j}, \quad \hat{H}' = e_M^2 \left(\frac{1}{r_{12}} + \frac{1}{r_{13}} + \frac{1}{r_{23}} \right), \quad e_M^2 = \frac{e^2}{4\pi\varepsilon_0}$$

where $\hat{H}_{0j}$ is the unperturbed hydrogenic component and $\hat{H}'$ accounts for the repulsive interactions among the three electrons. Since $\hat{H}$ acting on the eigenfunction yields

$$\left(\sum_{j=1}^{3} \hat{H}_{0j} + \hat{H}' \right) u_{100}(1)\, u_{100}(2)\, u_{200}(3)$$
$$= (2\,E_{100} + E_{200})\, u_{100}(1)\, u_{100}(2)\, u_{200}(3) + \hat{H}'\, u_{100}(1)\, u_{100}(2)\, u_{200}(3)$$

the total energy is given by

$$E_{\text{total}} = \langle u_{100}(1)\, u_{100}(2)\, u_{200}(3) | \left(\sum_{j=1}^{3} \hat{H}_{0j} + \hat{H}' \right) | u_{100}(1)\, u_{100}(2)\, u_{200}(3) \rangle$$
$$= (2E_{100} + E_{200}) + E_1,$$
$$E_1 = \langle u_{100}(1)\, u_{100}(2)\, u_{200}(3) | \hat{H} | u_{100}(1)\, u_{100}(2)\, u_{200}(3) \rangle$$

with E_1 denoting the first order correction due to the perturbing Hamiltonian. Also the z-component of the spin for example can be evaluated by using the spin functions as

$$(\hat{s}_{1z}+\hat{s}_{2z}+\hat{s}_{3z})\,\chi_+(1)\chi_-(2)\chi_\pm(3)\rangle = \frac{1}{2}\hbar(1-1\pm 1)\,\chi_+(1)\chi_-(2)\chi_+(3)\rangle$$

$$= \pm\frac{1}{2}\hbar\;\chi_+(1)\chi_-(2)\chi_+(3)\rangle$$

$$\langle\chi_+(1)\chi_-(2)\chi_\pm(3)|(\hat{s}_{1z}+\hat{s}_{2z}+\hat{s}_{3z})|\chi_+(1)\chi_-(2)\chi_\pm(3)\rangle = \pm\frac{1}{2}\hbar$$

The expectation value is determined by the spin states in u_{200}.

10.5. (a) The 11 electrons in Na-atom are assigned the following quantum numbers:

$$100\alpha,\ 100\beta;\ 200\alpha,\ 200\beta$$

$$211\alpha,\ 211\beta,\ 210\alpha,\ 210\beta$$

$$21-1\alpha,\ 21-1\beta,\ 300\alpha \text{ or } \beta$$

with α, β denoting the spin up and spin.

(b) The charge associated with 11 protons in the nucleus is screened in part, hence one has to introduce the effective atomic number Z_{eff}. Then the ionization energy required to knock out the valence electron in the outer orbit u_{300} is specified in terms of the ionization energy of the H-atom by

$$IP_1 = \frac{\mu Z_{\text{eff}}^2 e_M^4}{2\hbar^2}\frac{1}{3^2} = 13.6\,\text{eV}\ \frac{Z_{\text{eff}}^2}{9} = 5.14\,\text{eV}$$

Therefore $Z_{\text{eff}} \simeq 1.84$ and the atomic orbital can likewise be found in terms of the Bohr radius and Z_{eff} as

$$\langle r\rangle \sim \frac{a_0 n}{Z_{\text{eff}}} \sim \frac{3}{1.84} \times 0.05\,\text{nm} \simeq 0.08\,\text{nm}$$

10.6. (a) The shift in the energy level due to the spin orbit coupling is given from Eq. (10.40) by

$$\delta E_{so} \equiv \langle j, m_j|\hat{H}'_{so}|j, m_j\rangle = \frac{\hbar g_s \mu_B}{2}\langle f(r)\rangle[j(j+1) - l(l+1) - s(s+1)],$$

$$f(r) = \frac{Ze^2\mu_0}{8\pi m_e r^3}$$

The average value of $f(r)$ can be found with the use of the radial wavefunction but $< r >$ can also be approximated by the Bohr radius associated with each state.

Since the sum rule is given by $j = l \pm s$ there is no shift for $s = 0$ while for $l \neq 0$ the shift follows the rule

$$[(l+1/2)(l+1/2+1)] - l(l+1) - (1/2)(3/2) = l$$
$$[(l-1/2)(l-1/2+1)] - l(l+1) - (1/2)(3/2) = -(l+1)$$

Therefore the initial single level is split into two, one raised and the other lowered:

$$\delta E_{so} = (\Lambda/r_n^3)l; \quad -(\Lambda/r_n^3)(l+1), \quad \Lambda = \frac{\hbar g_s \mu_B}{2} \times \frac{Ze^2\mu_0}{8\pi m_e r_n^3}, r_n = na_0$$

(b) The longest wavelength of the Balmer series corresponds to the transition between $n=3$ to $n=2$. Hence the modification of the wavelength is given by

$$h\frac{c}{\lambda} = \delta E_{so} = [E_3 - (\Lambda/r_{n=3}^3)(l+1)] - [E_2 + (\Lambda/r_{n=2}^3)l]$$

Chapter 11

11.1. (a) The energy eigenequation

$$\left[-\frac{\hbar^2}{2m}\nabla^2 - e_M^2\left(\frac{1}{r_a} + \frac{1}{r_b}\right)\right](c_a u_a + c_a u_a) = E(c_a u_a + c_b u_b);$$
$$\left(-\frac{\hbar^2}{2m}\nabla^2 - \frac{e_M^2}{r_\alpha}\right)u_\alpha = E_0 u_\alpha, \alpha = a, b$$

can be rearranged as

$$c_a\left(E_0 - \frac{e_M^2}{r_b}\right)u_a + c_b\left(E_0 - \frac{e_M^2}{r_a}\right)u_b = E(c_a u_a + c_b u_b)$$

or

$$c_a\left(\Delta E - \frac{e_M^2}{r_b}\right)u_a + c_b\left(\Delta E - \frac{e_M^2}{r_a}\right)u_b = 0, \ \Delta E \equiv E_0 - E$$

Hence by carrying out the inner product on both sides with respect to u_a, u_b one obtains the coupled equation (11.5):

$$(\Delta E + C)c_a + (\Delta ES + D)c_b = 0$$
$$(\Delta ES + D)c_a + (\Delta E + C)c_b = 0$$

(b) Since the equation is homogeneous, the solution of the expansion coefficients c_a, c_b will be trivial unless the secular equation is satisfied,

$$\begin{vmatrix} \Delta E + C & \Delta ES + D \\ \Delta ES + D & \Delta E + C \end{vmatrix} = 0, \quad (\Delta E + C)^2 - (\Delta ES + D)^2 = 0$$

and ΔE can be found from the resulting quadratic equation and the eigenenergies (Eq. (11.10b)) are specified as

$$E_{\mp} = E_0 + \frac{C \mp D}{1 \mp S}$$

When these two roots are inserted into either one of the coupled equations c_a, c_b get related with each other and one can therefore obtain the corresponding wavefunctions as

$$\varphi_{\mp}(\underline{r}_a, \underline{r}_b) = \frac{1}{\sqrt{2}}(u_a \mp u_b); \quad c_b = \mp c_a$$

with the eigenvalues $E_{\mp}$.

11.2. Given the eigenfunction

$$|\varphi_{\pm}\rangle \equiv |u_a(\underline{r_1})u_b(\underline{r_2}) \pm u_b(\underline{r_1})u_a(\underline{r_2})\rangle$$

one has to evaluate the integral

$$\langle\varphi_{\pm}|\hat{H}|\varphi_{\pm}\rangle; \quad \hat{H} = \hat{H}_1 + \hat{H}_2 + \frac{e_M^2}{R_{ab}} + \frac{(-e_M^2)}{r_{b1}} + \frac{(-e_M^2)}{r_{a2}} + \frac{e_M^2}{r_{12}}$$

Since $\hat{H}$ is invariant under the exchange of $\boldsymbol{r}_1$ and $\boldsymbol{r}_2$ the four combinations of u_a, u_b

$$u_a(1)u_b(2)u_a(1)u_b(2), \quad j = \underline{r}_j \tag{A}$$

$$\pm u_a(1)u_b(2)u_b(1)u_a(2) \tag{B}$$

$$\pm u_b(1)u_a(2)u_a(1)u_b(2) \tag{C}$$

$$u_b(1)u_a(2)u_b(1)u_a(2) \tag{D}$$

reduce to two, since $A = D$, $B = C$ under the exchange of r_1, r_2 and therefore one needs to perform the integrations involving only A and B:

$$\langle u_a(1)u_b(2)|\hat{H}|u_a(1)u_b(2)\rangle = 2E_0 + \frac{e_M^2}{R_{ab}} + 2C + E_{RI} \tag{11.1}$$

where

$$C \equiv \langle u_a(1)|\frac{-e_M^2}{r_{b1}}|u_a(1)\rangle = \langle u_b(2)|\frac{-e_M^2}{r_{a2}}|u_b(2)\rangle$$

$$E_{RI} \equiv \langle u_a(1)u_b(2)|\frac{e_M^2}{r_{12}}|u_a(1)u_b(2)\rangle$$

Also,

$$\pm\langle u_a(1)u_b(2)|\hat{H}|u_b(1)u_a(2)\rangle = \pm\left[2E_0S^2 + \frac{e_M^2}{R_{ab}}S^2 + 2DS + E_{CE}\right] \tag{11.2}$$

where

$$S \equiv \langle u_a(1)|u_b(1)\rangle = \langle u_a(2)|u_b(2)\rangle$$

$$D \equiv \langle u_b(1)|\frac{-e_M^2}{r_{b1}}|u_a(1)\rangle = \langle u_a(2)|\frac{-e_M^2}{r_{a2}}|u_b(2)\rangle$$

$$E_{CE} = \langle u_b(1)u_a(2)|\frac{e_M^2}{r_{12}}|u_a(1)u_b(2)\rangle$$

Hence by adding Eqs. (11.1) and 11.2 therefore results

$$E_\pm = \frac{\langle\varphi_\pm|\hat{H}|\varphi_\pm\rangle}{\langle\varphi_\pm|\varphi_\pm\rangle} = \frac{2\left[2E_0\left(1 \pm S^2\right) + \frac{e_M^2}{R_{ab}}(1 \pm S^2) + 2C + E_{RI} \pm 2DS \pm E_{CE}\right]}{2(1 \pm S^2)}$$

$$= 2E_0 + \frac{e_M^2}{R_{ab}} + \frac{2C + E_{RI}}{1 \pm S^2} \pm \frac{2DS + E_{CE}}{1 \pm S^2}$$

and one thus obtains

$$E_b \equiv E_\pm - 2E_0 = \frac{e_M^2}{R_{ab}} + \frac{2C + E_{RI}}{1 \pm S^2} \pm \frac{2DS + E_{CE}}{1 \pm S^2}$$

11.3. Given $R_e = 0.24$ nm, $\nu_e = 1.1 \times 10^{13}\,\mathrm{s}^{-1}$ and also the value of the reduced mass

$$\frac{1}{\mu} = \frac{1}{m_{\mathrm{Na}}} + \frac{1}{m_{\mathrm{Cl}}}, \quad \mu = 23.3 \times 10^{-27}\,\mathrm{kg}$$

the parameters A, α and the bonding energy can be found as follows. One can start with Eq. (11.24)

$$\Delta E(R) = Ae^{-\alpha R} - \frac{e_M^2}{R} + \Delta E(\infty), \quad \Delta E(\infty) = 1.49\,\mathrm{eV}, \quad e_M = \frac{e}{4\pi\varepsilon_0}$$

Since Re represents the intermolecular distance at which $\Delta E(R)$ is at its minimum value one can write

$$\left.\frac{\partial \Delta E(R)}{\partial R}\right|_{R_e} = 0 = -A\alpha e^{-\alpha R_e} + \frac{e_M^2}{R_e^2} \tag{11.3}$$

Also the effective spring constant is specified by definition as

$$k \equiv \mu\omega_e^2 = \mu(2\pi\nu_e)^2$$

$$\equiv \left.\frac{\partial^2 \Delta E(R)}{\partial R^2}\right|_{R_e} = A\alpha^2 e^{-\alpha R_e} - 2\frac{e_M^2}{R_e^3} \tag{11.4}$$

Hence by combining Eqs. (11.3) and (11.4) one can write

$$\alpha\frac{e_M^2}{R_e^2} - 2\frac{e_M^2}{R_e^3} = \mu\omega_e^2 \tag{11.5}$$

and find α by using given values of $R_e = 0.24\,\text{nm}$, $\nu_e = \omega_e/2\pi = 1.1\times 10^{13}$ as

$$\alpha = \frac{2}{R_e} + \frac{\mu\omega_e^2 R_e^2}{e_M^2} \simeq 3.6\times 10^{10}\,\text{m}^{-1}$$

With α thus found A can in turn be determined from Eq. (11.3) as

$$A = \frac{1}{\alpha}\frac{e_M^2}{R_e^2}e^{\alpha R_e} \simeq 6.28\times 10^{-16}\,\text{J} = 3.92\times 10^3\,\text{eV}$$

With A, α thus determined one can obtain the bonding energy by evaluating ΔE at R_e:

$$\Delta E(R_e) = Ae^{-\alpha R_e} - \frac{e_M^2}{R_e} + 1.49\,\text{eV} \simeq -3.82\,\text{eV}$$

11.4. (a) Given the Hamiltonian

$$\hat{H} = \hat{H}_0 + \text{E}er\cos\theta, \quad \hat{H}_0 = -\frac{\hbar^2}{2m}\nabla^2 - \frac{e_M^2}{r}$$

and the wavefunction consisting of the eigenfunctions of $\hat{H}_0$

$$\varphi = c_1|u_{100}\rangle + c_2|u_{210}\rangle$$

the energy eigenequation reads as

$$\begin{aligned}\hat{H}\varphi &= E_1c_1|u_{100}\rangle + \text{E}er\cos\theta c_1|u_{100}\rangle + E_2c_2|u_{210}\rangle + \text{E}er\cos\theta c_2|u_{210}\rangle \\ &= E(c_1|u_{100}\rangle + c_2|u_{210}\rangle)\end{aligned}$$

or

$$(E_1 - E + \text{E}er\cos\theta)c_1|u_{100}\rangle + (E_2 - E + \text{E}er\cos\theta)c_2|u_{210}\rangle = 0$$

One can take the inner product on both sides with respect to u_{100}, u_{210} by making use of the orthonomality of the two eigenfunctions, and obtain the coupled equations

$$(E_1 - E)c_1 + m_{12}c_2 = 0, \quad m_{12} = \langle u_{100}|\text{E}er\cos\theta|u_{210}\rangle$$

$$m_{12}{}^*c_1 + (E_2 - E)c_2 = 0, \quad m_{12}{}^* = \langle u_{210}|\text{E}er\cos\theta|u_{100}\rangle$$

(b) The coupled equations are homogeneous, that is, the right hand sides are zero. Thus to avoid the trivial solution the secular equation has to be satisfied:

$$\begin{vmatrix} E_1 - E & m_{12} \\ m_{12}^* & E_2 - E \end{vmatrix} = 0, \quad E^2 - (E_1 + E_2)E + E_1E_2 - |m_{12}|^2 = 0$$

The roots of the quadratic equation constitutes the eigenenergy, which can be found as

$$\begin{aligned} E_\pm &= \frac{1}{2}\{(E_1 + E_2) \pm [(E_1 + E_2)^2 - 4(E_1E_2 - |m_{12}|^2)]^{1/2}\} \\ &= \frac{1}{2}\{(E_1 + E_2) \pm [(E_2 - E_1)^2 + 4|m_{12}|^2]^{1/2}\} \\ &= \frac{1}{2}\{(E_1 + E_2) \pm [\Delta E^2 + 4|m_{12}|^2]^{1/2}\}, \quad \Delta E = E_2 - E_1 \end{aligned}$$

With E_+, E_- inserted in either one of the coupled equations c_1, c_2 are interrelated as

$$\begin{aligned} \alpha_+ &\equiv \frac{c_2}{c_1} = -\frac{m_{12}^*}{E_2 - E_+} = -\frac{2m_{12}^*}{[\Delta E - (\Delta E^2 + 4|m_{12}|^2)^{1/2}]} \\ \alpha_- &\equiv \frac{c_2}{c_1} = -\frac{2m_{12}^*}{[\Delta E + (\Delta E^2 + 4|m_{12}|^2)^{1/2}]} \end{aligned}$$

Hence one can write the wavefunction as

$$\varphi_\pm = c_1(|u_{100}\rangle + \alpha|u_{210}\rangle), \quad \alpha = \alpha_+ = \alpha_-$$

where c_1 can be used for normalizing the eigenfunction.

(c) With c_1 determined from the normalization, that is,

$$1 = \langle\varphi|\varphi\rangle = |c_1|^2\langle u_{100} + \alpha u_{210}|u_{100} + \alpha u_{200}\rangle = |c_1|^2(1 + |\alpha|^2)$$

one can specify the atom dipole as

$$\begin{aligned} \langle \underline{r}\rangle &= \langle\varphi|\underline{r}|\varphi\rangle = |c_1|^2\langle u_{100} + \alpha u_{210}|\underline{r}|u_{100} + \alpha u_{210}\rangle \\ &= [\alpha\langle u_{100}|\underline{r}|u_{210}\rangle + c.c.]/(1 + |\alpha|^2), \quad \langle u_{100}|\underline{r}|u_{100}\rangle = \langle u_{210}|\underline{r}|u_{210}\rangle = 0 \\ &\propto \mathrm{E} \end{aligned}$$

Clearly the atomic dipole is driven by and is proportional to the electric field and the proportionality constant is the atomic susceptibility.

11.5. Given the Hamiltonian

$$\hat{H} = -\frac{\hbar^2}{2m}\frac{\partial^2}{\partial x_1{}^2} + \frac{1}{2}kx_1{}^2 - \frac{\hbar^2}{2m}\frac{\partial^2}{\partial x_2{}^2} + \frac{1}{2}kx_2{}^2 - \frac{2e_M^2x_1x_2}{R^3}$$

one may introduce new variables

$$\xi = x_1 + x_2, \quad \eta = x_2 - x_1$$

By inverting one can write

$$x_1 = \frac{1}{2}(\xi - \eta),\ x_2 = \frac{1}{2}(\xi + \eta)$$

Thus the Hamiltonian can be expressed in terms of new variables as follows.

$$\frac{\partial}{\partial x_1} = \frac{\partial}{\partial \xi}\frac{\partial \xi}{\partial x_1} + \frac{\partial}{\partial \eta}\frac{\partial \eta}{\partial x_1} = \frac{\partial}{\partial \xi} - \frac{\partial}{\partial \eta}$$

$$\frac{\partial^2}{\partial {x_1}^2} = \frac{\partial}{\partial \xi}\left(\frac{\partial}{\partial \xi} - \frac{\partial}{\partial \eta}\right)\frac{\partial x_1}{\partial \xi} + \frac{\partial}{\partial \eta}\left(\frac{\partial}{\partial \xi} - \frac{\partial}{\partial \eta}\right)\frac{\partial \eta}{\partial x_1} = \frac{\partial^2}{\partial \xi^2} - 2\frac{\partial^2}{\partial \xi \partial \eta} + \frac{\partial^2}{\partial \eta^2}$$

Similarly

$$\frac{\partial^2}{\partial {x_2}^2} = \frac{\partial^2}{\partial \xi^2} + 2\frac{\partial^2}{\partial \xi \partial \eta} + \frac{\partial^2}{\partial \eta^2}$$

Hence one can write

$$-\frac{\hbar^2}{2m}\left(\frac{\partial^2}{\partial {x_1}^2} + \frac{\partial^2}{\partial {x_2}^2}\right) = -\frac{\hbar^2}{2m}\cdot 2\left(\frac{\partial^2}{\partial \xi^2} + \frac{\partial^2}{\partial \eta^2}\right)$$

The potential energy is also expressed as

$$\frac{1}{2}k({x_1}^2 + {x_2}^2) = \frac{1}{2}k\left[\frac{1}{4}(\xi - \eta)^2 + \frac{1}{4}(\xi + \mu)^2\right] = \frac{1}{4}k(\xi^2 + \eta^2)$$

$$-\frac{2e_M^2}{R^3}x_1 x_2 = -\frac{2e_M^2}{R^3}\frac{1}{4}(\xi - \eta)(\xi + \eta) = -\frac{e_M^2}{2R^3}(\xi^2 - \eta^2)$$

Therefore by summing the kinetic and potential energy terms one obtains

$$\hat{H} = -\frac{\hbar^2}{2\mu}\frac{\partial^2}{\partial \xi^2} - \frac{\hbar^2}{2\mu}\frac{\partial^2}{\partial \eta^2} + \frac{1}{2}k_-\xi^2 + \frac{1}{2}k_+\eta^2$$

where

$$\frac{1}{\mu} = \frac{1}{m} + \frac{1}{m},\quad \mu = \frac{m}{2},\quad k_\mp = \frac{1}{2}k \mp \frac{e_M^2}{R^3}$$

Chapter 12

12.1. Given the energy eigenequation (12.4) one can use the Born approximation (Eq. (12.5)) and divide both sides by φ_e, obtaining

$$\left[-\frac{\hbar^2}{2m}\nabla_R^2 + E(R)\right]\chi(\underline{R}) = E\chi(\underline{R})$$

where $E(R)$ is the eigenenergy of electrons in the molecule and should now be treated as a function of R since the nuclei are not fixed but undergo the motion. Also, one can look for the solution in the form

$$\chi(R, \theta, \varphi) = \rho(R)Y_L^M(\theta, \varphi)$$

and insert into the equation and divide both sides with the spherical harmonics, obtaining

$$-\frac{\hbar^2}{2\mu R^2}\frac{\partial}{\partial R}\left(R^2\frac{\rho(R)}{\partial R}\right)+\frac{\hbar^2 L(L+1)}{2\mu R^2}\rho(R)+E(R)\rho(R)=E\rho(R)$$

Here use has been made of the fact that (i) the angular momentum operator naturally enters into the Laplacian when it is expressed in spherical coordinates (see (Eqs. (8.16) and (8.34)), (ii) the spherical harmonics are the eigenfunctions of the angular momentum operator and is cancelled from both sides, and (iii) the reduced mass of the two nuclei enters naturally in describing the internal motion of the two nuclei (see Eqs. (8.28) and (8.29)).
The total energy consists of the energy eigenvalue of the electrons attached to the nuclei and the vibrational and rotational energies of the two nuclei (see Eq. (12.9a)). Now the minimum value $E(R_e)$ represents the eigenenergy of the electrons with R_e denoting the equilibrium distance between the two nuclei. Consequently the radial equation (12.10) ensues.

12.2. (a) The moment of inertia of the two particle system with mass m_1, m_2 and displacement x_1 and x_2 from the fixed center is given by

$$I = m_1x_1^2 + m_2x_2^2$$

By introducing the center of mass and relative coordinates as

$$MX = m_1x_1 + m_2x_2, \quad x = x_1 - x_2, \qquad M = m_1 + m_2$$

and inverting the relation one finds

$$x_1 = X + \frac{m_2x}{M}, \quad x_2 = X - \frac{m_1x}{M}$$

Hence the moment of inertia can be specified in terms of X and x as

$$\begin{aligned} m_1x_1^2 + m_2x_2^2 &= m_1\left(X+\frac{m_2x}{M}\right)^2 + m_2\left(X-\frac{m_1x}{M}\right)^2 \\ &= MX^2 + \frac{m_1m_2}{m_1+m_2}x^2 \\ &= MX^2 + \mu x^2; \quad \frac{1}{\mu} = \frac{1}{m_1} + \frac{1}{m_2} \end{aligned}$$

with μ denoting the reduced mass. The first term accounts for the two nuclei moving together as a free particle while the second term represents the moment of inertia.

(b) To find the moments of inertia of H_2, HCl the respective reduced mass has to be found first:

$$\frac{1}{\mu_{H_2}} = \frac{1}{m_p} + \frac{1}{m_p} = \frac{2}{m_p}, \quad \mu_{H_2} = 8.4\times10^{-28}\,\text{kg}$$

$$\frac{1}{\mu_{HCl}} = \frac{1}{m_p} + \frac{1}{35.5m_p} = \frac{1}{m_p}\times 1.03, \quad \mu_{HCl} = 1.62\times10^{-27}\,\text{kg}$$

Here the mass of the atom was taken as the proton mass m_p times the mass number. Hence, $I(H_2) = 4.6 \times 10^{-48}$ kgm^2, $I(HCL) = 2.6 \times 10^{-47}$ kgm^2

(c) The rotational energy is given from Eq. (12.11a) by

$$E_r = \frac{\hbar^2 L(L+1)}{2I}; \quad I = \mu R_e^2, \ L = 0, 1, 2, \ldots$$

with I denoting the moment of inertia. The difference in energy involved in the transition from $L = 2$ to $L = 1$ is given by

$$E_r = \frac{\hbar^2}{2I}[2(2+1) - 1(1+1)] = \frac{2\hbar^2}{I}$$

The transition frequency is therefore found via the relation $h\nu = 2 \times \hbar^2/2\mu R_e^2$, that is,

$$\nu = \frac{2\hbar^2}{Ih} = \frac{2}{(2\pi)^2}\left(\frac{h}{I}\right)$$

Hence upon inserting the moment of inertia found in (b) one finds

$$\nu(H_2) = \frac{6.626 \times 10^{-34}\,(\mathrm{J\,s})}{2 \times \pi^2 \times 4.6 \times 10^{-48}\,(\mathrm{kg\,m^2})} = 7.3 \times 10^{12}\,\mathrm{Hz}$$

Likewise one finds

$$\nu(HCl) = 1.31 \times 10^{12}\,\mathrm{Hz}$$

12.3. (a) The reduced mass of each molecule has been found in the previous problem.

(b) The effective spring constants k can therefore be found as follows. By definition one can write from Eq. (7.1)

$$\frac{k}{\mu} = \omega^2 = (2\pi\nu)^2 = 4\pi^2 \times \left(\frac{c}{\lambda}\right)^2 = 4\pi^2 c^2 \times \tilde{\nu}^2; \quad \tilde{\nu} \equiv \frac{1}{\lambda}$$

Thus k can be specified in terms of the reduced mass μ, the velocity of light c and the wave number $\tilde{\nu}$. By inserting all the values given and converting them into the MKS unit one finds

$$k_{\mathrm{eff}}(H_2) = 580\,\mathrm{N\,m^{-1}}, \quad k_{\mathrm{eff}}(\mathrm{HCl}) = 990\,\mathrm{N\,m^{-1}}$$

12.4. The energy eigenequation of the 2D HO reads read as

$$\hat{H}\varphi(x,y) = E\varphi(x,y)$$

with the Hamiltonian given by

$$\hat{H} = \hat{H}_x + \hat{H}_y = \left[-\frac{\hbar^2}{2m}\frac{\partial^2}{\partial x^2} + \frac{1}{2}k_x x^2\right] + \left[-\frac{\hbar^2}{2m}\frac{\partial^2}{\partial y^2} + \frac{1}{2}k_y y^2\right]$$

As usual one can look for the solution in the form

$$\varphi(x,y) = u(x)u(y)$$

and insert into the eigenequation and divide both sides by $\varphi(x, y)$, obtaining

$$\frac{\hat{H}_x u(x)}{u(x)} + \frac{\hat{H}_y u(y)}{u(y)} = E$$

Since each term on the left hand side depends solely on x, y the only way to satisfy the equation is to put each term to a constant. In this case one obtains two independent ID harmonic oscillator eigenequations:

$$\hat{H}_x u(x) = E_x u(x), \quad \hat{H}_y u(y) = E_y u(y), \quad E_x + E_y = E$$

Hence one can transcribe all of the results obtained for analyzing the 1D HO in Chapter 7. Specifically the eigenenergy is given by

$$E(n_x, n_y) = \hbar\omega_x\left(n_x + \frac{1}{2}\right) + \hbar\omega_y\left(n_y + \frac{1}{2}\right); \quad \omega_\alpha = \left(\frac{k_\alpha}{m}\right)^{1/2}, \quad \alpha = x, y$$

and the corresponding eigenfunctions are given in Eq. (7.25). If $k_x = k_y = k$ the total energy reads as

$$E(n_x, n_y) = \hbar\omega(n_x + n_y + 1); \quad \omega = \left(\frac{k}{m}\right)^{1/2}$$

and is degenerate aside from the ground state, in which $n_x = n_y = 0$. For the first excited state there is two-fold degeneracy corresponding to $n_x = 1$ and $n_y = 0$ and vice versa. The degree of degeneracy increases in higher lying states.

12.5. (a) One can find the effective masses required by following the steps presented in 12.2. The reduced masses thus found are listed below:

$$\mu_{D2} = 1.67 \times 10^{-27}\,\text{kg}, \quad \mu_{CO} = 1.15 \times 10^{-26}\,\text{kg}$$
$$\mu_{O_2} = 1.33 \times 10^{-26}\,\text{kg}, \quad \mu_{NaCl} = 2.3 \times 10^{-26}\,\text{kg}$$

(b) With the use of the reduced mass and the wave number the effective spring constant given by

$$\frac{k}{\mu} = \omega^2 = (2\pi\nu)^2 = 4\pi^2 \times \left(\frac{c}{\lambda}\right)^2 = 4\pi^2 c^2 \times \tilde{\nu}^2; \quad \tilde{\nu} \equiv \frac{1}{\lambda}$$

can be evaluated as

$$k_{\text{eff D2}} = 5.96 \times 10^1\,\text{N m}^{-1}, \quad k_{\text{eff CO}} = 2.08 \times 10^2\,\text{N m}^{-1}$$
$$k_{\text{eff O}_2} = 1.27 \times 10^2\,\text{N m}^{-1}, \quad k_{\text{NaCl}} = 1.29 \times 10^1\,\text{N m}^{-1}$$

(c) Now that μ and k_{eff} have been found the respective zero point energy

$$E_0 = \frac{1}{2}h\nu = \frac{1}{2}h\left[\left(\frac{1}{2\pi}\right)\left(\frac{k_{\text{eff}}}{\mu}\right)^{1/2}\right]$$

can also be evaluated as

$$E_{0\,\mathrm{D2}} = 9.9 \times 10^{-21}\,\mathrm{J} = 6.14 \times 10^{-2}\,\mathrm{eV},\ E_{0\,\mathrm{CO}} = 7.08 \times 10^{-21}\,\mathrm{J} = 4.4 \times 10^{-2}\,\mathrm{eV}$$
$$E_{\mathrm{O}_2} = 5.15 \times 10^{-21}\,\mathrm{J} = 3.19 \times 10^{-2}\,\mathrm{eV},\ E_{0\,\mathrm{NaCl}} = 1.77 \times 10^{-21}\,\mathrm{J} = 1.1 \times 10^{-2}\,\mathrm{eV}$$

12.6 (a) One has to find the reduced mass first, which is given in this case by

$$\frac{1}{\mu} = \frac{1}{1.67 \times 10^{-27}\,\mathrm{kg}} + \frac{1}{58.066 \times 10^{-27}\,\mathrm{kg}}$$

Thus one finds $\mu = 1.62 \times 10^{-27}$ kg. The vibrational frequency can then be evaluated as

$$\nu_{\mathrm{vib}} = \frac{1}{2\pi}\omega = \frac{1}{2\pi}\left(\frac{k}{\mu}\right)^{1/2} = \frac{1}{2\pi}\left(\frac{516.3\,\mathrm{N\,m^{-1}}}{1.62 \times 10^{-27}}\right)^{1/2} = 8.9 \times 10^{13}\,\mathrm{Hz}$$

(b) The P and R branches are given from Eqs. (12.23) and (12.24) by

$$\widetilde{\nu}_P = \widetilde{\nu}_{\mathrm{vib}} - 2\widetilde{\beta}L_i\ :\quad \widetilde{\nu}_R = \widetilde{\nu}_{\mathrm{vib}} + 2\widetilde{\beta}(L_i + 1),\quad L_i = 1, 2, 3, \ldots$$

with

$$\widetilde{\beta} = \left(\frac{\hbar^2}{2\mu R_e^2}\right)\left(\frac{1}{ch}\right) = \frac{1}{16\pi^2}\frac{6.626 \times 10^{-34}\,\mathrm{J\,s}}{1.62 \times 10^{-27}\mathrm{kg} \times (1.27 \times 10^{-10})\,\mathrm{m}^2 \times 3 \times 10^8\mathrm{m\,s^{-1}}}$$
$$= 5.4 \times 10^2\,\mathrm{m}^{-1}$$

Therefore the associated frequencies are given by

$$\nu_p \equiv c \times \widetilde{\nu}_p = \nu_{\mathrm{vib}} - 2c\widetilde{\beta}L_i = (8.9 \times 10^{13} - 3.2 \times 10^{11}L_i)\,\mathrm{Hz}$$
$$\nu_R \equiv c \times \widetilde{\nu}_R = \nu_{\mathrm{vib}} + 2c\widetilde{\beta}(L_i + 1) = (8.9 \times 10^{13} + 3.2 \times 10^{11}(L_i + 1))\,\mathrm{Hz}$$

The frequencies of the three innermost P and R lines can be found by putting $L_i = 1, 2, 3$. Indeed the rotational frequencies are lower than the vibrational frequencies by about 2 orders of magnitudes. The zero point energy can be obtained in the usual manner.

12.7. One can treat the flipping of the proton spin in a manner identical to the paramagnetic electron spin resonance. Thus the spin function can be represented in analogy with Eq. (10.51) in terms of the nuclear spin up and spin down states:

$$\chi_N(t) = c_+(t)e^{-i(\omega_0/2)t}\chi_+ + c_-(t)e^{i(\omega_0/2)t}\chi_-;\quad \frac{\hbar\omega_0}{2} = \mu_{BN}B_0 \tag{12.1}$$

where B_0 is the static magnetic field applied and the nuclear magnetic moment μ_N is smaller than that of electron by three orders of magnitudes. The Hamiltonian of the proton interacting with the magnetic field is given by (see Eq. (10.47))

$$\hat{H} = \hat{H}_0 + \hat{H}' = \frac{g_N \mu_B}{\hbar}(B_0\hat{s}_{zN} + B_1(t)\hat{s}_{xN}),\quad B_1(t) = \frac{B_1}{2}(e^{i\omega t} + e^{-i\omega t})$$

Thus the Schrödinger equation reads as

$$i\hbar\frac{\partial}{\partial t}\chi_N(t) = \hbar\sum_{\pm}[\pm(\omega_0/2)c_{\pm}(t) + i\dot{c}_{\pm}(t)]e^{\mp(i\omega_0 t/2)}\chi_{\pm}$$
$$= (\widehat{H}_0 + \widehat{H}')\chi(t) = \sum_{\pm}\left[\pm g_N\mu_B B_0 + \frac{g_N\mu_B}{\hbar}B_1(t)\widehat{s}_{xN}\right]c_{\pm}(t)e^{\mp i(\omega_0 t/2)}\chi_{\pm} \tag{12.2}$$

where $g_{\rm N}$ is the gyromagnetic ratio of the proton and the eigenequation of $\chi_{\pm}$ is given by

$$\widehat{H}_0\chi_{\pm} = \pm g_N\mu_B B\chi_{\pm}, \quad g_N\mu_B B = \frac{1}{2}\hbar\omega$$

Evidently the first terms of the both sides of Eq. (12.2) cancel out, so that Eq. (12.2) reduces to

$$i\hbar\sum_{\pm}\dot{c}_{\pm}(t)e^{\mp(i\omega_0 t/2)}\chi_{\pm}\dot{c}_{\pm}(t) = \frac{g_N\mu_B}{\hbar}\sum_{\pm}B_1(t)\widehat{s}_{xN}c_{\pm}(t)e^{\mp(i\omega_0 t/2)}\chi_{\pm} \tag{12.3}$$

Hence by taking inner products on both sides of Eq. (12.3) with respect to χ_+ and χ_- and making use of the orthonormality of the spin functions one obtains the coupled equations

$$i\hbar\dot{c}_+ = \frac{g_N\mu_{BN}}{2}B_1 e^{-i\Delta t}c_-, \quad \Delta = \omega - \omega_0$$
$$i\hbar\dot{c}_- = \frac{g_N\mu_{BN}}{2}B_1 e^{i\Delta t}c_+$$

where $\hbar\omega_0$ is the difference in energy between the spin up and down states, Delta the frequency detuning between the driving frequency ω and ω_0. Also the coupling is caused by the spin flip operators

$$\langle\chi_+|\widehat{s}_{xN}|\chi_-\rangle = \langle\chi_+|\frac{1}{2}(\widehat{s}_{+N} + \widehat{s}_{-N})|\chi_-\rangle = \frac{1}{2}\hbar$$

(see Eqs. (10.1)–(10.3)). The coupled equation can be treated in a manner identical to the treatment of the electron paramagnetic resonance. Therefore the results obtained in Eqs. (10.55)–(10.57) can be directly used.

Chapter 13

13.1. (a) When the wavefunction (13.12) is inserted into the wave equation (13.11) there results

$$i\hbar\sum_{j=1}^{2}[-i\omega_j a_j(t) + \dot{a}_j(t)]e^{-i(E_j/\hbar)t}|u_j\rangle$$
$$= \sum_{j=1}^{2}E_j a_j(t)e^{-i(E_j/\hbar)t}|u_j\rangle + \widehat{H}'(t)\sum_{j=1}^{2}a_j(t)e^{-i(E_j/\hbar)t}|u_j\rangle, \quad \omega_j = E_j/\hbar$$

Obviously the first terms on both sides are identical and cancel each other out and equation reduces to

$$i\hbar\sum_{j=1}^{2}\dot{a}_j(t)\mathrm{e}^{-i(E_j/\hbar)t}|u_j\rangle = \hat{H}'(t)\sum_{j=1}^{2}a_j(t)\mathrm{e}^{-i(E_j/\hbar)t}|u_j\rangle,\ \hat{H}'(t) = -e\mathrm{E}(t)(\hat{e}_{\mathrm{f}}\cdot r)$$

One can perform the inner product on both sides with respect to u_1, u_2, obtaining

$$\begin{aligned} i\hbar\dot{a}_1(t)e^{-i\omega_1 t} &= -\widetilde{\mu}\mathrm{E}(t)a_2(t)e^{-i\omega_2 t},\ \widetilde{\mu}\equiv e\langle 1|\hat{e}_f\cdot \underline{r}|2\rangle,\ \hbar\omega_j = E_j \\ i\hbar\dot{a}_2(t)e^{-i\omega_2 t} &= -\widetilde{\mu}\mathrm{E}(t)a_1(t)e^{-i\omega_1 t} \end{aligned}$$

where the orthonormality of u_1, u_2 has been used and also the fact that $\langle u_j|\hat{H}'|u_j\rangle = 0$. The coupled equations can be rearranged as

$$\begin{aligned} \dot{a}_1 &= i\frac{\widetilde{\mu}\mathrm{E}(t)}{\hbar}a_2e^{-i\omega_0 t},\ \omega_0 \equiv \frac{E_2-E_1}{\hbar} \\ \dot{a}_2 &= i\frac{\widetilde{\mu}\mathrm{E}(t)}{\hbar}a_1e^{-i\omega_0 t} \end{aligned} \tag{13.1}$$

(b) The solutions of Eq. (13.1) has been obtained in the text and one can use Eq. (13.15) and write

$$a_1(t) = \cos\Omega t;\ \ a_2(t) = i\sin\Omega t$$

and describe the evolution in time of the atom dipole moment as

$$\begin{aligned} \langle\hat{\mu}\rangle &= \langle\psi|\hat{\mu}|\Psi\rangle;\ \Psi(r,t) = \cos(\Omega t)e^{-i\omega_1 t}u_1(r) + i\sin(\Omega t)e^{-i\omega_2 t}u_2(r) \\ &= \widetilde{\mu}i\{\sin(\Omega t)\cos(\Omega t)[e^{-i\omega_0 t} - e^{i\omega_0 t}],\ \ \widetilde{\mu}\equiv e\langle 1|\hat{e}_f\cdot\underline{r}|2\rangle \\ &= \widetilde{\mu}\sin(2\Omega t)\sin\omega_0 t \end{aligned}$$

where the dipole moment is specified by the integration

$$\langle\psi|\hat{\mu}|\psi\rangle \equiv \int_{-\infty}^{\infty}\mathrm{d}\underline{r}(\psi^*\hat{\mu}\psi)$$

Thus the atom dipole moment oscillates with the atomic transition frequency ω_0 while the magnitude of the moment evolves in time with the transition frequency Ω.

13.2. (a) With the wavefunction

$$\psi(\underline{r},t) = \sum_{j=1}^{2}a_{js}(t)|u_j\rangle;\ \ \hat{H}_0|u_j\rangle = E_j|u_j\rangle,\ j=1,2 \tag{13.1}$$

used in the wave equation (13.11), it reads as

$$i\hbar\sum_{j=1}^{2}\dot{a}_{js}(t)|u_j\rangle = \sum_{j=1}^{2}E_ja_{js}(t)|u_j\rangle + \hat{H}'(t)\sum_{j=1}^{2}a_{js}(t)|u_j\rangle$$

By performing the inner product on both sides with respect to u_1, u_2, one obtains

$$
\begin{aligned}
i\hbar\dot{a}_{1s}(t) &= E_1 a_{1s} - \tilde{\mu}\mathrm{E}(t)a_{2s}(t), \ \ \tilde{\mu} \equiv e\langle 1|\hat{e}_f \cdot \underline{r}|2\rangle \\
i\hbar\dot{a}_{2s}(t) &= E_2 a_{2s} - \tilde{\mu}\mathrm{E}(t)a_{1s}(t)
\end{aligned}
\tag{13.2}
$$

Here the orthonormality of the two eigenfunctions has been used together with the fact that the diagonal matrix elements of $\hat{H}'$ vanish due to the odd parity of the integrand involved.

(b) Since $a_{js}(t) = a_j(t) \exp -(i\omega_j t)$, one can write

$$
\dot{a}_{js} = \frac{d}{dt} a_j e^{-i\omega_j t} = \dot{a}_j e^{-i\omega_j t} + (-i\omega_j) a_j e^{-i\omega_j t}, \ \ j = 1, 2
$$

so that when $a_{js}(t)$ is replaced by $a_j(t)$ the coupled equations (13.2) are reduced to Eq. (13.13) in the text.

(c) Now that the two sets of the coupled equations are shown identical, one can use the solution obtained for $a_j(t)$ under the same initial condition and write

$$
\begin{aligned}
a_{1s}(t) &= a_1(t)e^{-i\omega_1 t} = \cos\Omega t e^{-i\omega_1 t} \\
a_{2s}(t) &= a_2(t)e^{-i\omega_2 t} = i\sin\Omega t e^{-i\omega_2 t}
\end{aligned}
\tag{13.3}
$$

When Eq. (13.3) is used in the representation of the wavefunction (13.1), it becomes identical to the wavefunction expressed in terms of $a_j(t)$ and therefore the description of the atom dipole moment should be the same (see Eq. (13.8)).

13.3 (a) With the electric and magnetic field given in Eq. (13.19)

$$
\underline{\mathrm{E}}_l = \hat{y}\sqrt{\frac{2}{V\varepsilon}} p_l(t) \sin k_l z, \ \ \underline{H}_l = \hat{x}\sqrt{\frac{2}{V\mu}} q_l(t)\omega_l \cos k_l z
$$

one can express the field energy residing in the lth mode as

$$
\begin{aligned}
\hat{H}_l &= \int_0^L A dz \left[\frac{\varepsilon|\mathrm{E}|^2}{2} + \frac{\mu|H|^2}{2}\right] \\
&= \int_0^L A dz \left[\frac{1}{AL}\sin^2(k_l z)\, p_l^2(t) + \frac{1}{AL}\omega_l^2 \cos^2(k_l z) q_l^2(t)\right] \\
&= \frac{1}{L}\int_0^L dz \left[\frac{1-\cos(2k_l z)}{2} p_l^2(t) + \omega_l^2 \frac{1+\cos(2k_l z)}{2} q_l^2(t)\right] \\
&= \frac{1}{2} p_l^2(t) + \frac{1}{2}\omega_l^2 q_l^2(t)
\end{aligned}
$$

where A, L denote the cross-sectional area and length of the cavity. In performing the integral a well known trigonometric identity has been used together with the boundary condition of the standing wave.

(b) The commutation relation of the creation and annihilation operators reads from Eq. (13.25) as

$$\begin{aligned}[a_l, a_l^+] &= \frac{1}{2\hbar\omega_l}[(\omega_l q_l + ip_l), (\omega_l q_l - ip_l)] \\ &= \frac{1}{2\hbar\omega_l}\{[\omega_l q_l, \omega_l q_l] - i\omega_l[q_l, p_l] + i\omega_l[p_l, q_l] + [p_l, p_l]\} \\ &= \frac{1}{2\hbar\omega_l}\{-i\omega_l[q_l, p_l] + i\omega_l[p_l, q_l]\} \\ &= \frac{1}{2\hbar\omega_l}\{-i\omega_l \times i\hbar + i\omega_l \times (-i\hbar)\} = 1\end{aligned}$$

where use has been made of the commutation relation

$$[q_l, p_l] = i\hbar, \quad [q_l, q_l] = 0, \quad [q_l, p_l] = 0$$

13.4. (a) One can invert the relation given in Eq. (13.25) and express q_l, p_l in terms of the creation and annihilation operators:

$$q_l = \left(\frac{\hbar}{2\omega_l}\right)^{1/2}(a_l + a_l^+), \quad p_l = -i\left(\frac{\hbar\omega_l}{2}\right)^{1/2}(a_l - a_l^+)$$

Now q_l, p_l corresponds to x, p_x of the harmonic oscillator, so that one can write the Hamiltonian as

$$\begin{aligned}H &= \frac{1}{2}kq_l^2 + \frac{1}{2}p_l^2, \quad m = 1 \\ &= \frac{1}{2}k \times \left(\frac{\hbar}{2\omega_l}\right)(a_l + a_l^+)(a_l + a_l^+) - \frac{1}{2} \times \frac{\hbar\omega_l}{2}(a_l - a_l^+)(a_l - a_l^+) \\ &= \frac{\hbar\omega_l}{4}[(a_l + a_l^+)(a_l + a_l^+) - (a_l - a_l^+)(a_l - a_l^+)], \quad k = \omega_l^2 \\ &= \frac{\hbar\omega_l}{4}[2a_l a_l^+ + 2a_l^+ a_l] = \frac{\hbar\omega_l}{4}2[a_l a_l^+ + a_l^+ a_l] \\ &= \hbar\omega_l\left[a_l^+ a_l + \frac{1}{2}\right], \quad a_l a_l^+ = a_l^+ a_l + 1\end{aligned}$$

(b) The standing wave mode representation of the electric field given in Eq. (13.33a) can be recast into the traveling wave mode as follows:

$$\begin{aligned}\underline{E}_l &= \hat{y}i\sqrt{\frac{\hbar\omega_l}{V\varepsilon}}[a_l^+(t) - a_l(t)]\sin k_l z \\ &= \hat{y}i\sqrt{\frac{\hbar\omega_l}{V\varepsilon}}[a_l^+(0)e^{i\omega_l t} - a_l(0)e^{-i\omega_l t}]\frac{(e^{ik_l z} - e^{-ik_l z})}{2i}\end{aligned}$$

Since any combination of the product $\exp(\pm i\omega t) \times \exp(\pm ikz)$ is the solution of the wave equation of E one can choose those combination describing the propagation in a desired direction. Thus one can utilize the representation given in Eq. (13.33a):

$$\underline{E}_k = i\underline{e}_{k\lambda}\sqrt{\frac{\hbar\omega_k}{2V\varepsilon}}[a^+_{k\lambda}(t)e^{-i\underline{k}\cdot\underline{r}} - a(t)_{k\lambda}e^{i\underline{k}\cdot\underline{r}}]$$

where $\hat{y}$ can be replaced by the polarization vector. One can likewise construct the traveling H field.

(c) The transition rate in Eq. (13.36) involves the transition matrix element given by

$$W \propto |\langle u_1, n_l + 1|e(\hat{e}_{l\lambda}\cdot \underline{r})(a^+_{k\lambda}(t)e^{-i\underline{k}\cdot\underline{r}} - a_{k\lambda}(t)e^{i\underline{k}\cdot\underline{r}})|u_2, n_l\rangle|^2$$

Now the interaction Hamiltonian operating on the state $|u_2, n_l\rangle$ yields

$$(a^+_{k\lambda}(t)e^{-i\underline{k}\cdot\underline{r}} - a_{k\lambda}(t)e^{i\underline{k}\cdot\underline{r}})|u_2, n_l\rangle = e^{-i\underline{k}\cdot\underline{r}}(n_l + 1)^{1/2}|u_1, n_l + 1\rangle - n^{1/2}e^{i\underline{k}\cdot\underline{r}}|u_2, n_l - 1\rangle$$

Obviously the first term can be connected to the state $|u_1, n_l + 1\rangle$ accounting for the electron making the transition from the upper to lower state, while emitting a photon. Hence the resulting transition rate

$$W \propto n_l + 1$$

naturally incorporates both the induced and spontaneous emission of radiation.

Chapter 14

14.1. (a) The coupled equations

$$\dot{a}_{1s}(t) = -i\omega_1 a_{1s} + i\frac{\tilde{\mu}E(t)}{\hbar}a_{2s}, \quad \dot{a}_{2s}(t) = -i\omega_2 a_{2s} + i\frac{\tilde{\mu}E(t)}{\hbar}a_{1s} \tag{14.1}$$

have been derived already in Eq. (13.2).

(b) By using Eq. (14.1) one can obtain

$$\begin{aligned}\frac{d}{dt}(a^*_{1s}a_{1s}) &= \dot{a}^*_{1s}a_{1s} + a^*_{1s}\dot{a}_{1s}\\ &= \left(i\omega_1 a^*_{1s} - i\frac{\tilde{\mu}E(t)}{\hbar}a^*_{2s}\right)a_{1s} + a^*_{1s}\left(-i\omega_1 a_{1s} + i\frac{\tilde{\mu}E(t)}{\hbar}a_{2s}\right)\\ &= -i\frac{\tilde{\mu}E(t)}{\hbar}a^*_{2s}a_{1s} + i\frac{\tilde{\mu}E(t)}{\hbar}a_{2s}a^*_{1s}\end{aligned}$$

$$\begin{aligned}\frac{d}{dt}(a^*_{2s}a_{2s}) &= \dot{a}^*_{2s}a_{2s} + a^*_{2s}\dot{a}_{2s}\\ &= \left(i\omega_2 a^*_{2s} - i\frac{\tilde{\mu}E(t)}{\hbar}a^*_{1s}\right)a_{2s} + a^*_{2s}\left(-i\omega_2 a_{2s} + i\frac{\tilde{\mu}E(t)}{\hbar}a_{1s}\right)\\ &= -i\frac{\tilde{\mu}E(t)}{\hbar}a^*_{1s}a_{2s} + i\frac{\tilde{\mu}E(t)}{\hbar}a^*_{2s}a_{1s}\end{aligned}$$

Therefore by subtracting the latter equation from the former one obtains

$$\frac{d}{dt}(\rho_{11} - \rho_{22}) = \frac{2i\widetilde{\mu}\mathrm{E}(t)}{\hbar}(\rho_{21} - \rho_{21}^{*})$$

(see Eq. (14.14)). One can likewise derive Eq. (14.15a), that is,

$$\frac{d}{dt}\rho_{21} = \frac{d}{dt}a_{2s}a_{1s}^{*} = -i\omega_0\rho_{21} + i\frac{\widetilde{\mu}\mathrm{E}(t)}{\hbar}(\rho_{11} - \rho_{22}), \quad \omega_0 = \frac{E_2 - E_1}{\hbar}$$

14.2. (a) In the absence of the electric field Eq. (14.15b) in the text is reduced to

$$\frac{d}{dt}(\rho_{11} - \rho_{22}) = -\frac{(\rho_{11} - \rho_{22}) - (\rho_{11}^{(0)} - \rho_{22}^{(0)})}{\tau}$$

or equivalently

$$\frac{d(\rho_{11} - \rho_{22})}{dt} + \frac{(\rho_{11} - \rho_{22})}{\tau} = \frac{(\rho_{11}^{(0)} - \rho_{22}^{(0)})}{\tau} \tag{14.2}$$

Now one may introduce a function

$$\chi = (\rho_{11} - \rho_{22})\exp(t/\tau)$$

in which case the left hand side of Eq. (14.2) can be expressed as

$$\frac{d(\rho_{11} - \rho_{22})}{dt} + \frac{(\rho_{11} - \rho_{22})}{\tau} = e^{-t/\tau}\frac{d}{dt}\chi \tag{14.3}$$

Therefore by equating Eqs. (14.2) and (14.3) one can write

$$e^{-t/\tau}\frac{d}{dt}\chi = \frac{(\rho_{11}^{(0)} - \rho_{22}^{(0)})}{\tau}$$

that is,

$$\frac{d}{dt}\chi = \frac{d}{dt}[(\rho_{11} - \rho_{22})e^{t/\tau}] = \frac{(\rho_{11}^{(0)} - \rho_{22}^{(0)})}{\tau}e^{t/\tau} \tag{14.4}$$

One can likewise introduce the function

$$\chi = e^{(i\omega_0 t + t/T_2)}\rho_{21}$$

and obtain

$$\frac{d}{dt}\chi = \frac{d}{dt}(e^{i\omega_0 t + t/T_2}\rho_{21}) = 0 \tag{14.5}$$

A straightforward integration of Eqs. (14.4) and (14.5) leads to Eq. (14.16) in the text.

14.3. (a) The complex equations (14.20a) and (14.20b) read in the steady state as

$$0 = i(\omega - \omega_0)(\sigma_{21}^{(r)} + i\sigma_{21}^{(i)}) + i\frac{\widetilde{\mu}\mathrm{E}_0}{2\hbar}(\rho_{11} - \rho_{22}) - \frac{(\sigma_{21}^{(r)} + i\sigma_{21}^{(i)})}{T_2}$$

$$0 = \frac{i\widetilde{\mu}\mathrm{E}_0}{\hbar}(2i\sigma_{21}^{(i)}) - \frac{(\rho_{11} - \rho_{22}) - (\rho_{11}^{(0)} - \rho_{22}^{(0)})}{\tau}$$

Thus by singling out the real and imaginary parts from both sides of these two equations one can write

$$(\omega - \omega_0)\sigma_{21}^{(i)} + \frac{\sigma_{21}^{(r)}}{T_2} = 0$$

$$-\frac{\sigma_{21}^{(i)}}{T_2} + (\omega - \omega_0)\sigma_{21}^{(r)} + \frac{\widetilde{\mu}\mathrm{E}_0}{2\hbar}(\rho_{11} - \rho_{22}) = 0$$

$$-\frac{2\widetilde{\mu}\mathrm{E}_0}{\hbar}\sigma_{21}^{(i)} - \frac{(\rho_{11} - \rho_{22})}{\tau} = -\frac{(\rho_{11}^{(0)} - \rho_{22}^{(0)})}{\tau}$$

and find the three unknowns from the three equations and obtain Eq. (14.21).

(b) With $\sigma_{21}^{(r)}, \sigma_{21}^{(i)}$, and $(\rho_{11} - \rho_{22})$ thus found one can specify the atomic susceptibility by relating the polarization vector P to the atom dipole as

$$\begin{aligned} P(t) &\equiv \mathrm{Re}N\langle\mu(t)\rangle \equiv \varepsilon_0\chi_a'\underline{\mathrm{E}}_0\cos\omega t + \varepsilon_0\chi_a''\underline{\mathrm{E}}_0\sin\omega t \\ &= N\widetilde{\mu}(\rho_{21} + \rho_{12}) = 2\mathrm{Re}[\widetilde{\mu}(\sigma_{21}^{(r)} + i\sigma_{21}^{(i)})e^{-i\omega t}] = 2\widetilde{\mu}[\sigma_{21}^{(r)}\cos\omega t + \sigma_{21}^{(i)}\sin\omega t] \end{aligned}$$

and the results agree with Eq. (14.23).

14.4. At the steady state in which ρ_{11}, ρ_{22} are independent of time the rate equation (14.29) reduces to

$$0 = \lambda_2 - \frac{1}{\tau_2}\rho_{22} - W_i(\rho_{22} - \rho_{11})$$

$$0 = \lambda_1 - \frac{1}{\tau_1}\rho_{11} + W_i(\rho_{22} - \rho_{11})$$

where the spontaneous emission lifetime τ_{sp} is in general much longer than τ_1, τ_2 and discarded. One can rewrite the equations as

$$m_{11}\rho_{11} + m_{12}\rho_{22} = \lambda_1, \quad m_{11} = W_i + 1/\tau_1, \quad m_{12} = -W_i$$

$$m_{21}\rho_{11} + m_{22}\rho_{22} = \lambda_2, \quad m_{21} = -W_i, \quad m_{22} = W_i + 1/\tau_2$$

and find ρ_{11}, ρ_{22} via the Kramer's rule as

$$\rho_{11} = \frac{\begin{vmatrix} \lambda_1 & m_{12} \\ \lambda_2 & m_{22} \end{vmatrix}}{\begin{vmatrix} m_{11} & m_{12} \\ m_{21} & m_{22} \end{vmatrix}}, \quad \rho_{22} = \frac{\begin{vmatrix} m_{11} & \lambda_1 \\ m_{21} & \lambda_2 \end{vmatrix}}{\begin{vmatrix} m_{11} & m_{12} \\ m_{21} & m_{22} \end{vmatrix}}$$

The explicit expansion of the determinants leads to the results given in Eq. (14.30).

14.5. (a) The wavelength, frequency and the frequency spacing of the standing wave modes in a cavity with length L are given from Eq. (14.26) as

$$\frac{\lambda_l l}{2} = L, \ \lambda_l = \frac{2L}{l}; \ \nu_l = \frac{c}{\lambda_l} = l\frac{c}{2L}; \ \Delta\nu_c = \frac{c}{2L} \quad l = 1, 2, 3 \ldots$$

where the optical index of refraction has been taken unity. Thus the fundamental wavelength and frequency versus L are:

L(m)	λ(m)	ν(Hz)
1	2	1.5×10^8
10^{-2}	2×10^{-2}	1.5×10^{10}
10^{-4}	2×10^{-4}	1.5×10^{12}

(b) The frequency at 500 nm wavelength is 0.6×10^{15} Hz. The frequency spacing of the standing wave modes for $L = 0.5$ m is 3×10^8 Hz. The bandwidth of a picosecond pulse is roughly given by $\Delta\nu \approx 1/\Delta t = 10^{12}$ Hz and the carrier frequency of the pulse centered at 500 nm wavelength is 6×10^{14} Hz. Thus the number of standing waves mode-locked is about 2×10^6 centered around the carrier frequency.

Chapter 15

15.1. (a) The degenerate and the non-degenerate representation of n

$$n = \frac{2}{\sqrt{\pi}} N_c F_{1/2}(\eta_{Fn}), \quad n = N_C e^{-(E_C - E_F)/k_B T}$$

can be explicitly compared by considering

$$F_{1/2}(\eta_{Fn}), \quad e^{\eta_{Fn}} \frac{\sqrt{\pi}}{2}$$

(see Eq. (15.6–15.8)). As clear from the plot shown, the two quantities are essentially identical when the Fermi level E_F is a few $k_B T$ below the conduction band. This indicates the range of the validity of the non-degenerate and analytical expression of n. But when E_F approaches the conduction band edge E_C or is raised above E_C, the analytical expression progressively over-estimates the actual value, hence should not be used.

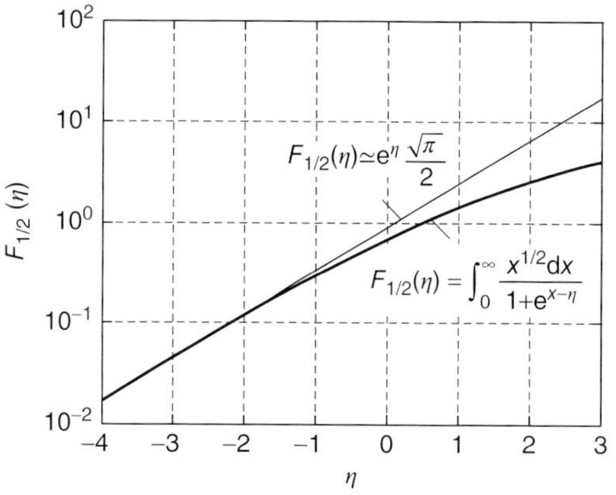

(b) The hole concentration is generally represented by

$$p = \int_{E_V-\Delta E_V}^{E_V} dEg_p(E)f_p(E), \quad g_p(E) = \frac{1}{2\pi^2}\left(\frac{2m_p}{\hbar^2}\right)^{3/2}(E_V - E)^{1/2} \tag{15.1}$$

where the density of states g_p is strictly analogous with g_n except that the hole effective mass m_p replaces m_n and $(E - E_C)$ is replaced by $(E_V - E)$ which is equivalent to $(E - E_C)$. The electrons move up the conduction band with increasing energy, while holes move down the valence band with increasing energy.
The hole occupation factor is by definition the probability that the state is not occupied by the electron:

$$f_p(E) \equiv 1 - \frac{1}{1 + e^{(E-E_F)/k_BT}} = \frac{1}{1 + e^{(E_F-E)/k_BT}} \tag{15.2}$$

By inserting Eq. (15.2) into Eq. (15.1) and precisely following the steps used for n one finds

$$p = \frac{2}{\sqrt{\pi}}N_V F_{1/2}(\eta_{Fp}); \quad \eta_{Fp} \equiv \frac{(E_V - E_F)}{k_BT}, \quad N_V \equiv 2\left(\frac{2\pi m_p k_B T}{h^2}\right)^{3/2}$$

For E_F lying above E_V a few thermal energy k_BT or more the same approximation can be made as in the case of n and one can obtain

$$F_{1/2}(\eta_{Fp}) \simeq e^{\eta_{Fp}} \int_0^{\infty} d\eta e^{-\eta}\eta^{1/2} = e^{\eta_{Fp}}\frac{\sqrt{\pi}}{2}$$

and the analytical expression for p in the non-degenerate regime

$$p = N_V e^{-(E_F-E_V)/k_BT}$$

15.2. (a) In the thermal equilibrium the law of mass action holds and in Si the intrinsic concentration is given at the room temperature by

$$np = n_i^2, \quad n_i = 1.45 \times 10^{10}\text{cm}^{-3}$$

Thus one can find n for given p as

$p\,\text{cm}^{-3}$	10	10^2	10^5	10^8
$n\,\text{cm}^{-3}$	2.1×10^{19}	2.1×10^{18}	2.1×10^{15}	2.1×10^{12}

One can also find n as a function of T from n_i given in Eq. (15.12) as a function T.

(b) To find N_D one has to find the location of E_F in the band gap. Confining to the case of non-degenerate n, p one can find E_F by using

$$n = n_i \exp \frac{E_F - E_i}{k_B T}$$

with E_i denoting the midgap corresponding to E_F of n_i.

$n\,\text{cm}^{-3}$	2.1×10^{19}	2.1×10^{18}	2.1×10^{15}	2.1×10^{12}
$E_F - E_i$(eV)	0.53	0.47	0.30	0.12

(c) Since $E_\text{C} - E_i = 0.56$ eV, and the non-degenerate statistics holds true for $E_\text{C} - E_\text{F} > 2k_\text{B}T$ and $k_\text{B}T = 0.025$ eV, the non-degenerate statistics can be used for all cases considered except for the case of 0.53 eV.

15.3. With $T \to 0$ the freeze out effect is operative and all electrons in the donor level E_D as well as in the valence band cannot be thermally excited into the conduction band in the n-type semiconductor. Thus the donor state should be occupied by the electron, which is ensured by E_F raised above E_D. By the same token E_F should be lowered below E_A level in the p-type semiconductor, so that no electron can be in the acceptor state. Then the holes cannot be generated in the valence band by electrons being promoted to the acceptor level E_A from the valence band.

15.4. (a) With $E_F - E_C = 0.1$ eV n is in a strongly non-degenerate regime, so that one has to use Eq. (15.6) and write

$$n = N_C \frac{2}{\sqrt{\pi}} F_{1/2}\left(\frac{0.1}{0.025}\right), \quad k_B T = 0.025\,\text{eV at } T = 300\,\text{K}$$

Since

$$N_C = 2.8 \times 10^{19}\,\text{cm}^{-3}, \quad F_{1/2}(4) = 6.5115 \text{ at } T = 300\,\text{K}$$

one finds $n = 1.82 \times 10^{20}\,\text{cm}^{-3}$. The N_D level should be higher than n, since E_F in this case is near or above E_D and not all of the electrons in the donor state are promoted to the conduction band.

(b) Similarly one has to use Eq. (15.10) for p and write

$$p = N_V \frac{2}{\sqrt{\pi}} F_{1/2}\left(\frac{0.15}{0.025}\right), \quad k_B T = 0.025\,\text{eV at } T = 300\,\text{K}$$

Since

$$N_V = 1.04 \times 10^{19}\,\text{cm}^{-3}, \quad F_{1/2}(6) = 11.447$$

one finds $p = 1.2 \times 10^{20}/\text{cm}^3$. Again N_A should be larger than p since E_F in this case is near or below E_A, so that not all of the acceptor atoms can accept the electron and generate hole in the valence band. The quantitative analysis requires the donor and acceptor statistics.

15.5. (a) The electrons in the quantum wire are confined in say y, z directions while moving freely along the x-direction. One can therefore utilize the quantized energy level from Eq. (4.27) as

$$E_{n,m} = \frac{\hbar^2 k_x^2}{2m_x} + E_0(n^2 + m^2); \; E_0 = \frac{\hbar^2 \pi^2}{2mW^2}; \quad n, m = 1, 2, \ldots$$

Hence the ground and first excited state levels are

$$E_{11} = 2E_0 \simeq 15\text{ meV}, \quad E_{12} = E_{21} = 5E_0 \simeq 37.5\,\text{meV at } W = 10\,\text{nm}, \; m_n = 0.9m_0$$

The 1D electron density is given by

$$n_{1\text{D}} = \sum_{n,m} \Delta n_{nm}$$

where Δn_{nm} can be evaluated in several steps as

$$\begin{aligned}
\Delta n_{nm} &= \int_{E_{nm}}^{\infty} dE g_{1\text{D}} f_n(E), \quad g_{1\text{D}}(E) = \frac{[(\sqrt{2}m^{1/2})/\pi\hbar]}{E^{1/2}}, \quad E \geq E_{nm} \\
&= \frac{\sqrt{2}m_n^{1/2}}{\pi\hbar} \int_{E_{nm}}^{\infty} dE \frac{1}{\varepsilon^{1/2}} \frac{1}{1 + e^{(\varepsilon + E_{nm} - E_F)/k_B T}}; \; \varepsilon = E - E_{nm}, \\
&\simeq \frac{\sqrt{2}m_n^{1/2}}{\pi\hbar} e^{-(E_{nm} - E_F)/k_B T} \int_{E_{nm}}^{\infty} \frac{dE\, e^{-(E - E_{nm})/k_B T}}{(E - E_{nm})^{1/2}}; \; e^{(\varepsilon + E_{nm} - E_F)/k_B T} >> 1
\end{aligned}$$

To perform the integration one can introduce a new variable

$$\xi = \left(\frac{E - E_{nm}}{k_B T}\right)^{1/2}$$

and put the integral in the form and evaluate it as

$$\int_{E_{nm}}^{\infty} \frac{dE\, e^{-(E - E_{nm})/k_B T}}{(E - E_{nm})^{1/2}} = 2\sqrt{k_B T} \int_0^{\infty} d\xi\, e^{-\xi^2} = \sqrt{k_B T \pi}$$

Therefore n_{1D} is given by

$$n_{1D} = \left(\frac{2m_n k_B T}{\pi}\right)^{1/2} \frac{1}{\hbar} e^{-(E_C - E_F)/k_B T} \sum_{nm} g_{nm}\, e^{-E_0(n^2+m^2)/k_B T}$$

where g_{nm} denotes the degeneracy of the n, m state.

(b) One can quote the result obtained in Eq. (4.26) and write the subband energy as

$$E_n = \frac{\hbar^2 \pi^2}{2m_n W^2} n^2, \quad n = 1, 2, 3, \ldots$$

and specify the 2D electron density as

$$n_{2D} = \sum_{s=1}^{\infty} \Delta n_s$$

where

$$\begin{aligned}
\Delta n_s &= \int_{E_s}^{\infty} dE\, g_{2D} f_n(E), \quad g_{2D} = m_n/\pi\hbar^2 \\
&= \frac{m_n}{\pi\hbar^2} \int_{E_n}^{\infty} \mathrm{d}(E - E_n) \frac{1}{1 + e^{(E-E_F)/k_B T}}, \quad E \geq E_n \\
&= \frac{m_n}{\pi\hbar^2} \int_{0}^{\infty} d\varepsilon \frac{1}{1 + e^{(\varepsilon+E_n-E_F)/k_B T}}, \quad \varepsilon = E - E_n \\
&\simeq \frac{m_n}{\pi\hbar^2} e^{-(E_n-E_F)/k_B T} \int_{E_n}^{\infty} d\varepsilon\, e^{-\varepsilon/k_B T}, \quad e^{(\varepsilon+E_n-E_F)/k_B T} >> 1 \\
&= \frac{m_n}{\pi\hbar^2} e^{-(E_n-E_F)/k_B T} k_B T
\end{aligned}$$

and n_{2D} is represented by

$$n_{2D} = \sum_n \Delta n_n = \frac{m_n}{\pi\hbar^2} k_B T \sum_n e^{-(E_n-E_F)/k_B T}$$

Chapter 16

16.1. (a) The mobility and diffusion coefficient of electrons for example are given from Eqs. (16.5)) and (16.7) by

$$\mu_n = \frac{q\tau_n}{m_n}, \quad D_n \equiv \frac{l_n^2}{\tau_n}$$

where τ_n, l_n are the mean collision time and the mean diffusion length, respectively. Now l_n is specified by the distance covered by the electron moving with the thermal speed v_T in the mean collision time, that is,

$$l_n = v_T \tau_n = \left(\frac{k_B T}{m_n}\right)^{1/2} \tau_n; \quad \frac{m_n v_T^2}{2} = \frac{k_B T}{2}$$

Here v_T has been found from the thermal energy via the equipartition theorem (see Eq. (1.15)). Hence one can derive the Einstein relation by

$$\frac{D_n}{\mu_n} = \frac{l_n^2/\tau_n}{q\tau_n/m_n} = \frac{\left(\frac{k_BT}{m_n}\right)\tau_n^2/\tau_n}{q\tau_n/m_n} = \frac{k_BT}{q}$$

(b) Under the illumination n, p are given from Eq. (16.16) by

$$n = n_i + n_{ph} = n_i + (\alpha I\tau_n/h\nu)$$
$$p = n_i + p_{ph} = n_i + (\alpha I\tau_p/h\nu) \tag{16.1}$$

where the second term in each expression denotes the photo-generated electron and hole concentrations. Under the illumination n, p are specified by the quasi-Fermi levels and are given by

$$n = n_i e^{(E_{Fn}-E_i)/k_BT}, \quad p = n_i e^{(E_i-E_{Fp})/k_BT} \tag{16.2}$$

Hence by combining Eqs. (16.1) and (16.2) one can write

$$e^{(E_{Fn}-E_i)/k_BT} = [1 + (\alpha I\tau_n/n_ih\nu)]$$
$$e^{(E_i-E_{Fp})/k_BT} = [1 + (\alpha I\tau_p/n_ih\nu)]$$

and by taking the logarithms on both sides one obtains

$$\frac{E_{Fn}-E_i}{k_BT} = \ln[1 + (\alpha I\tau_n/n_ih\nu)]$$
$$\frac{E_i-E_{Fp}}{k_BT} = \ln[1 + (\alpha I\tau_p/n_ih\nu)]$$

Therefore by adding the two equations one finds the splitting of E_{Fn} and E_{Fp} as

$$E_{Fn} - E_{Fp} = \ln[1 + (\alpha I\tau_n/n_ih\nu)] + \ln[1 + (\alpha I\tau_p/n_ih\nu)]$$
$$= \ln\{[1 + (\alpha I\tau_n/n_ih\nu)] \times [1 + (\alpha I\tau_p/n_ih\nu)]\}$$

16.2. (a) The resistivity is the inverse conductivity and is given by

$$\rho_n \equiv \frac{1}{\sigma_n} = \frac{1}{q\mu_n n} = \frac{1}{1.6\times10^{-19}\text{C}\cdot 800(\text{cm}^2(\text{V s})^{-1})\cdot n(\text{cm}^{-3})}$$
$$= \frac{1}{1.6\times10^{-19}\text{C}\cdot 800\cdot(\text{cm}^2(\text{V s})^{-1})n(\text{cm}^{-3})} = 10\,\Omega\,\text{cm}$$

One can thus find n as

$$n_n = \frac{1}{1.6\times10^{-19}\cdot 800\cdot 10} = 7.8\times10^{14}\,\text{cm}^{-3}$$

Once n_n is known p_n is to be found from the law of mass action at the room temperature as

$$p_n = \frac{n_i^2}{n_n} = \frac{(1.45\times 10^{10})^2}{7.8\times 10^{14}} = 2.79\times 10^5\ \mathrm{cm}^{-3}$$

Also the photogenerated density of e–h pairs is given by

$$n_{ph} = p_{ph} = g\tau = 10^{21}\cdot 10^{-6} = 10^{15}\ \mathrm{cm}^{-3}$$

The total conductivity $\sigma_T = \sigma_D + \sigma_{ph}$ is contributed by the dark and photo conductivities. The former component is given by

$$\begin{aligned}\sigma_D &= q\mu_n n + q\mu_p p \simeq q\mu_n n\\ &= 1.6\times 10^{-19}\,\mathrm{C}\cdot 800\,\mathrm{cm}^2\,(\mathrm{V\,s})^{-1}\cdot 7.8\times 10^{14}\,\mathrm{cm}^{-3} = 1.0\times 10^{-1}\,\mho\,\mathrm{cm}^{-1}\end{aligned}$$

Likewise the photoconductivity is specified as

$$\begin{aligned}\sigma_{ph} &= q(\mu_n n_{ph} + \mu_p p_{ph})\\ &= 1.6\times 10^{-19}\cdot 10^{15}\,\mathrm{cm}^{-3}\cdot (800+400)\,\mathrm{cm}^2\,(\mathrm{V\,s})^{-1}\\ &= 1.92\times 10^{-1}\,\mho\,\mathrm{cm}^{-1}\end{aligned}$$

(b) Since the total conductivity is mainly contributed by the photoconductivity the ratio is given by

$$\frac{\sigma_n}{\sigma_p} = \frac{\mu_n}{\mu_p} = \frac{1}{2}$$

16.3 (a) The light intensity is specified by the density n_{ph}, energy and the velocity of photon as

$$I = n_{ph}\cdot h\frac{c}{\lambda}\cdot c = h\frac{c}{\lambda}\cdot F_{ph},\ F_{ph} = n_{ph}\cdot c$$

where F_{ph} is the flux of photons. For $I = 10$ W, one finds the flux as

$$F_{ph} = \frac{10\,(\mathrm{J/s\,cm^2})\cdot 500\times 10^{-9}\,\mathrm{m}}{6.626\times 10^{-34}\,\mathrm{J\,s}\cdot 3\times 10^8\,\mathrm{m}} = 2.52\times 10^{19}\mathrm{cm}^{-2}\,\mathrm{s}^{-1}$$

(b) The generation rate is specified in terms of I and the attenuation coefficient α as

$$g = \alpha(I/h\nu) = 2.52\times 10^{20}\,\mathrm{cm}^{-3}\,\mathrm{s}^{-1}$$

16.4. The rate equation

$$\frac{\partial p_n}{\partial t} = g_L - \frac{p_n - p_{n0}}{\tau_p}$$

can be rearranged as

$$\frac{\partial p_n}{\partial t} + \frac{p_n}{\tau_p} = g_L + \frac{p_{n0}}{\tau_p} \quad \text{or} \quad \frac{\partial}{\partial t}(e^{t/\tau_p} p_n) = \left(g_L + \frac{p_{n0}}{\tau_p} \right) e^{t/\tau_p}$$

Upon integrating both sides there results

$$e^{t/\tau_p} p_n(t) - p_n(0) = (g_L \tau_p + p_{n0})(e^{t/\tau_p} - 1)$$

Or more specifically at t

$$p_n(t) = p_n(0) e^{-t/\tau_p} + (g_L \tau_p + p_{n0})(1 - e^{-t/\tau_p})$$

Thus for $t \gg \tau_p$ the initial value decays away and the steady state value is attained given by

$$p_n = (g_L \tau_p + p_{n0})$$

The first term of p_n is the photo-generated hole density while the second term is the equilibrium concentration. When the light is turned off at $t = T$ then the rate equation reads for $t \geq T$ as

$$\frac{\partial}{\partial t}(e^{t/\tau_p} p_n) = \frac{p_{n0}}{\tau_p} e^{t/\tau_p}$$

and by integrating both sides one finds

$$p_n(t) = p_n(T) e^{-t/\tau_p} + p_{n0}(1 - e^{-t/\tau_p})$$

Thus $p_n(T)$ decays away while $p_n(t)$ attains the equilibrium value in the long time limit.

16.5. (a) For $\sigma_n = \sigma_p = \sigma$ the recombination rate due to a single trap is given from Eq. (16.30) by

$$U = \frac{\sigma v_T N_t (pn - n_i^2)}{n + p + 2n_i \cosh \frac{(E_t - E_i)}{k_B T}}$$

For multilevel traps one can generalize U as

$$U = \sum_j U_j = \sum_j \frac{\sigma v_T N_t(E_j)(np - n_i^2)}{n + p + 2n_i \cosh\left(\frac{E_j - E_i}{k_B T}\right)}$$

with j denoting jth trap level. One can further introduce the trap profile

$$N_t(E_j) = f_t(E)$$

and recast U in terms of the distributed traps:

$$U = \sigma v_T (np - n_i^2) \int_{-E_G/2}^{E_G/2} \frac{f_t(E)}{n + p + 2n_i \cosh\left(\frac{E - E_i}{k_B T}\right)}$$

where E is taken centered at the midgap. For a flat distribution

$$f_t(E) = D_{ss}$$

U is obtained as

$$U = \frac{1}{\tau}(np - n_i^2) \int_{-E_G/2}^{E_G/2} \frac{dE}{n + p + 2n_i \cosh\left(\frac{E - E_i}{k_B T}\right)}, \quad \frac{1}{\tau} = \sigma v_T D_{ss}$$

One can carry out the integration incorporating the various cases in the denominator.

(b) For the Gaussian distributed trap profile

$$f(E) = N_t \exp - \frac{(E - E_{tc})^2}{2\sigma^2}$$

U reads as

$$U = \frac{1}{\tau}(np - n_i^2) \int_{-E_G/2}^{E_G/2} \frac{e^{-(E - E_{tc})^2/2\sigma^2} dE}{n + p + 2n_i \cosh\left(\frac{E - E_i}{k_B T}\right)}, \quad \frac{1}{\tau} = \sigma v_T N_t$$

In the n-type semiconductor for instance n is dominant so that U can be expressed in terms of the error function as

$$U \approx \frac{(np - n_i^2)}{\tau_n} \sqrt{2}\sigma \int_0^{\Lambda} e^{-\xi^2} d\xi, \quad \xi = \frac{E - E_{tc}}{\sqrt{2}\sigma}, \quad \Lambda = \frac{E_G/2 - E_{tc}}{\sqrt{2}\sigma}$$

$$= \left[\frac{(np - n_i^2)}{\tau_n} \sqrt{2}\sigma \right] \frac{\sqrt{\pi}}{2} erf\Lambda$$

Here the traps in the lower half of the bandgap has been discarded and the well known error function has been used,

$$erf\,\Lambda = \frac{2}{\sqrt{\pi}} \int_0^{\Lambda} e^{-\xi^2} d\xi$$

Chapter 17

17.1. (a) The junction parameters, for example, the built-in potential, depletion depth, the maximum electric field, and so on at given doping level can be specified explicitly by combining Eqs. (17.3)–17.5):

$$\varphi_{bi} = \frac{k_B T}{q} \ln\left(\frac{N_A N_D}{n_i^2}\right) = \frac{q}{2\varepsilon_s} \cdot \frac{N_A N_D}{N_A + N_D} \cdot W^2 = \frac{1}{2} E_{\max} W$$

$$W = x_n + x_p, \quad x_n = \frac{W}{(1 + N_D/N_A)}, \quad x_p = \frac{W}{(1 + N_A/N_D)}$$

Hence one can evaluate those parameters with the use of the formulae given above.

(b) Under the reverse bias $W, E_{\max}$ all increase and one can write from Eqs. (17.9)) and (17.10)) to

$$W(V_R) = \left[\frac{2\varepsilon_s \left(N_A + N_D\right)}{q \cdot N_A N_D} (\varphi_{bi} + |V_R|)\right]^{1/2}, \quad \varphi_{bi} \to \varphi_{bi} + |V_R|$$

$$E_{\max} = \frac{2(\varphi_{bi} + |V_R|)}{\left[\frac{2\varepsilon_s (N_A + N_D)}{q \cdot N_A N_D} (\varphi_{bi} + |V_R|)\right]^{1/2}} = \frac{2(\varphi_{bi} + |V_R|)^{1/2}}{\left[\frac{2\varepsilon_s (N_A + N_D)}{q \cdot N_A N_D}\right]^{1/2}}$$

and find the breakdown voltage $V_{\rm BR}$ for given breakdown field and doping level. For example for $E_{BR} = 3 \times 10^5\ \mathrm{V\,cm^{-1}}$, $N_A = 2 \times 10^{18}\ \mathrm{cm^{-3}}$ and $N_D = 10^{15}\ \mathrm{cm^{-3}}$, $\varphi_{bi} \simeq 0.78$ V and $V_{BR} \simeq 17$ V. For the same N_A but for different N_D of $2 \times 10^{17}\ \mathrm{cm^{-3}}$ $\varphi_{bi} \simeq 0.9$ V, and $V_{BR} \simeq 1.2$ V, and is reduced by a factor of about 10. This points to $V_{\rm BR}$ depending sensitively on the doping level.

17.2. The junction band bending is specified by the built-in barrier potential via

$$\varphi_{bi} = \varphi_{Fn} + \varphi_{Fp}$$

where the electron and hole Fermi potentials depend on doping level N_D and N_A, respectively. Thus, for sufficiently high doping level it is possible for E_F to be raised above E_C in the n bulk and E_F to be lowered below E_V in the p bulk regions, thereby making it possible to induce the band bending larger than the bandgap.

(b) To analyze the junction band bending larger than E_G the statistics of the donor and acceptor atoms are essential. For the degenerate doping level E_F can be raised above E_D in the n-bulk. By the same token E_F can be lowered below E_A in the p-bulk regions. Therefore all of the electrons in the donor atoms are not necessarily donated to the conduction band. Likewise all of the acceptor atoms can accept the electrons to create the holes in the valence band. Consequently the ionized donor and acceptor atoms constitute a fraction of N_D, N_A.

To estimate such N_D, N_A one may choose for instance N_D, N_A such that E_F coincides with E_C, E_V in n and p bulk regions, respectively and write

$$n = \frac{2}{\sqrt{\pi}} N_C F_{1/2}(0) = N_D^+ = \frac{N_D}{1 + g_D e^{(E_F - E_D)/k_B T}}, \quad E_C = E_F \tag{17.1a}$$

$$p = N_A^- = \frac{N_A}{1 + g_A e^{(E_A - E_F)/k_B T}} = \frac{2}{\sqrt{\pi}} N_V F_{1/2}(0), \quad E_V = E_F \tag{17.1b}$$

Here, N_D^+, N_A^- represent the ionized donor and acceptor atoms by donating electrons to the conduction band and holes to the valence band, respectively and constitute a fraction of N_D, N_A. Also g_D and g_A are the degeneracy factors for E_D and E_A states. By using the values of N_C, N_V in silicon and germanium into Eqs. (17.1) and (17.2) one can estimate the required levels of N_D, N_A.

17.3. In the completely depleted approximation the space charge ρ is taken constant at the level N_D, N_A in n and p regions, respectively. Consequently the E-field is linear in x. In this case the built in potential φ_{bi} is obtained by finding the triangular area under the E$-x$ curve, as discussed. The area is in turn decomposed into two triangular areas and one can write

$$\varphi_{bi} \equiv \frac{1}{2} \mathrm{E}_{\max} W = \frac{qN_A}{2\varepsilon_s} x_p^2 + \frac{qN_D}{2\varepsilon_s} x_n^2, \quad W = x_n + x_p$$

(see Eq. 17.5a) and Figure 17.2). At Δx distance inside W from the edge x_n, that is,

$$\Delta x = x_n - x$$

the space charge potential is less than φ_{bi} by an amount given by

$$\varphi(x) = \varphi_{bi} - \Delta\varphi(x), \quad \Delta\varphi(x) = \frac{qN_D}{\varepsilon_s} \Delta x \times \frac{1}{2} \Delta x = \frac{qN_D}{2\varepsilon_s} \Delta x^2$$

Here $\Delta\varphi(x)$ was approximated by the triangular area of height and base all specified in terms of Δx. Therefore n at Δx decreases and is given by

$$n(x) = n_{n0} \exp - \left(\frac{q^2 N_D \Delta x^2}{2\varepsilon_s k_B T} \right) = n_{n0} e^{-\xi^2}, \quad \xi \equiv \left(\frac{q^2 N_D}{2\varepsilon_s k_B T} \right)^{1/2} \cdot \Delta x$$

since $E_C(x) - E_F$ increases at Δx because of the band bending (see Figure 17.3)). Consequently one can estimate Δx at which n reduces to a negligible level, say 2% of N_D at x_n, that is, $\xi = 2$. Therefore one can assess Δx from ξ as

$$2 = \xi = \left(\frac{q^2 N_D}{2\varepsilon_s k_B T} \right)^{1/2} \Delta\widetilde{x}_n$$

For $N_A = 10^{17}\,\text{cm}^{-3}$ and $N_D = 10^{16}\,\text{cm}^{-3}$ for example one finds

$$\left(\frac{q^2 N_D}{2\varepsilon_s k_B T}\right)^{\frac{1}{2}} = \left[\frac{\left(1.6\times10^{-19}\right)^2 \text{C}^{-2}\cdot 10^{16+6}(m^{-3})\times 4\pi\times 8.988\times10^9(\text{N}\,\text{m}^2\,\text{C}^{-2})}{2\times 11.9\times 1.381\times 10^{-23}\,\text{J}\,\text{K}^{-1}\cdot 300\,\text{K}}\right]^{\frac{1}{2}}$$
$$\simeq 1.71\times10^7\,\text{m}^{-1} \simeq 1.71\times 10\,\mu\text{m}^{-1}$$

where the Coulomb constant has been used:

$$\varepsilon_S = \varepsilon_r \frac{1}{4\pi\varepsilon_0} = 11.9\times 8.988\times 10^9\ \text{m}^2\,\text{C}^{-2}$$

Therefore

$$\Delta\widetilde{x}_n = \frac{2}{\left(\frac{q^2 N_D}{2\varepsilon_s k_B T}\right)^{1/2}} \simeq 0.12\ \mu\text{m}$$

and one can likewise find

$$\Delta\widetilde{x}_p = \frac{2}{\left(\frac{q^2 N_A}{2\varepsilon_s k_B T}\right)^{1/2}} \simeq 0.04\ \mu\text{m}$$

For the same given doping level one can find the depletion depth W as

$$W = \left(\frac{2\varepsilon_s\left(N_A + N_D\right)}{qN_A N_D}\varphi_{bi}\right)^{1/2} \simeq 11\ \mu\text{m}$$

Therefore the sum of $\Delta\widetilde{x}_n$ and $\Delta\widetilde{x}_n$ is a mere fraction of W and the completely depleted approximation is shown a good approximation to make.

17.4. (a) Given the diffusion equation

$$\frac{d^2 p_n}{dx^2} - \frac{p_n - p_{n0}}{L_p^2} = 0, \quad x \geq x_n \tag{17.2}$$

one should first treat the homogeneous part, namely

$$\frac{d^2 p_n}{dx^2} - \frac{p_n}{L_p^2} = 0$$

The solutions of the homogeneous equation are given by $p_n \propto \exp\pm[(x - x_n)/L_p]$. Obviously the solution with the positive exponent should be discarded since it diverges at large x and one can write the solution of Eq. (17.2) as

$$p_n(x) = Ae^{-(x-x_n)/L_p} + p_{n0}$$

Now the constant of integration A should be used to satisfy the boundary conditions, namely

$$p_n(x_n) = p_{n0}e^{qV/k_BT} \text{ at } x = x_n, \ \ p_n(x \to \infty) = p_{n0}$$

Obviously these conditions are satisfied with the choice of A such that

$$p_n(x) = p_{n0}(e^{qV/k_BT} - 1)e^{-(x-x_n)/L_p} + p_{n0}$$

(b) Given the recombination rate given from Eq. (17.20) by

$$U = \frac{1}{\tau}\frac{n_i^2(e^{qV/k_BT} - 1)}{n + p + 2n_i \cosh(E_t - E_i)/k_BT}, \ \ np = n_i^2 e^{qV/k_BT}$$

the maximum U ensues with the minimum value of $n + p$, which can be found by putting the first derivative equal to zero, that is,

$$d(n + p) = 0$$

subject to the condition

$$pn = n_i^2 e^{qV/k_BT}$$

One can thus write

$$dn = -dp = -d\left(\frac{n_i^2 e^{qV/k_BT}}{n}\right) = \left(\frac{n_i^2 e^{qV/k_BT}}{n^2}\right) dn = \left(\frac{pn}{n^2}\right) dn = \frac{p}{n} dn$$

and therefore

$$n = p = n_i e^{qV/2k_BT}$$

17.5. The current flowing under a bias is generally specified in terms of the gradient of the quasi-Fermi level

$$J_n = \mu_n n \frac{d}{dx} E_{Fn}, \ \ J_p = \mu_p p \frac{d}{dx} E_{Fp} \tag{17.3}$$

(see Eq. (16.19)). Also the forward current is given from Eq. (17.17) by

$$J_n \simeq \frac{qD_n n_{p0}}{L_n} e^{qV/k_BT} \ \ \text{at } x = -x_p, \ \ J_p \simeq \frac{qD_p p_{n0}}{L_p} e^{qV/k_BT} \ \ at\ x = x_n \tag{17.4}$$

Thus by combining Eqs. (17.3) and (17.4) one can write

$$\frac{dE_{Fn}}{dx} = \frac{(qD_n n_{p0}/L_n)\, e^{qV/k_BT}}{\mu_n n_{n0}}, \quad \frac{dE_{Fp}}{dx} = \frac{(qD_p p_{n0}/L_p) e^{qV/k_BT}}{\mu_p p_{p0}}$$

Once injected into the depletion region, J_n, J_p are maintained at the injection level. For $N_A = 10^{17}\,\text{cm}^{-3}$ and $N_D = 10^{16}\,\text{cm}^{-3}$ for example the slope at the injection plane can be re-expressed by using the Einstein relation as

$$\frac{dE_{Fn}}{dx} = \frac{k_BT\,\mu_n n_{p0}\, e^{qV/k_BT}}{L_n \mu_n n_{n0}} = \frac{k_BT}{L_n}\frac{n_i^2}{p_{p0} n_{n0}} e^{qV/k_BT}, \quad \frac{D_n}{\mu_n} = \frac{k_BT}{q}$$

Also with the use of $\mu_n \approx 800\,\text{cm}^2\,(\text{V s})^{-1}$ and $\tau_n \approx 0.1\,\mu\text{s}$ the diffusion length is estimated as

$$\begin{aligned} L_n = (D_n\tau_n)^{1/2} &= \left(\frac{k_BT\mu_n\tau_n}{q}\right)^{1/2} \\ &= \left(\frac{1.381\times 10^{-23}\,\text{J}\cdot 300\,\text{K}\cdot 8\times 10-4\,\text{m}^2\cdot 10^{-7}\,\text{s}}{1.6\times 10^{-19}\,\text{C}}\right)^{1/2} \\ &\simeq 1.46\times 10^{-5}\,\text{m} = 14.6\,\mu\text{m} \end{aligned}$$

Hence by using of L_n thus found one can estimate the slope of E_{Fn} for the given doping level

$$\begin{aligned} \frac{dE_{Fn}}{dx} &= \frac{1.381\times 10^{-23}\,\text{J K}^{-1}\cdot 300\,\text{K}}{1.46\times 10^{-5}\,\text{m}}\frac{(1.45\times 10^{10})^2\,\text{cm}^{-6}}{10^{17}\times 10^{16}\,\text{cm}^{-6}}\cdot e^{qV/k_BT} \\ &\simeq 6\times 10^{-29}\, e^{qV/k_BT}\,\text{J m}^{-1} \\ &= 3.7\times 10^{-16}\, e^{qV/k_BT}\,\text{eV}\,\mu\text{m}^{-1} \end{aligned}$$

Thus for the forward voltage of 0.6 V for example one finds

$$\frac{dE_{Fn}}{dx} \simeq 4.8\times 10^{-6}\,\text{eV}\,\mu\text{m}^{-1}$$

Therefore for $W \simeq 11\,\mu\text{m}$ and for given doping level the total change of E_{Fn} is $\sim 5\times 10^{-5}$ eV and can therefore be neglected and E_{Fn} can be taken flat in W. One can likewise show that E_{Fp} can be taken flat.
For the case of the reverse bias

$$J_n \simeq \frac{qD_n n_{p0}}{L_n} = \frac{k_BT\mu_n n_{p0}}{L_n}$$

and therefore E_{Fn}, E_{Fp} can also be shown to be nearly flat. However for a large forward voltage the approximation ceases to be valid.

17.6. The Zener breakdown is caused by the Fowler-Nordheim tunneling given from Eq. (5.26) by

$$T \sim \exp\left[-\frac{4(2m_n)^{1/2}}{3q\mathrm{E}\hbar}E_G^{3/2}\right]$$

where $V - E = E_G$. Under a reverse bias V_R the junction electric field E is given from Eq. 17.10) by

$$E_{\max} \sim \frac{2(\varphi_{bi} + |V_R|)^{1/2}}{\left[\frac{2\varepsilon_s(N_A+N_D)}{q \cdot N_A N_D}\right]^{1/2}}$$

The critical field for the onset of the Zener breakdown is determined by the condition

$$q\mathrm{E}_{\max} \cdot W \approx \mathrm{E}_G - q\varphi_{bi}$$

The condition states that the band bending induced by V_R plus the intrinsic band bending exceeds E_G so that the valence band on the p side lines up with the conduction band on the n side. In this case the electrons in the valence band can tunnel through the triangular barrier into the conduction band on the n side, giving rise to the breakdown current.
Clearly this points to the fact that the small E_G enhances the F−N tunneling probability and also requires a smaller V_R for lining up the valence band to the conduction, causing the breakdown. Since W is determined by the doping level regardless of E_G the breakdown is more likely to occur in a smaller bandgap material for given doping level. The V_R responsible for the breakdown can be easily estimated with the use of the equations given above.

Chapter 18

18.1. The steady state diffusion equation

$$p_n'' - \frac{p_n - p_{n0}}{L_p^2} + \frac{\tilde{g}_D}{D_p} = 0, \quad \text{or } p_n'' - \frac{p_n}{L_p^2} = -\left(\frac{p_{n0}}{L_p^2} + \frac{\tilde{g}_D}{D_p}\right)$$

with the boundary conditions

$$p_n(x_n) = 0, \ \ p_n(x \to \infty) = p_{no} + \tilde{g}_D\tau_p$$

can be solved by first considering the homogeneous equation

$$p_n'' - \frac{p_n}{L_p^2} = 0$$

The solution of the homogeneous equation is given by $p_n \sim \exp \pm (x - x_n)/L_p$ and the positive branch should be discarded since it diverges at $x \to \infty$. Also the particular solution is obtained by inspection as

$$p_n = p_{no} + \tilde{g}_0 \tau_p, \quad L_p^2/D_p \equiv \tau_p$$

Therefore the solution of the diffusion equation is given by

$$p_n(x) = Ae^{-(x-x_n)/L_p} + p_{no} + \tilde{g}_D \tau_p$$

where A can be used for satisfying the boundary condition, namely $p_n(x_n) = 0$. Thus the solution reads as

$$p_n(x) = (p_{no} + \tilde{g}_D \tau_p)(1 - e^{-(x-x_n)/L_p})$$

(b) The diffusion equation of the electron in the p sides is given by

$$n_p'' - \frac{n_p - n_{p0}}{L_n^2} + \frac{\tilde{g}_D}{D_n} = 0, \quad \text{or} \quad n_p'' - \frac{n_p}{L_p^2} = -\left(\frac{n_{p0}}{L_n^2} + \frac{\tilde{g}_D}{D_n}\right)$$

The solution of the homogeneous equation is given by $n_p \sim \exp \pm (x + x_p)/L_n$. In this case the negative branch has to be discarded, since n_p diverges at $x \to -\infty$. Therefore by accounting for the boundary condition, $n_p(-x_p) = 0$ one finds

$$n_p(x) = (n_{po} + \tilde{g}_D \tau_p)(1 - e^{(x+x_p)/L_n})$$

18.2. The effect of R_s is to reduce the load voltage provided by the junction forward voltage. Therefore the power extracted is reduced as

$$P_L = V_L I_L = (V - I_L R_s)[I_l - I_s(e^{q(V_L + I_L R_s)/k_B T} - 1)]$$

with V denoting the open circuit voltage (see Eq. (18.17)). Thus the series resistance degrades the power extraction.

18.3. (a) The basic role of the solar cell is the power production

$$P = V_L \times I_L$$

and the junction solar cell is based on the band bending as occurs in the junction depletion depth W. The solar radiation is absorbed in W, generating the electron hole pairs therein. The e–h pairs thus generated should be separated and contribute to the load current I_L. The separation is naturally assisted by the junction band bending. Specifically the electrons roll down the junction potential hill while holes roll up the hill. Simultaneously the built in potential of the junction provides the load voltage V_L. Thus, the junction band bending is the main driving force for the operation of the solar cell.

(b) There are a few key factors involved in the efficiency of the solar cell. To optimize the power extraction I_l should me made as large as possible. In addition V_{OC} should be large, which is given from Eq. (18.14) by

$$V_{OC} = \frac{k_B T}{q} \ln\left(\frac{I_l}{I_S}\right)$$

where I_S denotes the saturated current of the junction (see Eq. (17.18)). Thus the large I_l also increases V_{OC}. Now I_l is commensurate with the linear attenuation coefficient given by

$$I_l \propto g \propto \alpha, \quad \alpha = A^*(\hbar\omega - E_G)^{1/2}$$

Therefore to increase α the bandgap of the material should be small so that the absorption could occur over a broader range of the solar spectrum. However small E_G increases I_S via the increased intrinsic carrier concentration

$$I_S \propto n_i^2 \propto \exp -E_G/k_B T$$

and decreases V_{OC}. It is therefore clear that to attain large V_{OC} wider bandgap is desirable.
In view of the merits and demerits of the wider and narrower bandgap an optimal compromise is in order. More important the optimal combination of materials to achieve large I_l and V_{OC} is an important issue.

18.4. (a) The overlap of E_C, E_V by 0.2 eV can be achieved by raising E_F above E_C by say 0.1 V in the n bulk, while lowering E_F below E_V by 0.1 eV in the p bulk. In this strongly degenerate regime the degenerate statistics has to be used:

$$\begin{aligned} n &= \frac{2}{\sqrt{\pi}} N_c F_{1/2}\left(\frac{E_F - E_c}{k_B T}\right) \\ &= N_D^+ = \frac{N_D}{1 + g_D e^{(E_F - E_D)/k_B T}}, \quad E_F - E_C = 0.1\,\text{eV} \end{aligned}$$

$$\begin{aligned} p &== \frac{2}{\sqrt{\pi}} N_V F_{1/2}\left(\frac{E_V - E_F}{k_B T}\right) \\ &= N_A^- = \frac{N_A}{1 + g_A e^{(E_A - E_F)/k_B T}}, \quad E_V - E_F = 0.1\,\text{eV} \end{aligned}$$

By inserting the values of N_C, N_V of Si and GaAa and evaluating the Fermi 1/2 integral at given argument one can find the required N_D, N_A.

(b) The flux of electrons under a forward bias is given from Eq. (17.17) by

$$F_n = \frac{J_n}{q} = \frac{D_n n_{p0}}{L_n} e^{qV/k_B T} = \left(\frac{k_B T \cdot \mu_n}{q}\right) \frac{1}{L_n} \cdot \left(\frac{n_i^2}{p_{p0}}\right) e^{qV/k_B T}$$

with

$$p_{p0} = \frac{2}{\sqrt{\pi}} N_V F_{1/2} \left[\frac{0.1}{k_B T\,(\text{eV})}\right]$$

Likewise one can specify F_p as

$$F_p = \frac{J_p}{q} = \frac{D_p p_{n0}}{L_p} e^{qV/k_B T} = \left(\frac{k_B T \cdot \mu_p}{q}\right) \frac{1}{L_p} \cdot \left(\frac{n_i^2}{n_{n0}}\right) e^{qV/k_B T}$$

By using the respective values of N_C, N_V, and the mobilities involved one can evaluate the electron and hole fluxes.

18.5. (a) The condition

$$f_v(E_a) > f_c(E_b)$$

is specified explicitly as

$$\frac{1}{1 + e^{(E_a - E_{Fv})/k_B T}} > \frac{1}{1 + e^{(E_b - E_{FC})/k_B T}}$$

and is equivalent to

$$e^{(E_a - E_{Fv})/k_B T} < e^{(E_b - E_{FC})/k_B T} \text{ or } E_a - E_{Fv} < E_b - E_{FC}$$

That is

$$E_{FC} - E_{Fv} < E_b - E_a = \hbar\omega$$

Similarly the condition

$$f_v(E_a) < f_c(E_b)$$

leads to

$$E_{FC} - E_{Fv} > E_b - E_a = \hbar\omega$$

(b) The gain coefficient as given by

$$\gamma(\omega) = \alpha(\omega)[f_c(E_b) - f_v(E_a)]$$

represents the probability of electron being in the conduction band at E_b level greater than that of being in the valence band at E_a level. In view of the fact that electrons are excited from the valence band to conduction band, the condition is analogous to the population inversion in laser devices by exciting the electrons from the ground state to the upper lasing level.

Chapter 19

19.1. (a) Before the contact E_F in the p$^+$ poly gate is lower than E_F in the n substrate. To keep E_F flat in equilibrium contact the band bending ensues. Since E_F in the n-substrate is higher than E_F in the electrode the electrons are injected into the p$^+$ gate electrode from the substrate. Consequently a negative charge sheet is formed at the surface of the gate electrode. The electric field emanating from the surface charge pushes the electrons from near the interface into the bulk, leaving donor atoms uncompensated. As a result the space charge is formed out of uncompensated donor ions and band bends up as the surface is approached from the bulk and the channel is inverted with further increase of band bending by applying the negative gate voltage. These discussions are compactly summarized in the figure shown below.

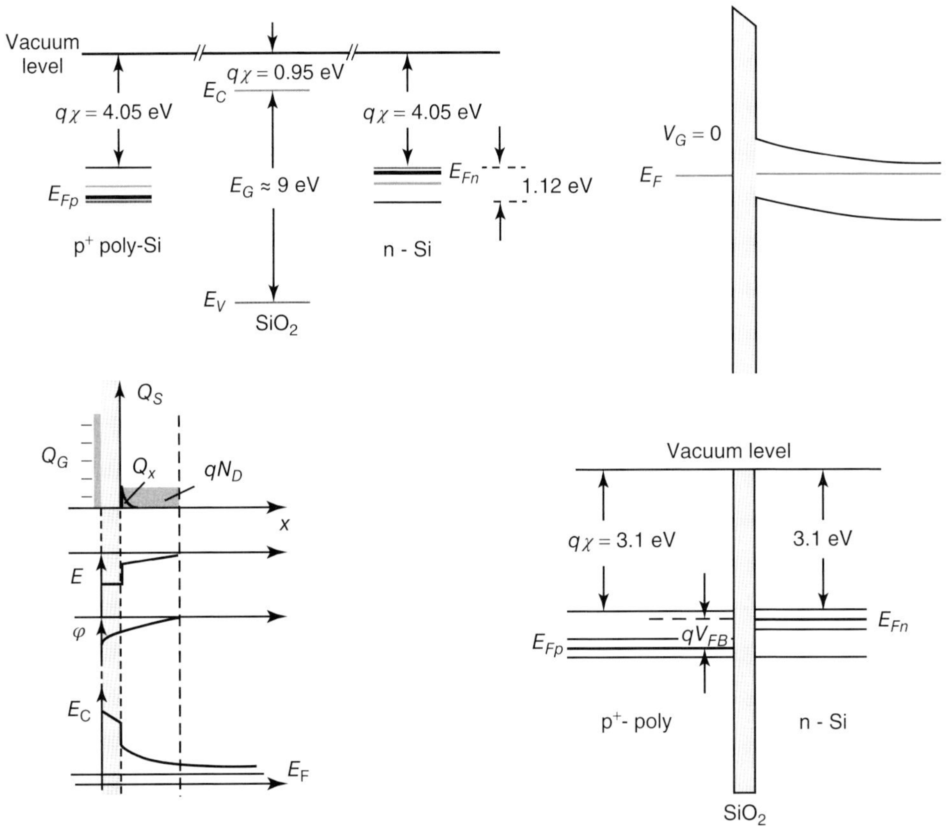

(b) The flat band voltage is defined as the difference of E_F between the substrate bulk and the gate electrode and is given in this case by

$$E_{Fn} - E_{Fp} = E_i + q\varphi_n - (E_i - q\varphi_p) = q\varphi_n + q\varphi_p$$
$$\approx 0.56\,\text{eV} + \frac{k_B T}{q} \ln\left(\frac{N_D}{n_i}\right), \quad \frac{E_G}{2} = 0.56\,\text{eV}$$

where $q\varphi_n$, $q\varphi_p$ denote respectively the Fermi potential of the electron and hole, respectively (see Eqs. (15.19) and (15.20)). Since E_{Fp} in the p^+ polysilicon gate practically coincides with E_V, $q\varphi_p$ was taken as $E_G/2$. For $N_D = 10^{16}\,\text{cm}^{-3}$

$$\varphi_{Fn} = \frac{k_B T}{q} \times \ln\left(\frac{N_D}{n_i}\right) = 0.025\,(\text{eV}) \times \ln\left(\frac{10^{16}}{1.45 \times 10^{10}}\right) = 0.336\,\text{eV}$$

Similarly one finds $\varphi_{Fn} = 0.39$ eV for $N_D = 10^{17}\,\text{cm}^{-3}$. Hence the value V_{FB} is 0.9 and 0.95 V respectively.

19.2. One can carry out the modeling of PMOS $I-V$ in strict analogy with the NMOS $I-V$ modeling by interchanging the roles of electrons and holes. When the p^+ poly-gate, SiO_2 and n substrate are in the equilibrium contact there is again the band bending to keep E_F flat. Since E_F in the n-substrate is higher than E_F of the gate electrode as clear from the band diagram shown the electrons are injected into the p^+ poly gate, forming thereby the negative charge sheet on the surface. Consequently the band bends up and the hole concentration is enhanced near the oxide interface. One can quantify the band bending by starting from the Poisson equation, which in this case is given in strict analogy with Eq. (19.4) by

$$\frac{d^2\varphi(x)}{dx^2} = -\frac{\rho(x)}{\varepsilon_S}, \quad \rho(x) = q[N_D^+ - n_n(x) + p_n(x)] \tag{19.1a}$$

Also the charge neutrality in the n bulk region prevails, that is

$$N_D^+ + p_{n0} = n_{n0} \tag{19.1b}$$

Because the band bends up in this case, that is, $\varphi \leq 0$ as clear from the band diagram shown in the previous problem p_n increases near the surface, while n_n decreases according to

$$p_n(x) = p_{n0}e^{-\beta\varphi(x)}, \quad n_n(x) = n_{n0}e^{\beta\varphi(x)}, \quad \beta = q/k_B T \tag{19.1c}$$

Thus by inserting Eqs. (19.1b) and (19.1c) in Eq. (19.1a) one can write

$$\frac{d^2}{dx^2}\varphi(x) = \frac{q}{\varepsilon_S}[n_{n0}(e^{\beta\varphi}-1) - p_{n0}(e^{-\beta\varphi}-1)]$$
$$= \frac{qn_{n0}}{\varepsilon_S}\left[(e^{\beta\varphi}-1) - \frac{p_{n0}}{n_{n0}}(e^{-\beta\varphi}-1)\right], \quad n_{n0} \simeq N_D \tag{19.2}$$

Since Eq. (19.2) is identical in form to Eq. (19.4) in the text one can repeat the same algebraic step and obtain the surface field E_s and surface potential Q_s as

$$Q_S = -\varepsilon_S E_S = \pm\sqrt{2}\frac{k_BT}{qL_D}F\left(\beta\varphi_s, \frac{p_{n0}}{n_{n0}}\right), \quad L_D = \left(\frac{k_BT}{q^2}\frac{\varepsilon_S}{n_{n0}}\right)^{1/2} \tag{19.3a}$$

where

$$F\left(\beta\varphi_s, \frac{p_{n0}}{n_{n0}}\right) = \left[(e^{\beta\varphi} - \beta\varphi - 1) + \frac{p_{n0}}{n_{n0}}(e^{-\beta\varphi} + \beta\varphi - 1)\right]^{1/2} \tag{19.3b}$$

Equation (19.3) is the PMOS counterpart of Eq. (19.7) in the text. The difference between the two consists of

$$\varphi_s \leftrightarrow -\varphi_s, \quad p_{p0} \leftrightarrow n_{n0}, \quad n_{p0} \leftrightarrow p_{n0}$$

With the use of Eq. (19.3) one can quantify the electron accumulation for $\varphi_s > 0$, the hole depletion for $0 > \varphi_s > -2\varphi_n$ and the strong inversion for $\varphi_s \le -2\varphi_n$. These different regimes are the mirror images of the n-type MOSFET on the axis of the surface potential φ_s. By using the expression thus obtained one can again repeat the algebraic steps used in NMOS and obtain

$$I_D = \frac{W}{L}\mu_p C_{OX}\left(|V_G| - |V_T| - \frac{1}{2}|V_D|\right)|V_D|$$

Here all of the biases are negative, so that the I_D expression is the mirror image of the NMOS I_D expression, as expected.

19.3. (a) The drain current is given in terms of the drain voltage V_D and channel length L as

$$I_D = \frac{W}{L}C_{OX}\mu_n\left(V_G - V_T - \frac{1}{2}V_D\right)V_D, \quad V_{DSAT} = V_G - V_T \tag{19.4}$$

By taking y as the channel length instead of L and the voltage therein $V(y)$ as the terminal voltage instead of V_D the same I_D can be expressed as

$$I_D = \frac{W}{y}C_{OX}\mu_n\left[V_G - V_T - \frac{1}{2}V(y)\right]V(y) \tag{19.5}$$

By equating Eqs. 19.4 and 19.5 one can write

$$\frac{1}{L}\left(V'_G - \frac{1}{2}V_D\right)V_D = \frac{1}{y}\left[V'_G - \frac{1}{2}V\left(y\right)\right]V(y), \quad V'_G \equiv V_G - V_T$$

and find $V(y)$ from the resulting quadratic equation in terms of V_G, V_D as

$$V(y) = V'_G \pm \left[V'^2_G - \frac{2y}{L}\left(V'_G - \frac{1}{2}V_D\right)V_D\right]^{1/2}$$

Since by definition the source voltage is zero, that is, $V(y = 0) = 0$ the positive branch of the solution should be discarded.

(b) With the channel voltage $V(y)$ thus found one can specify the channel field by

$$E(y) \equiv -\frac{\partial V(y)}{\partial y} = -\frac{\left(V'_G - \frac{1}{2}V_D\right)V_D}{L\left[V'^2_G - \frac{2y}{L}\left(V'_G - \frac{1}{2}V_D\right)V_D\right]^{1/2}} = -\frac{V_D}{L}\cdot\frac{1}{\left(\gamma - \alpha\frac{y}{L}\right)^{1/2}} \tag{19.6}$$

where

$$\alpha \equiv 2\frac{V_D}{V'_G - \frac{1}{2}V_D}, \quad \gamma = \frac{V'^2_G}{\left(V'_G - \frac{1}{2}V_D\right)^2}$$

At the device saturation $V_{DSAT} = V_G - V_T$ and therefore $\alpha = \gamma = 4$. Consequently the channel field at the drain terminal $y = L$ is diverges, as expected.

(c) The transit time of the electron across the channel is given with the use of Eq. (19.6) by

$$\tau_{tr} = \int_0^L \frac{dy}{v_d} = \int_0^L \frac{dy}{\mu_n|E(y)|} = \frac{L}{\mu_n V_D}\int_0^L dy\left(\gamma - \alpha\frac{y}{L}\right)^{1/2}$$
$$= \frac{L}{\mu_n V_D}\cdot\left(\frac{-2L}{3\alpha}\right)\left(\gamma - \alpha\frac{y}{L}\right)^{1/2}\Big|_0^L = \frac{2L^2}{3\mu_n V_D\alpha}[\gamma^{3/2} - (\gamma - \alpha)^{3/2}]$$

By inserting the values of α, γ in device saturation one obtains

$$\tau_{tr} = \frac{4}{3}\frac{L^2}{\mu_n V_{DSAT}}$$

in general agreement with the representation of τ_r, that is,

$$\tau_{tr} = \frac{L}{\langle v_d\rangle} \approx \frac{L}{\mu_n(V_D/L)} = \frac{L^2}{\mu_n V_D}$$

19.4. The backscattering coefficient is given from Eqs. (19.25) and (19.26) by

$$r_c = \frac{l}{l+\lambda} = \frac{1}{1+\lambda/l} = \frac{1}{1+v_T m_n \mu_n E_s/k_B T} = \frac{k_B T}{k_B T + v_T m_n v_D}, \quad v_D \equiv \mu_n E_s$$

where v_D is the drift velocity. Then one can write

$$\eta = \frac{1-r_c}{1+r_c} = \frac{v_T m_n v_D}{2k_B T + v_T m_n v_D} = \frac{v_T m_n v_D}{m_n[(2k_B T/m_n + v_T v_D]}$$
$$= \frac{v_T v_D}{[(2k_B T/v_T m_n) + v_D]} = \frac{v_T v_D}{[v_T + v_D]} = \frac{1}{(1/v_D)+(1/v_T)}, \quad k_B T \equiv \frac{m_n v_T^2}{2}$$

The insertion of η into Eq. (19.24a) leads to Eq. (19.27) in the text.

19.5. (a) The built-in voltage of the three junctions can be found with the use of the Fermi potentials as follows:The Fermi potentials in the n-bulk is given by

$$\varphi_{Fn} = \frac{E_{Fn} - E_i}{q} = \frac{k_B T}{q} \ln\left(\frac{N_D}{n_i}\right) = 0.025\,\mathrm{V} \times \ln\left(\frac{N_D}{1.45\times 10^{10}\,\mathrm{cm}^{-3}}\right) \text{ at } 300\,\mathrm{K}$$

so that

$$\varphi_{Fn} = 0.24\,\mathrm{V} \text{ at } N_D = 10^{16}\,\mathrm{cm}^{-3}, \quad \varphi_{Fn} = 0.39\,\mathrm{V} \text{ at } N_D = 10^{17}\,\mathrm{cm}^{-3}$$

The built-in voltages for the three junctions are summarized as follows:

	V_{bi} at $N_D = 10^{16}\,\mathrm{cm}^{-3}$	V_{bi} at $N_D = 10^{17}\,\mathrm{cm}^{-3}$
$\mathrm{p^+ - n}$	$0.56 + 0.24\,\mathrm{eV}$	$0.56 + 0.39\,\mathrm{eV}$
$\mathrm{n - i}$	$0.24\,\mathrm{eV}$	$0.39\,\mathrm{eV}$
$\mathrm{i - n^+}$	$0.56\,\mathrm{eV}$	$0.56\,\mathrm{eV}$

(b) The tunneling ensues and the tunnel current starts to flow when the conduction band in the n-region in $\mathrm{p^+}$-n junction is lowered and lines up with the valence band in the $\mathrm{p^+}$ region. Thus the minimum V_D required to induce the tunneling is found by the difference between the band gap of 1.12 eV for Si for instance and the V_{bi} therein, One therefore finds $V_D = 0.32$ V, $V_D = 0.17$ V for $N_D = 10^{16}\,\mathrm{cm}^{-3}$, $N_D = 10^{17}\,\mathrm{cm}^{-3}$, respectively. It is pointed out that when V_D is applied, it should be dropped in the three junctions. The exact partitioning of V_D among the three junctions is difficult to analyze. However since the $\mathrm{p^+}$-n junction should take up a large fraction of it because it has the largest V_{bi}.

Chapter 20

20.1. (a) If the barrier height of the quantum well is taken infinite for simplicity and also without sacrificing too much accuracy one can write from Eq. (4.4)

$$E_n = \frac{\hbar^2\pi^2}{2mW^2}n^2, \qquad n = 1, 2, \ldots$$

Thus for electrons with rest mass m_0 the ground state energy can be evaluated as

$$\begin{aligned} E_1 &= \frac{(1.055\times 10^{-34})^2(\mathrm{J\ s})^2(3.14)^2}{2\cdot 9.1\times 10^{-31}\,\mathrm{kg}\cdot(10^{-9})^2\ w^2(\mathrm{nm})^2}n^2[J],\ 1\,\mathrm{m} = 1^9\,\mathrm{nm} \\ &= (0.603/w^2)\times 10^{-19}[\mathrm{J}] = (0.603/w^2)\times 10^{-19}\times 6.23\times 10^{18}[\mathrm{eV}] \\ &= (0.376/w^2)[\mathrm{eV}] \end{aligned}$$

where W has been scaled in nm unit by putting

$$W = 10^{-9}\cdot w$$

(b) The W at which E_1 is equal to the thermal energy is thus determined by

$$\frac{0.376}{w^2}\,\mathrm{eV} = k_BT \cong 25\,\mathrm{meV}$$

That is,

$$w = \sqrt{\frac{0.376}{0.025}} = 3.88\,\mathrm{nm}$$

(c) W can be found by putting

$$E_1 = k_BT$$

(d) See Eq. (5.22) in the text.

20.2. (a) The electron lifetime in the well is determined by the condition

$$NT \simeq 1 \tag{20.1}$$

where T is the tunneling probability and N the average number the electron encounters the barrier before tunneling out. Since the barrier on the left of the well is thicker than the barrier on the right the lifetime is dictated by the latter barrier. The tunneling probability is given in this case from Eq. (5.22) by

$$T = \frac{1}{1+\Lambda},\quad \Lambda = \frac{V^2}{4E(V-E)}\sinh^2 d\sqrt{\frac{2m}{\hbar^2}(V-E)} \tag{20.2}$$

where the parameters involved are $V = 3.1$ eV, $d = 8$ nm. The kinetic energy E of the electron is dictated by the ground state energy E_1 or the thermal energy depending on whether E_1 is greater or less than k_BT. Since W given is larger than 3.88 nm as was found in the previous problem one has to use the thermal energy of 0.025 eV at 300 K. In this case $E = k_BT << V$. Therefore one can evaluate Λ in Eq. (20.2) as

$$\Lambda \simeq \frac{V}{2k_BT}\frac{1}{4}\exp\left[2d\left(\frac{2m}{\hbar^2}V\right)^{1/2}\right]$$
$$= \frac{3.1\,\text{eV}}{2\times 0.025\,\text{eV}}$$
$$\times\frac{1}{4}\exp\left\{2\cdot 8\times 10^{-9}m\left[\frac{2\times 9.1\times 10^{-31}\,\text{kg}\times 3.1\times 1.6\times 10^{-19}\,\text{J}}{\left(1.055\times 10^{-34}\right)^2(\text{J s})^2}\right]^{1/2}\right\}$$
$$= 5.2\times 10^{62}$$

Consequently the number of encountering the barrier N is given by

$$N \simeq \frac{1}{T} \sim 5.2\times 10^{62},\quad T = \Lambda^{-1}$$

Since the thermal velocity at room temperature is given by

$$v_T = \left(\frac{k_BT}{m}\right)^{1/2} \approx 6.8\times 10^4\,\text{m s}^{-1},\quad mv_T^2/2 = k_BT/2$$

the lifetime can be estimated by the round trip time of the electron in the quantum well times N,

$$\tau = \frac{2\times 10\times 10^{-9}(\text{m})}{6.8\times 10^4(\text{m s})^{-1}}\cdot 5.2\times 10^{62} \approx 2.94\times 10^{49}\,\text{s}$$

and is shown nearly infinite.

(b) Controlling the lifetime of the electron via the gate bias is the working principle of the flash EEPROM cell. Reducing τ to 1 μs requires the application of the electric field, so that the barrier potential is transformed into the triangular shape, thereby enabling the utilization of the F–N tunneling. Since the F–N tunneling probability is given from Eq. (5.26) by

$$T \simeq \exp -\frac{4\sqrt{2m}}{3q|\text{E}|\hbar}(V-E)^{3/2} \simeq \exp -\frac{4\sqrt{2m}}{3q|\text{E}|\hbar}V^{3/2},\qquad E \ll V = 3.1\,\text{eV}$$

τ is specified by

$$\tau = \frac{2W}{v_T}\frac{1}{T} = \left(\frac{2W}{v_T}\right)\exp\left[\frac{4\sqrt{2m}V^{3/2}}{3q\,|\text{E}|\,\hbar}\right],\quad \left(\frac{2W}{v_T}\right) = \frac{2\times 10\times 10^{-9}\,(\text{m})}{6.8\times 10^4\,(\text{m s}^{-1})}$$
$$= 2.9\times 10^{-13}\,\text{s}$$

Therefore for $\tau = 1$ μs the required E field is given by

$$|E| = \frac{4\sqrt{2m}V^{3/2}}{3q\hbar} \times \ln\left(\frac{2W}{\nu_T}\right), V = 3.1 \times 1.6 \times 10^{-19} J$$

$$= \frac{4 \cdot (2 \cdot 9.1 \times 10^{-31})^{1/2}(3.1 \times 1.6 \times 10^{-19})^{3/2}}{3 \times 1.6 \times 10^{-19} \times 1.055 \times 10^{-34} \times \ln(\nu_T \tau / 2W)}$$

$$= 7 \times 10^8\,\mathrm{V\,m^{-1}}$$

$$= 0.7\,\mathrm{V\,nm^{-1}}$$

20.3. The shift in V_T due to programming reads from Eq. (20.4) as

$$\Delta V_T = V_{TCGP} - V_{TCGE} = \frac{|Q_{FG}|}{C_{ONO}}$$

Since the capacitance per unit area of the ONO dielectric layer is approximately given by

$$C_{ONO} = \frac{\varepsilon_{OX}}{t_{OX}} = \frac{\varepsilon_r \times \varepsilon_0}{t_{OX}} = \frac{3.9}{15 \times 10^{-9}(\mathrm{m})} \times \frac{1}{4\pi \cdot 8.988 \times 10^9\,\mathrm{C}^{-1}}$$
$$= 2.3 \times 10^{-3}\,\mathrm{C\,V^{-1}\,m^{-2}}$$

one can find the number N of excess electrons for ΔV_T of 5 V as

$$5 \times C_{ONO} = |Q_{FG}| = q \times N$$

That is,

$$N = \frac{5 \times 2.3 \times 10^{-3}\,(\mathrm{C\,V^{-1}\,m^{-2}})}{1.5 \times 10^{-19}\,(\mathrm{C})} = 7.6 \times 10^{16}\,\mathrm{m^{-2}} = 7.6 \times 10^{12}\,\mathrm{cm^{-2}}$$

For the floating gate with the cross-sectional area 100 × 100 nm

$$N \approx 760$$

20.4.

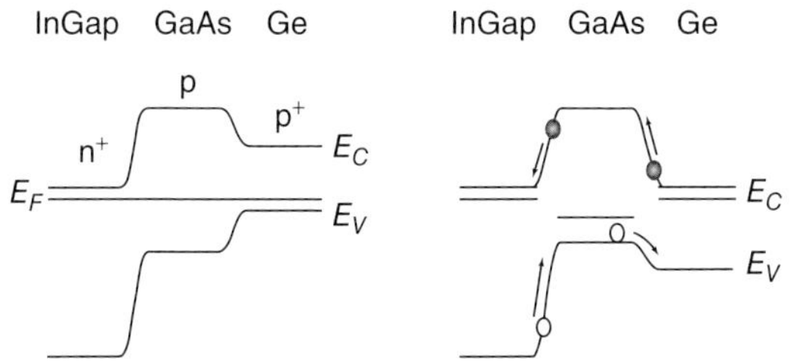

(a) Shown are the band diagrams of the stacked multi-junction solar cell in equilibrium (left) and under the illumination (right). In the former case the two junctions bend as usual to keep E_F flat. Naturally E_C in the p-region is highest, followed by

E_C in the p^+ and n^+ regions. This is because the difference $E_C - E_F$ should be preserved in respective bulk regions. Under the illumination the e–h pairs generated in the n^+-p junction are efficiently separated to contribute to the current. Also both junctions are forward biased due to the resulting space charge. The holes in the n^+-p junction naturally climb up the potential hill but have to diffuse into p^+ region by overcoming the junction barrier which is lowered by the forward bias developed. Likewise electrons generated in the $p-p^+$ junction region diffuse into the p-region by overcoming the barrier therein and roll down the potential hill in the n^+-p junction together with the electrons generated therein to contribute to the current.

20.5.

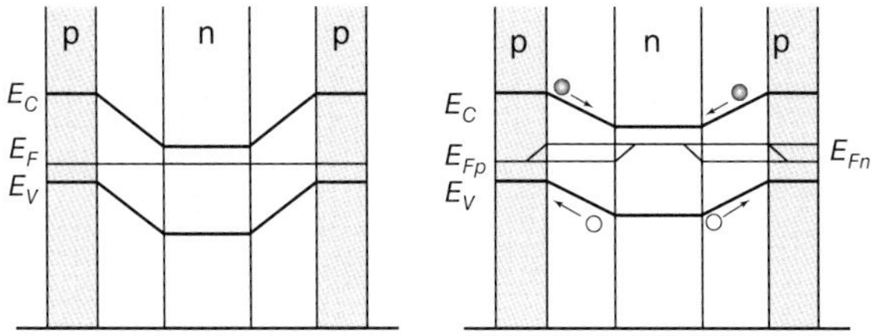

(b) Shown are the band diagrams of the vertical core–shell nanowire solar cell in equilibrium (left) and under illumination (right). In equilibrium the band bending reflects the two p–n junctions connected centered around the n-bulk. When the light is incident on the junction depletion depth, the e–h pairs are generated therein. Because of the junction band bending the generated e–h pairs are efficiently separated. Specifically holes roll up the potential hill into the p-region after traversing a short distance, while the electrons roll down the potential barrier into the region near the core of the wire. The resulting space charge, positive charge in the p- region and negative charge in the n-region induce the forward bias, thus providing the driving voltage of the cell. At the same time the junction band bending is reduced. Concurrently the separated electrons and holes drive the output current of the cell.

Index

Important Physical Numbers and Quantities

$1\ \text{cm} = 10^4\ \mu\text{m} = 10^8\ \text{Å} = 10^7\ \text{nm}$
Electron volt: $1\ \text{eV} = 1.602 \times 10^{-19}\ \text{J}$
Electron charge: $q = 1.602 \times 10^{-19}\ \text{C}$
Coulomb constant: $1/(4\pi\varepsilon_0) = 8.988 \times 10^9\ \text{N·m}^2\ \text{C}^{-2}$
Planck's constant: $h = 6.626 \times 10^{-34}\ \text{J s} = 4.136 \times 10^{-15}\ \text{eV s}$
$\hbar = h/2\pi = 1.055 \times 10^{-34}\ \text{J s} = 6.582 \times 10^{-16}\ \text{eV s}$
Boltzmann constant: $k_B = 1.381 \times 10^{-23}\ \text{J K}^{-1} = 8.617 \times 10^{-5}\ \text{eV K}^{-1}$
Bohr radius: $a = 0.529\ \text{Å} = 0.0529\ \text{nm}$
Avogadro's number: $N = 6.022 \times 10^{23}$ particles per mol
Electron mass in free space: $m_0 = 9.109 \times 10^{-31}\ \text{kg}$
Proton mass in free space: $m_p = 1.673 \times 10^{-27}\ \text{kg}$
Permeability of free space: $\mu_0 = 1.256 \times 10^{-8}\ \text{H cm}^{-1}$
Permittivity of free space: $\varepsilon_0 = 8.854 \times 10^{-14}\ \text{F cm}^{-1}$
Speed of light in free space: $c = 2.998 \times 10^8\ \text{m s}^{-1}$
Thermal voltage at room temperature (300 K): $k_B T/q = 0.0259\ \text{V}$
Wavelength of 1 eV photons: $1.24\ \mu\text{m}$

Important Electronic Properties of Semiconductors at Room Temperature

	Ge	Si	GaAs
Atoms (cm^3)	4.42×10^{22}	5.0×10^{22}	4.42×10^{22}
Breakdown field (V cm^{-1})	$\sim 10^5$	$\sim 3 \times 10^5$	$\sim 4 \times 10^5$
Dielectric constant	16.0	11.9	13.1
Effective density of states (cm^{-3})			
Conduction band N_c	1.04×10^{19}	2.8×10^{19}	4.7×10^{17}
Valance band N_v	6.0×10^{18}	1.04×10^{19}	7.0×10^{18}
Electron affinity, χ (V)	4.0	4.05	4.07
Energy gap (eV)	0.66	1.12	1.424
Intrinsic carrier concentration (cm^{-3})	2.4×10^{13}	1.45×10^{10}	1.79×10^6
Intrinsic Debye length (μm)	0.68	24	2250
Lattice constant (Å)	5.646	5.430	5.653
Lattice (intrinsic) mobility ($\text{cm}^2\ \text{V}^{-1}\text{·s}^{-1}$)			
Electrons	3900	1500	8500
Holes	1900	450	400